MICROSCOPY OF OXIDATION 2

MICROSCOPY OF OXIDATION 2

Proceedings of the
Second International Conference
on the Microscopy of Oxidation

held at Selwyn College, the University of Cambridge,
29–31 March 1993

Edited by
S. B. NEWCOMB and M. J. BENNETT

Sponsored and organised by the Materials Science Committee
of The Institute of Materials with the co-sponsorship of
The Institute of Physics and The Royal Microscopical Society

THE INSTITUTE OF MATERIALS

Book 552
First published in 1993 by
The Institute of Materials
1 Carlton House Terrace
London SW1Y 5DB

ISBN 0 901716 50 2

Typeset by
Keyset Composition, Colchester

Printed and bound at
The University Press, Cambridge

Contents

5.   Oxidation of Aluminium, Titanium, Zirconium and Manganese and Their Alloys

Preface

The Second International Conference of The Microscopy of Oxidation was held at Selwyn College, Cambridge from the 29th–31st March 1993. The conference was sponsored and organised by the Metal Science Committee of The Institute of Materials with the co-sponsorship of The Institute of Physics and The Royal Microscopical Society.

The meeting followed the general format of the very successful first meeting held in Cambridge in March 1990 on the theme of the application of microscopy to oxidation studies. In the second conference, this general theme was continued but broadened slightly to include some keynote papers on techniques which have not been used routinely in oxidation studies but which might hold some promise for the future in this area.

The high quality of all the oral papers and posters presented, and the lively discussion which followed each of the sessions, ensured an interesting and enjoyable meeting. My thanks go to all the members of the organising committee for their help and support and especially to Mike Bennett and Simon Newcomb for editing this volume.

It is hoped to hold the third meeting in this series in the Spring of 1996.

Gordon J. Tatlock
Chairman, Organising Committee

1

KEYNOTE PAPERS

Status and Potential of Newer Microscopy Techniques in Oxidation Studies

High-Resolution Transmission Electron Microscopy of Metal/Metal Oxide Interfaces

MANFRED RÜHLE, UTE SALZBERGER
and ECKART SCHUMANN

Max-Planck-Institut für Metallforschung Institut für Werkstoffwissenschaft
Seestr. 92, D-70174 Stuttgart 1, Germany

ABSTRACT

The structure and chemistry of heterophase boundaries, such as metal/metal oxide interfaces, can be resolved by dedicated techniques of transmission electron microscopy. High-resolution transmission electron microscopy (HREM) reveals the structure with high precision. The coordinates of atom columns can be detected with an accuracy of ±0.02 nm. However, only the projected structure of the interface can be revealed if (i) both crystals (adjacent to the interface) are oriented parallel to low-indexed Laue zones and (ii) the interface itself is also parallel to the incoming electron beam. Analytical electron microscopy (AEM) reveals the chemical composition of the interfaces with a spatial resolution of <1 nm (with dedicated systems). Successful HREM and AEM studies require the preparation of specimens with thicknesses < 100 nm. Often specimen preparation presents a major obstacle for applying advanced transmission electron microscopy techniques in materials science and also for the microscopy of oxidation. Specimen preparation techniques have been developed which allow the preparation of high-quality cross-sectional TEM specimens.

1. Introduction

The adhesion of an oxide scale to a metal substrate depends on the bonding of the two components across the interface. The bonding (adhesion) is strongly influenced by the atomistic structure of the interface and by the local chemical composition (segregation) of the interface. Therefore, it is desirable to determine both the atomistic structure as well as the chemical composition of such heterophase boundaries. Recently, much attention has been drawn to the identification of the structure and chemistry of heterophase boundaries, which play an important, sometimes controlling, role in the properties of modern material components. The results of the studies are published in proceedings of recent conferences.[1,2] The structure of the metal/ceramic (oxide) interface can be revealed by high-resolution transmission electron microscopy (HREM).[3,4] In

3

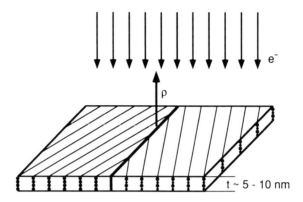

Fig. 1 Direct lattice imaging by HREM. The crystalline specimen must be adjusted so that the direction of the incoming electron beam coincides exactly with the orientation of atom rows. The schematic drawing shows a heterophase boundary. HREM can be successfully performed for pure tilt boundaries with tilt axis parallel to the direction of the incoming beam.

this short paper the HREM technique will be reviewed and applied to an interface formed after oxidation of a (100) surface of single crystalline Ni$_3$Al.[5] Atomistic structures and misfit dislocations were revealed. In addition, shear stresses parallel to the interface were analysed.

The imaging of heterophase boundaries requires that both crystals adjacent to the interface are oriented parallel to low-indexed Laue zones. In addition, the interface itself has also to be parallel to the incoming electron beam (see Fig. 1). Deviations from those conditions reduce the accuracy of the HREM results.

It should be emphasised that only the projection along *one* zone axis can be analysed with one set of micrographs. The three-dimensional analysis of an interface structure requires that, at least, a second projection is analysed. The analysis of one projection is, however, sufficient for the analysis of the three-dimensional structure of an interface if the interface possesses special symmetry properties.[6]

2. Experimental Studies of Interface Structure by High-Resolution Electron Microscopy

The point-to-point resolution[3] of HREM is now better than 0.17 nm for instruments with an acceleration voltage of 400 kV. The next generation of instruments pushes the resolution limit to ~0.1 nm. High-quality experimental images can be obtained. Nevertheless, the interpretation of the HREM micrographs is not possible on a naïve basis owing to the aberration of magnetic lenses. Therefore, the micrographs have to be analysed by comparing them to simulated images. Considerable advances have been made in this

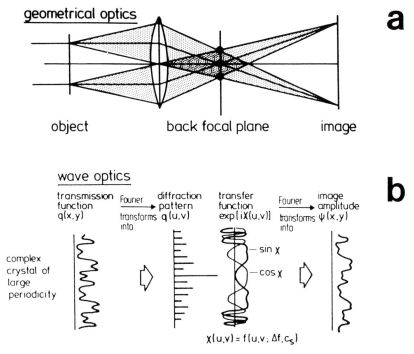

Fig. 2 Image formation by the objective lens of a transmission electron microscope (a) geometrical optical path diagram (b) wave optical description. (See text for explanation.)

analysis.[3,7] Methods and programs have been developed[8,9] which allow the simulation of HREM images of any given atomic arrangement.

The geometric beam path through the objective lens of a TEM is shown in Fig. 2a. Beams from the lower side of the object travel in both the directions of the incoming and of the diffracted beams. All beams are focused by the objective lens in the back focal plane to form the diffraction pattern. The image of the object is produced in the image plane by interference of the transmitted and diffracted beams. Fig. 2b uses wave optics to describe physical processes which contribute to the image formation. From the lower side of the foil, a wave field emerges that can be described by a transmission function $q(x,y)$ where x and y are the coordinates at the lower surface of the object. For an undistorted lattice, $q(x,y)$ represents a simple periodic amplitude and intensity distribution. The transmission function of a complex lattice with a large periodicity (e.g., an interface with a periodic structure) is very complicated and $q(x,y)$ is a non-periodic function for the distorted region of a crystal.

Since spherical aberration cannot be avoided with rotationally symmetrical electromagnetic lenses,[3,7] the beams emerging from an object at a certain angle (Fig. 2a) undergo a phase shift relative to the direct beam. Imaging with a small defocusing distance Δf, leads also to a phase shift. This depends on the sign and magnitude of the defocusing distance Δf, which is defined as the distance between lower foil surface and imaging plane. The influence of the lens errors and the defocus on the amplitude $q(u,v)$ of the diffraction patterns is described by the contrast transfer function (CTF).[3,7]

The image (Fig. 2b) is formed by a second Fourier transformation of the amplitude distribution in the diffraction pattern multiplied by the contrast transfer function. The amplitude in the image plane, $\Psi(x,y)$, is not identical to the wave field in the object plane (transmission function $q(x,y)$). The image is severely modified if scattering to large angle occurs, since the influence of the spherical aberration increases strongly with increasing scattering angle. The modification is most severe if the components in the diffraction pattern coincide with the oscillating part of the CTF (large values of u,v). If, however, the wave vectors lie within the first wide maximum of the CTF, it can be assumed that characteristic features and properties of the object can be directly recognised in the image. Good HREM imaging requires that the first zero value of the CTF under optimum defocus must be at sufficiently large reciprocal lattice spacings. Good imaging conditions are fulfilled for lattices with large lattice parameters.[3,7] If, however, deviations in the periodicity exist, components of the diffraction pattern also appear at large diffraction vectors. It is then most likely that certain Fourier components possess reciprocal lattice distances larger than the Scherzer focus. Lens aberrations and defocusing cannot be neglected in this case.

This qualitative description of image formation in HREM suggests that experimental studies have always to be accompanied by image simulation (see Fig. 3). The determination of instrumental parameters, the constants of spherical aberration, C_s, and chromatic aberration, C_c, defocusing distance Δf, is performed by analysing amorphous regions of the specimen. Then the image for the perfect regions of the two crystals adjacent to the interface must be simulated for different foil thicknesses. The simulated images are matched (for various values of Δf) to the experimentally obtained micrographs resulting in foil thicknesses t_1 and t_2 of both crystals adjacent to the interface. The vector $\mathbf{T}$ of the translation state of the two lattices with respect to each other can then be determined.[6] Finally, relaxations of atomic columns can be identified.

The quality of the HREM analysis depends sensitively on the quality of the specimen (Fig. 4a). For the image simulation an 'ideal' specimen is assumed which possesses an exact orientation relationship and a constant specimen thickness over the analysed region of the foil. In a 'real' specimen (Fig. 4b) the thickness may not be constant close to the interface (preferential etching) and the crystals adjacent to the interface may be slightly misoriented. The coverage of the specimen with an amorphous layer leads to an increase of the background noise in the images. Frequently, this 'specimen effect' limits the accuracy of the HREM evaluation. Many activities must be devoted to the best possible specimen preparation.

3. Specimen Preparation

Quantitative high-resolution electron microscopy requires that specimens with constant thickness ($t = \sim 5$–10 nm) are available with well-defined orientation relationships (Fig. 4).

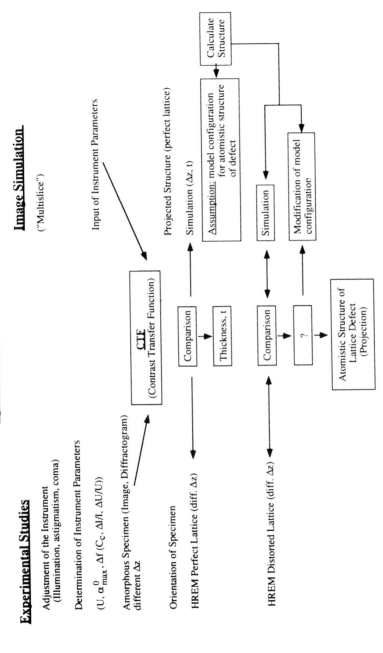

Fig. 3 Flow chart of quantitative high-resolution electron microscopy.

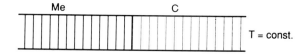

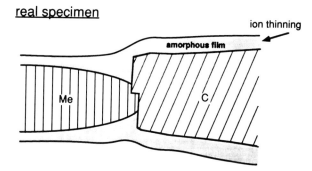

Fig. 4 Schematic drawing of (a) an 'ideal' specimen and (b) a 'real' specimen.

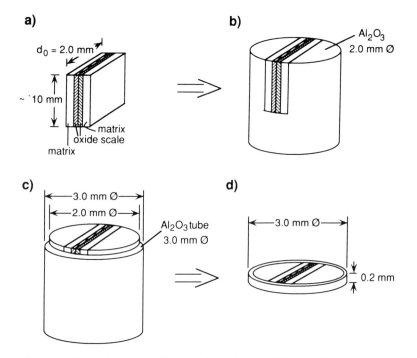

Fig. 5 Preparation of cross-section TEM specimens. (See text for explanation.)

The study of oxide scales requires that cross-sections are available. In our laboratory special efforts were made to develop a technique which allows the reproducible preparation of TEM cross-section with a high yield. The applied steps are indicated in Fig. 5.[10,11] Single crystals of Ni_3Al (surface parallel to (100)) were oxidised in air for 1 h at 950°C. Strips (thickness <0.2 mm) were cut out of the crystals with sizes of about 2×10 mm^2. Two strips were put together with their oxidised surface face-to-face (Fig. 5a) and mounted in the slit of an Al_2O_3 cylinder (Fig. 5b). The width of the slit corresponded to the thickness of the sandwich. The specimen-containing cylinder was then glued into a cylinder of either Al_2O_3 or brass (for better thermal conductivity during ion beam thinning, Fig. 5c). Slices with a thickness of ~0.2 mm thickness were cut from the specimen (Fig. 5d). The slices were polished, carefully dimpleground until the remaining thickness in the centre of the slice was <0.05 mm. Ion thinning is usually the last step for specimen preparation. The angle of incidence of the ion beam (of the ion beam thinner) must be as small as possible and it must be possible to focus the ion beam on the specimen. Ion beam thinners which are now commercially available fulfil these conditions.

Recently, the cross-section preparation technique was so refined that excellent specimens can be obtained with a high yield.

4. Experimental Results and Interpretation

A TEM cross-section specimen was prepared of an oxidised Ni_3Al single crystal ((100) surface, 950°C, 1 h, air). The conventional TEM studies were performed in a 200 kV instrument (JEOL 2000 FX), whereas the HREM studies were performed in a dedicated HREM instrument at 400 kV acceleration voltage (JEOL 4000 EX). The point-to-point resolution of the latter instrument is 0.17 nm.

Figure 6 shows a TEM micrograph of a cross-section through such a specimen. Different layers can readily be identified. The matrix (lower part of Fig. 6) consists of ordered γ'-Ni_3Al. A phase boundary separates γ'-Ni_3Al from Ni(Al) solid solution followed by a γ-Al_2O_3 and a NiO scale[5,12]. The HREM studies were performed at the interface between Ni(Al) and γ-Al_2O_3. A well-defined cube-on-cube orientation relationship exists between both cubic components.

Figure 7a shows the HREM micrograph of the interface between γ-Al_2O_3 and Ni(Al). The lattices of the cubic Ni(Al) as well as the basic structure of the cubic γ-Al_2O_3 can easily be recognised. The semi-coherent interface is composed of coherent regions in perfect registry (coherency) and misfit dislocations (indicated by arrows in Fig. 7a). The misfit dislocations relieve strains caused by a misfit of ~11% between Ni(Al) and γ-Al_2O_3. The misfit dislocations are located in the metallic Ni(Al) with a small 'stand-off' distance of the dislocation core to the interface plane.[13,14] The distances between adjacent misfit dislocations vary from 7 to 10 lattice planes.

Strained regions in the lower left part adjacent to the misfit dislocation can be identified owing to the change in background contrast. An image processed

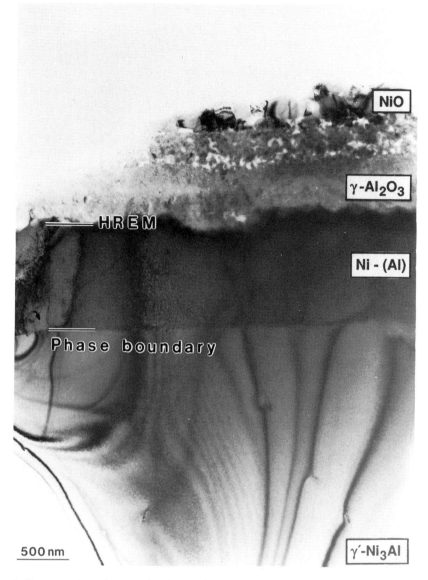

Fig. 6 Cross-section of an oxidised Ni₃Al single crystal (surface parallel to (100),
1 h at 950°C in air).

micrograph[15] is shown in Fig. 7b. Atom columns (bright spots) as well as misfit
dislocations can be identified more easily in this micrograph.

Image simulation of the regions of perfect matching (in between the lattice
defects) shows that bright spots represent the atomic columns of the atoms
imaged by the electron microscope (reverse contrast, first pass band).[3]

A preliminary quantitative evaluation of the HREM micrographs showed
that the interface between γ-Al₂O₃ and Ni(Al) is semi-coherent. Figure 7 clearly

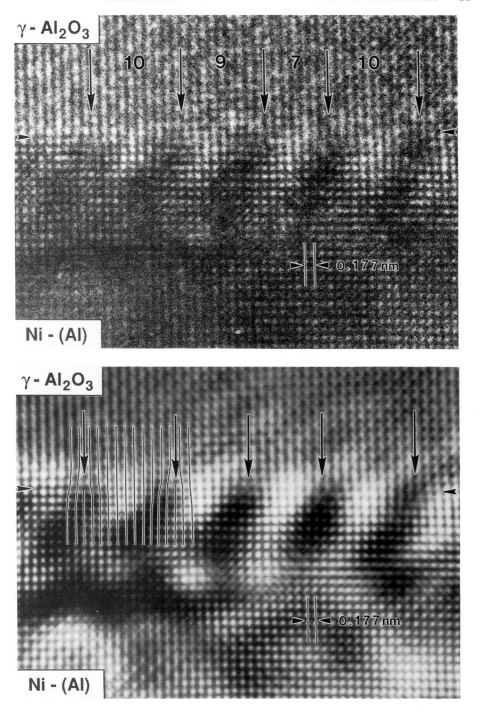

Fig. 7 High-resolution micrograph of the interface between Ni(Al) and γ-Al₂O₃. Both crystals possess a cubic structure and are oriented in a cube-on-cube relationship. (a) As received micrograph, (b) filtered micrograph.

indicates that the lattice adjacent to the core of the misfit dislocations is bent, indicating that large tangential strains exist parallel to the interface between the metal and the oxide. Raj *et al.*[14] developed a method which allows an evaluation of the tangential strains parallel to the interface. This method uses (i) the 'stand-off' distance of the misfit dislocations from the interface and (ii) the atomic coordinates of the columns close to the misfit dislocations for the identification of the shear tractions existing at the interfaces.

The evaluation requires that the position of the interface with respect to the core of the misfit dislocation is known with high accuracy. This is, however, not exactly known for the present micrograph (Fig. 7). It is difficult to locate the exact position of the interface. The following elastic properties of the two components were used: Ni(Al): shear modulus, $G = 76$ GPa, Poisson's ratio $\nu = 0.31$), Al$_2$O$_3$: $G = 150$ GPa, $\nu = 0.25$), misfit: 11%. The composite value of the shear compliance can be calculated[14] and the shear stresses parallel to interfaces were evaluated resulting in a shear strength of the interface between 60 and 120 GPa. This indicates an extremely high shear stiffness of the interface. This high value is expected owing to the large bending of the plane. It is most remarkable that the large stresses occurring due to this misfit of 11% are indeed accommodated by misfit dislocations.

5. Summary and Conclusion

High-resolution electron microscopy (HREM) is an important technique for analysing the atomistic structure of interfaces. HREM can be applied to interfaces between metal matrices and oxide scales. First HREM results are reported in this paper. The HREM studies require excellent cross-section specimens which can be obtained by cumbersome and time-consuming preparation. However, the HREM studies reveal the atomistic structure of an interface and result in values of the shear tractions existing at such interfaces. All those values are important quantities for the modelling of the mechanical behaviour (including spallation of the scale). Additional studies revealed that HREM studies in combination with analytical electron microscopy with high spatial resolution give important information which will contribute to an understanding of the fundamental processes occurring during high-tempera-ture oxidation and of the mechanical properties of the interfaces.

Acknowledgement

This work was supported by the Deutsche Forschungsgemeinschaft.

References

1. M. Rühle, A. G. Evans, M. F. Ashby and J. P. Hirth (eds), *Proc. Int. Workshop on Metal-Ceramic Interfaces*, Pergamon Press, Oxford 1990.

2. M. Rühle, A. G. Evans, A. H. Heuer and M. F. Ashby (eds), *Acta metall. Mater.*, **40**, 1992, Supplement S1-S368.
3. J. C. H. Spence, *Experimental High-Resolution Electron Microscopy*, 2nd ed. Oxford University Press, Oxford, 1988.
4. M. W. Finnis and M. Rühle, in: *Materials Science and Technology* (R. W. Cahn, P. Haasen and E. J. Kramer, eds), Vol. 1: Structure of Solids (V. Gerold, ed.) VCH-Verlag, Weinheim, 1992, pp. 533–605.
5. E. Schumann and M. Rühle, *Acta metall. mater.*, to be published.
6. J. Mayer, G. Gutekunst, G. Möbus, J. Dura, C. P. Flynn and M. Rühle, *Acta metall. mater.*, **40**, 1992, S217–S225.
7. P. Busek, J. Cowley, L. Eyring (eds), *High-Resolution Transmission Electron Microscopy*, Oxford University Press, Oxford, 1988.
8. M. A. O'Keefe, 'Electron Image Simulation: A Complimentary Processing Technique', in: *Electron Optical Systems* (O. Johavi, ed.) Chicago, SEO Inc. 1985, 209–220.
9. P. A. Stadelmann, *Ultramicroscopy*, **21**, 1987, 131–146.
10. E. Schumann, G. Schnotz, K. P. Trumble and M. Rühle, *Acta metall. mater.*, **40**, 1992, 1311–1319.
11. A. Strecker, U. Salzberger and J. Mayer, in preparation.
12. E. Schumann and M. Rühle, these proceedings.
13. W. Mader, *MRS Symp. Proc.*, **82**, 1987, 403–408.
14. R. Raj, F. Ernst, G. Gutekunst, W. Mader, J. Mayer, A. Trampert and M. Rühle, to be published.
15. G. Möbus, G. Necker and M. Rühle, *Ultramicroscopy*, **49**, 1993, 46–65.

Application of Dynamic Secondary Ion Mass Spectrometry to the Study of Oxidation

H. E. BISHOP

*Materials Characterisation Service, AEA Technology, Harwell Laboratory,
Oxfordshire OX11 0RA, United Kingdom*

ABSTRACT

Over the past few years Secondary Ion Mass Spectrometry (SIMS) has been playing an increasing role in the characterisation of attack resulting in oxidation and in determining oxidation mechanisms. For relatively planar oxide films elemental distributions may be obtained by depth profiling, but for more complex, thicker scales, imaging of polished cross-sections using high-resolution liquid metal ion probes gives the best results. SIMS offers a combination of high sensitivity (hence speed) with good lateral and depth resolution and of isotopic sensitivity. Limitations to SIMS are specimen charging and poor quantification. Secondary Neutral Mass Spectrometry may overcome these problems but usually at the cost of reduced sensitivity.

1. Introduction

The understanding of oxidation processes over the past 30 years has been greatly enhanced by the emergence of a family of surface and near-surface analytical techniques. Of these Secondary Ion Mass Spectrometry, SIMS, has evolved rapidly over the past few years and is now in a position to take a major role in characterising oxide scales and in elucidating oxidation mechanisms. The aim of this review is to identify the areas where SIMS may be applied to oxidation studies drawing on examples taken from work in our laboratory.

In SIMS a beam of primary ions is directed onto the surface of a sample and the secondary ions sputtered from its surface are analysed using a mass spectrometer.[1] As an analytical technique SIMS has a number of important advantages:

High sensitivity (with the consequent potential for high speed).
The capability for high depth and lateral resolution.
Isotopic sensitivity.
Ability to detect all elements, including hydrogen.

These advantages must be offset against the following problems:

SIMS is an inherently destructive technique.
Quantification is difficult as elemental sensitivities vary over many orders of magnitude and matrix effects for a given element may vary by up to three orders of magnitude.
Charging effects may occur for highly insulating samples.

The disadvantages may be largely overcome, at the expense of a loss in sensitivity and lateral resolution, by employing the closely related technique Secondary Neutral Mass Spectrometry (SNMS), for which there is a much smaller spread in sensitivities between elements and the measured signal is insensitive to the sample potential.

SIMS has two distinct modes of operation, static and dynamic. In the former the primary ion beam density is kept to a minimum so that the surface is essentially unchanged during the course of the measurement. In the dynamic mode much higher primary beam densities are used so that the sample is continuously eroded. Static SIMS is concerned with the composition of the surface monolayer, whereas dynamic SIMS measures near-surface composition and variations in composition with depth as the sample surface is eroded. In this paper attention will be confined to the applications of dynamic SIMS.

In the study of oxide films the strong matrix effects, that generally pose such problems in interpreting SIMS data, are actually beneficial. Positive ion yields of most species are enhanced by as much as three orders of magnitude, with the result that if an inert primary beam is used it is very easy to distinguish the oxide–metal interface both in depth profiles and in images.

2. Instrumentation

A SIMS facility requires an ion source and a mass spectrometer. The commonly used types of source and spectrometer are listed in Table 1 with their main attributes. The combination of a thermal ionisation source with a quadrupole mass spectrometer provides a relatively cheap add-on option to a multi-technique surface analytical instrument. A dedicated high-performance dynamic SIMS intrument will typically consist of a magnetic sector spectrometer together with a duoplasmatron and a caesium thermal ionisation source. Time-of-flight instruments with their high collection efficiency and ability to record the whole mass spectrum at once makes them ideally suited to static SIMS, but their low duty cycle makes them less useful for dynamic applications.

SIMS data may be collected in a variety of modes. The most common of these in oxidation studies is depth profiling, where the variations in composition with depth are determined by following changes in the SIMS signal as the sample is eroded by the primary ion beam. Other modes are simple mass spectra, imaging, line profiles and 3D imaging, where sequences of images are recorded as the sample is eroded. SIMS generates a large body of data and it is essential to have a good data acquisition and processing system to take full advantage of the technique.

Ion Sources	
Electron impact	cheap low brightness gas source
	probe typically $>100 \mu m$
Duoplasmatron	high-performance gas source
	probes down to 500 nm
Thermal ionisation	alkali metal source
	enhances negative ion yields
	probes down to 200 nm
Liquid Metal field emission	high brightness sources
	probes down to 20 nm
Mass Spectrometers	
Quadrupole	inexpensive, UHV compatible
	low transmission, narrow energy window
	unit mass resolution
Magnetic Sector	high transmission, wide energy window
	high mass resolution
	stigmatic imaging possible
Time-of-Flight	high transmission, wide energy window
	high mass resolution
	whole mass spectrum recorded in parallel
	low duty cycle
	stigmatic imaging possible

Table 1

3. Depth Profiling

Many oxides take the form of a uniform film over the surface of the metallic substrate. For such films the first requirement is to determine their composition and how this varies through the thickness of the film. For quantitative analysis of the major elements in thin films, <50 nm, electron spectroscopy techniques such as XPS and AES are to be preferred. Interpretation of the first 10 nm or so of any profile is complicated in SIMS because of the gradual build up of a concentration of the implanted primary ion beam in the surface. Until an equilibrium is established ion yields and sputtering rates are continuously varying. Once an equilibrium is established SIMS comes into its own as AES and XPS are relatively slow. SIMS profiling is particularly suited to the following situations:

The properties of a number of similar samples are to be compared.
Trace elements are of interest.
Hydrogen or lithium profiles are required.
Isotopic tracers are used.

An example of the first situation has arisen from a long term project measuring oxide scales produced in simulated Pressurised Water Reactor (PWR) water

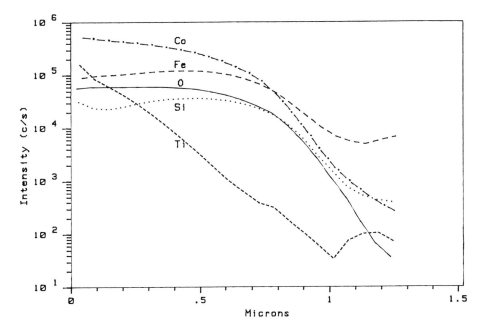

Fig. 1 Profile through reference oxide formed on a steel.

loops to determine the mechanisms of cobalt incorporation.[2] Scale thickness ranged from 50 nm to 3 μm. Using the Cameca IMS 3F and an argon primary beam, standard conditions have been established where the depth distribution of 10 elements is recorded using a sputtering rate of either 100 or 25 nm per minute, depending on the film thickness. A set of relative sensitivity factors was established using EPMA on thicker samples. An important feature of this approach was that only a 500 μm square of the sample was affected so that samples could be replaced in the loop for further exposure. A reference sample was run for each batch of specimens to check for consistency. A profile from this 1 μm reference sample is shown in Figure 1. Note how the strong matrix effect clearly delineates the oxide–metal interface.

An area where SIMS has proved particularly valuable has been in the study of the role of reactive elements in high-temperature corrosion studies. In Figure 2 we see depth profiles recorded from a set of scales produced on nickel under identical conditions after different thickness ceria coatings had been applied.[3] Above a critical ceria layer thickness a beneficial ceria-rich layer remains within the scale. The protection mechanism may be investigated using $^{18}O_2$ in the final stages of the oxidation. Figure 3 shows the profile through a chromia scale containing cerium.[4] The profile shows that new oxide is mainly formed at the oxide–metal interface by oxygen in diffusion, but in addition there is some new oxide formation together with oxygen exchange in the ceria layer.

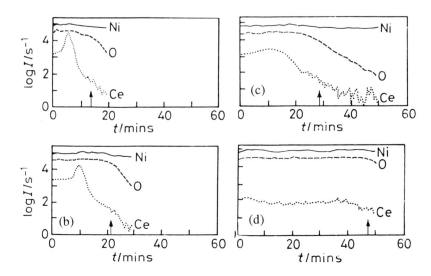

Fig. 2 Depth profiles for oxide scales formed on CeO_2 sol-coated Ni at 900°C, 14 h. Sol concentrations: (a) 2 g l^{-1}, (b) 0.8 g l^{-1}, (c) 0.4 g l^{-1}, (d) 0.01 g l^{-1}. $\wedge$ marks the scale–metal interface

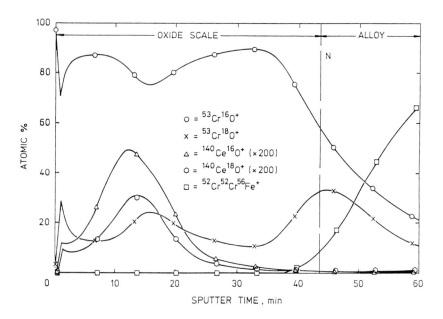

Fig. 3 Depth profile through the Cr_2O_3 scale formed on Fe-25 Cr, coated with 40 Å CeO_2, after oxidation at 1100°C first in $^{16}O_2$ and then in $^{18}O_2$.

4. Imaging

Depth profiles only give one-dimensional information, and, unless the oxide film is very smooth, depth resolution degrades rapidly with increasing depth. SIMS imaging may be used to investigate variations in composition across the surface of an oxide. It may be used also to map the in-depth variations in composition of thicker scales by preparing taper cross-sections. The high spatial resolution available using liquid metal ion guns makes the sectioning approach a particularly effective way of characterising scales.[5]

The extra information that may be obtained is illustrated in an ^{18}O tracer experiment on nickel oxide scales of the type profiled in Figure 3. Figure 4 shows cross-sections through two scales, one without, 4(a)-(c), and one with, 4(d)-(f), a protective ceria layer. In the secondary electron image of the untreated sample, the grain structure of the NiO is clearly seen, Figure 4(a). Figure 4(b) shows how the new ^{18}O rich oxide is formed at the scale-gas interface. Note however that there is some penetration of ^{18}O down the NiO boundaries in the direction of the oxide metal interface. On the ceria treated sample, new oxide still forms on the outer surface of the oxide, Figure 4(f), but there is also oxygen exchange taking place within the ceria layer. The ceria is also present in the grain boundaries of the inner NiO layer. In general, the time required to record a set of SIMS images is much less than that for comparable images obtained using EPMA or AES. The images presented here were recorded typically in one or two minutes, one to two orders of magnitude less than the time required for X-ray or Auger images.

The location of trace elements within an oxide scale is illustrated in Figure 5, which shows a section through an oxide scale produced at 700°C on an Ni_3Al alloy containing some boron. A duplex oxide is formed with an NiO outer layer and an inner oxide containing Ni and Al.[5,6] The depth profile showed an apparent uniform concentration of boron in this inner oxide. In the alloy the boron is present as a grain boundary phase and the SIMS images indicate that it retains a similar distribution within the inner oxide.

Tracer techniques are also of great value in studying the corrosion of composite materials. Both silicon carbide reinforced ceramic composites and nickel aluminium, reinforced with alumina fibres, have been studied using an ^{18}O tracer. In both cases the fibre–matrix interface proves to be a rapid diffusion path for oxygen, leading to anomalously high oxidation rates for these materials.

5. Secondary Neutral Mass Spectrometry

A major limitation to SIMS of many oxides is specimen charging. The oxides of Fe, Cr and Ni do not usually present problems, but alumina scales can be very troublesome. The majority of atoms sputtered from a surface actually leave as neutral species and are unaffected by charging. In SNMS these neutrals are ionised and mass analysed. Although, in general, much less sensitive than SIMS the technique is much easier to use and interpret in cases where charging

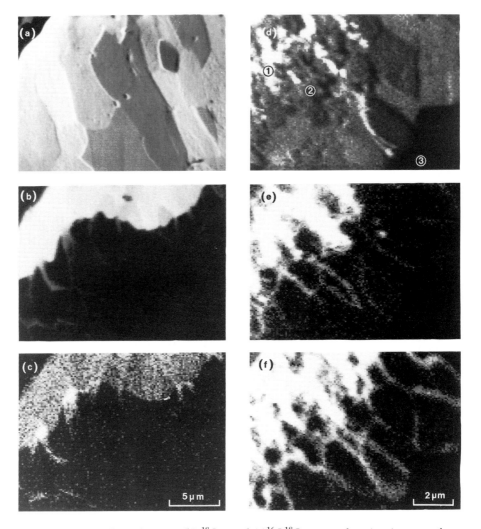

Fig. 4 (a) Secondary electron, (b) $^{18}O^-$ and (c)$^{16}O^{18}O^-$ secondary ion images of a transverse section of an NiO scale formed during sequential oxidation in $^{16}O_2{}^{18}O_2$ at 900°C. (d) $^{16}O^-$, (e) $^{140}Ce^{16}O^+$ and (f)$^{18}O^-$ ion images of transverse section through a scale formed on CeO$_2$ sol-coated Ni at 900 °C. Position 1, CeO dispersed NiO layer; 2, inner NiO layer; 3, Ni substrate.

in SIMS is severe. The technique has proved particularly useful in the study of the alumina scales formed on high-temperature alloys.[7]

6. Summary

SIMS has an important part to play both in the routine evaluation of corrosion scales and in more detailed mechanistic studies. Its chief advantages are high

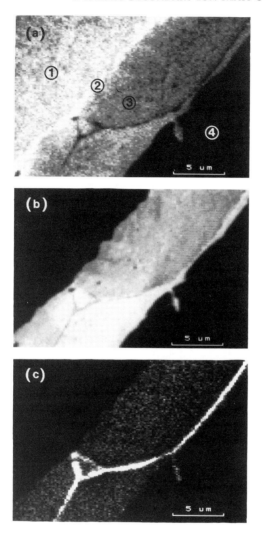

Fig. 5 (a) Ni$^+$, (b) Al$^+$ and (c) B$^+$ secondary ion images of a taper section of a scale formed on Ni$_3$Al at 700°C. Position 1, scale surface; 2, outer NiO layer; 3, inner oxide; 4, alloy substrate.

sensitivity, speed and isotopic sensitivity for both depth profiling and for imaging applications. The main limitations are difficulties with quantification, and, for highly insulating oxides such as alumina and silica, specimen charging. The associated technique of SNMS may overcome these problems but at a cost of either lower sensitivity or lower rates of data acquisition or both. Both SIMS and SNMS are at a stage of rapid instrumental development so that significant improvements in performance may be anticipated over the next few years.[8]

Acknowledgements

It is a pleasure to acknowledge the contributions of David Moon and Mike Bennett to this work. Figure 3 was kindly supplied by Dr. R. J. Hussey (NRC, Canada).

References

1. For a general introduction to SIMS see: A. Benninghoven, F. G. Rudenauer and H. W. Werner, *Secondary Ion Mass Spectrometry*, Wiley, New York, 1987.
2. M. D. H. Amey, P. Campion, G. C. W. Comley, W. J. Symons, and R. K. Butter, *Water Chem. Nucl. React. Syst.* **5**, 55–61, British Nuclear Energy Society, London, 1989.
3. D. P. Moon, *Oxid. Met.*, 1989, **32**, 47–66.
4. M. J. Bennett and D. P. Moon, *The Role of Active Elements in the Oxidation Behaviour of High Temperature Metals and Alloys*, ed. E. Lang, Elsevier Applied Science, 1989, 111–129.
5. H. E. Bishop, D. P. Moon, P. Marriott and P. R. Chalker, *Vacuum*, 1989, **39**, 929–939.
6. J. H. Devan, P. F. Tortorelli, H. E. Bishop and M. J. Bennett, to be published.
7. H. E. Bishop and S. J. Greenwood, *Surface and Interface Analysis*, 1988, **12**, 27–34.
8. See, for instance, *Proceedings of SIMS VIII*, ed. A. Benninghoven, K. T. F. Janssen, J. Tumpner and H. W. Werner, John Wiley and Sons, Chichester 1992.

The Use of a Hot-Stage Microscope in High-Temperature Corrosion Studies

J.F. NORTON, S. CANETOLI and P. PEX*

*Commission of the European Communities, Institute for Advanced Materials,
J.R.C.-Petten Establishment, 1755 ZG Petten, The Netherlands
Now with Energy Centrum Nederlands (ECN), Petten.

ABSTRACT

A hot-stage microscope has been specially adapted to enable *in-situ* observations to be made during high-temperature corrosion experiments using aggressive gas mixtures. In addition to studying the corrosion events as they occur, this facility also enables decisions to be taken whereby specimens can be quenched at critical stages during the nucleation and growth process and subsequently examined in much greater detail using a sequential SEM/EDX technique.

Events occurring during exposure of an Fe-Cr-Ni alloy to a carburising H_2-CH_4 gas mixture at 825°C have been monitored and recorded using a video camera. Observations show that, due to the presence of relatively low levels of oxygen-bearing species in the H_2-CH_4 gas mixture, an oxide layer was formed on the alloy surface during heating, prior to the commencement of the carburising reactions. At the test temperature carbides nucleated whilst at the same time the surface oxide was gradually being reduced by the gas mixture. It appears that the reduction of the Cr-oxide may also provide a reservoir of Cr which contributed to the growth of the carbide nuclei. The influence of prior surface finish and condition has been shown to be an important factor in this nucleation and growth process.

During similar higher-temperature investigations concerned with the oxidation behaviour of an Si_3N_4 material it was observed that at between 920°C and 950°C the surface layer which had formed became liquid and this was accompanied by the vigorous evolution of N_2. At higher temperatures silica glass again formed along with a Y_2SiO_5 phase.

1. Introduction

Engineering components operating at high temperatures often suffer corrosion in the service environment leading to a loss of load-bearing cross-section with a consequent reduction in plant performance and efficiency. Over the years, a great deal of materials research has therefore been directed towards improving the long-term performance and reliability of materials exposed to aggressive high-temperature process atmospheres. An important requirement for these studies is to establish a deeper understanding of the mechanisms involved and the factors governing the early stages of the corrosion process.

Although hot-stage microscopy is a well-known and established technique, its application has often been limited to use with simple environments such as air, argon or nitrogen. This paper summarises observations from in-situ nucleation and growth experiments carried out on materials exposed to high-temperature gaseous environments which, in some cases, contained explosive and toxic species. A special gas-tight environmental chamber containing a small furnace and fitted with a quartz viewing window has been attached to a conventional optical microscope and a video camera used to record the events.

From these dynamic *in-situ* observations it has been possible to interrupt experiments periodically at critical stages as the corrosion process develops in order to carry out more detailed examinations of rapidly quenched specimens. Such sequential studies have provided valuable complementary information on the morphological and compositional development of the phases formed.

In this paper, the practical application of the technique for high-temperature nucleation and growth studies is illustrated by examples taken from results obtained during short-term exposure of alloys and ceramics to various corrosive environments. Results are presented of observations of the carburisation of Fe-Cr-Ni alloy samples exposed to an H_2-CH_4 gas mixture at 825°C. In addition, a brief summary of the oxidation of a commercial silicon nitride exposed at temperatures of between 750°C and 1150°C is presented. More detailed results from this latter study are contained in another paper given at this conference.[1]

2. Experimental Technique

2.1 The hot-stage and gas supply system

In order to study the nucleation and growth of the corrosion products as they form when the gas first comes into contact with the hot surface of the material, a conventional optical microscope has been fitted with an environmental chamber sealed with a quartz viewing window.

Using special lenses, magnifications of up to ×320 are possible with the present system. Further, image enhancement has been obtained using a video camera enabling apparent magnifications of greater than ×1500 to be achieved whilst retaining excellent definition and image quality.

The corrosion processes are observed as they occur in the hot stage via an on-line TV monitor. In addition, video recording the sequence of events using time-lapse techniques is carried out which offers significant advantages since during high-temperature corrosive attack many changes occur very rapidly. Hard-copy prints of the more important features can be readily selected from the video film for accurate comparison purposes using an image-grabbing technique.

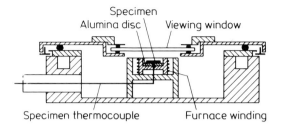

Fig. 1a Schematic diagram of hot-stage environmental chamber.

Complementary, more detailed examinations of the important changes in composition, morphology and distribution of the surface corrosion products have been made using XRD, and SEM/EDX analysis, as well as conventional cross-sectional microscopy. This has been achieved by rapidly quenching samples during the *in-situ* studies so as to 'freeze-in' key events, the development of which is then followed by the sequential re-examination of the selected areas.

The environmental chamber contains a small Pt-20%Rh-wound Al_2O_3 furnace in which sits a disc-type specimen, 5 mm dia. $\times 0.5$ mm thick. The temperature is monitored by a Pt/Pt–10%Rh thermocouple placed on the lower, back face of the specimen. There is a temperature difference between the upper and lower faces of the specimen and a calibration is carried out using pure metal standards of known melting point in order to establish the actual temperature of the viewed top face. The hot-stage has a temperature capability of 1500°C, is controlled by a power unit capable of generating complex temperature cycles and it is possible to apply high linear heating rates of up to 100 K min^{-1}. The environmental chamber was manufactured by Linkam Scientific Instruments Ltd.[2] A schematic diagram of the chamber showing the specimen and furnace configuration is presented in Figure 1a.

High purity pre-mixed gas mixtures with compositions designed specifically for the corrosion studies (in terms of the activity of reactants such as C, O and S) are used. The gas supply system has been designed and built within the Institute's laboratories and incorporates several necessary safety features which enable potentially explosive and toxic gas mixtures (containing species such as H_2, CH_4, CO, H_2S) to be used, in addition to the more commonly used innocuous atmospheres such as air, argon and nitrogen. The gas supply system is shown schematically in Figure 1b. It is worth pointing out that for studies involving sulphidising species a completely separate gas supply system and second environmental chamber are used.

2.2 Procedure

Disc specimens were machined from a ternary Fe-25Cr-35Ni alloy with a composition approximately that of the HP40 class of tube material used in petrochemical methane reformation plant. The chemical composition of the alloy is given in Table 1. These specimens were prepared with one of two types

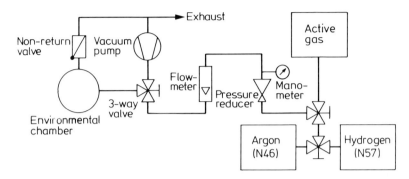

Fig. 1b Flow diagram of gas supply circuit to hot-stage.

Fe–Cr–Ni Alloy:

Cr	Ni	Fe	C	Si	Mn	S
24.8	35.1	Bal.	0.013	0.02	0.01	0.056

Si_3N_4:

	Y_2O_3	MgO	Al_2O_3	Fe_2O_3	Si_3N_4
	9.3	0.01	0.12	1.71	Bal.

Table 1 Chemical Analyses of Materials Studied (Wt %)

of surface finish: i.e. electropolished (smooth and work-free) or ground on SiC paper (180 grade, worked surface).

The specimens were exposed to an H_2-based gas containing nominally 4%CH_4 at 825°C at atmospheric pressure. Taking into account the presence of minor amounts of oxygen-bearing species in the component gases and assuming that equilibration of the mixture takes place at temperature, the mixture has a carbon activity of just less than 1, i.e. below the level at which carbon deposition from the gas phase will occur, and a very low oxygen partial pressure of 10^{-28} bar.

A commercial grade of silicon nitride produced by hot pressing at 1700–1800°C with the aid of a Y_2O_3 sintering agent was studied. The composition is shown in Table 1. Disc specimens were also exposed in static laboratory air at temperatures up to 1150°C during a series of experiments under isothermal-dwell and ramp conditions.

Before heating the specimens, the correct test atmosphere was established in the environmental chamber by evacuating and back-filling several times with argon followed by a similar procedure using the test gas. The specimen was then rapidly heated to the selected temperature using rates of up to 90 K min^{-1}. When the set-point temperature was reached the exhaust valve was opened thereby maintaining a constant flow of test gas over the specimen in order to ensure that the calculated reactant activities were achieved. In the case of the oxidising atmosphere used in the ceramics study, the procedure was slightly different, stagnant laboratory air being used.

3. Results

3.1 Carburisation studies of an Fe-Cr-Ni alloy in H_2-CH_4.

Figure 2 summarises the observations of the main surface changes occurring during the early stages of exposure to the carburising gas mixture. When viewed under the microscope in normal light, during the heating cycle the grains themselves did not all exhibit the same colour, thus reflecting variations in the thickness of the initial oxide film which had formed. A green colouration indicative of the presence of a chromia film was clearly evident. Carbides also nucleated from a very early stage, initially primarily at grain boundaries and at, or adjacent to, inclusions and finally within the grains themselves.

The sequence of micrographs shown in Figure 2 (a–d) illustrates how individual carbide nuclei, which had formed from a very early stage, i.e. 2.5 minutes, Figure 2b, grew laterally along the grain boundaries linking up with others, Figure 2c, prior to the precipitation of discernible amounts of intra-granular carbide after 45 minutes, Figure 2d. The influence of indigenous sulphide inclusions in promoting carbide precipitation is clearly evident, comparing Figure 2c with Figure 2a.

The benefits of extending the *in-situ* optical examination to a sequential scanning electron microscope (SSEM) technique during the intermittent quen-ching and re-heating of the specimens as key events were observed are illustrated in Figure 3. Figure 3a summarises the surface changes observed on an electropolished sample and clearly confirms that carbide nucleation oc-curred in the chromium-rich areas at or near the grain boundaries and around the indigenous sulphide inclusions, as seen in the micrograph of the specimen exposed for 30 minutes. Extended studies showed that after 55 minutes a phase having a needle-like morphology had grown on top of the existing carbide nuclei. Carbide nucleation within the grains was observed to increase considerably with continuing exposure with many discrete carbides being evident after 75 minutes, as shown in the figure.

The role of surface working in carbide nucleation and growth is evident when the observations presented in Figure 3a are compared with those made on a ground specimen exposed under identical conditions, Figure 3b. In this case, carbide precipitation is clearly influenced by the original grinding of the surface with the nucleation rate appearing fairly constant with increasing exposure. The carbides which nucleated along the grinding marks exhibited a needle-like morphology from a very early stage, as shown in the micrographs of the specimen exposed for 60 minutes and 90 minutes.

3.2 Oxidation studies of a hot-pressed Si_3N_4

Observations made during the heating of an Si_3N_4 specimen to 950°C followed by an isothermal-dwell period of 120 minutes are summarised in Figure 4. At

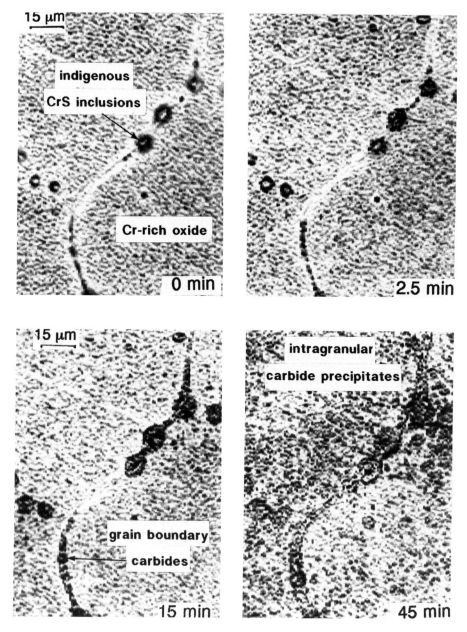

Fig. 2 (a–d) Micrographs of surface changes observed during *in-situ* study of carburisation of an Fe-Cr-Ni alloy at 825°C.

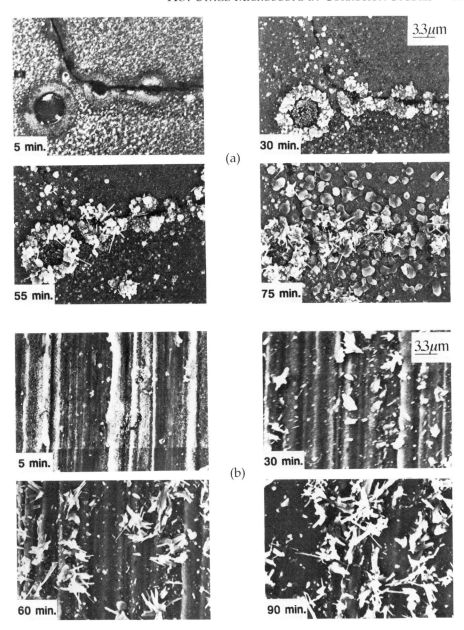

Fig. 3 Sequential SEM study of nucleation and growth of carbides on Fe-Cr-Ni alloy exposed to H_2-4%CH_4 atmosphere at 825°C. (a) electropolished surface, (b) ground surface.

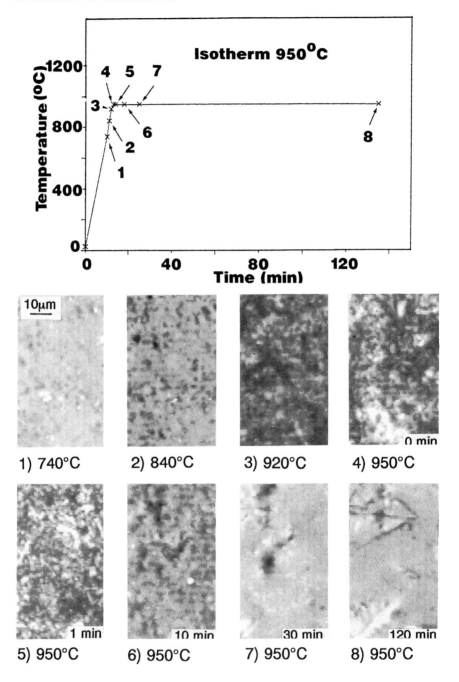

Fig. 4 Micrographs of surface changes observed during *in-situ* study of the air oxidation of Si_3N_4 at 950°C.

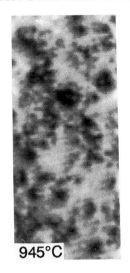

Fig. 5 Summary of main events occurring on the surface of Si_3N_4 heated slowly at 1 K min^{-1} from 830°C to 1050°C.

approximately 740°C, (point 1), the first visible reaction took place; this appeared as a black phase which nucleated and grew on the specimen surface with increasing temperature, (point 2). At about 920°C, (point 3) vigorous bubbling and gas evolution occurred. As the temperature reached 950°C, (point 4) a glassy phase started to nucleate and grow until after 10 minutes, (point 6) it covered almost the entire surface of the specimen. This appeared to thicken with time, (point 7) and a white phase was observed to form in places on top of the glassy layer by the end of the experiment (point 8).

Results obtained during a key experiment in which a specimen was very rapidly heated (90 K min^{-1}) to 830°C and then at a much slower rate of 1 K min^{-1} up to approximately 1050°C have enabled the dynamic changes occurring within this temperature range to be more easily observed. The growths of both the black phase below 920°C and the glassy layer plus needle-like white phase above this temperature were confirmed. This growth and the 'bubble phenomenon' are summarised in Figure 5 for selected temperatures of 845°C, 945°C and 960°C. The latter micrograph captures a bubble bursting (in the lower right-hand corner of the picture).

SEM/EDX examinations carried out on a rapidly quenched specimen after the termination of the experiment showed that the white phase was rich in yttrium (Y_2SiO_5) whilst the glassy phase contained mainly silicon.

4. Discussion

Carburisation studies of an Fe-Cr-Ni alloy have shown that due to trace amounts of oxygen-containing species in the H_2–CH_4 gas mixture a chromia

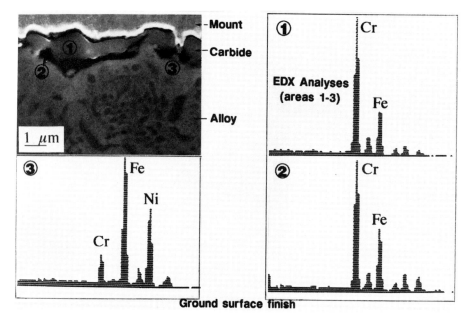

Fig. 6 Cross-sectional SEM examination of carburised specimen discontinued after 300 minutes' exposure at 825°C.

layer forms whilst the specimen is being heated to the test temperature. When the sample is at temperature (825°C) this oxide is thermodynamically unstable in the very low oxygen partial pressure of the gaseous atmosphere used for these studies. Therefore, it is possible that either this oxide may be reduced before carburisation commences or, alternatively, carbides may nucleate on the surface by direct *in-situ* conversion of the oxide.

In a previous study, Smith[3] considered that oxide-to-carbide conversion was likely to be slow at these temperatures and concluded that carbide nuclei will only grow slowly by oxide conversion, if they were isolated from the alloy substrate. Once nucleated, their penetration to the oxide-metal interface and subsequent growth to a clearly visible size depends on the thickness of the initial chromia layer. In a complementary experiment, exposure of identical specimens of this alloy for several hours supported this proposal when it was observed that carbides became discernible earlier on the surfaces of electro-polished specimens than on ground specimens, due to the fact that in the latter case, the associated surface working had promoted the growth of a thicker surface oxide layer.

Limited cross-sectional examinations of selected quenched specimens showed that in places carbides were present on top of the oxide and there was an apparent absence of any contact with the alloy substrate. Figure 6 shows that after 300 minutes' exposure of a ground specimen, in places a continuous oxide layer was evident beneath the carbides. EDX analyses showed high Cr peaks associated with both the outer carbides and the oxide present at the carbide/alloy interface whilst immediately under the oxide the substrate was

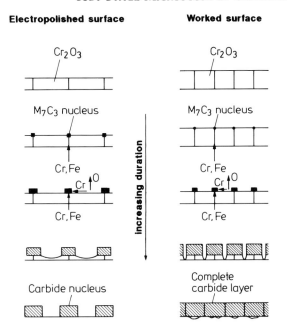

Fig. 7 Illustration of influence of surface condition on carburisation model originally proposed by Smith.[3]

significantly depleted in Cr. On the equivalent electropolished surfaces the oxide had almost disappeared in places with the thinnest areas of oxide tending to lie between the carbides rather than underneath them. This may be an indication that during the early stages of corrosion the carbides are not necessarily growing only by direct conversion of the oxide but are being supplied with chromium diffusing partly from the reduced oxide between the carbides (surface diffusion) and partly from the bulk alloy (probably via oxide grain boundaries).

Thus the nucleation and growth mechanism proposed in Smith's simple model remains largely valid but as a result of these *in-situ* studies may be modified as indicated in Figure 7. It is postulated that at the start of the carburisation, M_7C_3 carbides nucleate on the grain boundaries by direct conversion of oxide, as described in Smith's model. These nuclei will subsequently grow at a rate principally governed by the rate of chromium (and iron) diffusion from the alloy substrate along the oxide grain boundaries. This will occur more rapidly through the thinner oxide formed on electropolished specimens and, as a result, the carbides will become clearly visible earlier and continue to grow at a faster rate than those on the ground surface. At the same time, the oxide between the particles is reduced and the chromium liberated becomes incorporated in the carbide particles. As carbide nuclei grow laterally and begin to impinge, the underlying oxide may become completely isolated from the gas. This was especially noticeable on the ground surface, where the higher carbide nucleation density meant that a continuous carbide layer was formed earlier.

Later on in the corrosion process, after most or all of the oxide layer has disappeared, the carburisation rate is limited by the relative diffusion rates of carbon and carbide-forming elements within the bulk alloy. In the surface regions these rates are faster for the ground specimen because surface working induces grain refinement locally which enhances the contribution of grain boundary diffusion. Therefore, with continuing exposure, the rate of carburisation of the ground specimens may, for a time, overtake that of the electro-polished specimens.

Oxidation studies of an Si_3N_4 ceramic have enabled the surface phenomena occurring over the temperature range 750°C to 1150°C to be observed *in-situ* and related to those predicted by an existing model.[4]

The material appeared to remain chemically inert at temperatures up to 750°C with no visible oxidation evident. As the temperature was raised to 920°C, two phases were observed to nucleate which subsequently grew to cover the surface, a Y-containing O-apatite and a silicate glass.

A further increase in temperature resulted in dramatic changes due to the glassy phase becoming liquid, thereby enabling N_2, resulting from the oxidation of the indigenous H-phase, i.e. $Y_5(SiO_4)_3N$, to form as bubbles and be liberated from the outer surface. As higher temperatures were reached, silicon nitride was also oxidised which produced both N_2-release and SiO_2 formation. The glassy phase was thus enriched in silica and after some time its viscosity increased, although some bubbling was still apparent on the surface.

5. Concluding Remarks

An optical microscope fitted with an environmental chamber containing a Pt-20%Rh-wound Al_2O_3 furnace has been used to carry out *in-situ* corrosion experiments in gaseous atmospheres at high temperatures. The present hot-stage has a temperature capability of 1500°C and can be used with hazardous gases, e.g. H_2, CH_4, CO, H_2S, SO_2, and SO_3 in addition to the more commonly used air, nitrogen or argon atmospheres. A video camera enables electronically enhanced images of the events to be recorded as they occur.

In addition to its use for real-time, *in-situ* studies, valuable complementary information can be obtained indirectly by quenching samples in the hot-stage, effectively 'freezing' key events as they occur. This has enabled more detailed morphological and chemical data to be derived using an intermittent surface SEM/EDX technique whereby precisely the same area is re-examined after repeated exposures in the hot-stage, i.e. by sequential scanning electron microscopy (SSEM).

This paper presents examples which demonstrate the use of this technique to study the early-stage nucleation and growth of corrosion products on the surfaces of advanced alloys and engineering ceramics. A study of an Fe-25Cr-35Ni alloy exposed at 825°C to an H_2-CH_4 gas mixture has highlighted the role of thin surface oxides in influencing the nucleation of chromium-rich carbides thereby contributing to the mechanistic understanding of the carburisation process.

Similar types of *in-situ* study involving the oxidation of a commercial Si_3N_4 in air at temperatures between 750°C and 1150°C have enabled remarkable changes to be observed. The vigorous release of N_2 bubbles through a liquid surface corrosion product, formed as a result of the oxidation of phases within the Si_3N_4, at temperatures as low as 950°C, has been a noteworthy observation.

6. Acknowledgements

Studies concerned with behaviour of ceramic materials have been carried out under the direction of R. J. Fordham and reference has been made to a more detailed paper presented by him at this conference. J. F. Coste, a summer student from ENSM (St Etienne), and M. Spreij, a final-year student from the Technical University of Delft, have also contributed to various aspects of this work. The SSEM studies were carried out by K. Schuster and his valuable support is gratefully acknowledged.

References

1. R. J. Fordham, J. F. Norton, S. Canetoli, and J. F. Coste. 'In-situ studies of the surface changes occurring during the high-temperature oxidation of a silicon nitride,' Proceedings of this conference.
2. *TH 1500 high temperature stage*, Brochure of Linkam Scientific Instruments Ltd.
3. P. J. Smith, Ph.D. Thesis, The National University of Ireland, 1984.
4. J. B. Veyret and M. Billy, 'Oxidation of hot-pressed silicon nitride: modelling' in *Euro-Ceramics Vol. 3: Engineering Ceramics*, Proceedings of 1st European Ceramic Society Conference, Maastricht (NL), edited by G. de With, R. A. Terpstra and R. Metselaar, Elsevier Applied Science, 1989, p. 512.

Developments in Raman Microscopy and Applications to Oxidation Studies

DEREK J. GARDINER

Department of Chemical and Life Sciences, University of Northumbria at Newcastle, Ellison Place, Newcastle upon Tyne, UK, NE1 8ST

ABSTRACT

The technique of Raman microscopy is described along with a discussion of recent advances in Raman mapping and imaging. Single point, point-by-point mapping and Raman Microline Focus Spectrometry are discussed and application examples are drawn from: corrosion scale stress measurements, species mapping and PVC degradation.

Introduction

Raman spectrometry measures molecular vibrations by analysing laser radiation scattered from a sample.[1] Typically, though by no means exclusively, visible radiation from a cw gas laser is used to excite the spectra. This situation is shown schematically in Figure 1 where the different scattering events occurring at the sample are illustrated. Scattered photons arise from both elastic (Rayleigh) and inelastic (Stokes and anti-Stokes) scattering. It is the red shifted, Stokes Raman scattering which is normally analysed and a plot of intensity against Raman shift expressed in wavenumbers (cm^{-1}) is used to display the Raman spectra. When microscope optics are used to focus the laser light onto the sample and to collect the scattered light, the resulting Raman microscope enables Raman spectra to be observed from points on the sample limited in size only by the spatial resolution of the optics and the wavelength of the laser light.

Instrumentation

A single point Raman microscope may be realised by coupling a modified optical microscope to a standard Raman spectrometer. The modification

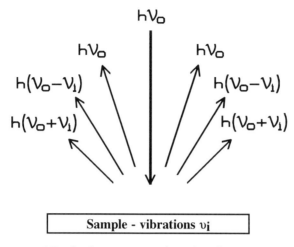

$$hV_o$$

$$hV_o \qquad hV_o$$

$$h(V_o-V_i) \qquad h(V_o-V_i)$$

$$h(V_o+V_i) \qquad h(V_o+V_i)$$

Sample - vibrations v_i

Fig. 1 Raman and Rayleigh scattering of incident laser light of energy $h\nu_0$.

comprises the use of a beamsplitter and coupling optics to introduce the laser into the microscope and to collect the scattered light.[2] Figure 2a shows a diagram of such a system which would incorporate a scanning double monochromator and photomultiplier detection.

Speed of acquisition of Raman data may be considerably improved by the use of linear diode array detectors which will collect all the spectral data simultaneously. If this is combined with computer-controlled movement of the sample, then point-by-point Raman mapping and imaging is possible. This arrangement is shown in Figure 2b.

A breakthrough in Raman microscope surface imaging and profiling came with the advent of ccd (charge coupled device) detectors. These two-dimensional detectors used in a slow scan mode and cooled to reduce dark current, comprise typically 400×600 or 256×1024 silicon sensors arranged as potential wells which trap photoelectrons from interaction with the impinging light. The electrons can be read out row by row and column by column to provide a digital image (intensity and position) of the detected light.

When used in conjunction with a line focus on the sample, the entire spatial distribution of Raman spectra along the micro-line on the surface is collected in a single exposure. This approach to Raman imaging and profiling is referred to as Raman Microline Focus Spectrometry (MiFS) and is shown schematically in Figure 2c.

Alternatively, it is possible to generate Raman images using global illumination of the sample and utilising a tuneable interference filter assembly to select the Raman shift range from which the image is to be formed.[3]

Applications

This paper is not intended as a comprehensive review but as an overview of the applicability of Raman microscopy to oxidation studies. Whilst recognising

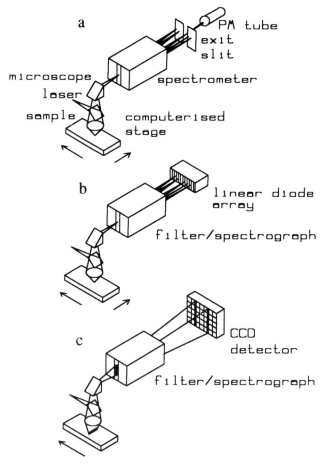

Fig. 2 Raman microscope configurations: A – scanning monochromator and Photomultiplier Tube (PMT) detection; B – spectrograph and Linear Diode Array (LDA) detector; C – Microline Focus Spectrometry (MiFS) using a ccd detector.

the research being carried out elsewhere, all of the application examples are taken from the author's own laboratory.

Single point studies

The nature and distribution of corrosion products formed on iron and iron-chromium alloys in air at high temperatures has been studied by Raman microscopy. Fe-Cr alloys, containing 2%, 5%, 14% and 18% chromium, were oxidised for 2 hours in air at 400, 600 and 850°C. Examination of the surface revealed varying distributions of Fe_2O_3, Fe_3O_4, Cr_2O_3 and $FeCr_2O_4$ by comparison with reference spectra obtained from the pure oxides.[4] Evidence was also obtained of the presence of 10 μm sized regions of Fe_2O_3 in the corrosion scale which, it was suggested, may indicate breakaway corrosion initiation sites.

A diamond anvil cell (DAC) used in conjunction with a Raman microscope, allows the effects of pressures up to 100 kbar on solids to be studied with ease.[5] In a study of the effects of hydrostatic stress on the Raman spectrum of α-Cr_2O_3, a DAC which employed an upper diamond anvil and a lower tungsten anvil was used. The sample was held in a 0.3 mm diameter hole in a 0.2 mm thick inconel gasket pressed between the two anvils and was positioned in the laser spot by viewing through the diamond anvil using the microscope. A 4:1 by volume, methanol:ethanol mixture was used as the pressure transmitting fluid around the sample to ensure a hydrostatic environment. Pressure was applied to the cell via a lever system and a hydraulic pump. Pressure measurements were realised using the ruby fluorescence method[6] and the resulting pressures were reported accurate to $\pm$0.5 kbar. The stress-induced shift of the v_1 A_{1g} vibration mode of α-Cr_2-O_3 was determined[7] as 2.8$\pm$0.8 kbar cm^{-1}.

In-situ oxidation studies can be undertaken using a microscope hot-stage.[6] This enables samples to be held at a pre-programmed temperature, in a controlled atmosphere, whilst at the same time allowing micro-Raman data to be collected from the surface. The build-up of an α-Cr_2O_3 scale on hydrogen annealed and polished, pure chromium metal samples was studied in this way. Measurements of the Raman shift of the $v_1(A_{1g})$ band were made on three different samples and at different positions on the samples as oxidation proceeded. The chromium was oxidised by heating at 800 °C in the hot stage whilst purging with argon containing a trace of oxygen. As a result, oxidation was controlled at a slow rate compared to the time required to make a Raman measurement. For two of the samples an initial decrease in Raman shift was evident during early development of the oxide scale. All samples exhibited essentially constant Raman shift values from *ca.* 80–230 minutes and thereafter the Raman shift was seen to decrease. Using the relationship between applied stress and Raman shift for this band, this behaviour was interpreted in terms of the stress changes in the surface of the α-Cr_2O_3 film during growth.

Point-by-point studies

Automated point-by-point Raman mapping is capable of generating species-specific images and profiles of a surface. In a recent study the phase-change response of zirconia ceramic to applied stress has been mapped.[8,9] An early example of Raman profile data obtained using the photomultiplier tube-based instrument is shown in Figure 3. The spectra were recorded from the surface of Fe-14%Cr alloy which had been corroded in air at high temperature.[10] The spectra, recorded every 10 μm along a 150 μm analysis line using a stepper motor driven stage, reveal a distinct species variation on the surface. The Raman band around 550 cm^{-1} is due to α-Cr_2O_3 whilst the broader bands arise from Fe_3O_4 and $FeCr_2O_4$. In the region between 40 and 80 μm the protective α-Cr_2O_3 film has been lost and a corrosion pit is developing.

Build-up of stress in protective oxide films results in cracking and eventual spallation of the film rendering the underlying metal open to further corrosive attack. The mechanisms of film growth and stress distribution are the subject

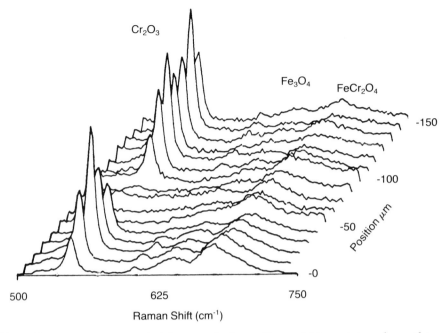

Fig. 3 Raman spectra recorded from analysis positions at 10 μm intervals on the surface of Fe-14%Cr alloy, oxidised in air at high temperature.

of considerable research effort by materials scientists. A detailed study[6] has been undertaken of the stress distribution in the α-Cr_2O_3 formed on Fe-20%Cr-25%Ni-Nb,Mn,Si(trace) stainless steels at high temperatures. The scanning, PMT instrument, which allows precise measurements of cm^{-1} shifts to be made, was used to obtain the results shown in Figure 4. The corrosion scale forms with low ridges over the grain boundaries of the substrate metal. The grain size was of the order of 20 μm and the analysis was made along a line crossing a complete grain at 1 μm intervals. The stress-induced shift in the Raman band of α-Cr_2O_3 is plotted against position in Figure 4 and reveals significant stress relief at the grain boundaries. This can be interpreted in terms of specific oxide scale growth mechanisms.

MIFS studies

Microline focus spectrometry has not been applied to oxidation studies to date. However, studies of polyvinylchloride degradation provide an insight into the potential of the technique. Polyvinylchloride is known to degrade at elevated temperatures and by exposure to UV radiation and chemical attack, through loss of hydrogen chloride. This results in the formation of chains of doubly bonded carbon atoms in the polymer, which can be detected at low concentration by resonance Raman spectroscopy. It is of interest to determine the rate at which the degradation proceeds, the extent to which the effect will ingress into the polymer sheet and the effect of the presence of polyurethane backing.

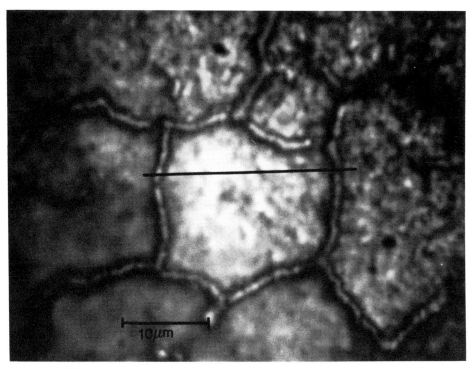

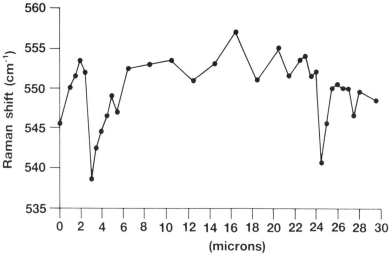

Fig. 4 Stress profile, derived from micro Raman data, of α-Cr_2O_3 scale formed on a 20Cr-25Ni-Nb steel, at high temperature.

Experiments were devised[11,12] in which polyurethane backed, polyvinylchloride sheet, was thermally degraded. The sheet was then sectioned and analysed by Raman MiFS. Figbure 5 is a MiFS profile of a resonance Raman band arising from the polyene degradation products across the sheet thickness. The result provides clear evidence of the extent of degradation ingress and of the

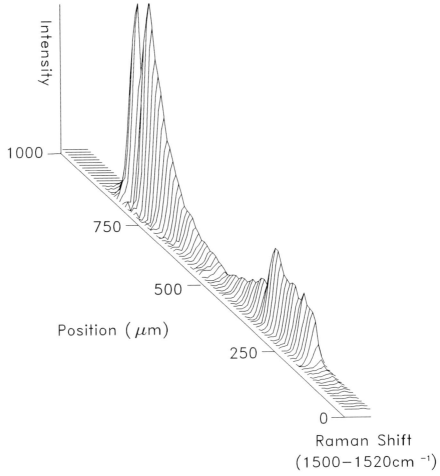

Fig. 5 Raman MiFS profile of polyene band intensity arising from dehydrochlorination of polyurethane-backed polyvinylchloride. The profile extends across the thickness of the sheet, from the polyurethane interface (800 μm position) to the air interface (200 μm position).

additional degradation occurring at the polyurethane foam side as a result of attack from residual amines in the foam.

Prospect

Raman microscopy is now firmly established as an important micro-analytical tool, although the instrumentation remains confined to academic research groups and to the larger industrial laboratories. The technique is limited in terms of spatial resolution, by the wavelength of the exciting light, to around 0.5 μm. Highly fluorescent samples also pose difficulties as the intense fluorescence may mask weaker Raman signals. Also many amorphous and non-stoichiometric compounds produce ill-defined spectra which are difficult to use for species identification. However, the power of the technique to tackle samples *in-situ* under a variety of environmental conditions often outweighs the disadvantages. Developments, such as the use of low-noise, high-sensitivity ccd detectors and small low-cost air-cooled lasers, as in the Raman Microline Focus Spectrometer, may be expected to extend the potential application of the technique. There are two distinct avenues for future instrument development. In the first place, improved general research instrumentation would combine the ability of multi-element detectors to acquire high-quality micrographic data sets in a reasonable time, with convenient access to the full range of spectroscopic measurements needed to determine all the available physical parameters exemplified in this article. In contrast, the identification of specific measurement applications having widespread commercial or research relevance could lead to dedicated instruments designed for mapping a single predetermined spectroscopic parameter.

Acknowledgements

The author wishes to thank the Ministry of Defence and BP Research, Sunbury-on-Thames, UK for financial support.

References

1. D. A. Long, *Raman Spectroscopy*, McGraw Hill International, 1977.
2. D. J. Gardiner, M. Bowden and P. R. Graves, *Phil. Trans. R. Soc. Lond.*, 1986, **A320**, 295.
3. D. Batchelder, *Meas. Sci. Technol.*, 1992, **3**, 561.
4. D. J. Gardiner, C. J. Littleton, K. M. Thomas and K. N. Strafford, *Oxidation of Metals*, 1987, **27**, 59.
5. D. J. Gardiner, M. Bowden, J. Daymond, A. C. Gorvin and M. P. Dare-Edwards, *Appl. Spectrosc.*, 1984, 38, 282.
6. G. J. Piermarini and S. Block, *Rev. Sci. Instrum.*, 1975, 46, 973.
7. J. Birnie, C. Craggs, D. J. Gardiner and P. R. Graves, *Corrosion Sci.*, 1992, **33**, 1.
8. M. Bowden, G. R. Dickson, D. J. Gardiner and D. J. Wood, *Appl. Spectrosc.*, 1990, **44**, 1679.
9. M. Bowden, G. D. Dickson, D. J. Gardiner and D. J. Wood, *J. Materials Sci.*, 1993, **28**, 1031.
10. D. J. Gardiner, C. J. Littleton and M. Bowden, *Microbeam Anal.*, 1987, 131.
11. M. Bowden, P. Donaldson, D. J. Gardiner, J. Birnie and D. L. Gerrard, *Anal. Chem.*, 1991, **63**, 2915.
12. J. W. Bradley, D. J. Gardiner, J. Birnie, N. M. Dixon, D. L. Gerrard and A. S. Wilson, *Proc. XIIIth Int. Conf. Raman Spectrosc.*, Wurzburg, Germany, 1048, 1992.

Acoustic Microscopy of Oxidation

G. DESPAUX, R. J. M. DA FONSECA and J. ATTAL

Laboratoire de Microacoustique de Montpellier – Université des Sciences et Technique du Languedoc Place Eugène Bataillon – 34095 – Montpellier – CEDEX 05 – France

ABSTRACT

Acoustic microscopy has been receiving great attention as a new technology applicable to materials' characterisation on the microscopic scale. It has demonstrated a large variety of applications in many fields: microelectronics, biology, micrometallurgy, industrial applications, etc. Recently, the acoustic microscope has been applied to detect quantitatively elastic parameter changes via a significant variation in leaky surface acoustic wave velocities. Moreover, high-frequency microechography is used as a complementary technique to measure, principally, the longitudinal wave velocity, thickness and density of materials, and acoustic wave phase measurements improve considerably the resolution of the acoustic microscope. So, acoustic microscopy can reveal inhomogeneities introduced by undesirable problems in the specimens.

1. Introduction

In the last few years, the manufacture of new materials has made considerable progress and this has largely been due to the development of control methods and nondestructive analysis techniques. Acoustics, pre-eminently a non-destructive characterisation method, is used via traditional echography. Nevertheless, among the new observational instruments, the scanning acoustic microscope (SAM) has become a particularly interesting analysis and microcharacterisation tool.[1] Its principle, based on the use of a convergent beam of ultrasonic waves at very high frequencies, allows one to obtain a resolution of the order of a micrometre for depths of about a millimetre. The principal interest of this technique is that most solids have acoustic absorptions several orders of magnitude lower than optical absorption. They can thus be considered as transparent to ultrasonic waves so that the structures located inside the material can be visualised. The complexity of the obtained images has led research groups to model the observed phenomena. By improving the lens shape[2] or studying the nature of the coupling liquids,[3] theoretical models have allowed a better use of the acoustic phenomena.

The first acoustic microscopes used to operate only in imaging mode, observing uniquely the qualitative information, which only specialists could interpret. Nowadays, it is possible to get significant quantitative results by means of the so-called acoustic material signature, also known as $V(z)$.[4,5] These quantitative results are easily interpreted and are directly related to physical aspects. Indeed, the measurement of the surface wave parameters results in the characterisation of the studied sample. Thus, Young's modulus, attenuation coefficient, adhesion quality, porosity, anisotropy, influence of different preparation method, etc. can be investigated non-destructively and with excellent accuracy. More recently, these quantitative and qualitative approaches were enriched with the information that may be obtained by investigating the phase of the acoustic signal.[6] Also, high-frequency microechography has advanced considerably in these last years thanks to new network analysers that were introduced[7,8] which enable interface topography investigation (thickness measurement) with resolutions that can reach the wavelength. Therefore, the aim of this paper is to describe acoustic techniques and to illustrate it with some applications.

2. Description of the acoustic microscope

The essential part of the SAM is a piezoelectric thin film transducer (typically ZnO or LiNbO$_3$) the thickness of which is a function of the applied frequencies. It transforms a high-frequency electrical signal into an ultrasound wave which propagates at the same frequency and is localised on a small rod of quartz or sapphire that acts as an acoustic delay line (Fig. 1). To focus the acoustic beam a spherical, aspherical, cylindrical or conical cavity is cut in the other face of the bar. The type of lens depends on the object or phenomenon to be observed and its development for quantitative elastic measurements is an area of great activity in acoustic microscopy. At high frequencies, the acoustic waves do not propagate in the air and a coupling liquid is necessary to transmit the acoustical signal from the lens to the sample. This fluid plays a fundamental part in the resolution of the acoustic microscope. It is the attenuation in the liquid that determines the highest frequency and the shortest wavelength that can be used, and therefore the best resolution that can be obtained. Water and mercury are employed in most of the applications.

The acoustic microscope can work in two modes: reflection and transmission. In reflection mode, the SAM operates with a single transducer for transmitting and receiving the acoustic signal and resembles a miniature radar or sonar system. In transmission, two transducers are symmetrically arranged with respect to the sample to be studied. The object acoustic image, which is in a parallel plane to its surface, is obtained by mechanically scanning the sample in two perpendicular directions (x-y) relative to the transducer. The amplitude of the transmitted or reflected beam is measured and displayed on the screen of the microscope mapping apparatus. The magnification of the scanning field

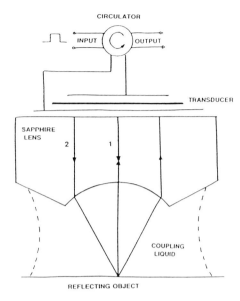

Fig. 1 Schematic configuration of the SAM in relection operating mode. Rays 1 and 2 represent axial and Rayleigh waves.

can be varied from 3 to 3000. The contrast on the acoustic images is due to the variations of elasticity, density as well as acoustic attenuation in the material. So, this microscope is particularly suitable for the observation of optically opaque materials, just where optical characterisation techniques fail.

3. Acoustic signature $V(z)$

The acoustic signal that originates from the sample is the result of the superposition of several types of waves which interfere at the transducer. It is possible, according to the aperture lens θ, to generate simultaneously in the specimen one or several modes that propagate with their specific velocities. So we can vary the difference of phase among these waves changing the relative distance z between the lens and the surface of the specimen. As z varies, the phases of these waves change at different rates, so that they will alternate between constructive and destructive interference. Under these conditions, the output voltage $V(z)$, provided by the transducer, presents pseudo-oscillations which constitute the acoustic material signature. In Figure 2 we show an example of an acoustic signature $V(z)$ at 544 MHz for a silicon sample, using water as coupling liquid.

An important surface acoustic wave which contributes to the $V(z)$ interference curve is the Rayleigh wave. It results in a superposition of the longitudinal and shear waves travelling along the surface with a common phase velocity V_r, which is slower than the velocity of either kind of wave in the bulk. In $V(z)$ measurements, the velocity V_r of the Rayleigh waves in the

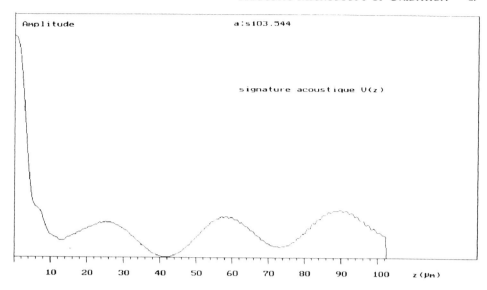

Fig. 2 Acoustic signature $V(z)$ for a silicon sample. The operating frequency was 544 MHz.

specimen surface is related to the operating frequency, f, of the SAM, the periodicity Δz of the resulting oscillations in $V(z)$ and the wave velocity in the coupling liquid V_0 by the following relation:[9]

$$V_R = V_0\left[1 - \left(1 - \frac{V_0}{2f\Delta z}\right)^2\right]^{-1/2}. \tag{1}$$

$V(z)$ can also be used in a defocus position to characterise and analyse the acoustic properties of bonding in layered materials. Actually, leaky Lamb waves are very sensitive to the quality of the bonding and layer thickness, so that the dispersion curves are highly affected by the adhesion conditions. Thus $V(z)$ techniques allow us to get information from these leaky Lamb modes and, consequently, to evaluate the layer adherence.

4. Acoustic imaging

4.1 Surface acoustic imaging

Surface imaging is used when we want to observe microstructures located at less than one wavelength under the surface of a sample. High resolution is therefore required and the operating frequency is typically higher than 500 MHz and normally reaches the GHz range. Figure 3 was obtained for an aluminium sample at 560 MHz. Here, we can clearly distinguish the boundaries between adjacent grains with a contrast which is related to the crystalline

Fig. 3 Surface acoustic image of an aluminium plate.

orientations, since aluminium is an anisotropic material. We note that the contrasts improve noticeably when the transducer is lightly defocused and this provides interesting results. In this case, at each focus depth, each orientation has its own contrasts. When the material presents grains with transverse dimensions higher than λ_0 (wavelength in the coupling liquid), we can correlate the obtained acoustic image and $V(z)$ simulation for each aluminium crystalline orientation. We notice that the lateral resolution is hardly affected by defocus, even for z different values corresponding to different wavelengths λ_0. This can be explained by the fact that the contrast is essentially dominated by Rayleigh waves. All the changes in the elastic properties give contrast because they affect the propagation of Rayleigh and other surface and pseudo-surface waves. Only for large values of z (about ten microns at 600 MHz), do we begin to observe the diffraction effects that disturb the image. Rubbed steels were investigated by SAM and scanning electronic microscopy. The results show the high quality of the acoustic images (see Figures 4a and b).

4.2 Bulk imaging

This technique is implemented for materials having a thickness about 5 to 10 wavelengths λ_0. In general, we use a strong defocus in order to obtain a good separation between the interior and surface echoes. For investigations deep into the material, the SAM employs frequencies ranging from 100 MHz to 500 MHz due to the acoustic wave attenuation that occurs principally in the coupling fluid. In this frequency range a depth resolution between 50 and 10

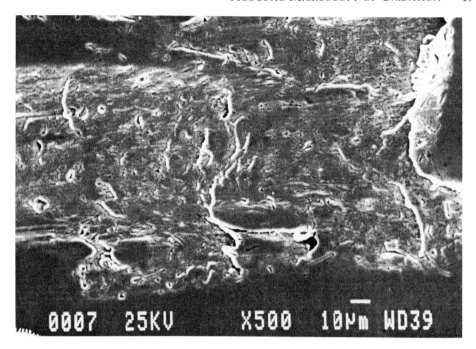

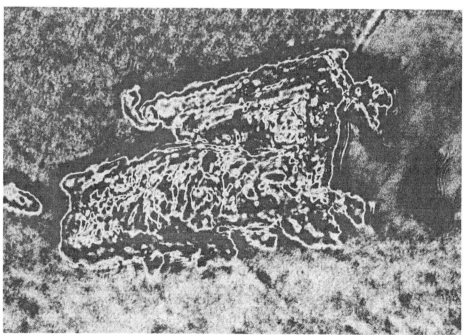

Fig. 4 Images of a rubbed steel obtained by (a) scanning electronic microscope and (b) scanning acoustic microscope at 600 MHz.

micrometres is achieved. Figure 5 is an example of subsurface imaging at 130 MHz. It was obtained with an aspherical sapphire lens designed in our laboratory and can be focused down 0.4 mm.[3] We can see the adherence defects in an Si-Mo interface of a thyristor. Specific theoretical models are adapted to analyse these images and consequently characterise the layer adherence.

4.3 Phase imaging

Up to now we have described the imaging techniques that only use the amplitude of the $V(z)$ signal. However, at all times, the phase can be measured to increase the precision of the experiments in many applications. For example, in optics, the interferometric methods determine the phase changes improving considerably the resolution. In acoustics, the phase also makes a contribution: for measurements using surface waves, it augments the precision of the measured velocity and allows materials which have very high Rayleigh velocities to be analysed. Moreover, the quantitative data obtained by the phase technique leads to the image acquisition.

  Phase contrast acoustic images are obtained which display the phase difference between the reflected wave and a reference one, instead of the wave amplitude. This technique is very sensitive to the surface topography and can also be used to take images of buried interfaces, parallel to the specimen surface. The phase images are very useful for 3D reconstruction and Young's modulus mapping. Moreover, it allows one to appreciably augment the experimental resolution. In Figures 6a and b, we present two images of an old five-franc coin taken at 60 MHz[6] in phase and amplitude respectively. The phase image was carried out through a thickness of 2 mm in the aluminium coin. The colour code is defined for each phase rotation: 0° and 360° are respectively represented by black and white colours and intermediate phase rotations by different grey levels. The measurement of the phase has a precision of the order of 1.4°, which is equivalent to λ/256. In this specific example of aluminium, a resolution of 390 nm was got at this operating frequency, which is about 125 times better than amplitude imaging.

5. Microechography

Echography has been employed in medical diagnostics [10,11] where the operating frequencies are low ($f < 20$ MHz). However, for microstructures and bonded material analysis, high frequencies are required and this makes the technique more complicated because the time delay among the different echoes is less than 10 ns and, moreover, the reflected signal amplitude decreases drastically with frequency. Recently, Kulik *et al.*[7,8] obtained good results in high-frequency microechography using a new network analyser model. This equipment can completely measure reflection and transmission parameters as a function of frequency and is equipped with a time domain option that has the capability of displaying the time domain response of a signal by computing the

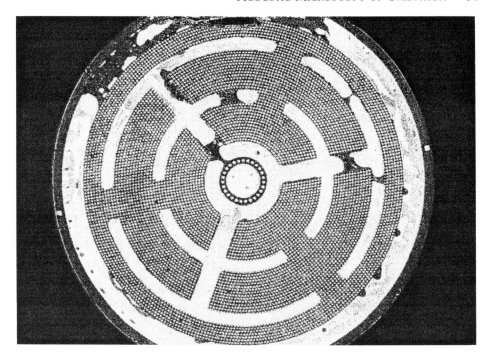

Fig. 5 Acoustic image (600 MHz) of a thyristor with some defects in its interface taken in two different defocussings: (a) small defocussing and (b) great defocussing.

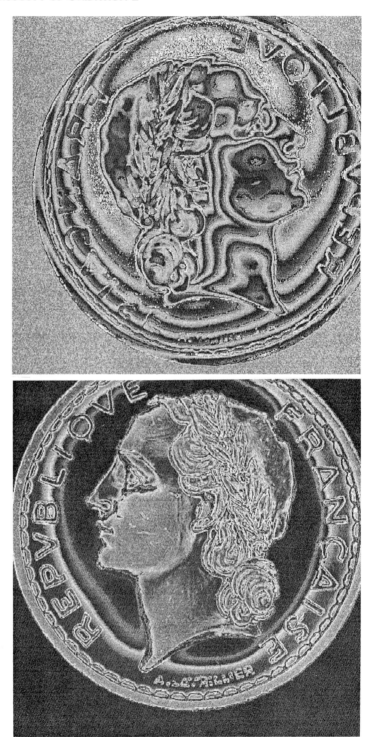

Fig. 6 Acoustic images of an old French coin in a) phase and b) amplitude.

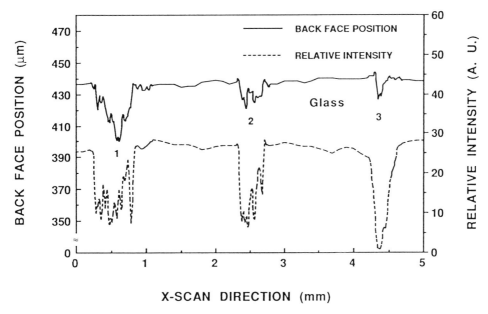

Fig. 7 Back side profile and relative echo intensity in the X-scan direction for a glass sample with grooves. This curve was measured by microechography at 200 MHz.

inverse Fourier transform of the frequency domain response. Consequently, it can directly measure the echo positions corresponding to the covered distance by longitudinal and Rayleigh waves. In addition the network analyser can also supply the acoustic signature $V(z)$.

In our laboratory, promising results were got by using this kind of equipment to implement the high-frequency microechography. A glass sample containing some grooves in its back polished surface was analysed at 200 MHz to evaluate the performance of the implemented experimental system. Figure 7 shows the back side profile and the relative echo intensity that were measured in the x-scan direction with a 5 μm scanning step. The thickness can be automatically calculated by a Network Analyser from the time delay, Δt, between the reflected signals on the surface and back face or flaw of the material, assuming that the longitudinal wave velocity in glass is known. In this example, the average thickness of the glass deduced from the data was 440 μm. The reflection coefficient undergoes changes when the acoustic beam finds the flaw while different scattering processes of the acoustic energy at the surface and back face explain the variations in the relative echo intensity. These results point to a lateral resolution of about 40 μm. The transducer frequency bandwidth, Δf, influences directly the accuracy of the groove height measurements. For glass, at a frequency of 200 MHz, this accuracy is about 15 μm. In addition, these results can be improved by a factor of 5 or more at the GHz range.

When the material velocity is unknown, this parameter can be found from a reference echo. This echo is initially measured from the substrate at a determined defocus, z_0, and with nothing on it except the coupling fluid. The lens is then moved to a region of the substrate where the specimen to be studied is located. In this new position, two echoes will be detected by the transducer, one from the top surface of the specimen, t_1, and another from the specimen-substrate interface, t_2. Designating the timing of the reference echo as t_0 and knowing the wave velocity in the coupling fluid, the sample longitudinal velocity is[12]

$$V_L = V_0 \frac{(t_0 - t_1)}{(t_1 - t_2)}.$$ (2)

Likewise, the acoustic impedance, density and all the elastic properties of the specimen can be measured from the relative echo amplitudes by using similar equations. Moreover, 3D imaging can be also performed by compounding the x and y positions, the time delay among the echoes and their relative amplitude. High-frequency microechography is thus a powerful technique for the quantitative identification and characterisation of materials.

6. Conclusions

We have shown that the acoustic microscope is a powerful tool for nondestructive materials testing. Young's modulus mapping can be done and the regions where it changes can be detected qualitatively and quantitatively with a high resolution (in Table 1 we give a summary of some typical values of resolution that can be obtained in acoustic microscopy). Moreover, high-frequency microechography is a complementary technique that enables measurement of the material thickness and the elastic properties in the x and y directions. Oxidation of samples can thus be detected using acoustic techniques as well as its influence on the elastic constants.

Frequency (MHz)	100	250	600	1000
Surface resolution (μm)	10	5.0	2.0	1.0
Depth resolution (μm)	40	20	8.0	4.0
Penetration (mm)	1.5	0.2	0.03	0.01

Table 1 Typical values of resolution and penetration of the acoustic microscope. These values were calculated for steel and using water as coupling liquid.

References

1. J. Attal, L. Robert, G. Despaux, R. Caplain and J. M. Saurel, *Acoustic Imaging*, 1992, **19**, 607–616.

2. A. Atalar and H. Köymen, *IEEE Trans.*, 1987, UFFC **34**, 2568–2576.
3. J. Attal, A. Saied, J. M. Saurel and C. C. Ly, *Acoustic Imaging*, 1989, **17**, 121–130.
4. J. Attal and C. F. Quate, *J. Acoust. Soc. Am.*, 1976, **59**, 69–73.
5. C. J. R. Sheppard and T. Wilson, *Appl. Phys. Lett.*, 1981, **38**, 858–859.
6. G. Despaux, 1993 Thesis, 'Microscopie acoustique à champ proche en utilisant la phase comme technique de super-résolution,' Montpellier (France).
7. A. Kulik, G. Gremaud and S. Sathish, *Acoustic Imaging*, 1989, **17**, 71–78.
8. A. Kulik, P. Richard, S. Sathish and G. Gremaud, *Acoustic Imaging*, 1992, **19**, 697–701.
9. J. Attal, R. Caplain, H. Coelho-Mandes, K. Alami and A. Saied, *Mechanics of Coatings*, 1990, **17**, 315–322.
10. L. Landini, F. Santarelli, M. Paterni and L. Verrazzani, *Acoustic Imaging*, 1992, **19**, 387–391.
11. J. C. Bamber, C. C. Harland, B. A. Gusterson and P. S. Mortimer, *Acoustic Imaging*, 1992, **19**, 369–374.
12. A. Briggs, *Acoustic Microscopy*, Clarendon Press, Oxford, 1992.

2

OXIDATION OF MILD AND LOW ALLOY STEELS

Study of the Corrosion of Steel in Concrete by Backscattered Electron Imaging in the Scanning Electron Microscope

ANASTASIA G. CONSTANTINOU and KAREN L. SCRIVENER

Department of Materials, Imperial College, London, England

ABSTRACT

The corrosion of steel reinforcement in concrete was studied by imaging polished sections with backscattered electrons in the scanning electron microscope (SEM). Three samples were examined, one which had been carbonated, one to which $CaCl_2$ had been added during mixing and, for comparison, one without chlorides or carbonation. In the case of the carbonated sample, the corrosion products migrated through the cement paste and there appeared to be some interaction with the surrounding hydrates. In the presence of chlorides corrosion occurred in localised areas and there appeared to be profound effects on the microstructure of the cement paste in the immediate vicinity.

1. Introduction

Corrosion of the steel reinforcement in concrete is normally inhibited because the protective oxide layer on the surface of the steel is chemically stable in the usual alkaline environment within the concrete. The composition of this protective iron oxide layer has been suggested to be within the γ Fe_2O_3–Fe_3O_4 solid-solution range.[1] Its effectiveness depends largely on maintaining conditions for its thermodynamic stability. However, the protective layer can be destroyed if the concrete becomes less alkaline as a result of carbonation or if chloride ions are present in the surrounding concrete. Then, given the right conditions, the steel reinforcement will actively corrode. The corrosion of the steel reinforcement results in the cracking and eventual spalling of the cover due to the internal stresses created as a result of the increase in volume associated with the transformation of steel to rust.[2]

Loss of protection due to carbonation occurs on exposure to the atmosphere, when air slowly permeates the concrete. Atmospheric CO_2 reacts with and decomposes the accessible hydrated cement compounds causing a significant reduction in the alkalinity of the concrete. The principal carbonation reaction which occurs initially is conversion of calcium hydroxide (Ca(OH)$_2$) into calcium carbonate (CaCO$_3$) and water. As carbonation proceeds, the remaining cement hydration products consisting of hydrated calcium silicates, aluminates, ferrites, or related complex hydrated salts, are attacked and decomposed with the ultimate formation of calcium carbonate, hydrated silica, alumina, ferric oxide and hydrated calcium sulphate, the last being derived from the setting regulator added to Portland cement.[3] If the concrete becomes carbonated to the depth of the reinforcement, the passive layer will no longer be stable and active corrosion will start.

The corrosion process is thought to be generalised and homogeneous, producing, over the long term, a reduction in the cross-sectional area of the steel and a significant amount of oxides which may crack the cover or diffuse through the pores to the surface of the concrete. However, the spatial distribution of corrosion and the deposition of the corrosion products are not well understood.

Loss of protection can also occur due to chloride penetration. Chloride ions may be introduced into concrete during manufacture or service. Within the former category, there is the possibility of deliberate inclusion of admixtures containing calcium chloride to accelerate the early stages of hydration, a practice that is now forbidden in many countries. Admixtures containing small quantities of calcium chloride may also be introduced as water-reducing agents. There are also many cases of the adventitious introduction of chlorides as contaminants of the aggregates or mixing water which may be almost unavoidable in certain circumstances where the use of locally available resources is necessary. More commonly chlorides permeate the concrete during service, either through contact with solutions containing chlorides, such as sea water, or through the use of de-icing salts. It is generally believed that the chloride ions become incorporated in the passive film, replacing some of the oxygen and increasing its solubility. The chlorides progressively migrate into the crevices where acidification of the crevices occurs due to hydrolysis of the resultant compounds.[4] With the progressive acidification and the increasing concentration of the chlorides, a critical solution composition is reached which destroys the passivity of the film which in turn loses its protective character. The breakdown of the passive film is a local phenomenon and results in the creation of microgalvanic cells. The local active areas will act as anodes where the iron will readily dissolve at a relatively low potential while the remaining passive areas will act as cathodes where oxygen reduction takes place at a higher potential.

The objective of the current study was to investigate the potential of backscattered electron imaging for examining these processes in order to further mechanistic understanding.

2. Experimental

Three reinforced laboratory samples were studied microscopically. The first sample was cast with a water/cement ratio of 0.6 and a sand/cement ratio of 3. The second sample was cast with 2%CaCl$_2$ (by weight of cement), with a

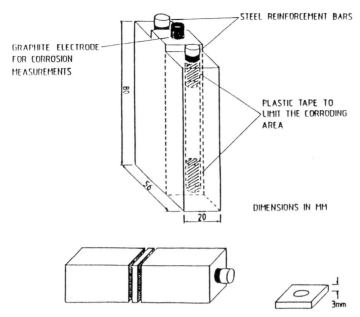

Fig. 1 The diagram shows the sample that was used for the electrochemical measurements and the section which was used for the SEM examination.

water/cement ratio of 0.6 and a sand/cement ratio of 3. After casting, both samples were subjected to 100% relative humidity at 20°C for one year. The third sample was cast with a water/cement ratio of 0.5 and a sand/cement ratio of 1. After casting, it was placed in a chamber with 60% relative humidity, and 100% CO_2. When the carbonation reached the reinforcement, the sample was subjected to 100% relative humidity; however, electrochemical measurements indicated that the sample was not corroding. The sample was then fully immersed in water for the last month in order to assist the initiation of active corrosion.

The samples were used originally for monitoring the corrosion rate in the reinforcement, employing the graphite rod as the counter electrode. They were cut, as indicated by the dotted line (Fig. 1), using a diamond-tipped blade. Sections of the reinforced samples were sliced with a Buhler Isomet low-speed saw with a high-concentration wafering blade. The sections were then resin-impregnated with 9 μm alumina powder and polished down to 0.25 μm grit size. Finally, the sections were coated with carbon. A JEOL-35CF scanning electron microscope (in the backscattered mode) linked to a Kontron SEM Image Processing System was used to examine the samples. An energy-dispersive X-ray detector (EDS) was also used to obtain a chemical analysis of the various phases present in the sample.

3. Results and Discussion

The sample without any additives or aggressives displayed the expected features of a one-year-old mortar sample. This provides a comparison for the

Fig. 2 The microstructure of a one-year-old sample, cast without any additives and subjected to 100% relative humidity after casting. The white grains are the unreacted cement grains and the grey rings surrounding them are the 'inner' product. The calcium hydroxide has a light grey appearance and the 'outer' product is the dark grey matrix.

changes in the microstructure due to chloride attack and carbonation. Since the sample was only one year old, the cement grains had not fully reacted with the water present to form the 'inner' product, a form of calcium silicate hydrate (C–S–H). The white grains shown in Fig. 2 are the unreacted cement grains and the grey rings surrounding them are the 'inner' product. The calcium hydroxide produced by the reaction of the silicate phases is deposited in the space originally occupied by water and can be recognised by its light grey appearance. The dark grey matrix is the 'outer' product formed through solution in the originally water-filled space. A major feature of this microstructure is its high porosity (black).

In the case of the carbonated sample, almost no corrosion was observed in the one-year-old sample. Figure 3 shows the typical microstructure of this sample. The uniform 'outer' product can still be observed, as well as the unreacted grains and the 'inner' product. The porosity, on the other hand, was greatly reduced due to the carbonation products (mainly calcium carbonate) depositing in the pores.

One of the sides of the sample had a small gap which was visible to the naked eye. This gap was probably introduced during casting and was filled with water when the sample was immersed in water. These conditions were sufficient to promote active corrosion. Gaps like this are also found in buildings, due to the sedimentation of the aggregate in the concrete under-

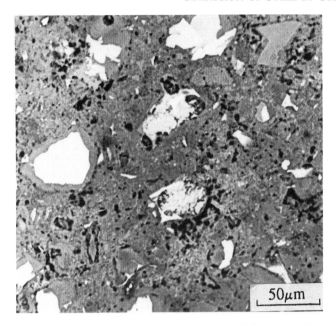

Fig. 3 The interface between the steel (white area) and the paste (dark area) in the one-year-old carbonated sample.

neath horizontal reinforcement. Rising bleed water becomes trapped under the bar resulting in the corrosion of the bottom surface of the steel.[5]

The corrosion products (identifiable by their white appearance) moved across the gap and into the paste, filling up the pores and the spaces around the aggregates. The area of the sample where the gap was present is depicted in Fig. 4, from which it is evident that changes in the microstructure have occurred which can be attributed to the corrosion of the reinforcement. The 'inner' product is darker.

These particles were studied at high magnification and mapped using the image analyser X-ray dot mapping facility. This indicated that the new product was now silicon oxide with little calcium present.

The process of corrosion in the carbonated sample is depicted schematically in Fig. 5. Only a process of corrosion for the area of the gap can be suggested in this case since corrosion was only observed there. Carbon dioxide reacts with the water present in the gap to form carbonic acid (H_2CO_3). The carbonic acid attacks the steel reinforcement and iron carbonate is formed. Since iron carbonate ($FeCO_3$) is slightly soluble in cold water (20–25°C) as well as hydrophilic, it is immediately hydrated and moves away from the interface between the reinforcement and the concrete paste. Carbonate ions will form continuously from the reaction of the water with the carbon dioxide and hence more corrosion products are produced which move towards the surface of the sample down the concentration gradient. Hydroxyl ions are also formed in the gap, when water reacts with oxygen (which is present in small quantities). These in turn combine with iron ions to form insoluble $Fe(OH)_2$ (iron

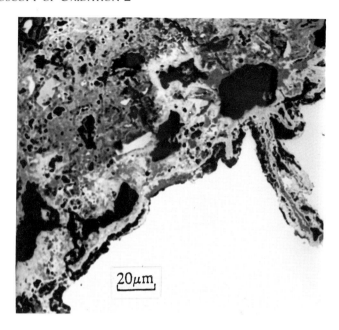

Fig. 4 The interface between the steel (white area) and the paste (dark area) in the carbonated sample (in the region of the gap). The corrosion products can be identified by their whitish appearance.

hydroxide) which deposits in the pores nearest to the reinforcement and can be observed in Fig. 4. The depletion of calcium in the 'inner' product can be attributed to the fact that CO_2 in solution is slightly acidic and it attacks the calcium silicate hydrate.

The sample cast with 2% $CaCl_2$ showed localised attack (which is a characteristic of chloride attack) and some pits were formed. These pits might have originated at imperfections in the steel. Figure 6 shows a typical pit found in the chloride-attacked sample. A layer of corrosion product has formed in the pit and has spread into the paste but not laterally. This layer consisted of at least two different oxides which are identifiable by the difference in grey level. EDS analysis, however, failed to differentiate between the two oxides. Alternative techniques should be employed to identify the two oxides.

The particles incorporated in the oxide layer (Fig. 6) were thought to be the same as the ones found in the severely corroded areas of the carbonated sample (Fig. 4). However, an X-ray dot map conducted on these particles showed that these particles were in fact hollow.

This difference in the level of attack between the carbonated and the chloride-attacked sample can be related to the level of corrosion in each sample. It is well documented[6,7,8,9] that hydrogen ions are evolved during the process of pitting corrosion, lowering the pH of the anodic area of the pit. On the other hand CO_2 forms a weaker acid. Hence, the level of attack in the case of the chloride-containing sample is higher.

CONCRETE

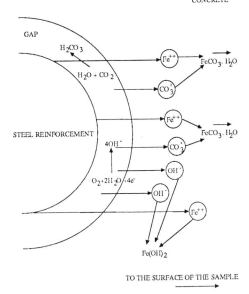

GAP

H_2CO_3

$H_2O + CO_2$

STEEL REINFORCEMENT

$4OH^-$

$O_2 + 2H_2O + 4e^-$

Fe^{++} → $FeCO_3 . H_2O$

CO_3^{2-}

Fe^{++} → $FeCO_3 . H_2O$

CO_3^{2-}

OH^-

OH^-

Fe^{++}

$Fe(OH)_2$

TO THE SURFACE OF THE SAMPLE

Fig. 5 The proposed process of corrosion of the reinforcement in the carbonated sample.

50μm

Fig. 6 One of the pits developed in the steel (of the 2% $CaCl_2$ sample) and the corrosion products which penetrate the paste. Two layers of oxides can clearly be distinguished by the difference in their grey levels.

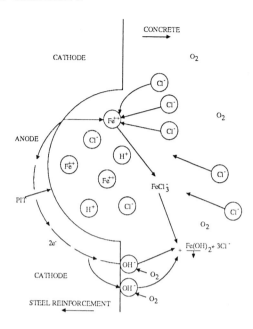

Fig. 7 The proposed process of corrosion of the reinforcement in the chloride-containing sample.

The process of corrosion in the chloride-induced sample is depicted schematically in Fig. 7. The chloride ions migrate to the newly formed pit, their concentration increases and microgalvanic cells are created. The anode is at the bottom of the pit and the cathode is at the area surrounding the pit. Oxidation of iron takes place at the anode (with the evolution of hydrogen ions), whereas the production of hydroxyl ions occurs at the cathode. Because of the local drop in pH at the anode, a soluble complex ion may form. This ion will diffuse away from the anode, provoking more corrosion. Some distance from the anode towards the cathode where both the pH and the concentration of oxygen are higher, the complex ion will break down, iron hydroxide will precipitate, thereby making available chloride ions which will then react further with ferrous ions at the anode. Since the chloride ions are not used up in this process and the corrosion is not stifled by a high concentration of iron ions in the vicinity of the steel, the process can continue with iron ions migrating away from the steel and reacting further with oxygen to form higher oxides or hydroxides. The corrosion process continues at the local anodic areas, causing the development of deep pits. Since hydroxyl ions are continuously consumed by the formation of insoluble iron hydroxide, hydroxyl-containing compounds such as the 'inner' product will dissolve in order to form more hydroxyl ions to enable the restoration of equilibrium in the area.

4. Conclusions

1. The use of backscattered electron imaging has proved successful in providing information about the spatial distribution of the corrosion and the deposition of the corrosion products.
2. In the case of the carbonated sample, one year's exposure in 100% RH at 20°C was not sufficient to initiate active corrosion. However, when the sample was fully immersed in water, the region where a gap was found started to corrode. The corrosion products were deposited in the pores and started to move out towards the surface of the sample.
3. The chloride-containing sample showed attack at localised areas and the corrosion products deposited in the pores around the pits did not extend laterally.
4. A major difference between the samples is the level of attack on the hydrates and can be attributed to the fact that more acidic conditions are created during the corrosion of the chloride-containing sample due to the evolution of hydrogen ions.

Acknowledgements

The author would like to thank Dr M.-A. Sanjuan and Dr M. C. Andrade (Instituto de Ciencias de la Contruccion Eduardo Torroja, Madrid, Spain) for providing the specimens and for monitoring the corrosion of the reinforcement. The authors would also like to thank Dr P. Sidky (Imperial College, London, UK) and Dr G. Constantinou (Geological Survey Department, Nicosia, Cyprus) for helpful discussions.

References

1. K. K. Sagoe-Crentsil and F. P. Glasser, 3rd Symp., *Corrosion of Reinforcement in Concrete* (ed. C. L. Page *et al.*), pp. 74–86, Elsevier Applied Science, 1990.
2. A. Rosenberg, C. M. Hansson and C. Andrade, *Materials Science of Concrete*. Vol. 1 (ed. J. S. Scalny and S. Mindess), pp. 285–313, The American Ceramic Society, 1989.
3. M. H. Roberts, 'Carbonation of concrete made with dense natural aggregates', *Information Paper IP6/81*, Building Research Establishment, 1981.
4. J. A. Gonzalez, A. Molina, E. Otero and W. Lopez, *Mag. Con. Res.*, **42**(150), 23–27, 1990.
5. T. P. Lees, in *Durability of Concrete Structures: investigation, repair, protection* (ed. G. Mays), pp. 10–36, 1992.
6. T. P. Hoar, D. C. Mears and G. P. Rothwell, *Cor. Sci.*, **5**, 279–289, 1965.
7. T. P. Hoar, *Cor. Sci.*, **7**, 341–355, 1967.
8. J. R. Galvele, *Cor. Sci.*, **21**(8), 551–579, 1981.
9. M. G. Alvarez and J. R. Galvel, *Cor. Sci.*, **24**(1), 27–48, 1984.

The Study of Atmospheric Corrosion in Steel by Backscattered Electron Imaging

J. SIMANCAS* and KAREN L. SCRIVENER†

*Department of Corrosion and Protection, Centro Nacional Investigaciones
Metalurgicas, Madrid, Spain
†Department of Materials, Imperial College, London, England

ABSTRACT

The corrosion of steel in the atmosphere is important in many different structures such as buildings, bridges, cars, ships, etc. This work describes a study of corrosion layers built up on steel exposed in different field locations. Specimens were subjected to natural weathering in several locations in Spain for more than 13 years. Backscattered electron (bse) imaging was used to analyse the corrosion layers. Many of these were observed to have a complex structure with layers of different porosity and grey level contrast. Some impurities, especially sulphur and chloride, were detected in the oxide layers. This paper presents some preliminary results from this study and tries to link differences in the observed corrosion layers to the atmospheric conditions at the different exposure sites.

1. Introduction

Atmospheric corrosion of metals has been studied all over the world for many decades; in many countries it is the first subject chosen when starting up corrosion research. The reason is the extensive use of metals for outdoor structures. Most of the studies have been field tests at sites in different types of climate, and much useful information has been obtained from them. Atmospheric corrosion proceeds in a relatively complicated system consisting of metal, corrosion products, surface electrolyte, and atmosphere. The electrolyte film on the surface will contain various species deposited from the atmosphere or originating from the corroding metal. For the thermodynamics and kinetics of the corrosion process the composition of the electrolyte is often of decisive importance.

In atmospheric corrosion the most important pollutants are SO_2 and Cl. The main part of SO_2 pollution is caused by combustion of fossil fuels, i.e. oil and coal in industrialised regions. The sulphur dioxide is oxidised on most particles or in droplets of water to sulphuric acid.[1] Chlorides are deposited mainly in marine atmospheres as droplets or as crystals formed by evaporation of spray carried by the wind from the sea. Other sources of chloride emission are coal burning and municipal incinerators. In marine environments chloride deposition usually decreases strongly with increasing distance from the shore, as the droplets and crystals settle by gravitation or may be filtered off when the wind passes through vegetation. The initiation of corrosion on a clean metal surface in a non-contaminated atmosphere is a very slow process even in environments saturated with water vapour. A more important factor for the initiation of corrosion is the presence of solid particles on the surface, specially particles of hygroscopic salts, such as chlorides or sulphates, which form a corrosive electrolyte on the surface. Carbonaceous particles can also start the corrosion process as they may form cathodes in microcells with the steel surface.

The objective of this research was to investigate the possibility of the application of backscattered electron (bse) imaging and energy dispersive (EDS) X-ray analysis to a steel exposed at different test sites in Spain for different exposure times.

2. Experimental

The main network of testing sites in Spain (Fig. 1) comprises several types of atmospheres: rural (El Escorial), urban (Madrid, Zaragoza), urban–marine (Barcelona, Alicante) and industrial–marine (Bilbao). These corrosion stations are provided with temperature and humidity recorders, rain gauges, and instrumentation for measuring the duration of the visible wetting of a metallic surface and levels of atmospheric pollution by SO_2 and chlorides. Long-term research programmes are being carried out at these sites.

The research programme which provided the specimens in this study was

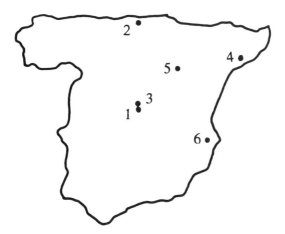

Fig. 1 Location of corrosion testing stations in Spain: 1, Madrid; 2, Bilbao; 3, Escorial; 4, Barcelona; 5, Zaragoza; 6, Alicante.

Fig. 2 Measurement of corrosion layer thickness. ×300.

initiated to determine the effect of long-term exposure on the rate of atmospheric corrosion.[2] Panels of mild steel, zinc, copper, and aluminium have generally been exposed, in triplicate, on racks at an angle of 45° to the horizontal. In this research only the results from the steel are discussed. The samples were prepared for microscopy as follows. The specimens were mounted on edge in epoxy resin and polished so that the steel substrate and corrosion products could be viewed in cross-section. Samples were polished to a $\frac{1}{4}$ μm diamond finish. The specimens were coated with carbon and examined in a JEOL 35 CF scanning electron microscope, equipped with a solid-state backscattered electron detector and a Link AN 10000 energy dispersive X-ray analysis system. The SEM was also interfaced to a Kontron IBAS image analyser which could collect the SEM image and also the X-ray counts from up to four element windows. The images collected on the image analyser are digitised into a 512×512 pixel array and 256 grey levels.[3] To measure the thickness of the oxide layers 20 images were collected from each side of the specimen at a magnification of 300×. Five measurements were made on each image (Fig. 2). To quantify the compactness of the layer the percentage area of pores and cracks was measured on 20 images at 1500×.

3. Results and Discussion

Backscattered electron images (Fig. 3) show that the oxide layers have a complex structure. Near the outer surface of the layers alternating light and dark bands were observed. It is thought that these arise from the segregation of contaminants during wetting and drying of the sample. However, any variations in chemical composition were on too fine a scale to be determined by

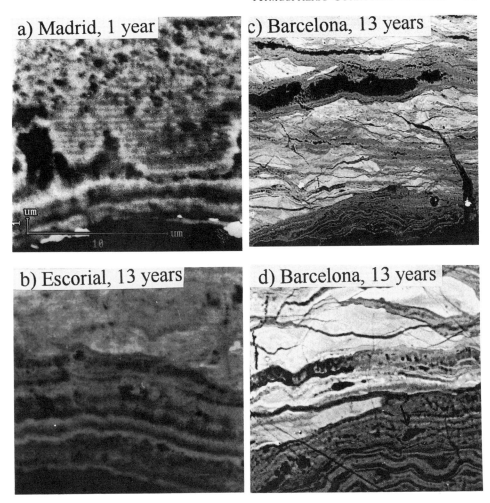

Fig. 3 Images of corrosion layer showing complex structure: (a) Madrid, 1 year; (b) Escorial, 13 years (c) Barcelona, 13 years; (d) Barcelona, 13 years.

EDS analysis. X-ray mapping did however reveal that some of the dark areas originally thought to be pores (e.g. that arrowed in Fig. 3d) were in fact inclusions of rock minerals which had been deposited from the atmosphere.

In the specimen with little corrosion (Madrid 1 year, Fig. 3a, and Escorial 13 years, Fig. 3b), the corrosion products appeared more homogeneous, below the striated surface region. In some of the older samples a mosaic of light and dark regions can be observed. It appears that the dark regions have formed within cracks in the light regions. Figure 3c and d shows the appearance of the oxide layer from the specimens exposed in Barcelona. In some areas, X-ray mapping indicates that contrast corresponded to inhomogeneous distribution of chlorine. However, in most cases the spatial variation of the microstructure was too fine and the chemical differences too slight to be detected by EDS.

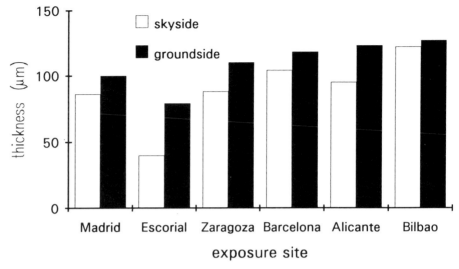

Fig. 4 Thickness of corrosion layers on steel exposed in different localities for 13 years.

The corrosion layers produced by atmospheric exposure are very fragile and some cases material is lost during preparation. The extent of loss is a function of the aggressive atmospheric and the exposure time. Figure 4 shows the average corroded layer thickness for the samples exposed for 13 years at different sites. In all cases the layer on the groundside of the specimen was thicker than that on the skyside. This is attributed to the different lengths of time of wetness and dryness. On the groundside, the times of wetness are longer. It is clear that the specimen exposed in a rural environment (Escorial) has undergone by far the least corrosion. Parallel weight loss measurement,[4] of the amount of metal reacted, showed that the industrial–marine environment of Bilbao led to the greatest amount of corrosion followed by urban–marine environments and then by non-marine–urban environments.

Figure 5 compares the layer thickness measured in the SEM and the amount of reacted metal calculated from weight loss measurements. The gross deviation of the measurements at greater degrees of reaction suggests that there has been significant loss of material from these specimens. Loss of material appears to occur when the corroded layer reaches about 100 μm in thickness. It is notable that the specimens which have lost the largest amount of material have been exposed in marine environments, where the presence of chlorine accelerates corrosion. At lower degrees of corrosion there is a linear relationship between the amount of metal reacted and the measured thickness of the corrosion layer. The relationship appears to be 1 : 1, which is not expected, as a much greater volume of oxide should be formed from a given volume of metal. This suggests that some material may have been lost even from these specimens.

Measurements of the amount of metal reacted by weight loss indicated that the specimens exposed at Madrid and Escorial showed a deviation to the

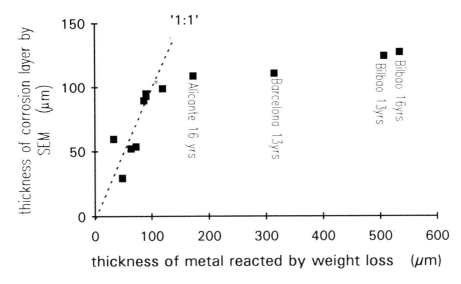

Fig. 5 Comparison of the thickness of corrosion layers by SEM and the thickness of metal reacted by weight loss.

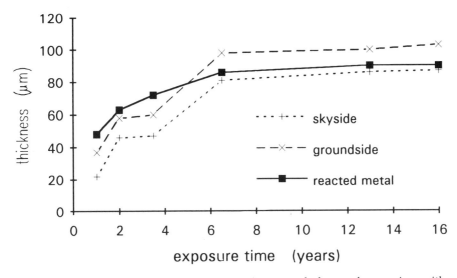

Fig. 6 Variation of the corrosion layer thickness and the total corrosion with exposure time for the samples exposed at Madrid.

usually observed bilogarithmic law of atmospheric corrosion rate versus time of exposure.[5] At these test stations the thickness increased quickly with exposure time up to 6 years after which the thickness increased very slowly with time. To investigate this phenomenon further, the corroded layer thickness and the compactness of this layer were measured for samples exposed at Madrid for different lengths of time. The measurements of thickness (Fig. 6) correspond

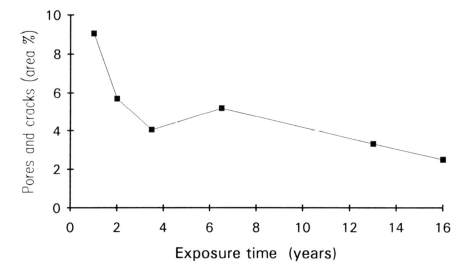

Fig. 7 Variation of cracks and pores (area %) in oxide film with exposure in Madrid.

closely to the weight loss measurements and confirm that there is very little thickening of the corrosion layer after some 5 or 6 years. The compactness of the layers was quantified by measuring the percentage area of pores and cracks in the layer. These measurements clearly show a reduction in the porosity of the corroded layer corresponding to the drop in corrosion rate (Fig. 7). The difference in the porosity of the corroded layer can be observed in the micrographs (Fig. 8). The sample exposed for 1 year is shown in Fig. 8a along with the porosity that can be discriminated in this image (around 10%). In the specimen exposed for 16 years (Fig. 8b) the discriminated porosity is only 3%.

4. Conclusions

1. Backscattered electron imaging is a useful technique for the study of atmospheric corrosion.
2. The corrosion product formed during atmospheric corrosion has a complex structure, the origins of which have not been fully elucidated.
3. The decrease in corrosion rate observed in some samples appears to be associated with a reduction in the porosity of the oxide layer.

Acknowledgements

We acknowledge the support of the Ramsay Memorial Fellowships Trust and Spanish Scientific Research Council (CSIC), for the visit of J.S. to Imperial College during which the work presented here was carried out.

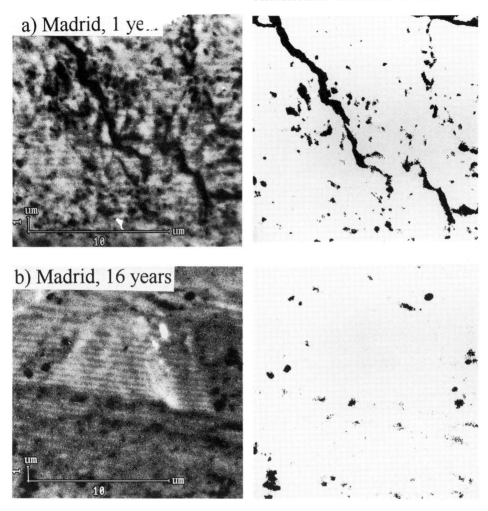

Fig. 8 Images of corrosion layer from: (a) Madrid, 1 year, ×600, bse image and discriminated cracks and pores, black; (b) Madrid, 16 years, bse image and discriminated cracks and pores, black.

References

1. V. Kucera and E. Mattsson, *Atmospheric Corrosion* (ed. W. H. Ailor), p. 218. Wiley, New York, 1982.
2. S. Feliu and M. Morcillo, *Atmospheric Corrosion Testing in Spain* (ed. W. H. Ailor), p. 913. Wiley, New York, 1982.
3. K. L. Scrivener and P. L. Pratt, *Proceedings First International RILEM Congress, Paris 1987.* Vol. 1 (ed. J. C. Maso), pp. 61–68, London, Chapman and Hall. 1987.
4. M. Morcillo. Private Communication.
5. M. Morcillo, S. Feliu and J. Simancas. *British Corrosion J.*, **28**(1.1), 1993.

Oxidation of a precipitation-strengthened die steel

W. J. R. NISBET,* J. D. H. PAUL, G. W. LORIMER,
D. G. LEES and N. RIDLEY

*Manchester Materials Science Centre, University of Manchester/UMIST,
Grosvenor Street, Manchester M1 7HS
Shell Research, Arnhem, The Netherlands

ABSTRACT

The effect of variations in carbon and nitrogen concentration, between 0.45 and 0.6 wt% and 0.3 and 0.5 wt%, respectively, and additions of cobalt up to 4 wt%, on the oxidation behaviour of a 22/4/9 die steel have been evaluated. Samples of the steel were isothermally oxidised at 900°C in O_2 at 0.1 atm. Steels with a high C/N content exhibited superior oxidation resistance to those with a low C/N concentration. Cobalt additions decreased the oxidation rate in low C/N steel, but also induced breakaway oxidation.

Metallographic examination of a die that had been in service for >1000 hours revealed a carbide/nitride-free zone adjacent to the surface. XRD and TEM identified a layer of ferrite between the surface spinel oxide and the precipitate-free zone. Both the precipitate free zone and the layer of ferrite were associated with Mn, C and N depletion.

1. Introduction

Precipitation-strengthened austenitic stainless steels based on a 22Cr–4Ni–9Mn (wt%) composition (22/4/9) are used as die steels for the diffusion bonding (DB) and superplastic forming (SPF) of titanium alloys at elevated temperatures (approximately 900°C). Prolonged exposure at the DB/SPF temperatures and thermal cycling results in pitting/oxidation of the surface which limits the life of the die.

The work described in this paper was part of a study into how to extend die life-time by careful choice of composition. This was done by investigating the isothermal and cyclic oxidation behaviour of 22/4/9 steel with C and N additions at the high and low specification levels, and with and without the addition of 4 wt% Co. A metallographic and electron optical study was also conducted on a die which had been in use for over 1000 hours.

wt%	C	Si	Mn	S	P	Ni	Cr	Co	Mo	N$_2$	Fe
22/4/9(L)	0.46	0.5	9.37	0.008	0.03	4.43	21.8	<0.02	<0.02	0.32	Bal
22/4/9(H)	0.60	0.6	9.27	0.007	0.032	4.37	22.2	<0.02	<0.02	0.51	Bal
22/4/9Co(L)	0.48	0.4	9.60	0.007	0.027	4.52	22.1	4.26	<0.02	0.32	Bal
22/4/9Co(H)	0.6	0.4	9.49	0.006	0.032	4.55	22.6	4.31	<0.02	0.52	Bal

Table 1 Composition of 22/4/9 steels.

2. Experimental

Evaluation of the effect of alloying elements on oxidation

Four steels were evaluated with the nominal compositions:

22/4/9(L) 0.45 C, 0.3 N (wt%)
22/4/9(H) 0.6 C, 0.5 N
22/4/9Co(L) 4 Co, 0.45 C, 0.3 N
22/4/9Co(H) 4 Co, 0.6 C, 0.5 N

The actual composition of the steels is shown in Table 1.

Metallographic specimens of as-cast 22/4/9 steels were prepared for optical examination in the conventional manner by grinding and polishing and then electrolytically etching in 10% chromic acid at 5 volts for 15 seconds.

Oxidation tests were conducted in a manometric oxidation test rig developed by Lees et al.[1] This method is based on measuring the difference in pressure developed between an empty reference tube and a tube containing the specimen. The pressure difference is measured using a pressure transducer which gives a voltage reading, this is then converted into a weight gain using a predetermined calibration coefficient. Samples with a total surface area of approximately 2 cm^2 were ground to P600 finish on all faces. Prior to testing the samples were ultrasonically cleaned, degreased in acetone and then dried. Both isothermal and cyclic tests were performed at 900°C in O$_2$ at 0.1 atm pressure for 20 hours. In the cyclic testing, the heating period was 60 min with a 15 min cooling period between cycles.

2.1 Examination of an in-service die

A transverse specimen was removed from a 22/4/9 die which had been in service for >1000 hours and was prepared for metallographic examination in the conventional manner. The specimen was electrolytically etched in 10% chromic acid and examined using optical and scanning electron microscopy. The composition profile at the surface of the die was determined using energy dispersive X-ray analysis and associated software that facilitated a series of analyses ≈1 μm apart. This produced a linescan from the surface into the parent material. Transverse sections of the die surface were prepared for transmission electron microscopy (TEM) by ion beam thinning by a method described by Nisbet et al.,[2] based on the work of Rowlands[3] and Newcomb et *al.*[4] The sections were examined using a Philips EM 430 TEM at 300 kV fitted with an energy-dispersive X-ray detector and the ancillary electronics.

The surface of the die was examined using X-ray diffraction (XRD) with Cu Kα radiation. Three surface conditions were examined: the as-received die surface, the die surface after wire brushing, and a specimen from the bulk material.

3. Results and discussion

3.1 Effect of alloying elements

The microstructures of the various as-cast 22/4/9 steels are shown in Fig. 1. It can be seen that the high C/N-containing steels contain a significant volume fraction of a pearlite-like lamellar decomposition product. This is in marked contrast to the low C/N steels which have very few precipitates. The addition of Co has a negligible effect on both the high and low C/N steel microstructure.

The results for the isothermal oxidation tests on the various 22/4/9 steels are shown in Fig. 2. In the early stages of oxidation all the steels exhibited parabolic growth which is associated with the formation of Cr_2O_3. The breakaway oxidation that occurred after a very short time in 22/4/9(L) and after longer times in 22/4/9Co(L) and 22/4/9Co(H) could be associated with the formation of less protective spinel. Breakaway oxidation may also occur if the oxide grows to a critical thickness at which cracking or rupturing of the oxide takes place and exposes metal denuded in Cr. Some tests on 22/4/9Co(H) showed breakaway oxidation while others did not. This probably demonstrates the poor adherence of the oxide on this steel. The isothermal oxidation tests on the as-cast steels demonstrate the beneficial effect of high C/N levels and, to a lesser extent, Co additions, on the oxidation resistance of the die materials.

The increased oxidation resistance of high C/N compared to low C/N levels is a surprising result because high C/N levels promote extensive formation of Cr rich carbides and carbo-nitrides which consequently deplete the matrix of Cr. Therefore the low C/N steels would be expected to have superior oxidation resistance. However, if the effect of S is considered, the beneficial effect of high C/N may be explained. Impurity sulphur can segregate to the scale/metal interface and reduce the adherence and promote spalling.[5-8] In a high C/N steel sulphur may segregate to the large number of precipitate/matrix inter-faces, leaving an adherent oxide/metal interface. In contrast, a low C/N steel has relatively few precipitate/matrix interfaces and thus oxide adherence is reduced. Lees[7] also suggests that S segregates to the oxide grain-boundaries and promotes cation transport, and that when no sulphur is present the grain-boundaries transport only oxygen. This may explain why the final oxidation rates of the high C/N steels are lower than those of the steels with low C/N levels.

The results for the cyclic oxidation tests are shown in Fig. 2. The weight gain after 20 hours for the non-cobalt-containing steels is very similar to that under isothermal testing conditions. However, for the cobalt-containing steels the

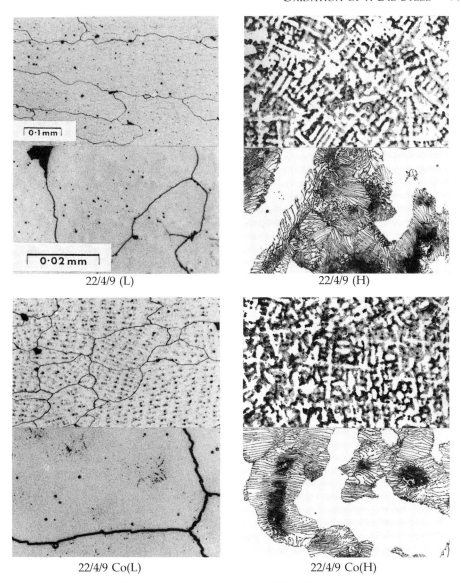

22/4/9 (L)   22/4/9 (H)

22/4/9 Co(L)   22/4/9 Co(H)

Fig. 1 Microstructure of as-cast 22/4/9 die steels.

situation is slightly different. The initial rate of oxidation is greater for the cyclic than for the isothermal oxidation mode. This could be explained by thermal stresses leading to increased spallation under cyclic conditions. For 22/4/9Co(H), although no significant breakaway oxidation was observed, the weight gain after 20 hours was greater than that under isothermal conditions. No significant breakaway oxidation was observed in the cyclic testing of 22/4/9Co(L).

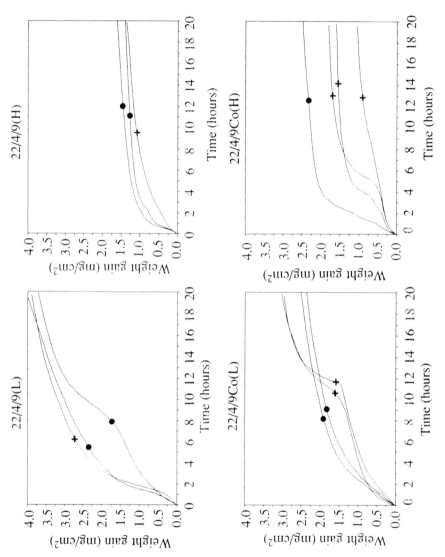

Fig. 2 Isothermal (+) and Cyclic (●) Oxidation runs on 22/4/9 steels.

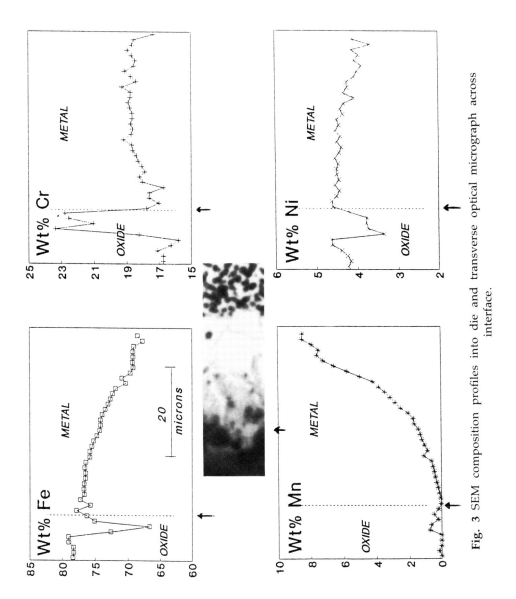

Fig. 3 SEM composition profiles into die and transverse optical micrograph across interface.

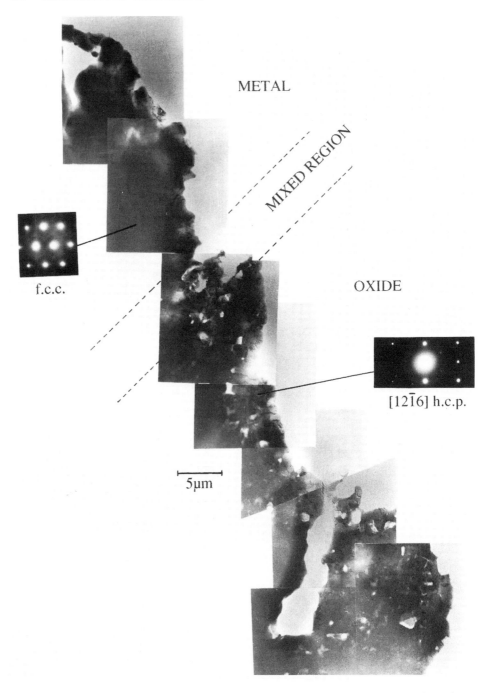

Fig. 4 TEM micrograph showing section from oxide to bulk material with diffraction patterns.

3.2 In-service die

Macroexamination of the in-service die showed a slightly pitted surface and a semi-adherent oxide scale. A transverse optical micrograph of this surface is shown in Fig. 3, and confirms the surface roughness of the die and also shows a precipitate-free zone which extends to a depth of 40–50 μm below the oxide–metal interface. Energy-dispersive X-ray analysis of the surface metallic region (precipitate-free zone), showed over a depth of 50 μm that the iron content decreased from $\approx$80wt% to $\approx$67–68wt%, the chromium content increased from $\approx$15wt% to $\approx$19wt% and the nickel content remained essentially constant at $\approx$4.4wt%. The precipitate-free zone was totally depleted in Mn up to a depth of $\approx$10 μm, after which the composition gradually increased to the bulk composition after $\approx$50 μm.

The structure across the metal/oxide interface at the surface of the SPF die, as revealed by TEM, is shown in Fig. 4. Three regions were identified: oxide, a mixed oxide/metal region, and the precipitate-free zone. Diffraction patterns from the oxide were indexed as hexagonal, and X-ray diffraction examination (see Table 2) of the as-received die surface identified the oxide as $(Cr,M)_2O_3$ which has a hexagonal structure. The composition of the outer oxide was approximately:

$$47wt\%Cr, 36wt\%Fe, 13wt\%Mn, 4wt\%Ni$$

and the composition of the inner oxide was approximately

$$66wt\%Cr, 3wt\%Fe, 30wt\%Mn, 1wt\%Ni.$$

Diffraction patterns from the precipitate-free zone were indexed as face-centred cubic. Diffraction patterns from the metal in the mixed region were indexed as body-centred cubic. X-ray diffraction analysis of the wire-brushed surface identified $(Cr,M)_2O_3$ and ferrite (see Table 2), while for the bulk material the structure was predominantly austenite although $Cr_{23}C_6$ was also detected. The typical composition of this precipitate was 68wt%Cr, 5wt%Mn, 1.5wt%Ni, 25.5wt%Fe. Other precipitates, richer in Cr, were also identified by TEM. These precipitates had a typical composition of: 93wt%Cr, 2wt%Mn, 5wt%Fe, and were probably $Cr_2(C,N)$.

In the initial stages of oxidation it is likely that a thin protective layer of Cr_2O_3 forms. As the oxide grows, the surface of the die will become depleted in Cr. This decrease in Cr content at the die surface promotes the formation of a spinel containing Cr, Mn and some Fe. Spinel formation promotes the

Sample	Phases identified
Bulk material	Austenite, $M_{23}C_6$
As-received die surface	$(Cr,M)_2O_3$
Wire-brushed die surface	$(Cr,M)_2O_3$, ferrite

Table 2 Results from X-ray diffraction.

depletion of Mn and Cr, and enrichment of Ni and Fe in the die at the oxide–metal interface. The depletion of Mn cannot account for the formation of the precipitate-free zone at the surface of the die. The absence of precipitates may indicate that N and C depletion is occurring at the surface. The denudation of Mn, N and C will be most marked at the metal-oxide interface, and the removal of these elements (which are γ stabilisers) may have promoted the formation of the ferrite that was detected in this region.

4. Conclusions

1. The 22/4/9 steels with high C/N levels have better oxidation resistance than the corresponding low C/N steel under isothermal conditions.
2. Cobalt additions tend to lead to breakaway oxidation under isothermal conditions. For the low C/N steel, cobalt addition improves the isothermal and cyclic oxidation resistance. For the high C/N steels, cobalt addition does not significantly affect the isothermal oxidation resistance but leads to a slight deterioration in cyclic oxidation resistance.
3. The surface of the 'as used die' had a precipitate-free zone extending to a depth of up to $\approx 50\,\mu m$ which was associated with depletion of Mn, C and N. The depletion of these γ stabilisers also led to the formation of ferrite between the surface spinel oxide and the precipitate-free zone.
4. The oxidation resistance can be ranked as follows:
   Isothermal: 22/4/9Co(H) $\approx$ 22/4/9(H) $>$ 22/4/9Co(L) $>$ 22/4/9(L)
   Cyclic: 22/4/9(H) $>$ 22/4/9Co(H) $\approx$ 22/4/9Co(L) $>$ 22/4/9(L)

Acknowledgements

This research was sponsored by British Aerospace, Salmesbury, whose financial support and supply of materials are gratefully acknowledged.

References

1. M. Skeldon, J. M. Calvert and D. G. Lees, *Oxid. Met.*, **28**, 109, 1987.
2. W. J. R. Nisbet, G. W. Lorimer, C. Sherhod and M. J. Stowell, *Mater. Sci. Technol.*, **6**, 182, 1990.
3. M. I. Manning and P. C. Rowlands, *Br. Corros. J.*, **15**(4), 184, 1980.
4. S. B. Newcomb, C. B. Boothroyd and W. M. Stobbs, *J. Microsc.*, **140**, 195, 1985.
5. Y. Ikeda, K. Nii and K. Yoshihara: *Proc. 3rd JIM Int. Symposium on High-Temperature Corrosion of Metals and Alloys, Trans. Jpn. Inst. Met.*, **24** (Suppl.), 207, 1983.
6. A. W. Funkenbusch, J. W. Smeggil and N. S. Bornstein: *Met. Trans.*, **16A**, 1164, 1985.
7. D. G. Lees, *Oxid. Met.*, **27**, 75, 1987.
8. P. Fox, D. G. Lees and G. W. Lorimer, *Oxid. Met.*, **36**, 491, 1991.

Mixed-Oxidant Corrosion in Non-equilibrium CO-CO$_2$-H$_2$-H$_2$S Gas Mixtures at 400 and 550°C

W. T. BAKKER* and J. A. BONVALLET†

*Electric Power Research Institute
†Lockheed Missiles and Space Company

ABSTRACT

Entrained slagging gasifiers are designed to gasify coal with little or no steam additions at 1300–1600°C. The resulting CO–CO$_2$–H$_2$–H$_2$S syngas is rapidly cooled in heat exchangers producing steam, without significant change in gas composition. Therefore, the syngas is not at equilibrium at the heat exchanger metal temperature, which ranges from 350 to 360°C. Non-equilibrium oxygen and sulphur partial pressure of the syngas can be calculated from the high-temperature CO$_2$/CO and H$_2$S/H$_2$ ratios and the gas equilibrium constants at the heat exchanger metal temperature. Calculations show that the PS$_2$/PO$_2$ ratio of a typical syngas is smaller than the PS$_2$/PO$_2$ ratio at the Cr$_2$S$_3$/Cr$_2$O$_3$ phase boundary, above 440°C, but higher at temperatures below 440°C. This is caused by a change in the equilibrium constant of the H$_2$S/H$_2$ reaction at the boiling point of sulphur. This increases the PS$_2$ of the syngas 1–3 orders of magnitude, depending on temperature.

Model alloys containing 20% Cr, 35% Ni, and 3% V or Ti, balance Fe, were exposed to a CO–CO$_2$–H$_2$–H$_2$S gas mixture simulating a dry feed, oxygen blown, entrained slagging gasifier at 400 and 550°C. Significant changes in scale development were observed. At 550°C the scale consisted of an outer (Fe, Ni) S Layer, which appeared to grow outward and was quite porous and friable, and a more compact mixed oxide/sulphide inner scale. The latter was inhomogeneous and usually enriched in sulphide on the gas side. At 400°C both inner and outer scales consisted of sulphides only, mainly (Fe, Ni) S on the outside and Fe Cr$_2$ S$_4$ and Cr$_2$ S$_3$ on the inside. Additions of Ti and V did not significantly change the scale composition, but markedly affected scale growth rates. Optical microscopy, SEM/EDS and XRD were used to analyse the scales.

1. Introduction

Coal gasification is close to commercial reality. Several large demonstration plants in the 100–300 MWe range are in operation or under construction.

Several more are being planned. The main incentive to use coal gasification technology instead of conventional pulverised coal burning boilers is a concern for the environment. Coal gasification combined cycle power plants (CGPPs) can remove 99+% of SO_x and NO_x emissions, even when using high-sulphur coals, without excessive costs. At present, conventional pulverised coal power stations can only remove 90–95%, at least at reasonable cost. Fully integrated CGPPs are also more efficient than conventional plants and thus emit less CO_2 per kilowatt of electricity produced.

At present, the most favoured processes operate above the melting-point of coal ash, in the 1300–1600°C range and are called entrained slagging gasifiers. They are refractory lined as the service life of bare metal alloys is too short at these temperatures to be practical. The sensible heat in the raw syngas must be recovered to increase the overall efficiency of the process. This is generally done by raising steam in both radiant and convective syngas coolers. These are similar to boilers, but generally operate at elevated pressures 1700–4000 kPa (~250–600 psi). In most systems, the syngas coolers are only used as evaporators, and thus operate in the 350–450°C (660–830°F) range. In some systems it is desirable to superheat or reheat some of the steam, usually at 500–600°C (~900–1100°F). Thus, the practical temperature range for high-temperature corrosion studies in coal gasification environments is 350–600°C. Temperatures of 550°C and 400°C were selected for this study.

Gas composition in syngas coolers of entrained slagging gasifiers depends on the coal gasification process and the type of coal used. Entrained slagging gasifiers with a dry pulverised coal feed system produce a CO-rich, dry syngas. A typical composition is (vol%) H_2 32, CO 64, CO_2 4, H_2S 0.2–1.2%, HCl at 200–600 ppm. The H_2S content of the syngas is mainly dependent on the coal composition with 0.2% H_2S representing a low sulphur coal (<1% S) and 1.0% a high sulphur coal (>2.5%S). Usually the theoretically dry syngas contains some steam, possibly up to 3%. Frequently some steam is injected in the coal gasification reactor to moderate its temperature, and may not entirely convert to H_2 and CO. In other designs, quenched, water-containing syngas is recirculated to reduce the gas temperature in the syngas coolers. In this study we will report on corrosion in nominally dry syngas at 400–600°C, the range of metal temperatures encountered in syngas coolers.

2. Oxygen and Sulphur Partial Pressures of Non-equilibrium H_2–CO–CO_2–H_2S Mixtures

The composition of the syngas produced in entrained slagging gasifiers at 1200–1600°C remains essentially unchanged, when cooling rapidly on its passage through the syngas coolers. Thus, the syngas in contact with the water or steam cooled heat exchanger surfaces is not in equilibrium at the metal temperature. In a previous paper,[1] a method to calculate the non-equilibrium pressure of oxygen and sulphur was presented, using the fixed high-temperature gas composition and the equilibrium constants of the appropriate gas phase reactions ($CO + \frac{1}{2}\ O_2 = CO_2$,[1] $H_2 + S = H_2S^2$), obtained from the

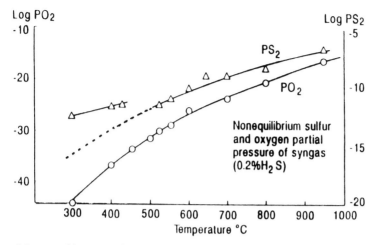

Fig. 1 Non-equilibrium sulphur and oxygen pressure of dry syngas (0.2%H₂S) as a function of temperature.

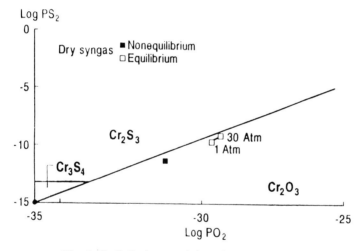

Fig. 2 Cr-O-S phase stability diagram at 500°C.

JANAFF tables. Figure 1 shows the nominal oxygen and sulphur pressures as a function of the temperature of the typical dry syngas, containing 0.2% H_2S. The logarithm of the oxygen pressure decreases monotonically with temperature as a result of the increase in equilibrium constant of reaction.[1] However, the log PS_2 changes at about 440°C, because of an abrupt change in the equilibrium constant of reaction (2) at the boiling point of sulphur. This increases the sulphur partial pressure of the gas below 440°C. Figure 2 illustrates that at temperatures above 440°C the gas composition falls just within the field where Cr_2O_3 is the stable phase when plotted on a Cr–O–S phase equilibrium diagram. With decreasing temperature, the position of the gas composition in the phase stability diagram remains unchanged until 440°C.

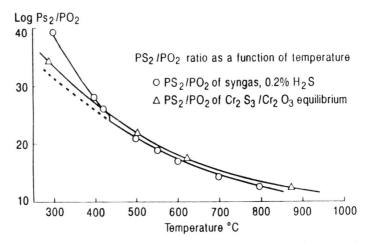

Fig. 3 PS_2/PO_2 ratios of dry syngas and Cr_2S_3/Cr_2O_3 phase boundary as a function of temperature.

At this point the PS_2 increases abruptly and at 400°C the gas composition is well within the stability field of $Cr_2 S_3$. This is shown in Fig. 3, where the PS_2/PO_2 ratio of the syngas is compared with the PS_2/PO_2 ratio of the $Cr_2 S_3/Cr_2 O_3$ phase boundary at the PO_2 of the syngas at each temperature.

3. Experimental Procedures

Corrosion tests were performed in electrically heated vertical tube furnaces in gas mixtures flowing at a rate of 1 litre/minute. Detailed test procedures have been given previously.[2] The flow rate of the gas mixture was chosen to be high enough to prevent changes in the gas composition in the high-temperature zone of the furnace. All tests were performed isothermally without interruption. At 550°C, 150, 600 and 1350 h tests were performed using the following gas mixture (vol.%): CO 64, CO_2 4, H_2 31.8, H_2S 0.2, HCl 400 ppm. At 400°C a single 600 h test was done using the same gas composition as at 550°C.

Exposed metal alloy specimens, measuring about $12 \times 6 \times 3$ mm, were mounted in bakelite or epoxy resin and polished using standard polishing procedures. Scale thickness and metal loss were measured by microscope to determine the corrosion loss. When the corrosion rate is high, the decrease in specimen thickness is generally higher than the scale thickness indicating scale spallation. At lower corrosion rates, there is a good agreement between the two measurements. When the two measurements differ, the highest measurement is considered the most valid one. Optical photomicrographs were taken to document scale morphology.

Selected alloys were studied further using a Jeol JXA-840A, scanning electron microscope, with a KEVEX 8000 computerised EDS/WDS microanalyser. This equipment is able to deconvolute the EDS spectrum and provide semi-quantitative scale analyses. From the elemental analyses, the local scale composition was calculated based on the relative stability of the various metal oxides and sulphides. Since the oxygen content of the scale could not be determined accurately, it was assumed that only sulphides and oxides were present in the scale. Cursory WDS analysis indicated this was a fair assumption. The stability of the oxides was assumed to increase in the following order: NiO, MnO, FeO, V_2O_3, Cr_2O_3, TiO_2, SiO_2. This ranking is based on the location of the gas composition in the M–O–S equilibrium diagrams of the various metals.[3] Thus, a scale composition with 2at% Si, 13% S, 60% Cr, 25% Fe would have the following calculated mineral composition (at%): FeS 13, FeO 12, $Cr_2 O_3$ 72, Si O_2 3. X-ray diffraction indicates that FeO and Cr_2O_3 combine to form $FeCr_2O_4$ spinel. Thus, the final mineral composition is FeS 13, $FeCr_2 O_4$ 24, Cr_2O_3 60, SiO_2 3. Similarly, NiS and MnS form solid solutions with FeS and are shown as such; FeS and Cr_2S_3 combine to form the sulphur spinel $FeCr_2 S_4$.

4. Alloys Investigated

Alloys investigated were prepared from high-purity raw materials in the laboratory. Compositions are given in Table 1.

5. Results

5.1 550°C exposures

Significant corrosion occurred at 550°C. Figure 4 shows a plot of the square of the alloy recession as a function of time. The data indicate a significant influence of the minor alloying elements on the recession rate, with the oxide-forming element Ti being detrimental. The data also indicate that the slow-growing scale on the V containing, alloy exhibits protective parabolic growth kinetics, while the scale growth rates of the other alloys exceed those predicted from parabolic kinetics, especially that of the Ti-containing alloy. Microscopic examination of the alloys indicated that all three alloys had similar scale morphologies, as illustrated in the optical photomicrograph in Fig. 5 for alloy 1. The scales consisted of a loose outward-growing scale and a somewhat denser, but inhomogeneous inward-growing scale. SEM/EDS analysis shows

	Nominal Composition wt%						
Alloy No.	Ni	Cr	V	Ti	Mn	Si	C
1	35	20	–	–	1	0.3	<0.1
2 (V)	35	20	3	–	1	0.3	<0.1
3 (Ti)	35	20	–	3	1	0.3	<0.1

Table 1 Alloys investigated.

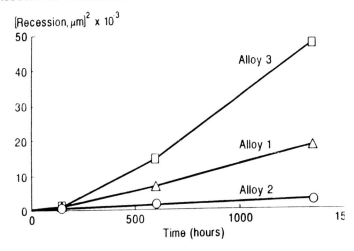

Fig. 4 Corrosion loss at 550°C.
1. 20Cr-35Ni-bal Fe
2. 20Cr-35Ni-3V-bal Fe
3. 20Cr-35Ni-3Ti-bal Fe

that the outward-growing scale consists of (Fe, Ni) sulphide with a metal to sulphur ratio close to one. Little or no chromium or other alloying elements are present in the outward-growing sulphides.

The inward growing scale consists of mixed oxides and sulphides, generally with an increasing sulphur content towards the gas side. The top layer of the scale often consists of almost stoichiometric $FeCr_2S_4$. V and Ti are present only in the inward-growing scale generally evenly distributed, although the sulphur content of the scale varies considerably, especially near the metal-scale interface. Table 2 shows the calculated mineral composition of the inner scales of all three alloys studied. Alloy 3 suffered from an irregular, pitting type of attack. Surprisingly the scale overlaying the slow-growing areas were sulphur rich, while oxide rich scales were present in the pits. This is clearly illustrated in the elemental maps shown in Fig. 6. The reason for the accelerated attack in oxygen rich areas is still under study.

5.2 400°C exposure

Corrosion at 400°C was minimal in the dry syngas, with a scale thickness ranging from 8–9 µm after 600 h exposure. The outward-growing scale was similar to that observed at 550°C, near stoichiometric Fe(Ni)S. The inward-growing scale consisted only of chrome iron sulphides, with some Ni present. Table 3 gives the mineral composition of the inner scales of each alloy. The Cr/S ratio of the Cr_2S_3 is surprisingly close to 1.5. Only the Ti-containing alloy has a minor amount of oxide in the inner scale. An oxygen analysis by WDS confirmed the absence of oxygen in the scale on alloy 1. Figure 7 shows the scale morphology of alloy 1, as well as elemental maps of the major scale constituents.

Alloy	Composition	Corrosion loss μm	Scale	FeNi(Mn)S	FeCr₂S₄	MS	FeCr₂O₄	Cr₂O₃	MO	SiO₂
1	20Cr-35Ni	134	sulphur rich	6	38	–	–	55	–	1
			sulphur poor	24	–		–	74		2
2	20Cr-15Ni-3V	49	general*	12	–		–	76	11V₂O₃	1
			sulphur rich	23	–	9V₂S₃	–	51	15V₂O₃	2
			oxide rich	4			40	40	13V₂O₃	3
3	20Cr-35Ni-3Ti	217	above peaks	–	54	15 Cr₂S₃	–	24	7TiO₂	–
			in pits	4	26	–		58	10TiO₂	2

Table 2 Calculated scale composition (at.%) from SEM/EDS spectra 1350 h, 550°C, dry syngas.

Alloy	Composition wt%	Corrosion loss μm	FeCr₂S₄	Cr₂S₃	S/Cr ratio	MS/MO	Cr₂O₃	SiO₂
1	20Cr-35Ni	8	40	59	1.7	–	–	1
2	20Cr-15Ni-3V	7	28	53	1.5	17V₂S₃	–	2
3	20Cr-35Ni-3Ti	9	46	26	1.5	14TiO₂	12	2

Table 3 Calculated scale composition (at.%) from SEM/EDS spectra 631 h, 400°C, dry syngas.

40 μm

Fig. 5 Optical photomicrograph of scale on alloy 1 550°C, 1350 hrs. exposure.

6. Discussion

Corrosion of simple Fe–Ni–Cr alloys in non-equilibrium H_2–H_2S–CO–CO_2 gas mixtures can be described with reference to the Cr–O–S phase diagrams. At 550°C oxygen and sulphur pressures of the gas mixture studied fall in the field, where Cr_2O_3 and FeS are the stable phases (Fig. 2). However, the gas composition is close to the Cr_2S_3–Cr_2O_3 phase boundary and, therefore, most likely to the left of the 'kinetic boundary', as defined by Perkins[4] and Natesan.[5] At higher oxygen pressures to the right of the kinetic boundary the growth rate of Cr_2O_3 is high enough to form a continuous, protective Cr_3O_3 scale. At lower oxygen pressures a mixed oxide/sulphide scale is formed which allows the diffusion of Fe and Ni to the surface to form FeNi sulphides. At high temperatures this leads to very high corrosion rates, when the FeS–NiS eutectic temperature is exceeded. At 550°C corrosion rates are still high, but reaction kinetics are close to parabolic, suggesting a diffusion-controlled process, which may lead to commercially acceptable corrosion rates in properly designed

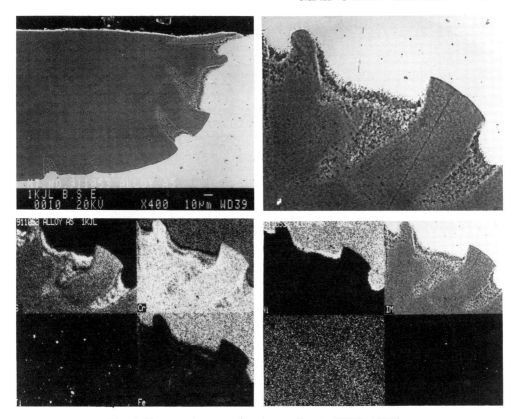

Fig. 6 Elemental maps of scale on alloy 3, 550°C, 1350 hrs.

alloys containing 20–25% chromium. It is further of interest to note that the sulphur content of the inner scale is quite variable and that Cr_2S_3 or $FeCr_2S_4$ is present especially near the surface of the inner scale. This confirms that the PS_2/PO_2 ratio of the gas is close to that of the Cr_2S_3/Cr_2O_3 phase boundary so that local variations in sulphur and oxygen pressure can lead to the formation of chromium sulphides.

At 400°C the gas mixture is clearly in the Cr_2S_3 stability field. Formation of Cr_2O_3 is, therefore, thermodynamically less likely. It is, therefore, not surprising that the inner scale consists almost entirely of Cr_2S_3, of near stoichiometric composition. It is of interest to note that the growth rate of Cr_2S_3 scales at 400°C is quite slow, about the same as that of Cr_2O_3 at 1000°C.[6] Thus parabolic reaction kinetics are likely and commercially acceptable corrosion rates may be expected.

The effect of alloying additions V and Ti, although significant, is not straightforward. It is generally thought that strong oxide-forming elements are beneficial. For instance, it was previously shown that addition of Ti to type 310 stainless steel significantly decreased its corrosion rate at 800–1000°C in H_2–H_2O–H_2S–CO–CO_2 gas mixtures.[7] The present data not only show that Ti additions are clearly detrimental, SEM analysis also shows increased corrosion

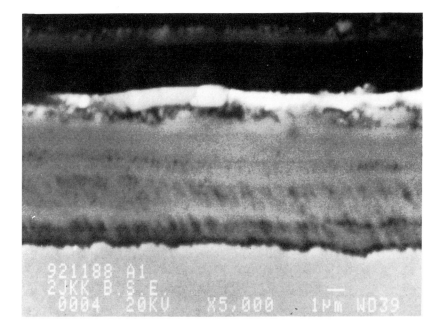

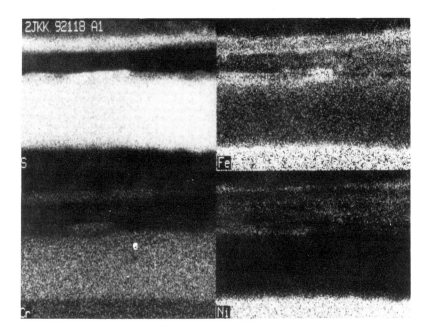

Fig. 7 Elemental maps of scale on alloy 1, 400°C, 600 hrs.

in areas where the scale adjacent to alloy is relatively oxygen rich (Fig. 6). This suggests that a high-sulphur content in the inner scale is not necessarily detrimental. It may also explain why vanadium, a strong sulphide former in the gas mixture studied, significantly reduces the corrosion rate. This was formerly suggested by Strafford[8] for high temperature conditions, where the formation of a $V_2 S_3$ subscale was demonstrated. It must be noted, however, that the formation of a $V_2 S_3$ subscale or even the formation of $V_2 S_3$ is not clearly shown in the present work. Routine scale composition calculations assuming FeS will form in preference to $V_2 S_3$ show that in most areas $V_2 O_3$ instead of $V_2 S_3$ will be stable. If we assume that $V_2 S_3$ is the more stable species, the 'general' scale composition of alloy 2 in Table 2 would be approximately (at%): $V_2 S_3$ 11, $Fe Cr_2 O_4$ 26, $Cr_2 O_3$ 50. Such a scale may well be reasonably protective.

7. Conclusions

1. Mixed oxidant corrosion in non-equilibrium, highly reducing and sulphidising gases at 400–550°C is generally similar to that found previously in more oxidising equilibrated gases at higher temperatures.
2. Sulphide-rich scales are not necessarily less protective than oxide-rich scales.
3. The change in Kp of the $H_2 + S = H_2S$ reaction at the boiling point of sulphur (440°C) increases the sulphur pressure of H_2–H_2S–CO–CO_2 gas mixtures below 440°C and prevents the formation of $Cr_2 O_3$ at 400°C in the gas mixture studied.

References

1. W. T. Bakker, J. A. Bonvallet and J. H. W. de Wit, 'Corrosion in coal gasification environments at 550°C', *Proc. High Temperature Materials Corrosion, Les Embiez (France)*, May 1992, in print.
2. R. A. Perkins *et al.*, *EPRI report GS-6971*, August 1990.
3. E. A. Gulbransen and G. H. Meier, *DOE report, FE 1354701 UC, 90H*, May 1980.
4. R. A. Perkins and S. J. Vonk, *EPRI Report FP-1280*, December 1979.
5. K. Natesan, *Proc. Corrosion-Erosion-Wear'* (ed. A. Levy) p. 100, NACE, 1982.
6. K. Natesan, *Proc. Materials for Coal Gasification* (ed. W. T. Bakker) p. 137, ASM, 1988.
7. R. W. Bradshaw *et al.*, *Sandia Report 78-8277*, March 1979.
8. K. N. Strafford *et al.*, *Corrosion Science*, **29**(6), 775, 1989.

3

OXIDATION OF NICKEL, CHROMIUM AND THEIR ALLOYS

Scanning Tunnelling Microscopy Study of the Atomic Structure of Thin Anodic Oxide Films Grown on Nickel Single Crystal Surfaces

V. MAURICE, H. TALAH and P. MARCUS

Laboratoire de Physico-Chimie des Surfaces, CNRS (URA 425) – Université Pierre et Marie Curie, Ecole Nationale Supérieure de Chimie de Paris, 11 rue Pierre et Marie Curie, 75231 Paris Cedex 05, France

ABSTRACT

Scanning tunnelling microscopy has been used to study the thin anodic oxide film grown on Ni(111) in 0.05 M H_2SO_4. *Ex situ* atomic resolution imaging demonstrates the crystalline character of the passive oxide film and the epitaxy with the substrate. Stepped surfaces are measured which indicate a tilt of the surface of the film with respect to the (111) orientation. Local variations of the film thickness are likely to result from this tilt. The chemical nature of the atomic structure which is resolved and the possible epitaxial relationships resulting from the surface tilt are discussed.

1. Introduction

The thin anodic oxide films grown on metal surfaces represent an important area of scientific investigation because of their properties of passivation of metal surfaces against corrosion. Numerous studies by means of depth sensitive spectroscopic techniques have shown that the distribution of the chemical species within these thin films is well described by a bilayer model.[1] The inner part of the film is made of an oxide component and the outer part is made of a hydroxide or oxihydroxide component.[1] The extent of ordering within passive films is a more controversial topic although the amorphous character has been emphasised.[2] The case of pure nickel is a typical example. The thickness of the oxide inner part of the film has been reported to range from 0.4 to 1.2 nm, and that of the hydroxide outer part has been reported to range from fractions of a monolayer to a complete monolayer (0.6 nm thick).[3–6] The extent of ordering within the passive film on nickel, which is possibly related to the *ex situ* or *in situ* conditions of investigation of the structure, is still a subject of debate.[7,8]

We have undertaken a Scanning Tunnelling Microscopy (STM) investigation

99

of the structure of the passive film formed on a pure nickel substrate in acid electrolyte. Our choice of a (111) oriented Ni single crystal substrate was mainly dictated by the concern of selecting a system for which the passive film had previously been reported as being crystalline[7] in order to facilitate atomic resolution imaging. Preliminary experiments confirmed the possibility of achieving lateral atomic resolution in *ex situ* investigations of passive films.[9] This paper reports further *ex situ* investigations of the atomic structure of the anodic oxide film grown on Ni(111) with the discussion being focussed on the epitaxy of the film with respect to the metal substrate.

2. Experimental

Sample preparation from a Ni single crystal rod (purity 99.999%) involved successively: orientation within ±1° by X-ray diffraction, spark machining, mechanical and electrochemical polishing and annealing at 1275 K for a few hours in a flow of purified hydrogen at atmosphere pressure. The sample was then transferred at room temperature and under the hydrogen atmosphere into a nitrogen-containing glove box where the electrochemical experiments were performed. Sulphuric acid (0.05 M) was prepared from reagent grade H_2SO_4 and ultrapure water (resistivity of 18 MΩ cm). After immersion of the electrode in the electrochemical cell, the potential was stepped from the open circuit value to the passivation potential. Three values were selected: +550, +650 and +750 mV/SHE. After 12 minutes of anodic polarisation and completion of the formation of the passive film, the experiment was stopped by emersing the Ni(111) electrode at the applied potential. The sample was then rinsed with ultrapure water, dried in nitrogen gas and finally transferred to air for STM imaging.

STM imaging was performed with the Nanoscope II and III from Digital Instruments (Santa Barbara, CA) operating in atmospheric conditions. Maximum scan range both in X and Y axis of the scanner used for atomic resolution imaging was about 0.7 μm. Images were recorded for the most part in the constant current topographic mode of the STM. All reported topographic images have had a least square plane subtracted in order to remove any tilt of the scanning head relative to the sample surface. Sample bias voltages used for atomic resolution imaging were in the range from 30 to 200 mV, positive or negative, and setpoint currents were in the range from 0.5 to 1 nA. Tunnelling probe tips were made from W wire etched in 1M KOH.

3. Results and Discussion

STM observations of the metal substrate before the passivation treatment revealed the presence of the steps and terraces of the surface. Steps of monolayer to multilayer height could be imaged which were all aligned along the <−101> directions of the substrate, as determined from a crystal alignment performed by X-ray back diffraction measurements. The width of

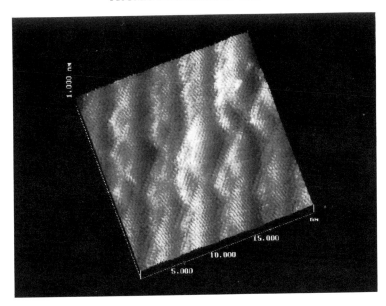

Fig. 1 (20×20) nm^2 topographic image of the stepped surface of the thin anodic oxide film grown on Ni(111) at +750 mV/SHE in 0.05 M H$_2$SO$_4$, sample bias voltage V$_t$ = +135 mV, and setpoint current I$_t$ = 0.8 nA. The kinked structure of the step edges and the atomic lattice on the terraces are evidenced.

the terraces was frequently found to exceed 100 nm. The roughness which was measured on the terraces is supposedly resulting from the formation of the native oxide. Atomic resolution was not achieved on these surfaces.

After the passivation treatment, two major levels of structural modifications were observed. On a submicroscopic scale (0.2 μm), a roughening dependent on the passivation potential was observed which probably results from the competition between metal dissolution and nucleation and growth of the passive film as reported elsewhere.[10] On the atomic scale (<20 nm), a corrugated lattice with stepped surfaces has been resolved. The parameters of this lattice and the characteristics of the stepped surfaces are found to be independent of the passivation potential.

Figure 1 shows a typical image recorded on the atomic scale after passivation which indicates that the passive film is crystalline. Stepped surfaces were imaged over large areas without noticeable defects. A corrugated lattice is resolved on the terraces which has the following parameters: 0.32 ± 0.02 nm and $117° \pm 5°$. The agreement is excellent with the lattice parameters of a (111) oriented NiO structure: 0.295 nm and 120°. The step edges of these passivated surfaces are disoriented within 5 to 10° from the main crystallographic directions: $\langle -101 \rangle$ and $\langle 1-21 \rangle$ with the presence of kinks. In Fig. 1, the step edge is facetted along the [-101] and [01-1] directions. Figure 2 shows a high magnification image of these stepped surfaces where the atomic lattice is clearly resolved on the terraces and where kinks oriented along the [-101] and [0-11] directions are visible. Figure 3 shows the image of the fine structure of a

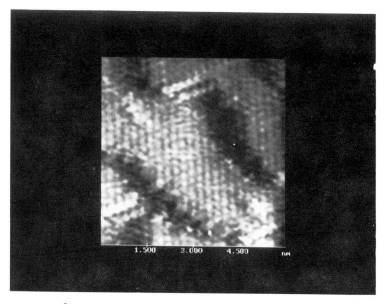

Fig. 2 $(6 \times 6)\,nm^2$ current image of the stepped surface of the thin anodic oxide film grown on Ni(111) at $+750\,mV/SHE$, $V_t = +85\,mV$, $I_t = 0.5\,nA$. The corrugation of the atomic lattice is resolved on the terraces and the step edges are kinked along the close packed $\langle -101 \rangle$ directions in the upper part and lower left part of the image.

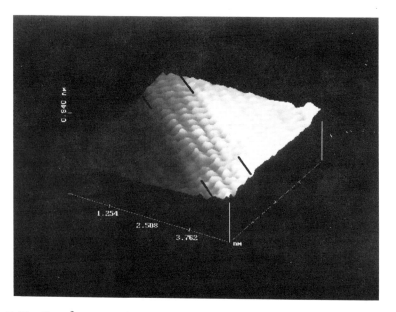

Fig. 3 $(5 \times 5)\,nm^2$ topographic image of the fine structure of a step edge along the $[-101]$ direction of the surface passivated at $+750\,mV/SHE$, $V_t = +113\,mV$, $I_t = 0.5\,nA$. The lines indicate a two-stage transition between the upper terrace (right) and the lower terrace (left).

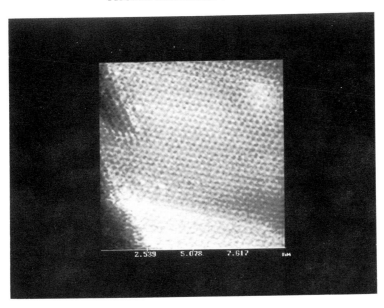

Fig. 4 (10×10) nm^2 topographic image of a large (111) terrace at the surface of the passive film formed on Ni(111) at +650 mV/SHE, V_t = +111 mV, I_t = 0.5 nA.

step locally oriented along the [–101] direction. It shows a two-stage transition between the upper and lower terraces with an intermediate terrace being resolved as indicated on the figure. This intermediate terrace is characterised by a higher corrugation. The width of the (111) oriented terraces that have been measured in most images varies locally and ranges from 2 to 5 nm. The step height also varies locally but remains consistent with monolayer or double layer NiO spacings (0.24 and 0.48 nm, respectively). This amounts to an average tilt of $8 \pm 5°$ for these stepped surfaces with respect to the (111) orientation of the terraces. Considering the value of 8° for the tilt, the orientation of these stepped surfaces is (433) for an average [0–11] step direction, and (765) for an average [1–21] step direction. Only one case of (111) terraces extended over larger distances has been found. It is shown in Fig. 4. The terrace recorded in this image was extending without any noticeable defects over a distance of about 20 nm. This suggests a tilt of less than 1° with respect to a (111) orientation.

The relation between the chemical nature of the passive film and the STM image is dependent on the tunnelling mechanism taking place for such thin film systems. A direct mechanism that would involve tunnelling from the metal substrate through the passive film is not likely. Such a mechanism would probably require the penetration of the tip into the passive film because the overall thickness of the film is large with respect to the usual width of the tunnelling barrier (about 0.5 nm).[11] In addition, the measured lattice parameters are consistent with those of nickel oxide (0.417 nm) but not with those of metallic nickel (0.352 nm). Two other mechanisms can be invoked which differ depending on the role of the hydroxide outer layer. In the case of the mechanism involving electron transfer from the metal substrate to the conduction band of the oxide inner part followed by tunnelling through the hydroxide outer part, the structural information in the images would be relative to the oxide inner part of the passive film. In such a case, the thickness of the hydroxide outer layer should not exceed the width of the tunnelling barrier. The image shown in Fig. 3, where a two-stage transition between upper and lower terraces has been resolved, could then be explained by a transition between the two O planes (or Ni) via a Ni plane (or O) in the (111) oriented NiO structure which alternates O planes and Ni planes. In the case of a mechanism involving electron transfer from the oxide inner part to surface hydroxyls groups which would act as centres for tunnelling to the tip, the structural information of the images would be also relevant to the hydroxide outer part of the passive film. The recorded images would then suggest that the surface hydroxyl groups form a complete ordered monolayer of (1×1) periodicity with respect to the underlying nickel oxide. The image shown in Fig. 3 could then be explained by a transition between two hydroxyl monolayers covering terraces of the underlying oxide. The intermediate area could correspond to a terrace not covered by hydroxyl groups in the vicinity of a step edge. The structural information contained in the recorded images does not allow us to favour either of these two possible mechanisms.

The interface between the metal substrate and the thin oxide film cannot be directly resolved in STM experiments and therefore, assumptions can only be made about the orientation of the metal substrate surface. The observation in Fig. 4 of a large (111) terrace at the surface of the passive film suggests the possibility of parallel surfaces for the film and the substrate over distances of about 20 nm. This possibility is shown on Fig. 5a and results in an NiO(111) // Ni(111) with an NiO[–101] // Ni[–101] epitaxial relationship. The lattice mismatch (expansion of 16.5% of bulk NiO with respect to bulk Ni) between the metal substrate and the oxide inner part of the film could possibly be accommodated by a site coincidence growth model with a large unit cell. A (6×6) NiO // (7×7) Ni superlattice has been observed previously[12] for NiO(100) expitaxially grown on Ni(100), demonstrating an interesting case of strained epitaxy for highly mismatched crystals achieved by large unit cells.

Still assuming large (111) oriented terraces at the metal surface but taking into account the average tilt of $8 \pm 5°$ recorded for the surface of the thin oxide film suggests the possibility of having the NiO(111) terraces tilted with respect to the Ni(111) terraces. This possibility is schematically shown in Fig. 5(b). Two different epitaxial relationships would result, considering that steps have been measured along directions close to [0–11] and [1–21]. Taking into account the value of 8° for the tilt, these epitaxial relationships would be: NiO(433) // Ni(111) with NiO[0–11] // Ni[0–11] and NiO(765) // Ni(111) with NiO[1–21] // Ni[1–21]. In such a case, the tilt could result from a partial relaxation in one direction of the strained epitaxy due to the large lattice mismatch. The recording in almost all areas of the passivated samples, of stepped surfaces indicates that the tilted epitaxial relationships would be more favourable than the parallel epitaxial relationship.

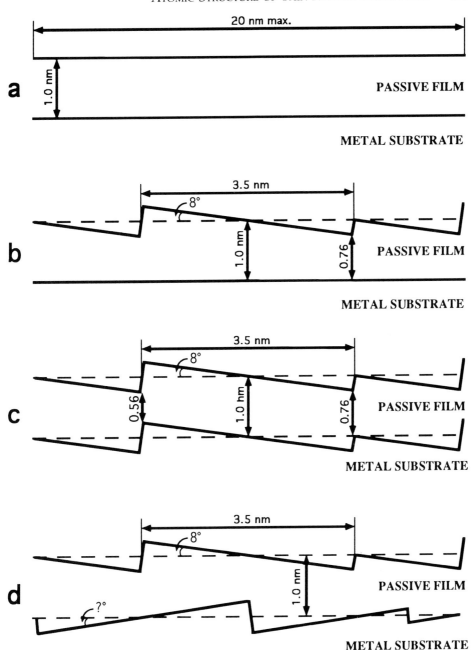

Fig. 5 Schematic representations of the orientation of the surface of the passive film with respect to the orientation of the surface of the metal substrate. A thickness of 1 nm and a tilt of 8° have been considered. Monolayer oxide steps (0.24 nm) and double layer oxide steps (0.48 nm) are represented at the surface of the film and at the surface of the substrate. (a): parallel surfaces, (b): non-parallel surfaces, (c): tilted parallel surfaces, (d): tilted non-parallel surfaces.

Two other cases can be considered if one assumes stepped surfaces also at the surface of the substrate. The first case is schematically shown on Fig. 5(c) where the orientation of the substrate surface is duplicated at the surface of the thin oxide film. Taking into account the value of 8° of the tilt, the resulting epitaxial relationships would be: NiO(433) // Ni(433) with NiO[0–11] // Ni[0–11] and NiO(765) // Ni(765) with NiO[1–21] // Ni[1–21]. Such a model would imply that the strained epitaxy of the (111) terraces would be partially relaxed by the presence of the steps. The second case taking into account a stepped substrate surface is schematically shown in Fig. 5(d). This possibility has been inferred from the previously reported results[13] of NiO grown on stepped Ni(100) which have shown that the presence of steps at the substrate surface induces a tilt of about 8° of NiO crystallites. The exact amount of disorientation at the substrate surface necessary to induce the tilt measured at the thin oxide film surface in our experiment as well as the mechanism by which this tilt could be generated are not elucidated. These two models of tilting suppose a disorientation of the substrate which was not observed before passivation. Surface facetting could possibly occur during the formation of the passive film mainly as the result of the strained epitaxy. The identification of facetting from the images recorded at low magnification is made difficult by the presence of numerous structural patterns which probably result from the metal dissolution (corrosion) occurring during the passivation experiment.

A major consequence of the presence of steps at the surface of the passive film would be local variations of the thickness of the film as illustrated in Fig. 5. Taking into account an average thickness of 1 nm, the steps at the film surface would induce a thickness decrease of about 0.24 nm. This decrease could even be about 0.44 nm in the case of two coinciding steps at the thin film surface and substrate surface. The bottom of the steps of the thin oxide film could constitute preferential sites for the breakdown of the passive film where the barrier effect of the film on cation diffusion would be drastically decreased.

4. Conclusions

STM measurements in air of thin anodic oxide films grown on Ni(111) single crystal surfaces in 0.05 M H_2SO_4 show the crystalline character of the films. On the atomic scale, stepped surfaces are imaged. Their characteristics are independent of the passivation potential and indicate a tilt of $8 \pm 5°$ of the surface of the film with respect to the (111) orientation of the terraces. The lattice parameters of 0.30 ± 0.02 nm and the $117 \pm 5°$ measured from the images fit those of NiO(111), the inner component of the passive film. Although the NiO(111) // Ni(111) with NiO [0–11] // Ni[0–11] epitaxial relationship can be deduced from a local measurement of extended (111) terraces, the tilt recorded in most areas suggests the following preferential epitaxial relationships: NiO(433) // Ni(111) with NiO[0–11] // Ni[0–11] and NiO(765) // Ni(111) with NiO[1–21] // Ni[1–21] or NiO(433) // Ni(433) with NiO[0–11] // Ni[0–11] and NiO(765) // Ni(765) with NiO[1–21] // Ni[1–21] depending on the possibility of a facetted substrate surface. In any case, the tilt is assigned to a partial relaxation of a strained epitaxy resulting from the large lattice mismatch between the oxide film and the metal substrate. A major consequence of this tilt would be local variations of the thickness of the film which would constitute preferential sites of breakdown.

References

1. See, e.g. P. Marcus, The role of alloyed elements and adsorbed impurities in passivation of metal surfaces, in *Electrochemistry at Well-Defined Surfaces*, (eds J. Oudar, P. Marcus and J. Clavilier), Special Volume of *J. Chimie Physique*, **88**, 1697–1711, 1991.
2. J. Kruger, Nature of passive film on iron: Does it affect breakdown?, in *Advances in Localized Corrosion* (eds H. Isaacs, U. Bertocci, J. Kruger and S. Smialowska), NACE, pp. 1–7, 1987, and references therein.
3. P. Marcus, J. Oudar and I. Olefjord, XPS study of the passive film on nickel, *J. Microsc. Spectrosc. Electron.*, **4**, 63–72, 1979.
4. B. P. Lochel and H.-H. Strehblow, Breakdown of passivity of nickel by fluorine, *J. Electrochem. Soc.*, **131**, 713–723, 1984.
5. F. T. Wagner and T. E. Moylan, Electrochemically and UHV grown passive layers on Ni(100): a comparison by AC impedance, XPS, LEED, and HREELS, *J. Electrochem. Soc.*, **136**, 2498–2506, 1989.
6. D. F. Mitchell, G. I. Sproule and M. J. Graham, Measurement of hydroxyl ions in the passive oxide films using SIMS, *Appl. Surf. Sci.*, **21**, 199–209, 1985.
7. J. Oudar and P. Marcus, Role of adsorbed sulphur in the dissolution and passivation of nickel and nickel-sulphur alloys, *Appl. Surf. Sci.*, **3**, 48–67, 1979.
8. R. Cortes, M. Froment, A. Hugot-Legoff and S. Joiret, Characterization of passive films on nickel and nickel alloys by ReflExafs and Raman spectroscopy, *Corrosion Sci.*, **31**, 121–127, 1990.
9. V. Maurice, H. Talah and P. Marcus, *Ex situ* STM imaging with atomic resolution of Ni(111) electrodes passivated in sulfuric acid, *Surface Sci.*, **284**, L431–L436, 1993.
10. V. Maurice, H. Talah and P. Marcus, to be published.
11. G. Binnig, H. Rohrer, Ch. Gerber and E. Weibel, Surface studies by STM, *Phys. Rev. Lett.*, **49**, 57–61, 1982.
12. R. S. Saiki, A. P. Kaduwela, M. Sagurton, J. Osterwalder, D. J. Friedman, C. S. Fadley and C. R. Brundle, X-ray photoelectron diffraction and LEED study of the interaction of oxygen with the Ni(001) surface: c(2 × 2) to saturated oxide, *Surf. Sci.*, **282**, 33–61, 1993, and references therein.
13. M. Bäumer, D. Cappus, H. Kuhlenbeck, H.-J. Freund, G. Wilhemi, A. Brodde and H. Neddermeyer, The structure of thin NiO(100) films grown on Ni(100) as determined by LEED and STM, *Surf. Sci.*, **253**, 116–128, 1991.

The Incorporation of Internal Oxide Precipitates into Growing Oxide and Sulphide External Scales

F. H. STOTT, A. STRAWBRIDGE* and G. C. WOOD

Corrosion and Protection Centre, University of Manchester Institute of Science and Technology, P.O. Box 88, Manchester M60, 1QD, UK
*Present address: Department of Materials Science and Engineering, Ohio State University, Columbus, Ohio 43210, USA

ABSTRACT

During oxidation of dilute nickel-base alloys containing chromium, aluminium, vanadium, molybdenum or tungsten, internal precipitates of the more stable second-element oxides are formed in the alloy beneath a thickening nickel oxide scale. A study has been made of the interactions between such precipitates and the scale as the scale/alloy interface encroaches on them. In particular, the precipitates have been preformed in a Ni/NiO Rhines pack at high temperature and, subsequently, the pretreated specimens have been oxidised in 1 atm oxygen at 1200°C, or sulphidised in H_2/H_2S at 600°C, in order to determine the parameters which influence the interactions. Significant differences have been observed, with some internal precipitates, such as $NiAl_2O_4$ and $NiCr_2O_4$, being incorporated easily into nickel oxide and nickel sulphide scales while others, such as MoO_2, are not incorporated at all into a nickel oxide scale. The results are discussed in terms of properties of the scales and precipitates.

1. Introduction

Most high-temperature alloys are based on nickel, cobalt or iron and contain elements such as chromium and/or aluminium that form more stable oxides than the base metal oxide. Upon oxidation, the latter often produces an external scale while the more active elements are oxidised internally to form particles at, or below, the scale/alloy interface. The resistance is determined, in part, by the effectiveness of the overall scale as a barrier to oxygen or metal ion transport. For instance, nickel-chromium alloys can have good resistance due to the development of a Cr_2O_3 layer at the base of the scale. However, this is

108

established only if the concentration of chromium exceeds about 15 wt%. At lower values, the internal oxide particles are incorporated into the thickening NiO scale and a complete layer is unable to form.[1] Moreover, for other alloys, such as nickel-molybdenum, the particles are apparently not incorporated into the scale and build up as a partial or complete layer at the scale/alloy interface at lower concentrations of the more active element.[2]

In the present paper, results are presented from an investigation of the factors that determine the interactions between internal oxide particles and an external scale as the scale/alloy interface encroaches on them, since this has consequences for protection of the alloy. The reasons why the particles sometimes pile up at the interface and sometimes are incorporated into the scale are also of more general scientific interest and application.

2. Materials and Methods

The alloys used were (in weight %) Ni–5.0%Cr, Ni–2.03%Al, Ni–4.18%V, Ni–4.15%Mo and Ni–4.0%W, corresponding to (in atomic%) Ni–5.6%Cr, Ni–4.4%Al, Ni–4.8%V, Ni–2.4%Mo and Ni–1.3%W. Specimens (5 × 10 × 1 mm) were polished to 1 μm finish and degreased.

The specimens were exposed in a Rhines pack, enabling the more active elements to be oxidised internally without formation of a surface scale. The pack was made by sealing one specimen in an evacuated, quartz capsule containing NiO (3g) and Ni (1g) powders. This was exposed for 48 h at 1100°C in a horizontal furnace, followed by furnace cooling. The treated specimens were ground on successively finer SiC papers and polished to 1 μm diamond finish. This removed the surface to a depth of about 50 μm. They were then ready for the oxidation or sulphidation tests.

Oxidation involved inserting the specimens in the reaction tube of a horizontal furnace. The sealed system was flushed with dried oxygen at 1 atm pressure and a flow rate of 40 cm^3 min^{-1}. The furnace was switched on and the specimens brought to 1200°C (±5)°C over a 2 h period and exposed for 12 h before being furnace cooled. Sulphidation involved locating the specimens in the inner part of a horizontal double-walled reaction tube. The reactive gas, H$_2$/2%H$_2$S, flowed between the inner and outer walls, through alumina particles to bring it to equilibrium at temperature, and then over the specimens. For a test, the system was sealed and flushed with argon before the reactive gas was introduced at a flow rate of 50 cm^3 min^{-1}. The specimens were brought to 600(±5)°C over a 30 min period and exposed for 10 h before being furnace cooled.

Following the tests, analyses of the specimens in cross-section were carried out using X-ray diffraction, optical and scanning electron microscopy and electron probe microanalysis. In some cases, the morphologies of internal oxide precipitates were revealed in more detail by deep etching in bromine/methanol to remove the metal substrate. Extracted internal oxide precipitates were identified using electron diffraction. Further details are given elsewhere.[3]

3. Experimental Results

3.1 Internal oxidation treatment

Following the internal oxidation treatment, there was a small increase in volume recorded for each specimen, of up to 5%. In every case, internal precipitates had formed in the alloy substrate, with depths of penetration from the surface of 300 μm (Ni–2%Al), 140 μm (Ni–4%V), 170 μm (Ni– 5%Cr), 170 μm (Ni–4%Mo) and 190 μm (Ni–4%W). In some cases, there was evidence for a precipitate-denuded zone at the surface, consistent with the hypothesis that creep of metal is induced by the stresses arising from internal oxide precipitation.[4]

The shapes and natures of the precipitates depended on the alloying element. They appeared to be acicular for Ni–2%Al (Fig. 1a) but were shown to be continuous, densely packed rods of small diameter ($\sim$1 μm), aligned perpendicular to the surface, following deep etching (Fig. 1b). X-ray and electron diffraction and microanalysis confirmed that they were Al_2O_3 adjacent to the internal oxide/alloy interface; however, they had transformed to $NiAl_2O_4$ nearer to the surface. The precipitates in the other alloys were discrete particles. They were acicular for Ni–5%Cr (Fig. 1c) and were Cr_2O_3 near the internal oxide front, but $NiCr_2O_4$ for most of the precipitate zone. They were relatively spherical for the other three alloys (Fig. 1d–f). In Ni–4%V, they were probably V_2O_3 at the internal oxide/alloy interface but contained both nickel and vanadium for much of the zone. Identification was difficult but diffraction patterns were consistent with $NiVO_3$. In Ni–4%Mo, only MoO_2 was detected throughout the zone while, in Ni–4%W, only $NiWO_4$ was indicated by analysis and diffraction. In addition to internal oxidation, for all the alloys except Ni–4%Mo, there was preferential precipitation of internal oxide in the alloy grain boundaries, with areas denuded of precipitates immediately adjacent to the boundaries.

3.2 Oxidation in 1 atm oxygen

Oxidation of pre-treated specimens of the five alloys resulted in development of an external NiO scale. However, some differences between the alloys were observed in terms of the interactions of the internal precipitates with the advancing scale/alloy interface.

For Ni–2%Al, the precipitates were incorporated into the inner layer of the scale (Fig. 2a) while the outer layer of the scale (not shown) was about 85 μm thick, and free of precipitates. There were three distinct morphologies in the inner layer; a thin ($\sim$8 μm) compact region at the scale/alloy interface, a very porous intermediate region ($\sim$50 μm thick) and a less porous outer region ($\sim$40 μm thick). The precipitates were present in all three regions, with similar spatial resolution as in the alloy. However, dissolution of the precipitates to saturate the NiO matrix with aluminium (to a level of 2%) usually resulted in break-up of the rods, leaving residual acicular precipitates. As in the alloy, there were areas denuded in particles adjacent to incorporated intergranular

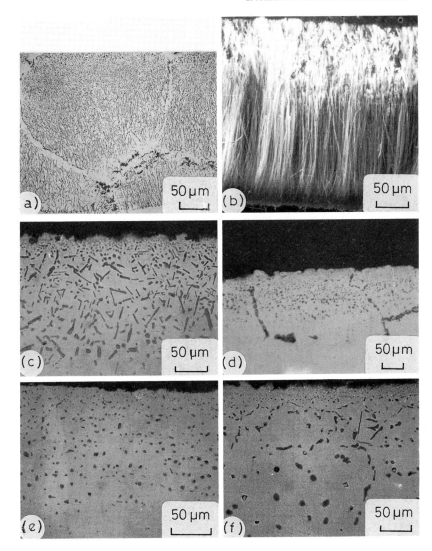

Fig. 1 Optical micrographs of cross-sections of nickel-base alloys after internal oxidation in a Rhines pack for 48 h at 1100°C. (a) Ni–2%Al; (b) Ni–2%Al, after deep etching (SEM); (c) Ni–5%Cr; (d) Ni–4%V; (e) Ni–4%Mo; (f) Ni–4%W.

oxide (Fig. 2b). The intermediate porous region was usually thinner (and the outer region thicker) in locations associated with such oxide (Fig. 2a). There was preferential inward penetration of NiO at the scale/alloy interface, along the intergranular oxide at the alloy grain boundaries.

The precipitates were also incorporated into the scale on Ni–5%Cr (Fig. 2c). The outer layer (not shown) was compact, being about 95 μm thick. The inner layer showed three morphologies. There was a compact region (8–10 μm) at the scale/alloy interface (Fig. 2d), a porous intermediate region (10 μm) and a

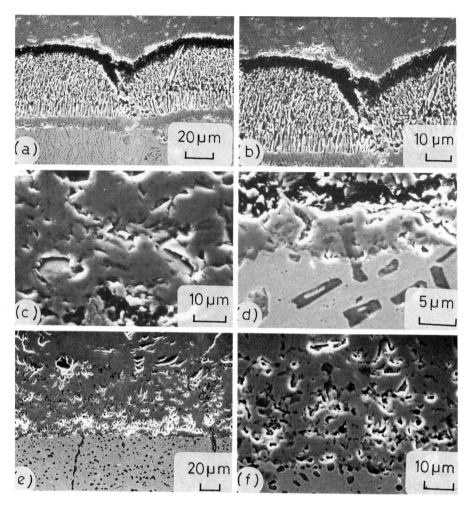

Fig. 2 Scanning electron micrographs of cross-sections of nickel-base alloys after internal oxidation and oxidation for 12 h at 1200°C in 1 atm oxygen, showing inner scale and internal oxide zone. (a) Ni–2%Al; (b) Ni–2%Al, detail of inner scale; (c) Ni–5%Cr; (d) Ni–5%Cr, detail of scale/alloy interface; (e) Ni–4%V; (f) Ni–4%V, detail of scale/alloy interface.

compact outer region (~40 μm) (Fig. 2c). The precipitates were incorporated into the inner region with a similar spatial distribution as in the alloy (Fig. 2d). However, they were smaller in the outer region (Fig. 2c), consistent with considerable dissolution of chromium into the NiO matrix, to the solubility limit of about 3%.

The scale on Ni–4%V consisted of a columnar outer layer (30 μm), an intermediate layer (70 μm) and an inner layer (75 μm). Incorporated particles were present in the inner layer only (Fig. 2f), although there was a vanadium concentration gradient across the intermediate layer, as observed previously,[2]

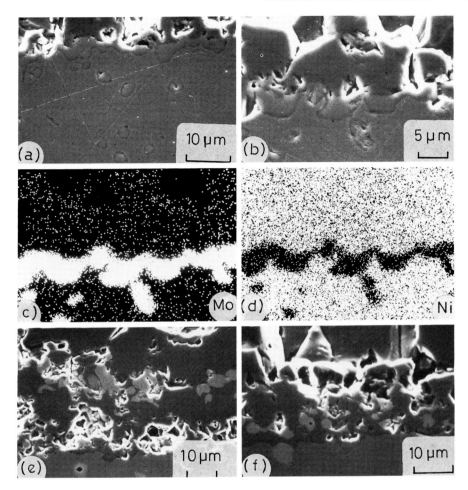

Fig. 3 Scanning electron micrographs of cross-sections of nickel-base alloys after internal oxidation and oxidation for 12 h at 1200°C in 1 atm oxygen, showing inner scale and internal oxide zone. (a) Ni–4%Mo; (b) Ni–4%Mo; (c) Mo X-ray map of (b); (d) Ni X-ray map of (b); (e) Ni–4%W; (f) Ni–4%W.

from the solubility limit (4%) at the inner interface to zero at the outer layer. The precipitates were more irregular in shape than in the alloy, with two or more sometimes having apparently merged together. In some areas, towards the outside of the inner layer, the particles were small or non-existent. Incorporated intergranular precipitates had generally broken up, consistent with partial dissolution in the NiO matrix (Fig. 2e).

There were no precipitates in the scale on Ni–4%Mo; instead, they accumulated as a discontinuous layer at the alloy/scale interface (Fig. 3a). Nonetheless, the scale was relatively thick, with a columnar outer layer (60 μm thick) and a smaller-grained inner layer (20 μm thick). The layer of precipitates covered 80% of the alloy/scale interface after 12 h and was irregular in thickness, up to 6 μm. Analysis indicated that it contained molybdenum but no nickel (Figs. 3b–d) and was probably MoO_2. No molybdenum was detected in the scale.

Although there was no apparent build-up of precipitates at the alloy/scale interface for Ni–4%W, the scale was of similar thickness to that on Ni–4%Mo and consisted of a columnar outer layer (60 μm) and a fine-grained inner layer (30 μm). The precipitates were incorporated into the inner layer (Figs 3e and f). In some areas, there were indications that the particles had initially accumulated at the interface (Fig. 3e) while, in others, they were of a similar size and distribution as in the alloy (Fig. 3f). There was no measurable tungsten in the NiO matrix, consistent with the very low solubility of this element in the oxide.

3.3 Sulphidation in H_2/2%H_2S

Relatively thick nickel sulphide scales, shown by diffraction to be Ni_3S_2, were developed on all the pretreated alloys during exposure to the sulphidising environment at 600°C.

The scale on Ni–2%Al consisted of an outer layer of variable thickness (100–500 μm), and an inner layer containing incorporated oxide precipitates. The latter could be separated into two regions, a compact outer part (25–50 μm) and a porous inner part (50–200 μm). The precipitates retained their rod-like form in the sulphide scale (Fig. 4a). There was no apparent dissolution of aluminium into the sulphide matrix. The scale/alloy interface was irregular, due to inward sulphide penetration being more rapid adjacent to internal oxide rods.

The outer scale on Ni–5%Cr was similar to that on Ni–2%Al, ranging in thickness from 75 to 300 μm. The inner layer consisted of two regions, both containing incorporated precipitates, of similar shape and size as in the alloy. The outer part was compact, about 50 μm thick, while the inner region was porous. Figure 4(b) shows that the precipitates were readily taken up into the scale at the alloy/scale interface. The interface penetrated deeper into the alloy adjacent to the particles and there was no obvious dissolution of chromium into the matrix.

The scale on Ni–4%V consisted of a ballooned, outer layer, 150–250 μm thick, which had detached from the inner layer. The latter was compact, 15–50 μm thick, and contained incorporated precipitates across its full width, but no dissolved vanadium in the sulphide matrix. However, there was a tendency for the particles to build up in the inner layer near the scale/alloy interface, resulting in a higher population density at that location than elsewhere (Fig. 4c). The scale/alloy interface was much flatter than for the previous two alloys (Fig. 4d). This figure also shows clearly the incorporation of precipitates into the scale.

A ballooned outer scale, 125–300 μm thick, had formed on Ni–4%Mo and detached from the inner layer which was compact, adherent and relatively thin (25 μm). It contained incorporated precipitates, located entirely near the scale/alloy interface (Fig. 4e). They had partially merged together, while the alloy/scale interface was relatively flat.

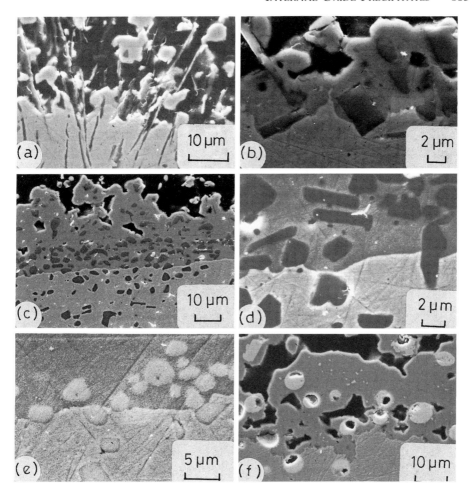

Fig. 4 Scanning electron micrographs of cross-sections of nickel-base alloys after internal oxidation and sulphidation for 10 h at 600°C in $H_2/2\%H_2S$, showing inner sulphide scale and internal oxide zone. (a) Ni–2%Al; (b) Ni–5%Cr; (c) Ni–4%V; (d) Ni–4%V, detail of interface; (e) Ni–4%Mo; (f) Ni–4%W.

The scale on Ni–4%W consisted of a compact outer layer, 150–300 μm thick, and an inner layer containing incorporated precipitates, but no dissolved tungsten. The latter consisted of a compact, outer region (50 μm) and a porous inner region. There was some build-up of incorporated particles in the inner scale, near the scale/alloy interface (Fig. 4f). The interface was irregular, with preferential inward penetration of scale adjacent to internal precipitates.

There was no transition of internal oxide into sulphide prior to incorporation into the Ni_3S_2 scale for any of the alloys, although, at least for the Ni–Mo case, there was some evidence for sulphidation of the incorporated precipitates. However, detailed analyses of such precipitates were not carried out since these did not influence the incorporation process.

4. Discussion

4.1 Interactions between particles and the scale

Although this study is concerned with dilute nickel-base alloys, the results have implications for the development of healing layers on high-temperature alloys during oxidation and for interactions between sulphide and oxide phases during exposures in mixed oxygen- and sulphur-containing environments. For instance, in some coal-gasification systems, the stable phases for such alloys may be oxides of the reactive elements, chromium or aluminium, and sulphides of the base metal elements, such as nickel. Interactions between surface sulphides and oxide precipitates, particularly in the early stages, may be important in determining the compositions of the longer-term steady-state scales.

In the present investigation, there are differences in the sizes, shapes and structures of the internal oxide precipitates in these alloys. Thus, on oxidation or sulphidation, the phases that contact the encroaching scale are $NiAl_2O_4$ rods in Ni–Al alloys, discrete acicular precipitates of $NiCr_2O_4$ in Ni–Cr alloys, regularly shaped precipitates, probably of $NiVO_3$, in Ni–V alloys, spherical precipitates of MoO_2 in Ni–Mo alloys and spherical precipitates of $NiWO_4$ in Ni–W alloys.

Interactions between the precipitates and the scale are complex. In several systems, notably Ni–2%Al and Ni–5%Cr during oxidation and sulphidation, the particles are incorporated into the scales with similar spatial resolution as in the alloy; some dissolution of the second element results in a decrease in volume and, in the case of $NiAl_2O_4$ particles in NiO, break-up of the continuous rods to leave residual acicular precipitates. This is particularly the case for NiO sales at 1200°C, although there is much less dissolution of particles into Ni_3S_2 scales at 600°C. In other systems, including Ni–4%W and, possibly, Ni–4%V during oxidation (Figs 3e and 2f) and Ni–4%V, Ni–4%Mo and Ni–4%W during sulphidation (Figs 4c, e and f), there is evidence for a partial build-up of particles at the interface but they are eventually incorporated into the scale, resulting in closely spaced or linked precipitates near to the scale/alloy interface. However, for Ni–4%Mo during oxidation, as the scale encroaches, it sweeps the particles with it, resulting in their build-up at the alloy/scale interface (Figs 3a and b).

It has been suggested[2,5] that incorporation of particles into an external scale involves cation vacancy coalescence at the particle/alloy interface which produces voids around the particles. This stimulates dissociation of the scale as it encroaches on them and oxidation of the alloy matrix beneath them; the particles are thereby taken up into the scale. However, in the present research, there was no evidence for voids around the particles. Another model for incorporation of internal oxide into Cr_2O_3 scales on TD Ni–Cr alloys[6] does not require formation of voids. Here, Cr_2O_3 is formed beneath the ThO_2 agglomerates by diffusion of oxygen through them. However, such a model is not applicable in the present systems as diffusion of oxygen through the particles is very slow and no scale has been detected on their underside.

The incorporation of particles into the scale involves the gradual encroach-ment of the scale on and around them, as confirmed by a time series of experiments not reported here. Although this encroachment may result from loss of metal at the alloy/scale interface following outward diffusion of nickel ions across the scale, it is likely that it also involves formation of new nickel oxide or nickel sulphide at that interface. There is evidence for preferential inward movement of scale along the particle/alloy interfaces, associated with enhanced inward diffusion of oxygen or sulphur in the interface (Figs 2a, 4a, b and f). More significant inward penetration of NiO was occasionally observed along intergranular oxides (Fig. 2a).

4.2 Factors influencing scale-particle interactions

There are significant differences in the shapes and sizes of the precipitates in the various systems, with some slight trends in this respect in terms of scale interactions. The rods of $NiAl_2O_4$ and the acicular precipitates of $NiCr_2O_4$ are incorporated easily into both NiO and Ni_3S_2 scales while the more spherical precipitates of $NiVO_3$ and $NiWO_4$ are also incorporated into such scales, although there is some indication of partial linking in the scale for the latter systems. Indeed, particularly for Ni–4%W during oxidation and Ni–4%Mo during sulphidation, the distribution of precipitates in the scale (Figs 3e and 4e) indicates that a partial build-up at the interface is followed by incorporation of the linked particles into the scale after some time. The spherical particles of MoO_2 build up at the NiO/alloy interface and are not incorporated at all into the scale (Figs 3a–d).

The solubility and diffusivity of the second element in the scale are probably not very important in determining the extent of precipitate incorporation, although these parameters do determine the extent of dissolution of such precipitates in the scale. There is very little dissolution of any internal oxide in the sulphide scales, consistent with low solubility at the reaction temperature.

The physical state of the particles is important in the sweeping of grain boundaries, in that liquid oxides are swept while solid oxides are not.[7] The present observations cannot be correlated to the melting points of the precipitates, but their physical state (e.g. plasticity) and that of the scale are important factors in determining the interactions. In addition, differ-ences in the degree of coherency characteristics of the interfaces may result in differences in interactions between the particles and the scale. Inert oxides such as MoO_2 are reported to have different interfacial char-acteristics, for instance in terms of availability as vacancy sinks, compared with more active interfaces.[8] Similarly, they may have different transport or coherency characteristics. However, insufficient is known about the interfacial properties of the internal oxide/nickel oxide (or nickel sulphide)/nickel matrix systems to comment precisely on these characteristics. It is a subject worthy of more detailed study since much greater understanding of the coherency characteristics of interfaces would benefit various processes, not just oxidation.

5. Conclusions

1. During oxidation or sulphidation of nickel containing internal oxide precipitates, the scale encroaches on the precipitates which are either incorporated into it, or are swept by the scale/alloy interface. In the former case, they may either be incorporated easily or may be swept for a period first, resulting in a different distribution from that in the alloy.
2. Rod-shaped $NiAl_2O_4$ precipitates and acicular $NiCr_2O_4$ precipitates are incorporated into NiO scales at 1200°C and Ni_3S_2 scales at 600°C with a similar distribution as in the substrate, while more regularly shaped $NiVO_3$ and $NiWO_4$ precipitates are swept for short periods before being incorporated. Spherical MoO_2 precipitates are swept by the NiO/alloy interface and are not incorporated at all while such particles are also swept by the Ni_3S_2/alloy interface but are eventually taken into the scale.
3. Factors determining such interactions include the sizes and shapes of the precipitates, the physical states of the precipitates and the scale and the coherency characteristics of the various interfaces.

Acknowledgements

The authors thank Nuclear Electric plc (Berkeley Laboratories) for financial support and Professor J. E. Harris for general discussions in this field.

References

1. G. C. Wood and T. Hodgkiess, *Nature*, **211**, 1358–1361, September 1966.
2. G. C Wood, F. H. Stott and J. E. Forrest, *Werkstoffe u. Korros*, **28**, 395–404, 1977.
3. A. Strawbridge, Ph.D. Thesis, University of Manchester, 1992.
4. S. Guruswamy, S. M. Park, J. P. Hirth and R. A. Rapp, *Oxidation Metals*, **26**, 77–100, 1986.
5. R. A. Rapp, *Corrosion*, **21**, 382–390, 1989.
6. C. G. Giggins and F. S. Pettit, *Metall. Trans.*, **2**, 1071–1079, 1971.
7. M. F. Ashby and R. M. A. Centamore, *Acta Metall.*, **16**, 1081–1086, 1968.
8. J. E. Harris, *Mater. Sci. Technol.*, **4**, 457–461, 1988.

The Analytical Electron Microscopy of Thin Oxide Films Formed on Cerium Implanted Nickel

F. CZERWINSKI and W. W. SMELTZER

Institute for Materials Research, McMaster University, Hamilton, Ontario, Canada, L8S 4M1

ABSTRACT

The microstructure of oxide films on cerium implanted, high-purity nickel has been investigated. The oxide films were formed in 5×10^{-3} Torr oxygen at 973 K for 4 h. The NiO with thickness of about 170 nm showed a relatively uniform surface morphology and a lack of an orientation relationship between the oxide and the substrate. Auger electron spectroscopy indicated that the cerium concentration maximum was located in the vicinity of the gas–oxide interface. Thin foils were prepared by ion milling to examine sections at different depths by transmission electron microscopy using selected area and convergent beam diffraction. The foil thickness and cerium concentrations were determined by electron energy loss spectroscopy. It was found that three sub-layers with different structure and cerium concentrations may be distinguished. Cerium was present as CeO_2 particles distributed both inside nickel oxide grains and at grain boundaries. The segregation of Ce in NiO grain boundaries was also detected by scanning transmission electron microscope and energy dispersive X-ray analysis. These results are discussed in terms of a reactive element effect on the development of the microstructure of oxide films.

1. Introduction

The ion implantation of reactive elements to improve high-temperature oxidation properties of metals and alloys is used with considerable success. As was stated by George et al.,[1] the physical damage has no essential effect on high-temperature oxidation properties of implanted metal, which depends on the kind of implanted species.

In our previous papers,[2,3] we analysed the effect of cerium on high-temperature oxidation of nickel by surface applied ceria sol coatings containing 5 nm size ceria particles. In this study we employed extensively analytical electron microscopy to study oxide growth on cerium implanted nickel during early stages of exposure to oxygen.

2. Experimental Procedure

Specimens with a thickness of 1 mm and a diameter of 9.5 mm were prepared from 99.99% purity polycrystalline nickel supplied by A. D. MacKay. After mechanical polishing to 1 μm diamond paste and annealing at 1173 K for 1 h in vacuum of 10^{-7} Torr, samples were chemically polished as a final step.[3] Nickel was implanted at a level of 2×10^{16} ion/cm^2 with Ce$^+$ at implantation energy of 150 keV. A small part of each sample was not implanted to have a reference surface for comparison purposes. Oxidation at 973 K in 5×10^{-3} Torr oxygen was performed in ultra-high vacuum manometric apparatus. The cold insertion to oxygen and furnace raised techniques were used. Some samples were annealed before oxidation at 1073 K for 1 h in vacuum of 10^{-9} Torr.

The composition of oxide was analysed using sputter depth profiling and Auger electron microscopy (AES). The surface morphology was observed by scanning electron microscopy (SEM) whereas oxide microstructure was analysed in details by transmission electron microscopy (TEM). Thin foils were prepared parallel to the oxide–metal interface by two different techniques. For conventional observations the oxide was stripped from the substrate in a saturated solution of iodine in methanol.[4] For high magnification analysis of the films on different depths, the foils were prepared by ion thinning using a technique described by Pint.[5] Ion thinning was conducted in a Gatan 600 duomill with an Ar beam at 4–4.5 keV. The individual ceria particles and the segregation of reactive element to oxide grain boundaries were analysed by a field emission gun scanning transmission electron microscope (STEM) with 3.5 nm electron probe and windowless Link detector (EDX).

3. Results and Discussion

3.1 Oxide growth kinetics and surface morphology

The oxidation kinetics of pure and cerium implanted nickel are shown in Fig. 1. After 4 h of exposure the oxide on implanted nickel has a thickness of 170 nm as estimated from oxygen uptake and is about three times thinner than that grown on pure nickel. The corresponding difference in parabolic rate constants is more than one order of magnitude. Ultra-high vacuum annealing at 1073 K for 1 h resulted in an increase of the nickel oxidation rate in comparison to specimens oxidised after implantation.

The morphology of oxide surfaces shown in Fig. 2 indicates that on the unimplanted part the selective oxidation occurs of differentially oriented nickel grains. On the implanted part the oxide thickness is rather uniform. The preferential oxide growth on nickel grain boundaries was practically inhibited. Annealing of implanted samples before oxidation promotes nodular growth

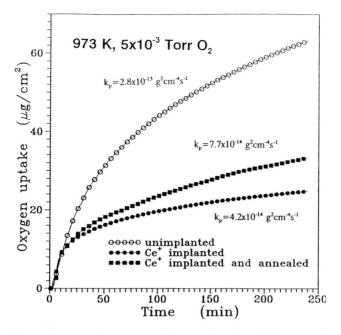

Fig. 1 Oxidation kinetics for pure and Ce implanted nickel. Annealing: 1073 K, 1 h, vacuum 10^{-9} Torr.

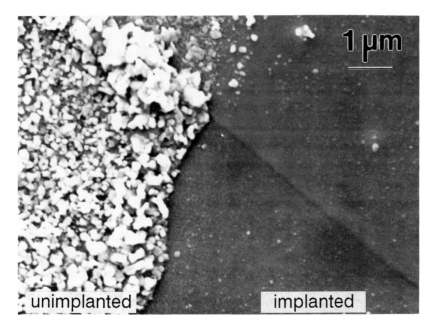

Fig. 2 Morphology of oxide surface formed on Ce implanted and unimplanted nickel after oxidation at 973 K for 4 h.

which suggests the existence of easy diffusion paths for Ni^{2+} ions in ceria modified nickel oxide.

3.2 Depth distribution of cerium

After implantation, as was estimated by computer TRIM-91 simulation,[6] the maximum cerium concentration is located about 20 nm beneath the nickel surface. After oxidation according to AES analysis, the cerium is concentrated in the near surface region of nickel oxide. Since the ceria particles can be considered as inert markers, the depth location of cerium after oxidation indicates the transport mechanism during oxide growth. A correlation exists between the presence of reactive element in the outer part of oxide and inhibition of oxidation rate, i.e. shift of the reactive element peak towards the oxide–metal interface always results in a higher oxidation rate. The presence of cerium in the vicinity of the oxide surface is similar as was detected by combination of Rutherford backscattering spectrometry and AES techniques after the same time of oxidation at 973 K for ceria sol coated nickel.[2,3] This ceria distribution suggests that oxidation occurs predominantly by inward oxygen diffusion as was proposed in reference 7.

3.3 Oxide microstructure and microchemistry

The TEM bright field image of 170 nm thick oxide formed on implanted nickel is shown in Fig. 3. The SAD pattern reveals randomly oriented grains of NiO, while the very strong continuous rings were built by fine particles of CeO_2. There is no orientation relationship between oxide and nickel in the substrate as was observed for ceria coated nickel with identical surface preparation. This difference is induced by physical damage due to ion implantation promoting nucleation of randomly oriented oxide grains.

The 170 nm thick film of nickel oxide was examined at different depths. During milling by Ar ions from gas or metal side, the portions of oxide were removed to allow observations at high magnifications of characteristic regions. The foil thickness was measured by electron energy loss spectroscopy (EELS).[8] In Fig. 4 is shown the bright field image of a 35 nm thick surface portion of oxide. The maximum cerium concentration is located in this part of the film. The very small and close-spaced ceria particles are incorporated within fine grained nickel oxide.

The TEM image of 65 nm thick foil obtained by simultaneous sputtering from both sides by Ar ions is shown in Fig. 5(a). The presence of ceria particles was supported by STEM/EDX measurements and convergent beam electron diffraction. No NiO–CeO_2 compounds were detected. Cerium segregation to nickel oxide grain boundaries was measured. The typical Ce distribution profile across an oxide grain boundary is shown in Fig. 5(b). In the analysed region, no ceria particles were observed at all under the magnification up to 2 million times. The similar Ce segregation has been found by Moon[9] in thick nickel oxide scale formed on ceria sol coated nickel after long time oxidation at 1173 K. Although the quantification of the EDX spectrum is not performed, the

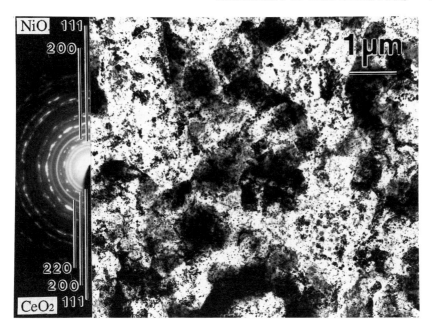

Fig. 3 The TEM bright field image of 170 nm thick oxide film after 4 h oxidation at 973 K.

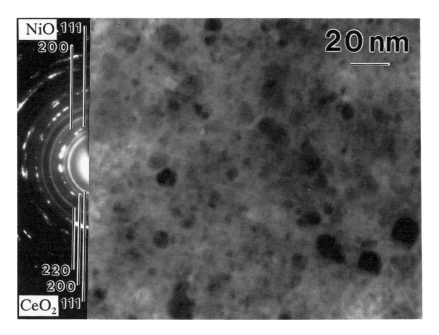

Fig. 4 The TEM bright field image from 35 nm thick surface portion of oxide.

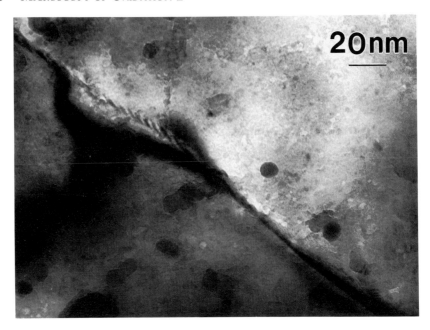

b

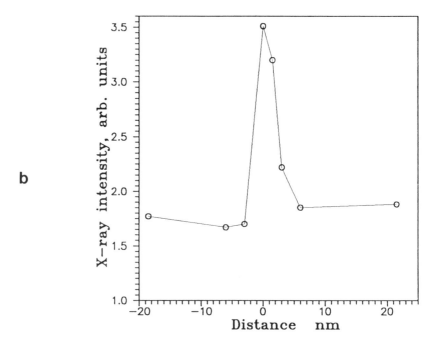

Fig. 5 The TEM bright field image of outer portion of nickel oxide film (a) and cerium concentration profile across nickel oxide grain boundary as detected by STEM/EDX (b).

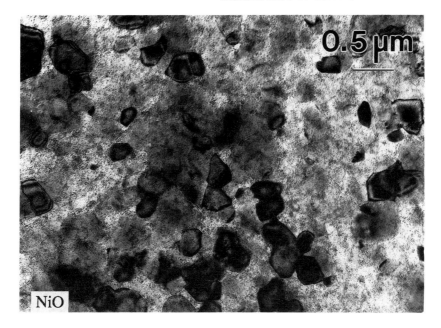

Fig. 6 The TEM microstructure of oxide beneath the ceria modified portion of nickel oxide film.

profile presented in Fig. 5(b) clearly demonstrates that the Ce ion segregants are already present in oxide boundaries during the initial stages of oxidation.

The microstructure of oxide formed in the vicinity of the oxide–metal interface is shown in Fig. 6 and is composed of NiO grains with diameter of 0.5μm. These oxide grains are formed during the steady stage of parabolic oxidation presumably by inward diffusion of oxygen.

Vacuum annealing after implantation is usually used to redistribute implanted species.[10] Cerium diffuses towards the nickel surface during annealing as was supported by AES analysis. Furthermore, annealing in an ultra-high vacuum of 10^{-9} Torr available in oxidation apparatus led to the presence of ceria on the surface as was detected by AES. The microstructure of the ceria-rich region of oxide formed on an annealed sample is shown in Fig. 7. There is an essential difference in size of ceria particles in comparison with the oxide shown in Fig. 4. The coarse ceria particles are the possible reason for the observed higher oxidation rate of an annealed specimen shown in Fig. 1. However, this effect is lower than that observed for ceria sol-coated nickel with particle size modified by vacuum annealing after deposition.[7] It is believed that in the implanted samples, additionally to the observed coarse particles much smaller particles which act as sources for outward diffusion of Ce segregations to oxide grain boundaries are also present; however, experimentally a proof of this is not straightforward.

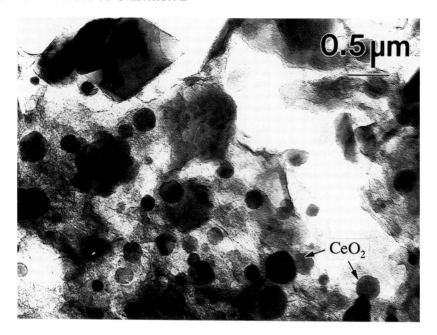

Fig. 7 The ceria particles in oxide film formed after annealing of implanted samples at 1073 K for 1 h in vacuum of 10^{-9} Torr. Oxidation: 973 K, 4 h.

4. Conclusions

Examinations of oxide formed on cerium implanted nickel demonstrate marked inhibition of nickel oxidation rate by implanted Ce at 973 K, i.e. in the generally accepted regime of short-circuit diffusion for nickel in the oxide film.

The implant of cerium is located after oxidation mainly in the outer part of the oxide film as CeO_2 particles and Ce ion segregants in oxide grain boundaries. Such location suggests a change in oxidation mechanism from predominant Ni^{2+} outward diffusion to predominant O^{2-} inward diffusion as was observed for ceria sol-coated nickel.[7]

Three sub-layers were distinguished with different microstructure and cerium concentration in the 170 nm thick oxide film formed after 4 h of oxidation at 973 K.

Acknowledgements

The authors express their appreciation to the Natural Sciences and Engineering Research Council of Canada for financial support of this research. We wish also to thank B. McClelland for ion implantation and Dr A. Perovic for STEM/EDX measurements.

References

1. P. J. George, M. J. Bennett, H. E. Bishop and G. Dearnaley, *Mat. Sci. Eng.*, **116**, 111–117, 1989.
2. F. Czerwinski and W. W. Smeltzer, *Proc. of the 182 Meeting of The Electrochemical Society, Toronto, Symposium Oxide Films on Metals and Alloys*, pp. 81–91, 1992.
3. F. Czerwinski and W. W. Smeltzer, *J. Electr. Soc.*, submitted.
4. J. M. Perrow, W. W. Smeltzer and J. D. Embury, *Acta Met.*, **16**, 1209–1218, 1968.
5. B. A. Pint, Ph.D. thesis, Massachusetts Institute of Technology, 1992.
6. J. F. Ziegler, TRIM-91, the transport of ions in matter, *IBM-Res.*, New York, 1991.
7. F. Czerwinski and W. W. Smeltzer, *Oxid. Met.*, submitted.
8. T. Malis, S. C. Cheng and R. F. Egerton, *J. Electr. Micros. Techn.*, **8**, 193–200, 1988.
9. D. P. Moon, *Oxid. Met.*, **32**(1/2), 47, 1989.
10. J. M. Hamplikian, O. F. Devereux and D. I. Potter, *Mat. Sci. Eng.*, **116**, 119–127, 1989.

The Effect of Ceria Coatings on Early-stage Oxidation of Nickel Single Crystals

F. CZERWINSKI and W. W. SMELTZER

Institute for Materials Research, McMaster University, Hamilton, Ontario, Canada, L8S 4M1

ABSTRACT

A study has been performed of the growth of nickel oxide films on the (100) and (111) crystal faces of nickel superficially modified by 14 nm thick ceria sol coatings.

Oxidation kinetics were determined by use of an ultra-high vacuum manometric apparatus at temperatures in the range of 873–1073 K, oxygen pressure of 0.005 Torr and exposure times up to 4 h. The ceria coatings reduced the oxidation rate; however, the extent of the effect depends on the nickel surface orientation. The (100) Ni face oxidised with a higher rate both before and after coating. The microstructure of oxide formed on both pure and ceria modified nickel faces has been studied by AES, SEM and TEM techniques. NiO grown on pure nickel shows a strong orientation relationship between the oxide and the nickel substrate. The same orientation relationship is also present for ceria coated specimens; however, the faint diffraction rings superimposed on the intense maxima indicated that there is also a thin film of fine grained, randomly oriented NiO formed on top of the ceria particles. After oxidation, the Auger electron spectroscopy sputter profiles revealed that the cerium concentration maximum was in the vicinity of the gas–oxide interface. The results are discussed in terms of the influence of the ceria particles on the development of oxide microstructure.

1. Introduction

The dependence of oxidation rate on crystallographic plane for nickel was a subject of numerous studies, for example.[1-3] The kinetics for nickel oxide growth on different crystal faces at temperatures less than about 1200 K are related to the microstructure of oxide and have been interpreted in terms of the character and number of easy diffusion paths at crystallite boundaries in the oxide.[3] Additions of reactive elements decrease markedly the oxidation rate of nickel in the short-circuit diffusion temperature regime. However, for chemically polished polycrystalline nickel coated with cerium oxide, the oxidation rate was found to be still dependent on substrate grain orientation.[4]

In this paper, the early-stage oxide growth on (100) and (111) faces of nickel coated superficially with thin ceria films has been investigated, to obtain a more complete understanding of this oxidation rate anisotropy.

2. Experimental Procedure

Specimens with thickness of 1 mm and surface area of about 1 cm^2 were prepared from 99.999 wt% purity nickel supplied by Research Crystals Inc. The wafers after spark cutting were sequentially electropolished in 60% sulphuric acid and lightly mechanically polished with 1 μm diamond paste. As a final step the chemical polishing was performed at a temperature of 298 K in solution consisting of 65 ml acetic acid, 35 ml nitric acid and 0.5 ml hydrochloric acid.[5] This technique removed the plastically deformed layer and produced a high-quality mirror-like surface. Crystallographic orientation of a specimen surface was determined by Laue back reflection X-ray technique. Ceria sol coatings with thickness of 14 nm were deposited on unpreoxidised substrate by cold dipping as described in reference 6.

Oxidation in the temperature range of 873–1073 K, oxygen pressure of 5×10^{-3} Torr and exposure time up to 4 h was conducted in ultra-high vacuum manometric system.[4] The cold insertion of the specimen to oxygen and furnace raise method with heating rate of 200 K/min were used to avoid coarsening of CeO$_2$ particles during heating in vacuum. The oxide growth morphology was observed by scanning electron microscopy (SEM). The coating and oxide microstructures were analysed by transmission electron microscopy (TEM) and selected area electron diffraction (SAD). Thin foils, parallel to the oxide–metal interface, were prepared by oxide stripping from the substrate in a saturated solution of iodine in methanol.[7] The depth composition of oxide was analysed by sputter profiling and Auger electron spectroscopy (AES).

3. Results and Discussion

3.1 Oxidation kinetics

The oxidation kinetics for blank and ceria coated nickel crystals, measured at temperature of 873 and 1073 K, are given in Fig. 1. An estimate of the instantaneous parabolic oxidation rate constant was calculated from oxygen uptake after 4 h of exposure. Results are shown in Table 1. For uncoated samples, the presented data support previous studies.[1,2] The (100)Ni face oxidised with higher rate at all temperatures examined, exhibiting a difference in parabolic rate constants of about two orders of magnitude. The inhibition of nickel oxidation by ceria coatings with the same thickness of 14 nm depends on the orientation of the crystal face.

For (100)Ni the presence of coating markedly decreased the nickel oxidation rate. The parabolic rate constants are approximately one order of magnitude lower in comparison to uncoated samples (Table 1). It is worth while to note that uncoated (100)Ni oxidised at rates similar to those observed for polycrys-

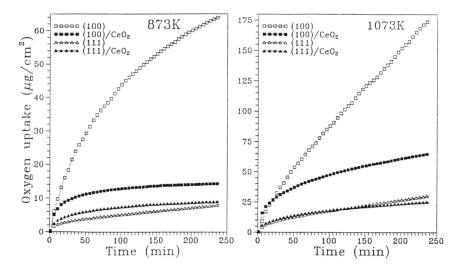

Fig. 1 Oxidation kinetics at 873 and 1073 K of pure and ceria sol coated nickel crystal faces.

talline nickel with the surface mechanically repolished prior to oxidation to promote the nucleation of randomly oriented oxide. Also inhibition of oxidation by ceria coating observed for this crystal face is similar to that measured for polycrystalline mechanically repolished and subsequently sol coated nickel.[4]

For (111)Ni the ceria coatings did not change markedly the nickel oxidation characteristics up to 4 h of exposure (Fig. 1). A small decrease of parabolic rate constants was observed during oxidation at 973 and 1073 K. At 873 K, the value of this rate constant for coated samples is even slightly higher than for pure nickel. As was stated in reference 4, ceria sol coatings do not act as barrier layers. The oxidation curves for (111)Ni illustrate that for the behaviour to be effective, the ceria particles must be incorporated into growing oxide to build the ceria modified layer of nickel oxide with cerium ions segregated to oxide

		Oxidation temperature (K)		
Crystal face	Surface treatment	873	973	1073
100	Uncoated	2.9×10^{-13}	1.4×10^{-12}	2.2×10^{-12}
	Coated	1.4×10^{-14}	2.8×10^{-13}	2.9×10^{-13}
111	Uncoated	4.6×10^{-15}	9.8×10^{-14}	6.4×10^{-14}
	Coated	5.7×10^{-15}	4.5×10^{-14}	5.3×10^{-14}

Table 1 Parabolic rate constants (*) k_p for oxidation of pure and ceria sol coated nickel crystal faces ($g^2 \, cm^{-4} \, s^{-1}$).

* Calculated as w^2/t where w and t are the oxygen uptake and oxidation time respectively.

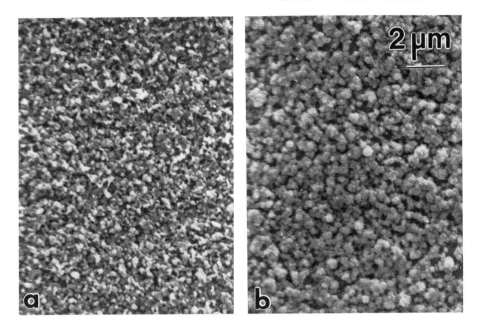

Fig. 2 SEM surface morphology of pure (a) and ceria sol coated (100) nickel face (b) after 4 h oxidation at 973 K.

grain boundaries.[4] The thickness of this layer, estimated from oxygen uptake, is about 30 nm. Before this protective layer is formed, the ceria coated (111)Ni oxidised with a higher rate than the uncoated sample because the small ceria particles present on the surface promote the nucleation of oxide crystallites, as was observed for Fe–Cr alloys by Rhys-Jones *et al.*[8]

3.2 Surface morphology and depth composition of oxide

The SEM surface morphology of oxide formed at 973 K on pure and ceria coated (100)Ni is shown in Fig. 2. Oxide formed on pure (100)Ni is fine grained and uniform. After applying ceria sol, a further reduction of grain size is observed. Observations of growth surface revealed that during very initial periods of exposure to oxygen, i.e. the first few minutes, the small grains of oxide nucleate on the top of coating. Furthermore, the changes of growth surface are rather small, suggesting that a growth process takes place under the coating. A similar effect was obtained for (111)Ni: however, due to a much lower oxidation rate for this face, the gradual increase in the number of oxide grains formed on the coating top was observed up to about 2 h of oxidation.

A typical composition depth profile for oxide grown on ceria coated (111)Ni, as revealed by AES, is shown in Fig. 3. After 4 h of oxidation the cerium is located close to the outer surface of 180 nm thick oxide film. For 435 nm thick oxide formed on (100)Ni the cerium was also present near the gas–oxide interface, whereas the essential portion of oxide is present beneath it.

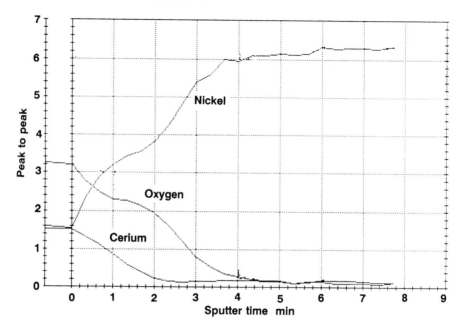

Fig. 3 AES sputter depth profile of oxide formed on ceria coated (111) nickel face after 4 h oxidation at 973 K.

3.3 Development of oxide microstructure

Using TEM–SAD technique, the oxide microstructure was examined at various stages of growth. The typical micrographs for (100)Ni and (111)Ni with corresponding electron diffraction patterns are shown in Figs. 4 and 5. All sections of oxide films examined were compact and did not exhibit any microcracks and pores.

The nickel oxide grown on both uncoated metal faces approximates single crystals containing crystallites slightly misoriented with respect to one another. The boundaries between crystallites represent the paths of easy diffusion. Higher density of these paths in the microstructure of oxide grown on (100)Ni than that grown on (111)Ni explain the measured difference in oxidation rate. The major oxide growth textures developed were $\langle 100 \rangle$NiO and $\langle 110 \rangle$NiO for (100)Ni and (111)Ni respectively. Several other orientations of NiO growing on these Ni faces are presented by Khoi et al.[3]

In this study the main attempt was focused on the analysis of oxidation of the ceria coated nickel faces. The examination of thin foils of stripped oxide formed during the very initial (transient) stage of reaction reveals that SAD patterns are composed of rings only for NiO and CeO_2. It means that during the transient stage the ceria modified layer of NiO is formed. The SEM detection of small NiO grains on top of the coating after the first minutes of oxidation suggests that formation of this layer takes place by outward Ni^{2+} diffusion. The microstructure, stereological analysis and microchemistry examinations of this portion of oxide, composed of small NiO grains, ceria

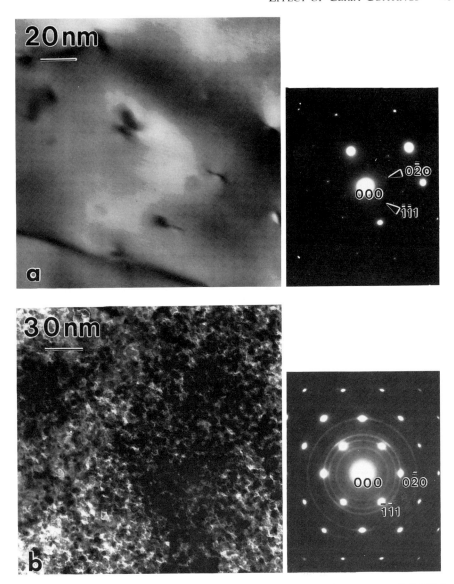

Fig. 4 TEM micrographs and SAD patterns of oxide formed on uncoated (111)Ni after 30 min of oxidation (a) and on ceria coated (111)Ni after 4 h oxidation at 973 K (b), (indicated spots for NiO).

particles and cerium segregations in grain boundaries, are given elsewhere.[9] SAD patterns of oxide formed during subsequent steady stage of oxidation consisted additionally of intense maxima for NiO. The diffraction analysis revealed that the primary growth directions for this oxide are the same as observed on uncoated nickel, i.e. $\langle 100 \rangle$NiO and $\langle 110 \rangle$NiO for (100)Ni and (111)Ni respectively. Thus, oxide formed during steady stage has an orientation relationship with the nickel substrate. Assuming that the ceria particles

Fig. 5 TEM micrographs and SAD patterns of oxide formed on uncoated (100)Ni after 10 min of oxidation (a) and on ceria coated (100)Ni after 4 h oxidation at 973 K (b) (indicated spots for NiO).

may be considered as inert markers, the above observations suggest the change of oxidation mechanism from outward Ni^{2+} diffusion to predominant inward O^{2-} diffusion. A similar position of CeO_2 particles was found for early-stage oxidation of polycrystalline ceria coated nickel.[9]

The comparison of oxidation kinetic measurements and microstructural observations reveals that the marked decrease of oxidation rate is achieved after the ceria modified portion of nickel oxide is formed. It is believed that this layer is transport controlling during steady stage of oxide growth. In reference 9 we proved that the diameter of ceria particles is a crucial factor affecting the oxidation kinetics of sol coated nickel. To be highly effective, the extremely small (5 nm in size) ceria particles should be relatively quickly incorporated into the growing oxide film. Ceria particles remaining on the metal surface, due to sintering, increase their size and become less effective. Such lower efficiency of reactive elements derived from sol coatings was observed during oxidation of nickel at a relatively low temperature.[4,10,11] Ceria coatings deposited on (111)Ni, due to low oxidation rate, remain for a longer time on the substrate surface. It may also affect the inhibition of oxidation of (111) nickel face by surface applied ceria sol coatings.

4. Conclusions

1. The effect of superficially applied 14 nm thick ceria sol coatings on nickel oxidation in the temperature range of 873–1073 K depends on crystal face orientation. (111)Ni oxidised at a lower rate in comparison to (100)Ni, both before and after coating. However, the higher reduction of oxidation rate was achieved for rapidly oxidised (100)Ni.
2. For both crystal orientations analysed, the oxide formed on coated samples is composed of an outer, cerium-rich layer and located beneath it, NiO growing in orientation relationship with the nickel substrate. The different rates of formation of the ceria-modified layers of NiO are responsible for observed differences in oxidation rates of sol-coated crystal faces of nickel.
3. The position of the CeO_2 particles in the outer region of the oxide suggests a change of oxidation mechanism from outward Ni^{2+} diffusion to predominant inward O^{2-} diffusion.

Acknowledgement

The authors are grateful to the Natural Sciences and Engineering Research Council of Canada for financial support of this research.

References

1. J. V. Cathart, G. F. Petersen and C. J. Sparks, *J. Electroch. Soc.*, **116**, 664, 1969.
2. R. Herchl, N. N. Khoi, T. Homma and W. W. Smeltzer, *Oxid. Met.*, **4/1**, 35–49, 1972.
3. N. N. Khoi, W. W. Smeltzer and J. D. Embury, *J. Electroch. Soc.*, **11**, 1495–1503, 1975.
4. F. Czerwinski and W. W. Smeltzer, *J. Electroch. Soc.*, submitted.
5. L. P. Fox, US Patent No 2,080,678, June 8, 1954.

6. F. Czerwinski and W. W. Smeltzer, *Proc. of the 182 Meeting of The Electroch. Soc., Oxide Films on Metals and Alloys, Toronto*, pp. 81–91, 1992.
7. J. M. Perrow, W. W. Smeltzer and J. D. Embury, *Acta Met.*, **16**, 1209–1218, 1968.
8. T. N. Rhys-Jones, H. J. Grabke and H. Kudielka, *Corr. Sci.*, **27**, 49, 1987.
9. F. Czerwinski and W. W. Smeltzer, *Oxid. Met.*, submitted.
10. D. P. Moon and M. J. Bennett, *Materials Science Forum*, **43**, 269–298, 1989.
11. A. T. Chadwick and R. I. Taylor, *J. Microscopy*, **140**, 221–242, 1985.

The Effect of Reactive Element Additions upon Alloy Depletion Profiles Resulting from the Preferential Removal of the Less Noble Metal During Alloy Oxidation

A. GREEN* and B. BASTOW[†]

*Department of Materials Science & Engineering, University of Liverpool
[†]British Nuclear Fuels Ltd., Sellafield, Seascale, Cumbria, England

ABSTRACT

Differences in alloy depletion profiles brought about by additions of reactive element additions (0.5 wt % yttrium, hafnium, cerium and zirconium) to a Ni–20wt % Cr alloy after oxidation at 1000°C have been measured using electron probe microanalysis. An explanation of the differences is proposed using arguments based upon a theoretical description obtained from a combination of Wagner's original solution for least noble metal depletion and the classical expression for scale growth. Use of this theoretical description is postulated to be a means of determining important intrinsic scale properties, for example scale cation diffusivity, when experimentally determined alloy profiles and alloy interdiffusion rates are known. In order to complete the description, it was necessary to experimentally determine any changes to alloy interdiffusion rate which might have been incurred by the reactive element addition. This was determined by electron microprobe microanalysis of concentration profiles formed in designed diffusion couples and a modified Matano–Boltzmann technique to calculate alloy interdiffusion rates. Using the determined values of alloy inter-diffusion rates and available intrinsic scale property data, theoretical depletion profiles were constructed reflecting a reduction in Cr_2O_3 scale diffusivity rates for alloys containing the reactive element additions.

1. Introduction

The effect of rare earth additions on the oxidation of alloys forming chromia and alumina scales has been studied extensively with reference to the changing kinetics, scale morphology[1] and stress generation.[2] However, although it is recognised that minor additions of rare earths or other reactive elements may also influence the transport properties of both the alloy and the scale, direct measurements of diffusion profiles to investigate the importance of these parameters have not been reported.

The present investigation is concerned with the measurement of diffusion profiles formed beneath oxide scales on alloy substrates containing the reactive elements Y, Ce, Zr and Hf, together with separate determinations of diffusion coefficients in the same nickel-base alloys using conventional pseudo-binary diffusion couples. These data allow more specific identification of the role of changes in alloy diffusion rates on oxidation behaviour.

2. Experimental procedures

The nickel-based alloys used in this study were prepared from high purity elements by vacuum melting in a high-frequency induction furnace. Both oxidation and diffusion couple specimens were prepared from as cast materials, in order to prevent the measurements being influenced by uncontrolled selective oxidation. The cast material had a grain size comparable to the depths of the diffusion profiles measured, minimising any influence of grain boundary diffusion in the alloy. Alloys used for interdiffusion studies were based on pure nickel, Ni–20wt%Cr or Ni–20wt%Cr–6% Al with additions of 0.5wt% of either yttrium or hafnium. Oxidation studies were carried out on the same alloys and also on other Ni–20wt%Cr-based alloys containing 0.5wt% of either cerium or zirconium. The oxidation of pure nickel containing reactive elements has been described elsewhere.[3]

Diffusion couples were produced from 1 cm square slices, 2–3 mm thick, cut from the alloy ingot with mating surfaces polished to a 1 μm diamond finish. The two slices forming a diffusion couple were held together by binding platinum wire around them and wrapping in tantalum foil to eliminate any residual oxygen when the couples were annealed in evacuated quartz tubes (10^{-5} Pa). Alloy compositions in a diffusion couple were chosen to give a step of 20% Cr across the interface, with the reactive element concentration equal on both sides for Ni–20%Cr-based alloys and a step of 6%Al for Ni–20%Cr–6%Al-based alloys, with both chromium and reactive element concentrations equal on both sides of the interface. Annealing was carried out in a vacuum furnace (10^{-5} Pa) for 240 hours at 1000 $\pm$ 5°C in all cases.

After annealing the couples were sectioned at right angles to the interface and then mounted for metallographic examination. Microprobe analysis of profiles in both diffusion couples and depletion zones was carried out using a JEOL JXA 50A Microprobe analyser, by point counting at intervals for chromium or aluminium. ZAF corrections to the original intensity ratio data were minimised by using one of the bulk alloy components as the concentration reference standard, rather than pure elements.

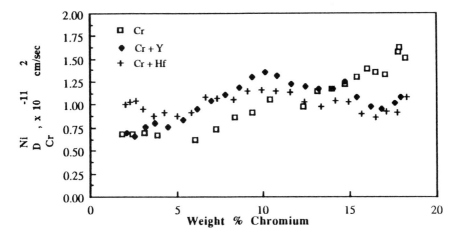

Fig. 1 Interdiffusion coefficients in binary Ni–Cr alloys showing the effect of chromium concentration and the effect of additions of 0.5wt% of either yttrium or hafnium.

3. Diffusion coefficients

The concentrations profiles measured in the pseudo-binary diffusion couples were interpreted using a modified form of the Boltzmann–Matano method[4,5] in order to overcome the tedious and very often inaccurate calculation of the location of the Matano interface. The profiles for couples formed between pure nickel or nickel containing either yttrium or hafnium on one side and the matching Ni–20%Cr alloy, with or without yttrium or hafnium additions as appropriate, on the other side, gave the diffusion coefficients shown in Fig. 1. The results from the interpretation of the monotonic aluminium profile as a pseudo-binary system are shown in Fig. 2, together with the equivalent data obtained from alloys containing either yttrium or hafnium on both sides of the interface.

4. Depletion profiles

The profiles formed in the binary Ni–20%Cr alloy are shown in Fig. 3 and are consistent with those reported previously by Hodgkiess *et al.*[6] The effects of reactive elements on the shape of this depletion profile are shown in Fig. 4.

All the scales on the Ni–20%Cr type alloys were chromia, with no immediate evidence for oxides of the reactive elements present as second phases within the scale. The major significant differences between scales on alloys with and without reactive element additions included improved adhesion, absence of convoluted morphologies and finer grain structures when any of these additional elements were present, a factor which possibly influenced the scale growth rate, as discussed below.

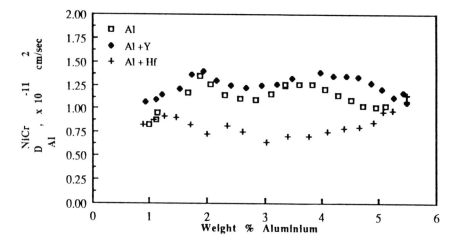

Fig. 2 Interdiffusion coefficients in ternary Ni–Cr–Al alloys showing the effect of aluminium concentration and the effect of additions of either 0.5wt% of yttrium or hafnium.

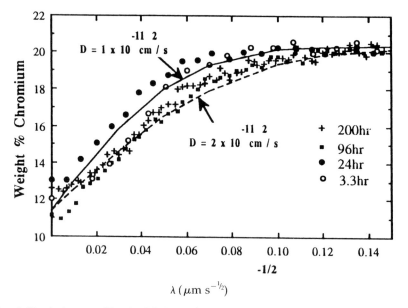

Fig. 3 Depletion profiles in Ni–20wt%Cr alloys oxidised in air to form Cr_2O_3 scales at 1273°K. Profiles for different oxidation periods are normalised by plotting data in terms of the coordinate $\lambda = x/t^{1/2}$. Solid and dotted lines are theoretical profiles calculated as described in the text using the values of interdiffusion coefficients shown. +200 h; ■ 96 h; ● 24 h; ○ 3.3 h.

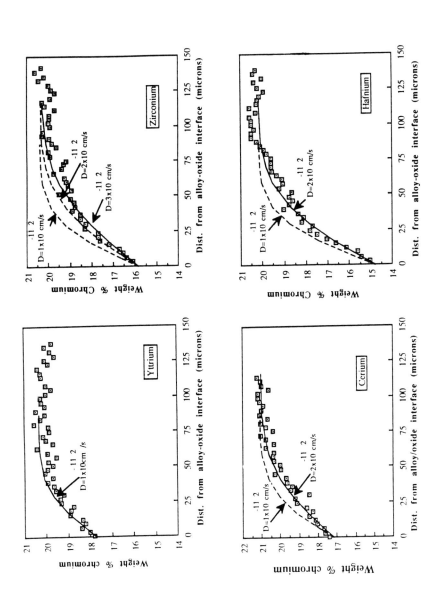

Fig. 4 Depletion profiles in Ni–20wt%Cr alloys oxidised in air to form Cr_2O_3 scales at 1273 K for 96 h, as influenced by the presence at 0.5wt% additions of reactive elements from Groups 3A and 4A of the Periodic Table: yttrium (top left), zirconium (top right), cerium (bottom left) and hafnium (bottom right). The lines are theoretical profiles calculated using the values of the interdiffusion coefficients shown, as described in the text.

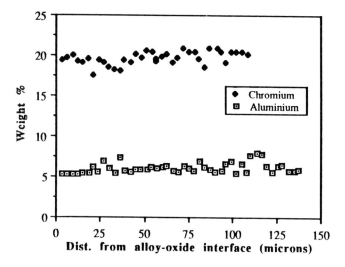

Fig. 5 Variation in the concentrations of aluminium and chromium in an Ni–20wt%Cr–6wt%Al alloy oxidised in air at 1273 K for 96 h to form an Al$_2$O$_3$ scale.

In contrast to the depletion profiles measured in the Ni–20%Cr alloys, those in the Ni–20%Cr-6%Al alloys were barely detectable, whether or not reactive elements were present. A typical alloying element distribution is shown in Fig. 5. The large fluctuations in the measured concentrations are due to the presence of Ni$_3$Al precipitates in the bulk alloy and reduced scatter in the measured aluminium concentrations only occurs in a narrow zone, less than 25 μm wide, immediately next to the alloy/scale interface. Here aluminium concentrations show slight evidence of depletion below the bulk alloy composition so that Ni$_3$Al is no longer stable, giving a precipitate-free zone.

The surface scale formed on these alloys were alumina, with internal oxidation of both aluminium and any reactive element present. As for the chromia scales on Ni–20%Cr alloys, there was no consistent evidence for the presence of reactive elements in the scales.

5. Interpretation of depletion profiles

Both the diffusion-controlled growth of scales on alloys and the associated depletion zone beneath the scale have been described quantitatively by Wagner.[7,8] The relevant equations can provide a self-consistent description of the oxidation process when they are combined with the appropriate thermodynamic description of the equilibrium condition at the alloy–scale interface. The applicability of this description to the oxidation of both Ni–Cr and Fe–Cr alloys has been demonstrated previously.[6,9] However, in both cases there remains a degree of uncertainty in the choice of values of the diffusion coefficients used in the calculations. This uncertainty cannot be avoided if the equations are used as a means of estimating interface compositions in the alloy and parabolic scale growth rate constants from the known properties of the oxide and alloy and the defined experimental conditions, i.e. the bulk alloy composition and the oxidising atmosphere.

However, the same description can be evaluated in a different sequence to estimate realistic scale diffusion rates, if the interfacial chromium compositions measured from depletion profiles are substituted into the relevant equations, rather than attempting to predict these values from first principles, as in earlier work.[6,9] Thus, knowing the chromium composition in the bulk alloy, N_{Cr}°, and at the alloy-scale interface, N_{Cr}',

$$\frac{N_{Cr}^{\circ} - N_{Cr}'}{1 - N_{Cr}'} = \pi^{1/2} \eta \exp \eta^2 . \mathrm{erfc}\, \eta = F(\eta) \tag{1}$$

according to Wagner,[8] with:

$$\eta = \frac{V_{\mathrm{alloy}}}{V_{\mathrm{oxide}}} \sqrt{\frac{k}{D^{\mathrm{alloy}}}} . \tag{2}$$

Here V_{alloy} and V_{oxide} are the molar volumes of the alloy and oxide respectively, k is the parabolic rate constant in terms of the increase of scale thickness and D^{alloy} is the alloy interdiffusion coefficient. Equation (2) allows estimation of k since D^{alloy} has been measured independently (Fig. 1) for the same alloys. From this value of k, the diffusion coefficient of the mobile ion species in the scale can be estimated, using the scale growth expression derived by Wagner.[7] Assuming that chromium diffusion is the rate controlling process in the oxidation of the Ni–20% Cr alloys:

$$k = \frac{z_{Cr}}{z_o} v D_{Cr}^{\mathrm{oxide}} [(a_o'')^{1/v} - (a_o')^{1/v}], \tag{3}$$

where $z_{Cr}/z_o = 3/2$, v ($=15$) is the oxygen activity dependence of the diffusion coefficient, a_o'' is the oxygen activity in the oxidising atmosphere (air, $a_o'' = 0.446$) and a_o' is the oxygen activity at the alloy scale interface. The equilibrium reaction at this interface is:

$$2/3Cr + 1/2O_2 = 1/3Cr_2O_3, \quad 1/3\Delta G^{\circ}Cr_2O_3 = -261 \text{ kJ mole}^{-1} \text{ at } 1273 \text{ K}, \tag{4}$$

and, since the Ni–Cr alloys considered are, on average, approximately ideal,[6] then:

$$\frac{1/3\Delta G_{Cr_2O_3}}{RT} - 2/3 \ln N_{Cr}' = \ln a_o'. \tag{5}$$

Values of D_{Cr}^{oxide} estimated in this way, for the scales formed on Ni–20% Cr alloys doped with different reactive elements, are given in Table 1, with other profile parameters. In each case the value of D_{Cr}^{oxide} given is the one which is consistent with a value of D^{alloy}, through equation (2), which gives the best agreement between the measured depletion profiles shown in Fig. 4 and theoretical profiles described in reference 8.

$$N_{Cr} = N'_{Cr} + (N^o_{Cr} - N'_{Cr}) \left[\frac{erf(\lambda/2D^{alloy})^{1/2} - erf(\alpha/2D^{alloy})^{1/2}}{erfc(\alpha/2D^{alloy})^{1/2}} \right]. \tag{6}$$

Here α, the rate constant for the recession rate of the alloy/oxide interface, is related to the oxide parabolic growth rate k by:

$$\frac{\alpha}{k} = \left(\frac{V_{alloy}}{V_{oxide}} \right)^2 = 4.75. \tag{7}$$

In using this interpretation, however, it is important to recognise its limitations for estimating values of D^{oxide}_{Cr} from the depletion profiles, since it relies firmly on the assumption that the oxidation process in these alloys is controlled by cation diffusion through the scale and thus the resulting diffusion coefficients are representative only of oxide scales on these Ni–Cr alloys.

6. Discussion

The measured values of interdiffusion coefficients in Fig. 1 are shown in Fig. 6 to be consistent with other data for interdiffusion Ni–Cr alloys,[11,12] as are the slightly higher values subsequently estimated in order to achieve improved agreement between experimental and theoretical depletion profiles. That such differences in the alloy interdiffusion coefficient do occur between the five alloys tested is also apparent from a qualitative comparison of depletion zone widths (Table 1), i.e. the narrower the zone, the smaller the alloy interdiffusion coefficient. Although the differences are relatively small, leaving consistency with other data unaffected (Fig. 6), the variation is still sufficient to change the shape of the depletion profile (Fig. 4). However, these differences between alloys are only apparent when they are undergoing oxidation and are not reproduced in the alloy in the absence of oxidation (Fig. 1). Speculation that the oxidation process might directly result in an increase in the alloy interdiffusion coefficient as a result of vacancy injection in some cases is not consistent with the relative values of the estimated parabolic scale growth constants given in Table 1, the narrowest depletion zones and hence the lowest values for D^{alloy} corresponding to both the highest (Ni–Cr) and lowest (Ni–Cr–Y) values for k. An alternative source of vacancies in the alloy may arise from the growth stresses generated in the scale.[13] If the scale is sufficiently well bonded to the alloy for the scale stresses to be transmitted across the alloy-scale interface, then creep may occur within the alloy, generating vacancies, and increasing the alloy interdiffusion coefficient. This mechanism is more consistent with the observed effect of reactive elements producing a more adherent scale, with a contiguous interface more capable of transferring stress but it still fails to interpret all the observed features in the profiles reported here.

The apparent effect of the reactive metals in reducing chromium diffusivity in the scale and hence the scale growth rate is also significant. It is also notable,

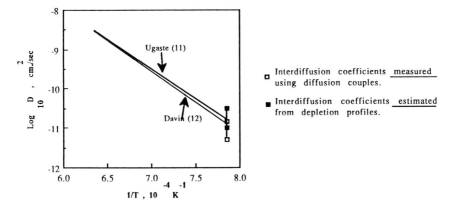

Fig. 6 Comparison of interdiffusion coefficients in Ni–Cr alloys measured in previous work with values measured and estimated in the present work. □ Interdiffusion coefficients *measured* using diffusion couples; ■ interdiffusion coefficients estimated from depletion profiles.

in the case of scale diffusion coefficients, that, as shown in Fig. 7, the values are comparable with other data[14–20] which is believed to be representative of Cr diffusion via grain boundaries in Cr_2O_3. From the data in Table 1 yttrium is clearly the most effective in reducing chromium diffusivity in the scale, approximately 0.32 atomic %, causing an elevenfold reduction compared to the yttrium-free Ni–20%Cr alloy. Cerium, which is chemically more similar to yttrium than either zirconium or hafnium, has a similar strong effect for a proportionately smaller addition of 0.2 atomic % while the two group IVA metals, zirconium and hafnium, are relatively less effective in reducing the scale diffusion rate and hence the scale growth rate. The mechanism by which this reduction is achieved for these scales is not clear since it is not compatible with the effects of doping, which conventionally would be expected to cause either no change (Y^{3+}, Ce^{3+}) or an increase (Zr^{4+}, Hf^{4+}, Ce^{4+}). Nor is it consistent with experimental observations that additions of reactive elements to the alloys cause a reduction in the grain size of the oxide scales, a change which would be expected to increase the availability of grain boundary short circuit paths, leading to an increase rather than the calculated decrease.

However, it is noted that the finer grain scales on the alloys containing reactive elements are less susceptible to severe deformation, sufficient to prevent convolutions, in comparison to the scales on the undoped Ni–20%Cr alloy. The reduced deformation will be associated with reduced defect generation within each oxide grain, with a consequent reduction in the number of short circuit diffusion paths and thus reduced diffusion rates, as implied by the calculated values. Such a mechanism, originally suggested by Stringer[21] for reducing diffusion rates in the oxide, relies ultimately on the oxide on alloys containing reactive elements having an increased resistance to deformation, relative to the oxide on binary Ni–Cr alloys, from the instant when oxide grains first form. The cause of any increase in strength is, however, not readily

Alloy	Reactive element concentration (atomic %)	Concentration depletion weight (%Cr)	Estimated depletion zone width (μm after 96 h)	Parabolic rate constant, k (cm^2 s^{-1})	D_{alloy} (cm^2 s^{-1})	Estimated diffusion coefficients D_{oxide} (cm^2 s^{-1})	D_{alloy}/D_{oxide}
Ni-Cr	0	7.3–9.8	90	5.13×10^{-13}	1×10^{-11}	3.1×10^{-14}	320
Ni-Cr-Y	0.32	2.5	60	4.49×10^{-14}	1×10^{-11}	2.7×10^{-15}	3700
Ni-Cr-Ce	0.20	3.8	105	1.91×10^{-13}	2×10^{-11}	1.1×10^{-14}	1750
Ni-Cr-Zr	0.31	4.4	130	4.75×10^{-13}	3×10^{-11}	2.4×10^{-14}	1270
Ni-Cr-Hf	0.16	5.1	110	3.44×10^{-13}	2×10^{-11}	2.1×10^{-14}	970

Table 1 Effect of reactive element additions on the extent of depletion profiles in Ni–20%Cr alloys and the values of alloy and oxide diffusion coefficients estimated from the profiles

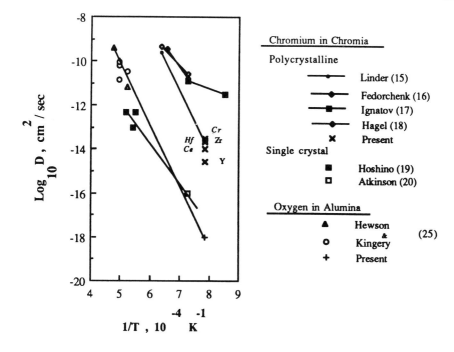

Fig. Comparison of diffusion coefficients in Cr_2O_3 and Al_2O_3 measured in previous work with values estimated from depletion profiles (Figs 3 and 4) in the alloys indicated.

apparent from examination of the scale microstructure. A probable alternative explanation for both grain refinement and reduced chromium transport through the scale is based upon studies by Yurek et al.[22,23] using SIMS and STEM, proposing that the presence of yttrium at the oxide grain boundaries produces a change in the relative rates of chromium and oxygen transport, as well as producing a grain boundary drag effect to change the microstructure of the oxide by inhibiting oxide grain growth.

Depletion and scale formation on the Ni–Cr–Al alloys differs significantly from that for the Ni–Cr alloys. Consistent with previous work the scale formed is alumina, with a growth rate much slower than that of chromia on Ni–Cr alloys, the generally accepted growth mechanism identifying inward diffusion of oxygen in the scale as the rate controlling process.[24] Also, while oxygen diffusion in alumina is much slower than that of chromium scales (Fig. 7), diffusion rates in the respective underlying alloys are very similar (Figs 1 and 2). Consequently, following the trend established in Table 1, showing that depletion becomes less significant as the ratio of the diffusion coefficients in the alloy and oxide, D^{alloy}/D^{oxide} increases, the absence of any measurable depletion in the Ni–20%Cr–6%Al alloy is consistent with the ratio $D^{alloy}/D^{Al_2O_3}$ being much higher than that for $D^{alloy}/D^{Cr_2O_3}$ for Ni–20% Cr alloys. A value for the maximum diffusion coefficient of oxygen in the alumina scale, consistent with the absence of any measurable depletion in the alloy, can be estimated using equation (1), which for small values of η (i.e. $k \ll D^{\text{alloy}}$) becomes:

$$N^o_{Al} - N'_{Al} = \pi^{1/2} \eta(1 - N'_{Al}). \tag{8}$$

If the lowest measurable depletion is $N^o_{Al} - N'_{Al}) \approx 0.001$ then equations (1), (2) and (3) give approximately:

$$\frac{0.001}{1 - N^o_{Al}} = \pi^{1/2} \frac{V_{\text{alloy}}}{V_{Al_2O_3}} \sqrt{\frac{z_{Al} v D^{\text{oxide}}_o (a''_o)^{1/v}}{z_o D^{\text{alloy}}}}, \tag{9}$$

since the term $(a'_o)^{1/v}$ is negligible, in view of the great stability of alumina. Substituting appropriate values for the known alloy and scale properties:

$$N_{Al} \text{ in Ni-20\% Cr-6Al} = 0.1206; \quad D^{\text{alloy}} = 1 \times 10^{-11} \text{ cm}^2 \text{ s}^{-1} \quad \text{(Figure 2)}$$

$$z_{Al}/z_o = 3/2; \quad V_{\text{alloy}}/V_{Al_2O_3} = 3.51$$

and, since the oxygen activity dependence of aluminium diffusion in alumina, v, is not known:

$$v D^{Al_2O_3}_o \approx 10^{-18} \text{ cm}^2 \text{ s}^{-1}.$$

Referring to Fig. 7, it is evident that, even allowing for an order of magnitude decrease in the estimate for $D^{Al_2O_3}$ if $v \sim 10$, this maximum value for the diffusion coefficient, consistent with the absence of any measurable depletion, is also consistent with extrapolated data for diffusion coefficients in alumina.[25]

7. Conclusions

1. The interdiffusion coefficient at 1273 K in alloys of nickel containing up to 20 %Cr and up to 6%Al lies in the range 0.5 to 1.5×10^{-11} cm^2 s^{-1}, with little or no consistent increase apparent as the chromium and aluminium contents increase.
2. The addition of 0.5% of either yttrium or hafnium has no significant effect on the interdiffusion rates in either the Ni–Cr or the Ni–Cr–Al alloys.
3. The addition of 0.5% of yttrium, cerium, zirconium or hafnium to the Ni–20 %Cr alloy causes reductions in the scale growth rates, improves the maintenance of contact between alloy and scale and reduces the degree of chromium depletion in the alloy beneath the growing scale.
4. Yttrium is the most effective addition for reducing depletion in the Ni–20% Cr alloy while the effect of cerium is smaller. Zirconium and hafnium show the least effect although they both still cause some reduction in chromium depletion compared to Ni–Cr alloys which are free of reactive elements.
5. The effect of reactive element additions can be correlated with a decrease in chromium ion diffusion rates in the chromia oxide scale.

6. Depletion of aluminium in Ni–Cr–Al alloys, with or without the addition of reactive elements, is negligible because of the much slower alumina scale growth rates in comparison to chromia scale growth rates on Ni-Cr alloys.

References

1. D. P. Moon, *Mat. Sci. and Technol.*, **5**, 754, 1989.
2. A. Rahmel and M. Schütze, *Oxid. of Metals*, **30**, 255, 1992.
3. A. Green, *Mat. Sci. and Technol.*, **8**, 159, 1992.
4. F. Sauer and V. Freise, *Z. Electrochem.*, **66**, 353, 1962.
5. A. Green and N. Swindells, *Mat. Sci. and Technol.*, **1**, 101, 1985.
6. T. Hodgkiess, G. C. Wood, D. P. Whittle and B. D. Bastow, *Oxid. of Metals*, **12**, 439, 1978.
7. C. Wagner, Atom movements, *Amer. Soc. Metals*, Cleveland 1951, p. 153.
8. C. Wagner, *J. Electrochem. Soc.*, **99**, 369, 1952.
9. B. D. Bastow, D. P. Whittle and G. C. Wood, *Oxid. of Metals*, **12**, 413, 1978.
10. Y. Adda and J. Philibert, *La Diffusion dans les Solides*, Presses Univ. de France, Paris, vol. II, 1966.
11. Y. E. Ugaste, *Fiz. Metall. Metalloved*, **24**, 442, 1967.
12. A. David, V. Leroy, D. Coutsouradis and L. Habraken, *Cobalt*, **19**, 51, 1963.
13. J. E. Harris, *Act. Metall.*, **26**, 1033, 1978.
14. P. J. Harrop, *J. Mat. Sci.*, **3**, 206, 1968.
15. R. Linder and A. Akerstrom, *Z. Phys. Chem.*, **6**, 162, 1965.
16. I. M. Fedorchenko and Y. B. Ermolovich, *Ukr. Chim. Zhur.*, **26**, 429, 1960.
17. D. V. Ignatov, I. L. Belokurova and I. N. Belgamin, *Proc. Conf. on the use of Radioactive and Stable Isotopes and Radiation in the National Economy and in Science*, USSR, p. 326, 1958; U.S. AEC Dept. NP-tr-448, 1958, p. 256; *Nucl. Sci. Abs.*, **14**, 1935.
18. W. C. Hagel and A. U. Seybolt, *J. Electrochem. Soc.*, **108**, 1146, 1961.
19. K. Hoshino and N. L. Peterson, *J. Amer. Cer. Soc.*, **66** (C-202), 1983.
20. A. Atkinson and R. I. Taylor, *NATO ASI*, Series B, vol. 129, Plenum Press, New York, 1985.
21. J. Stringer, B. A. Wilcox and R. I. Jaffee, *Oxid. of Metals*, **5**, 11, 1972.
22. C. M. Cotell, G. J. Yurek, R. J. Hussey and D. F. Mitchell, *Oxid. of Metals*, **34**, 173, 1990.
23. C. M. Cotell, G. J. Yurek, R. J. Hussey and D. F. Mitchell, *Oxid. of Metals*, **34**, 201, 1990.
24. R. Prescott and M. J. Graham, *Oxid. of Metals*, **38**, 233, 1992.
25. C. W. Hewson and W. D. Kingery, *J. Amer. Cer. Sol.*, **50**, 218, 1967.

The Oxidation of a Re-containing γ/γ' Alloy

A. H. DENT, S. B. NEWCOMB and W. M. STOBBS

Department of Materials Science and Metallurgy, University of Cambridge, Pembroke Street, Cambridge CB2 3QZ, U.K.

ABSTRACT

We report here a comparison of the differing scaling behaviours of a Re-containing γ/ι' alloy as oxidised at 750°C and at 950°C in air. The edge-on TEM approach used revealed changes in both the morphology and composition of the scales formed at the two temperatures and points to changes in mechanism. A thick outer oxide based on NiO/CoO is formed at the lower temperature while at 950°C the majority of the surface became covered by a thinner Al_2O_3/Cr_2O_3 scale.

1. Introduction

The current single crystal γ/γ' alloys used as turbine blades for high temperature applications are normally coated to maximise their service life but their oxidation behaviour has been extensively studied both as coated and as uncoated. The examination of the scaling behaviour of the uncoated alloys aids an understanding of the mechanisms of scale growth on the coated blade[1] and the oxidation of binary NiAl alloys has also been studied in the same context.[2]

Any differences in the scale development for coated and uncoated alloys during oxidation will be a function not only of the differences in composition of the coating and of the alloy but also of the developed differences in microstructure and composition in the sub-coat and sub-scale region respectively; it is for example well known both that the coating process can result in major microstructural changes in the sub-coat region and that this can alter the degradation behaviour.[3]

Here we describe and compare the scaling and behaviour of an uncoated Re-containing γ/γ' alloy at two different temperatures using the transmission electron microscopy (TEM) of edge-on thin foils to follow the development of both the surface oxides and the changes in the sub-scale region of the alloy beneath these. The use of the edge-on TEM approach allows us, in principle, to focus on the part of the developing microstructure which should give an insight into the oxidation mechanisms prevailing and in this context we examined in particular the rôle of the rhenium in the sub-scale zone noting, from our findings, how this would be expected to differ for coated blades.

2. Experimental

The nominal composition of the alloy studied is shown in Table 1, the material as oxidised having previously been fully solutionised.

C	Cr	Re	Al	Hf	Mo	Ni	Ti	W	Co	Ta
<0.06	6.5	3.0	5.6	0.1	0.6	bal	1.0	6.4	9.5	6.5

Table 1 CMSX4 nominal composition (wt%)

Two sets of coupons of the single crystal alloy were cut with (001), (011) and (111) surface normals. These were ground to a 1 μm finish and cleaned in methanol before they were oxidised in air. While we thus characterised the effect of the surface orientation on the oxidation behaviour, here we will concentrate on the characteristics of the scaling which were communally observed after, on the one hand, 200 h at 750°C and, on the other, 50 h at 950°C. The cross-sectional TEM foils of the oxidised surfaces, which were examined using conventional techniques including thin window EDX for their local analysis, were prepared using standard techniques[4] and the localised data obtained were compared with the results of an examination of mounted cross-sectional bulk samples. These were given a nitric acid γ' etch to enhance the visibility of this phase and examined optically, as well as using EDX in the SEM, to reveal both the more general trends in the oxide scaling behaviour as well as any tendency to heterogeneity.

3. Results

We first describe the results obtained optically and then relate these to the data obtained in our more detailed TEM study with the aid of the schematic representations of the more generally observed scale and sub-scale structures which are shown for 950°C in Fig. 1(c) and for 750°C in Fig. 1(d). In describing our TEM results we start with the sample oxidised at 950°C and then go on to describe the averaged composition profiles across the oxide scales as obtained in the SEM for both this temperature and 750°C. This allows us to emphasise the changes in the general trends, before detailing our TEM observations for the lower temperature.

3.1 Light microscopy of scales formed at 950°C and 750°C

Figure 1(a) shows an optical micrograph of a part of the specimen oxidised at 950°C where two types of oxidation have occurred. The more common scaling

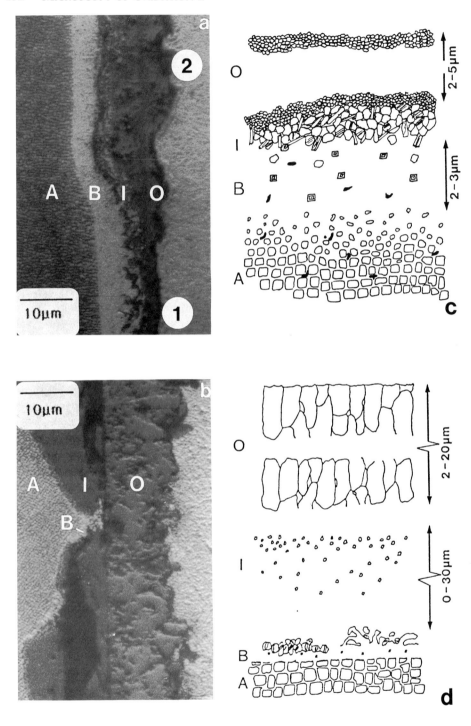

Fig. 1 Optical micrographs of cross sections of the oxides at (a) 950°C and (b) 750°C, and schematic summary diagrams of their respective microstructures as seen in edge on TEM.

morphology seen at this temperature, as found over about 90% of the surface, is marked in Fig. 1(a) at 1. Above the unmodified substrate zone A we see a modified zone B, apparently 5–8 μm in depth, where there has been dissolution of the ordered phase with inward oxidation above this in zone I. Here there are discrete particles within the γ indicating internal oxidation ahead of a very irregular inward-moving oxidation front. Outside this, the surface scale O varied in thickness between 2 and 5 μm. The less common form of scaling seen on this specimen, as in region 2, is clearly less protective though the width of zone B is apparently unchanged. Comparing the widths of zones I and B in the two regions it is clear that the rate of inward oxidation in region 1 is not controlled by the up-diffusion of the matrix constituents being oxidised. It would thus also appear that any differences in the inward formed oxides for the two regions are not such as would change the sub-scale matrix microstructure and composition to a level at which the local diffusivities might have altered. In region 2, the inward oxidation zone I is denser and has a better defined interface with the modified matrix than was the case in region 1, though unoxidised material is still apparently retained above the inward oxidation front. Both the inward and outward developing scales (up to 20 μm in total thickness) exhibit a banded morphology with apparent changes in the outward formed oxide above the position of the original metal surface and this has led to gross irregularities of the outer surface. The significance of these changes will become clear in section 3.2.2.

   The scale morphology seen for samples oxidised at 750°C is exemplified by Fig. 1(b). The thickness of the sub-scale region B, in which γ' is lost, is now much smaller than for the specimen oxidised at 950°C, as is indicative of the lower diffusion distances at the lower temperature. However, we can also see that there is now a more distinct interface between the zones of inward (I) and outward (O) oxidation. While there is now apparently less variation in the type of process occurring in each of the two zones at this lower temperature there are more gross localised changes in the depth of penetration of inward oxidation (0–30 μm) than at 905°C. The outward growing surface oxide is also very variably developed, ranging from 2 to 20 μm in thickness and thus, in this context, more closely resembles the higher temperature outer scale in region 2 than that in region 1.

3.2 The microstructure and composition of the scales

With the heterogeneities in scaling seen in both the low and the high temperature optical specimens in mind we turn to the TEM study of the oxidised alloy. Figures 1(c) and (d) are general schematic representations of the low and high temperature samples respectively, as derived from the TEM data, and are included here to facilitate an understanding of the local microstructure, as described below.

3.2.1 The 950°C, 50 h microstructure

The low magnification micrograph in Fig. 2(a) shows zones A, B, I and O of an edge-on foil prepared from a region of type 1 of a sample oxidised at 950°C and the different features in each of the different areas are best understood with reference to Fig. 1(c). The ordered γ' phase can be seen to be lost over a distance of about 2–3 μm in zone B, and it is clear that there has been plastic relief of the stresses associated with the scale formation in this zone in a way which will have been assisted by the reduction in local flow stress associated with the loss of the ordered phase. The dislocations present, as presumably generated at the base of the oxidation front where their density is highest, tend to be held up at the base of the modified zone at γ' particles still present there. It should be noted that we consistently found that the modified zone B proved to be less thick, as examined in the TEM, than was apparent optically. We conclude that the electrochemical nature of the γ' etch used is such that the ordered particles are not preferentially etched at the edge of the denuded zone. Cuboidal precipitates of TiN some 200–300 nm in size, as shown in Fig. 2(b), were found in the γ phase throughout zone B, cube–cube oriented with the matrix. The presence of these nitrides in the modified matrix layer beneath the scale demonstrate the way nitrogen will diffuse into an alloy ahead of oxygen, preferentially extracting nitride formers that might otherwise take a more active part in the oxidation process. While nitrides were observed in the modified zone B for all the specimens examined, whether at this temperature or at 750°C, TiC precipitates (Fig. 2c) were seen in the γ phase both within zone B and in areas apparently unaffected by γ' denudation from further back in the γ/γ' matrix but only at 950°C. Such carbides were probably thus formed from the <0.06 wt% of carbon present in the alloy.

The internal oxidation observed within the TEM foils existed mainly as an irregular (1–3 μm thick) 'band' (I) between the matrix and outer scale. Only a few isolated oxide precipitates of α-Al$_2$O$_3$ of an equiaxed 0.5–0.8 μm morphology (Fig. 2d) were found wholly within the denuded zone B. In this context internal oxidation was found to be less prevalent than was apparent from the optical micrographs of this type of region (Fig. 1a), cautioning too generalised a view of the oxide morphology based wholly upon the schematics. Those α-Al$_2$O$_3$ precipitates found at the interface between zones I and O tended to be faulted and of lath morphology and were larger than those formed internally within zone B. The continuous oxide grains here (Fig. 2e) were of a more highly faulted spinel structure, EDX traces showing that they contained Ni. With a measured lattice parameter of $a_o \approx 0.798$ nm, the structure of these oxides is based on the spinel NiAl$_2$O$_4$ so that the inward formed oxide can be expected to be more permeable than if α-Al$_2$O$_3$ alone were formed there. That the spinel is formed here is indicative of the reduced Al activity at the main inward oxidation front associated with the formation of the α-Al$_2$O$_3$ particles lower down.

The surface scale consisted of α-Al$_2$O$_3$ and Cr$_2$O$_3$. The 50–100 nm equiaxed form of the grain structure (Fig. 2f) suggests that the outward developing scale grows from its base. EDX traces of this lower region of the outward formed scale indicated the incorporation of significant amounts of Ta and W but not Re as well as smaller amounts of Ti at the base of the outward formed oxide.

The sub-scale region shown in Fig. 3(a) exhibits a very different type of morphology than we have described above for zone B and was probably beneath a type 2 region though the outer oxides had in this case spalled. The width of the denuded zone (~3 μm) suggests more active inward and outward

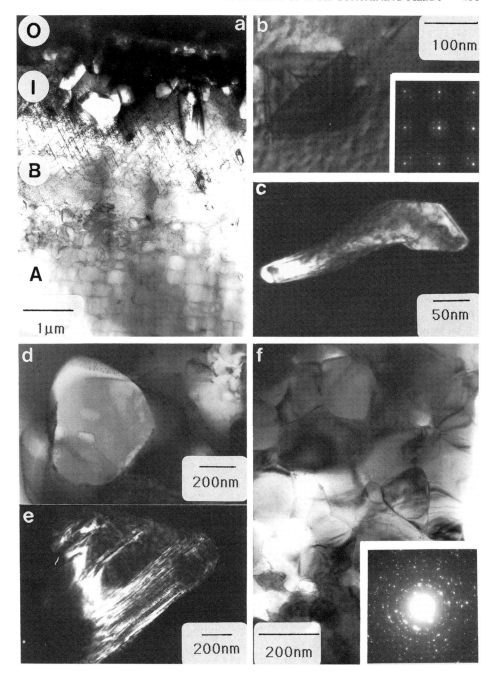

Fig. 2 TEM micrographs of the oxide and sub-scale region of the specimen oxidised at 950°C.

diffusion here and is thus concomitant with there having been a thicker scale. Consistently, the TiN precipitates formed in this particular area tended to be smaller (50–150 nm) and decorated the banded dislocation arrays indicative of creep assisted recrystallisation leading to accelerated nitridation and potentially differing outward oxidation processes. Indeed the only oxides retained above the denuded zone B were α-Al_2O_3 and Cr_2O_3 with an equiaxed fine (50–100 nm) morphology. That the spinels previously seen in zone I were not observed is thus probably associated with greater ease of outward Al and Cr diffusion through zone B given its increased dislocation content. This too is consistent with there being no apparent gradients of the alloy content in the dislocated sub-scale region.

3.2.2 Composition profiling

SEM EDX composition profiles of the bulk 950°C oxidised specimens, shown in Figs 3(b) and (c), as ratios relative to the local Ni content, further corroborate the distinct differences in process that can occur from place to place for this temperature (for the type 1 and type 2 regions respectively) even if the data obtained are necessarily on a coarser scale than was possible for the TEM foils. In Fig. 3(b), for a type 1 region, the different zones are clearly delineated and loss of Al, Cr and to some extent Ti as well as of Ta and W in zone B can be seen to give way to a rapid increase in Al and Cr as we reach the outward oxide with the Ta and W concentrated at the base of the outward formed scale. Interestingly Re, as a γ segregant, tends to remain in the γ' denuded B zone. Figure 3(c) is a profile of the similarly ratioed composition changes across the thicker, banded scaling of a type 2 region in Fig. 1(a). Aluminium can now be seen to be segregating mainly to the inward oxidation zone (I) whereas chromium was found to segregate to the darker band at the base of the outward formed oxide (Fig. 1a) with the heavier elements (other than Re which is retained within the γ' denuded sub-scale zone) again concentrating near to the original metal surface. More interestingly, as we progress outward across the scale, the reduction in the ratioed concentration of Cr reflects the way the scale changes to being based on NiO as was confirmed by an X-ray diffraction study of the oxide surface.

Turning to the EDX profile of the 750°C oxidised specimen, as shown in Fig. 3(d), we can still locate the now rather thin zone B, but the concentration gradients at the base of the inward grown zone I are fairly similar to those seen at the higher temperatures. Now we see the incorporation of both Ta and W at the top of this more clearly inward developing oxide (though again Re is retained in zone B) and a relatively sharper distinction with the outward oxide which, on the evidence of the composition profile, must be taken to be based on NiO with an increasing concentration of Co as we approach the outer, and actively oxidising surface. It remains to examine, using TEM, the more localised microstructure associated with these grosser trends for the 750°C oxidised specimens.

3.2.3 The 750°C, 200 h microstructure

The first sign of a morphological change in the sub-scale structure for the 750°C oxidised sample (Fig. 4a) as we approach the oxide is again the dissolution of

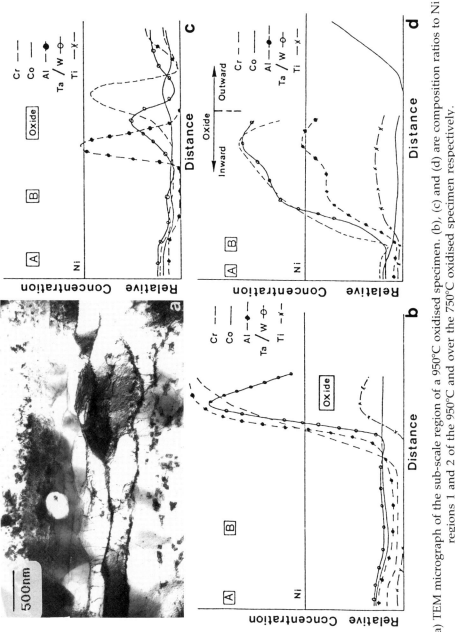

Fig. 3 (a) TEM micrograph of the sub-scale region of a 950°C oxidised specimen. (b), (c) and (d) are composition ratios to Ni over regions 1 and 2 of the 950°C and over the 750°C oxidised specimen respectively.

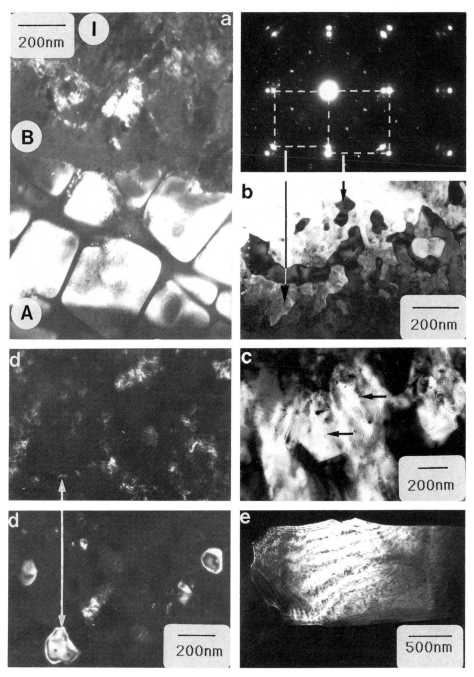

Fig. 4 TEM micrographs of the oxide and sub-scale region of the specimen oxidised at 750°C. (d) shows two images of the same region, the top taken with a spinel reflection, the bottom image using an α-alumina reflection.

the ordered γ' precipitates, though now over a much more localised, 0.3–0.5μm thick, region than at 950°C. Again however we see dislocation debris, as associated with stress relaxation due to the formation of the outer oxide, decorating the inner γ/γ' interfaces. TiN precipitates were also again found in zone B though now they were much smaller (~25 nm) than at the higher oxidising temperature. Two distinct types of interface between zones B and I were observed in different regions. In both cases the inward developing oxide was a Ni-containing aluminium-based spinel of highly faulted morphology oriented with the matrix. However, whereas in some cases the interface was rough and inter-convoluted as in Fig. 4(b), in others it was more discrete and locally voided exhibiting trans-granular cracking perpendicular to the oxidation front (Fresnel contrast-enhanced examples of such cracks are arrowed in Fig. 4c). The spinel oxides here were similar in chemistry to those observed at 950°C with lattice spacings $\approx$2% above that of γ-Al$_2$O$_3$ consistently with their demonstratedly higher Ni content. Their cube–cube orientation relation and large misfit with the matrix suggest how their progressive development via interface diffusion could lead to the inter-convoluted form of the interfaces with the substrate. On this basis the cracked spinels at flatter interfaces suggest a slightly later stage in the development of the inward oxidation, as would be affected more by changes in the oxidation behaviour above this. In the upper parts of the inward oxidised zone I small (100–200 nm) grains were seen at increasing density within the spinel oxide which could be demonstrated to be Cr$_2$O$_3$ and α-Al$_2$O$_3$ (the upper image in Fig. 4(d) was obtained using a spinel reflection and the lower using an α-alumina reflection). The contrast observed is similar to that exhibited by the subgrain structure in plan view oxide images from γ-Al$_2$O$_3$ found on NiCrAl alloys by Smialek and Gibala.[5] Above these regions the outward formed oxide (O) was found to be completely different from that generally formed at higher oxidising temperatures. Figure 4(e) is a dark field image from this outward oxide which consisted of 2–3 μm columnar grains of a NiO-based structure containing increasing amounts of Co in single phase solid solution towards the outer surface of the scale. Comparing the grossly different oxidation processes found at 750°C with those described earlier for higher oxidising temperatures, it would seem that several factors conspire to promote a mutually synergistic change in the predominant oxidation behaviour. The smaller energy barrier for nucleation of the spinel oxide, given its favourable 'cube–cube' oriented epitaxial growth, relative to that for the energetically preferred α-Al$_2$O$_3$ apparently promotes the development of the metastable spinel structures of relatively increased Ni content at these lower temperatures. The increased relative availability of Ni for outward oxidation as NiO is thus promoted relative to that of Al and Cr for the formation of alumina and chromia by the way in which the oxide thus formed allows ingress of oxygen for the further development of the scaling at the I/O interface. That the outward oxidation occurs in the main at the free surface is suggested by the outward increasing Co content of the developing outward formed scale (O). Co will be in increasing supply at the outer surface as inward oxidation proceeds given the thermodynamically preferential oxidation in the inward formed oxide of for example Al and Cr at the lower partial oxygen pressures prevalent there. There are however further indications, such as the trans-granular voiding of the inward formed spinels, that the Ni content of the outward formed scales could well be further enhanced at its inner front as oxidation proceeds by ion exchange in the initially inward formed spinels. The way small α-alumina and chromia grains are found in the upper parts of the inward formed spinel (Fig. 4d) is consistent with up-diffusing Al and Cr being exchanged with Ni from the spinel to form these thermodynamically more stable oxides. Such a transformation is indicative of a transition state in the way the oxidation will develop from this point onward and can be related to the form of oxidation development reported by Pettit[2] for a binary NiAl alloy. In this case at 750°C, again with the outward formation of NiO, inward spinel formation was observed to be a precursor to the stable inward formation of α-alumina, though only within the preformed spinel.

4. Conclusion

It has been demonstrated that a small change in oxidation temperature can lead to gross changes in both the morphology and the nature of the scaling process. Equally, while the predominant oxidation process at 950°C is essentially protective with the formation of α-alumina and chromia as the outer scales, it is apparent that perturbations in the initiation of this scaling behaviour can lead to the local stabilisation of non-protective scaling behaviour. It is well known[2] both that the non-protective behaviour is associated with a change in the outer scale formed to it being NiO based, and that this can be promoted by local development of internal oxidation with the formation of α-alumina ahead of the inward moving front. However, the insight provided by the detailed edge-on TEM examination of the scales allows further clarification of the mechanisms involved. For example, it is now clear that the way α-alumina forms at the higher temperature ahead of the continuous inward oxidation front enhances the tendency to form spinels there as the Al activity is thereby lowered. Similarly the forms of the changes in local microstructure are indicative of the way the greater diffusivity of these spinels relative to that of α-alumina would further stabilise the up-diffusion of Ni under the concentration gradient then created by its oxidation, in the main, at the outer surface of the outward forming scale once this had started. That a form of this latter type of oxidation behaviour is dominant at the lower oxidation temperature thus becomes unsurprising, the lower diffusion distances in the γ' denuded region of the sub-scale alloy reducing the availability of Al, while the reduced nucleation barrier for the formation of the epitaxially oriented spinels at this lower temperature further conspires both to prevent the formation of a protective α-alumina layer and to enhance the up-diffusion of Ni relative to Al. It is however interesting that the evidence of an exchange mechanism at this lower temperature leading to the localised formation of α-alumina and chromia within the developing inward forming spinel oxide will further enhance the availability of Ni and promote easier cation diffusion through the voiding thus created.

It is further interesting to note that on the basis of the work we have described, the scaling would be expected to be rather different if oxygen rather than air had been used for the oxidation. Ti is considered to be detrimental to the formation of alumina on Re-containing superalloys in that it has been associated with the development of more complex, less protective oxides.[6] That Ti is denuded from the sub-scale matrix to form nitrides, particularly in association with sub-scale recrystallisation, emphasises that in the absence of this type of process (as would be the case in pure oxygen) the incorporation of the Ti in the oxide could well decrease the tendency to the formation of a protective scale, at least at the higher temperatures.

In conclusion, this in turn leads us to note the rôles of the high atomic number elements such as Ta, W and Re in the oxidation processes examined. In this context it is significant that Ta and W, which do not segregate strongly to the disordered matrix phase, tend to become incorporated into the base of the scales. In contrast, Re which is strongly segregated to the γ phase formed for these alloys beneath the scale, when they are un-coated, does not appear to play an active part in the oxidation process itself. Nevertheless it could well develop a dynamic barrier for up-diffusion of other elements in the γ' denuded sub-scale region as it becomes more concentrated there. It is thus fascinating that when these alloys are aluminised or platinum aluminised the interface between the coating and the matrix becomes denuded in the γ phase and Re is incorporated into the coating.[7] Re can thus be expected to play a very different part in the oxidation of the alloy as coated by comparison with its behaviour when un-coated.

Acknowledgements

We would like to thank Professor C. J. Humphreys for the provision of laboratory facilities and are also grateful to Rolls-Royce both for the provision of the specimens examined as well as for, in association with the SERC, financial support. We particularly note our indebtedness to the late Dr T. Rhys-Jones for his encouragement to investigate the degradation behaviour of these interesting alloys.

References

1. J. L. Smialek and G. H. Meier, in *Superalloys 2* (eds. C. T. Sims, N. S. Stoloff and W. C. Hagel), pp. 293-326, Wiley-Interscience, New York, 1987.
2. F. S. Pettit, *Trans. AIME*, **239**, 1296–1305, 1967.
3. L. Singheiser, H. W. Grunling and K. Schneider, *Surf. Coat. Tech.*, **42**, 107–117, 1990.
4. S. B. Newcomb, C. B. Boothroyd and W. M. Stobbs, *J. Microsc.*, **140**, 195–207, 1985.
5. J. L. Smialek and R. Gibala, *Met. Trans. A*, **14A**, 2143–2161, 1983.
6. S. W. Yang, *Ox. Met.*, **15**, 375–397, 1981.
7. S. B. Newcomb and W. M. Stobbs, 'The TEM characterisation of the oxidation behaviour of coated blade alloys', Rolls-Royce plc *Report RR/AEP 78/2 BOC 074*, Cambridge, 1991.

TEM Investigation of the Oxidation of Nickel-based Superalloys and Ni–Cr Alloys

RÉGINE MOLINS and ERIC ANDRIEU

ENSMP, Centre des Matériaux, BP 87 91003 EVRY Cedex, FRANCE

ABSTRACT

We report an analytical TEM study of the oxidation behaviour of Ni-based superalloys and Ni–Cr alloys given short exposure times in air at 650°C. The resulting oxide scales and the microstructural changes in the bulk beneath the oxidised surface are described.

1. Introduction

Nickel-based superalloys are known for being subject to intergranular or interfacial embrittlement due to oxidation.[1] Earlier studies on fatigue crack growth at 650°C have shown the importance of environment in the crack propagation rate.[2,3] The high crack propagation rates currently measured emphasise the importance of the early stages of oxidation as a critical step in the embrittling process. Consequently, it is necessary to characterise the microstructural and chemical changes occurring in nickel-based superalloys and to identify the oxides growing on the alloy surfaces after short exposure times.

TEM equipped with analytical spectrometers permits us to investigate microstructural and chemical changes with a nanometric resolution.

Different nickel-based alloys, with differing microstructures (monocrystal, polycrystal, precipitates) and chemical compositions (alloying elements), have been used in this study and examined after short exposures in an air environment at 650°C.

2. Experimental

Two different binary Ni–Cr alloys containing different Cr contents have been studied, as well as two nickel-based superalloys: AM1 and Inconel 718. The

162

Ni	Co	Cr	Mo	W	Al	Ti	Ta	C	Zr
Bal	6.6	7.94	2.10	5.66	5.16	1.21	8.17	0.002	0.010

Table 1(a) AM1: composition of the alloy in wt%

Ni	Cr	Mo	Fe	Nb	Ti	Al	C
Bal	18.6	3.1	18.5	5	0.9	0.4	0.04

Table 1(b) Inconel 718: composition in wt%

compositions of the Ni–Cr alloys are respectively 20 and 30 wt% of Cr for Ni–20Cr and Ni–30Cr. The compositions of the superalloys are given in Table 1.

AM1 is a single crystal alloy and has an austenitic matrix in which a γ' hardening phase is precipitated in coherence with the matrix. The γ' precipitates are cubic and range in size from 0.5 to 0.7 μm. Their volume fraction is approximately 70%. Nickel, chromium, molybdenum, cobalt and tungsten are present in the γ matrix, and nickel, aluminium, titanium and tantalum in the form of $Ni_3(Ti,Al,Ta)$, are the main constituents of the γ' phase. This alloy has been chosen for our study because it provides a large number of γ/γ' interfaces to observe.

Inconel 718 is a polycrystalline alloy (150 μm grain size), remarkable for its high Nb and Fe contents. It is hardened in two different ways: solid solution hardening comes from Cr, Fe and Mo and structural hardening from γ' $Ni_3(Ti,Al)$ and γ'' Ni_3Nb precipitate phases. The γ' precipitates are quasi-spherical, 20 to 60 nm in diameter, and the γ'' precipitates are disc shaped, lying parallel to the (100) plane of the matrix. The volume fraction of the hardening phases does not exceed 20%. Intergranular precipitation is limited.

Ni–Cr alloys have been used as simplified alloy models to get a better understanding of the more complex superalloys matrix oxidation. These alloys have been studied in the literature in various oxygen pressure and generally for long times and at high temperatures.[4,5] Only a few studies deal with the transient stage of oxidation.[6,7]

TEM investigations were carried out on a 300 kV microscope equipped with EDS and EELS. The oxides and microstructural variations have been identified by using electron diffraction techniques and STEM elemental concentration linescans with a few nanometers probe size.

Two types of specimens have been used: thin foils electrolytically thinned and then oxidised for short times at 650°C; and cross-sectional foils prepared from bulk samples oxidised for longer times.[8]

3. Results

3.1 Ni–Cr alloys

We have compared the intragranular and intergranular oxidation mechanisms in Ni–Cr alloys of different Cr content.

The early stages of oxidation (1 min, at 650°C under air) on thin foils show clear differences. The homogeneous formation of NiO takes place on the surface of the Ni–20Cr alloy whereas an oxide scale consisting of Cr_2O_3 nuclei within a NiO layer forms on the Ni–30Cr alloy. Nuclei of Cr_2O_3 are present on the grain boundaries of both alloys.

For oxidation times longer than 1 min, the thin foils became completely oxidised. Then, cross-sectional foils have been prepared from half-cylinders oxidised at 650°C under air. Figures 1 and 2 show bright field micrographs of the Ni–30Cr and Ni–20Cr alloys after 2 min and 10 min oxidation respectively, as well as the elemental concentration linescans through the alloy/oxide interfaces. After 2 min oxidation, the elemental profile on Ni–30Cr alloy shows a chromium enrichment at the alloy surface. This area has been identified by electron diffraction technique as Cr_2O_3. No external outer scale of NiO has been detected. The Ni–30Cr alloy forms a continuous layer of Cr_2O_3 which was 30 nm in thickness. Beneath this oxide, the alloy was depleted in Cr (20%) over a depth of 200 nm, where the metal had a subgrained microstructure, crystallographically misorientationed with the bulk alloy grain.

For the Ni–20Cr, the same Cr_2O_3 layer appears under a NiO layer after 10 min oxidation. The NiO external scale contains Cr acting as a substitutional cation in the NiO lattice. Diffraction techniques allow us to identify NiO oxide type, without trace of $NiCr_2O_4$ spinel formation. The Cr_2O_3 oxide clearly penetrates into the grain boundary visible in Fig. 2. The Cr depleted area itself extends to 100 nm, with the same structural transformation as described above.

Observations after longer oxidation times have confirmed the presence of these 'recrystallised' and Cr depleted zones on both alloys with a grain size far below that of the bulk alloy: 0.5 μm instead of 100.

Our results have shown that the main difference between the two alloys is in the growth rate of the Cr_2O_3. A continuous layer is observable after 2 min on Ni–30Cr whereas 10 min are required for the Ni–20Cr. NiO is the first oxide appearing on Ni–20Cr, while Cr_2O_3 forms at the beginning of oxidation on Ni–30Cr. These results are consistent with our observations of oxidised foils where Cr_2O_3 appeared from the beginning of oxidation of the Ni–30Cr.

3.2 AM1

Earlier studies[9] of oxidised foils have shown the different scaling behaviours of the matrix and of the γ' precipitates from the first stages of oxidation. When the matrix is oxidised, it forms NiO, with coherent grains in epitaxy, while the γ' phase forms a polycrystalline Ni oxide as well as some nuclei of $NiAl_2O_4$ and $NiCr_2O_4$. A mixture of complex oxides containing the alloying elements are found at the γ/γ' interface.

Figure 3 shows a cross-sectional observation of the scale formed after 15 min exposure and the elemental concentration linescan through the oxide scale and 50 nm into the alloy from the metal/oxide interface. The profile in the alloy corresponds to the γ' phase $Ni_3(Ta)$, as confirmed by the high content of Ta detected.

Two oxide layers were identified at the bulk surface: an external scale

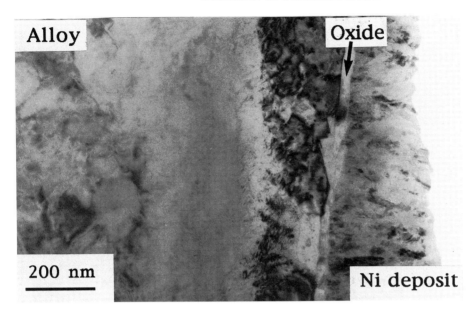

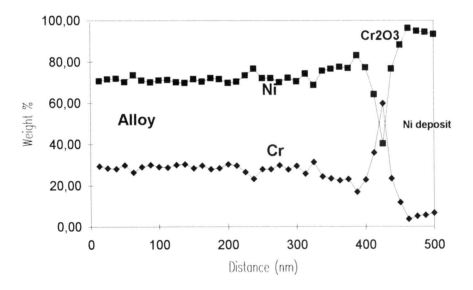

Fig. 1 Cross sectional micrograph of the oxide formed on Ni–30Cr and the elemental concentration linescan across the oxide and sub-scale alloy.

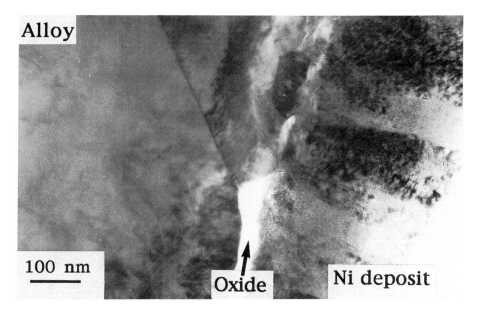

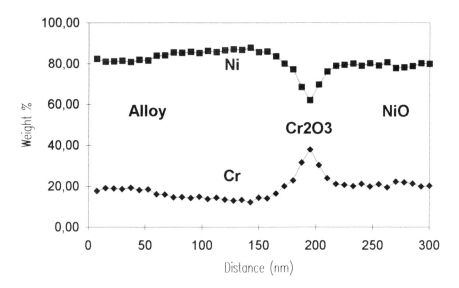

Fig. 2 Cross sectional micrograph of the oxide formed on Ni–20Cr and the elemental concentration linescan across the oxide and sub-scale alloy.

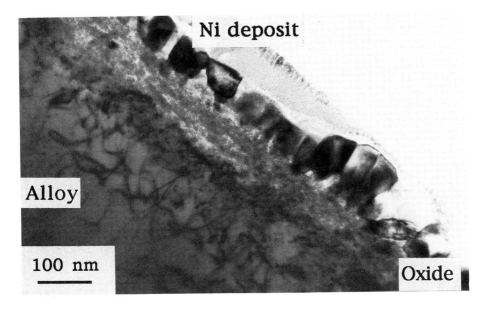

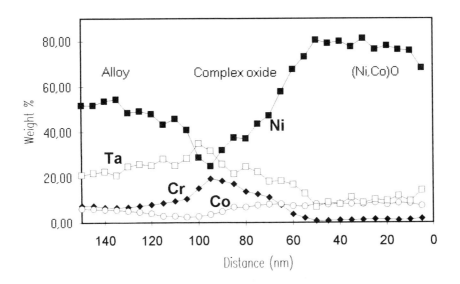

Fig. 3 Cross sectional micrograph of the oxide layers formed on the AM1 alloy and the elemental concentration linescan through the oxides and sub-scale alloy.

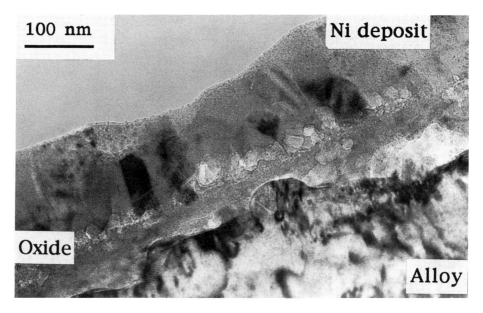

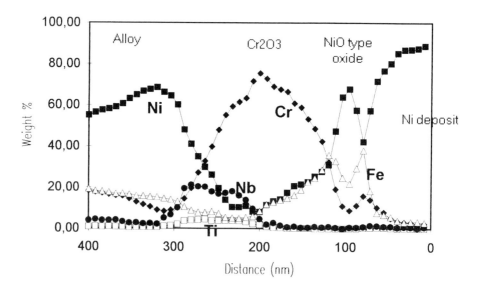

Fig. 4 Cross sectional micrograph of the oxide layers formed on Inconel 718 and the elemental concentration linescan through the oxides and sub-scale alloy.

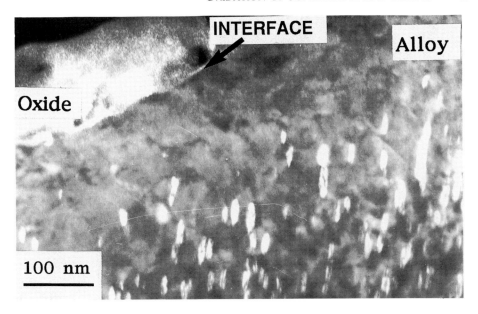

Fig. 5 Dark field micrograph imaged with a $\gamma' + \gamma''$ superlattice reflection showing the removal of precipitates in the metal beneath the oxide formed on Inconel 718.

(50 nm) of columnar grains of (Ni,Co)O and a nanocrystalline internal scale, 50 nm in thickness, of a complex oxide, rich in Cr, Al, Ta and Ti, as confirmed by X-ray analysis.

3.3 Inconel 718

Figure 4 shows a cross-sectional bright field micrograph of the superalloy and oxide layers formed after 15 min oxidation as well as an elemental concentration linescan to a depth of 400 nm from the surface of the oxide. Two oxide layers were identified to have been formed on the surface of the alloy: an external scale, which is approximately 70 nm thick, of NiO oxide containing various amounts of Cr and Fe, and an intermediate scale, of the same thickness. Its composition varies from Cr_2O_3 near the NiO interface to Cr(Nb, Ti)O_4 at the oxide/alloy interface. This layer is acting as a passivating coating, Cr_2O_3 avoiding further oxidation of the bulk material.

The concentration linescan reveals a decrease of Cr, Nb, Ti and Al contents in the alloy close to the alloy/oxide interface to values which are approximately one half the bulk alloy contents, 100 nm from the surface of the oxide. A dark field micrograph imaged using a $\gamma' + \gamma''$ superlattice spot shows the disappearance of these precipitates beneath the oxide (Fig. 5). This can be explained by the observed depletion of chromium in the matrix, which increases the solubility of Al and Nb, thus causing the dissolution of Ni_3Al and Ni_3Nb. This phenomenon has been demonstrated previously[10,11] for higher oxidation temperatures. The disappearance of hardening phases results in a local softening of the alloy.

4. Conclusions

TEM study in association with microanalysis techniques allows us to investigate the physical and chemical changes occurring during the early stages of oxidation of superalloys and nickel alloys, resulting in very thin scales and chemical variations.

TEM examination of the different superalloys and the Ni–20Cr alloy examined has revealed the presence of two distinct oxide layers:
1. An outer scale, consisting in a NiO type oxide, with columnar grains uniformly oriented, containing low amounts of the alloying elements.
2. An inner scale, a nanocrystalline oxide: Cr_2O_3 for the Ni–Cr alloys, and a mixture of complex oxides for the superalloys.

No external scale of NiO has been detected for the Ni–30Cr alloy. A continuous layer of Cr_2O_3 forms spontaneously.

Microstructural and chemical changes in the bulk beneath the oxide layers have been examined with the following features:
1. A significant depletion of chromium near the oxidised surface.
2. A disappearing of hardening precipitates associated with a recrystallised band, allowing local mechanical softening.

Acknowledgements

The authors would like to thank the DRET for financial support.

References

1. E. H. Bricknell and D. A. Woodford, *Met. Trans. A*, **12A**, 425–432, 1981.
2. A. Pineau, Environment induced cracking of metals (ed. R. P. Gangloff and M. B. Ives), *NACE*, **10**, 111–122, 1988.
3. E. Andrieu, G. Hochstetter, R. Molins and A. Pineau, *Corrosion-Deformation Interactions*, Fontainebleau, Oct. 1992.
4. C. S. Giggins and F. S. Pettit, *Trans. Metallurg. Soc. AIME*, **245**, 2495–2507, 1969.
5. P. Moulin, A. M. Huntz and P. Lacombe, *Acta Met.*, **28**, 745–756, 1980.
6. N. S. McIntyre, T. C. Chan and C. Chen, *Oxid. of Metals*, **33**, 457–479, 1990.
7. B. Chattopadhyay and G. C. Wood, *Oxid. of Metals*, **2**, 373–399, 1970.
8. S. B. Newcomb, C. B. Boothroyd and W. M. Stobbs, *J. Microscopy*, **140**, 194–207, 1985.

9. R. Molins and E. Andrieu, *3rd International Symposium on High Temperature Corrosion*, Les Embiez, May 1992.
10. B. Pieraggi and F. Dabosi, *Werkstoffe und Korrosion*, **38**, 584–590, 1987.
11. S. B. Newcomb and W. M. Stobbs, *Microscopy of Oxidation*, The Institute of Metals, 135–141, 1990.

The Oxidation of Nickel-based Alloys Containing Sulphur

G. J. TATLOCK*[†], P. G. BEAHAN*, J. RITHERDON* and
A. PARTRIDGE[†]

*Department of Materials Science and Engineering, and
[†]IRC in Surface Science, University of Liverpool, PO Box 147, Liverpool,
L69 3BX, UK

ABSTRACT

Scanning Auger microscopy and analysis have been used to study the *in-situ* oxidation of polycrystalline nickel samples, which have been heat treated to produce overlayers of carbon or sulphur impurities prior to oxidation. It is shown that the oxidation process varies with overlayer composition, temperature and grain orientation; and the results are compared with previous work on well-characterised single crystal nickel surfaces which have been dosed with impurities.

1. Introduction

The segregation of impurity elements to surfaces and grain boundaries in metals can have a profound effect on the physical properties of the alloy system.[1] Intergranular failure has often been ascribed to the presence of grain boundary impurities, while the presence of species such as sulphur at metal/oxide interfaces has been linked with reduced oxide adherence.[2] It has also been suggested that the presence of carbon or boron nitride on a surface can affect the initial stages of oxidation; and that the presence of these species on UHV chamber walls reduces oxygen adsorption significantly. This may have important practical applications in the design of vacuum chambers with superior outgassing characteristics.[3]

The surface segregation of carbon and sulphur has been extensively studied in various systems including nickel and nickel-based alloys. At low temperatures, carbon is observed to segregate to the surface while sulphur segregation dominates at higher temperatures.[4] A recent scanning Auger study of polycrystalline nickel[5] has shown that at intermediate temperatures (673–723 K) graphitic carbon forms over most of the surface, but sulphur occupies the regions round the emergent grain boundaries and is thought to diffuse onto the surface via these grain boundaries. Initially the sulphur is confined to a narrow region, but slowly spreads out with time at temperature to form a zone up to ~20 μm wide centred on the boundary. It has been suggested that site competition between the sulphur and carbon ensures that the zones remain well defined.

The oxidation of pure nickel has been studied widely.[6] At room temperature, the process can be described by three stages: rapid dissociative chemisorption, oxide nucleation and lateral growth to coalescence, followed by thickening of the continuous oxide film. Much of this work has been done on carefully prepared single crystal surfaces. For example, on Ni(111), p(2 × 2) and ($\sqrt{3} \times \sqrt{3}$)R30° ordered overlayers have been observed before oxide islands are formed at coverages above ~0.34 ML of oxygen.[7] The presence of impurities such as sulphur can have a major effect on the oxide formation and stability on nickel, although in recent Ni(111) single crystal studies (e.g. reference 8), it was shown that basic mechanisms remain unchanged.

In this paper we report a preliminary series of experiments with polycrystalline nickel substrates in which the early stages of oxidation in the presence of carbon and/ or sulphur overlayers have been examined. These results extend the previous single crystal studies and show how oxidation behaviour can vary markedly from area to area on a polycrystalline surface.

2. Experimental

High-purity polycrystalline nickel containing 100 wt ppm of carbon and approximately 10 wt ppm of sulphur, calcium, manganese and phosphorus was obtained in the form of 1 mm thick, temper annealed sheets (the content of other impurities totalled less than 100 ppm). These sheets were cold rolled to a thickness of 300 μm and annealed in Ar filled quartz tubes at 1373 K for 2 h and then oil quenched. The tarnished surface layer was then removed by gentle grinding and polishing before small sections of the sheets (7 × 7 × 0.3 mm) were mounted on a heatable stub and loaded into a scanning Auger microscope (V.G. Microlab II). In each case the samples were ion wiped with 10 keV Ar ions until free from contamination as measured by Auger analysis and then heated *in situ* using an isolated d.c. power supply. Sample temperatures were measured with an optical pyrometer which had previously been calibrated with a thermocouple mounted on a sample and temperatures were considered accurate to ±10°. Oxygen adsorption experiments were performed by leaking high-purity oxygen directly into the analysis chamber. The dosage was determined by recording the oxygen partial pressure using a mass spectrometer attached to the chamber. The microscope was operated throughout the experiments at a primary beam voltage of 30 kV which gave a high lateral spatial resolution of ~80 nm. The Auger spectra were collected by a 150° concentric hemispherical analyser and displayed in direct energy mode. The carbon (271 eV) sulphur (150 eV) and oxygen (507 eV) Auger peak heights were ratioed to the nickel (849 eV) signal after background stripping, and expressed in each case as a peak height ratio (PHR).

3. Results

3.1 Carbon-covered surfaces

When the polycrystalline nickel samples were heated to 625 K, first nitrogen and then carbon were observed to segregate to the surface. After heating for 60 min, only carbon was detected on the surface and prolonged heating for several hours led to a graphitic overlayer approximately 2 monolayers thick. When 300 Langmuirs (L) of oxygen were leaked into the chamber with the sample still at 625 K, all the carbon was removed from the surface of the sample and replaced by oxygen. However, the oxygen coverage varied markedly from grain to grain on the sample, as discussed in section 3.3. If the heating was maintained for a further 90 min, the oxygen disappeared from the surface (possibly into the bulk nickel substrate), and was replaced once more by carbon.

If the samples were cooled to room temperature prior to oxygen dosing, the uptake of oxygen was markedly reduced, in agreement with the work of Fujita and Homma,[3] and the carbon overlayer appeared to passivate the surface against oxygen uptake.

3.2 Sulphur-covered surfaces

Although there are some similarities between sulphur- and carbon-covered surfaces exposed to oxygen, there are also major differences. In order to produce a sulphur overlayer, the samples were heated to 875 K after ion cleaning at room temperature. By varying the time at temperature, a sulphur coverage of between 10 and 50% of a monolayer could be obtained. The samples were then allowed to cool back to room temperature before dosing with oxygen. Figure 1(a) shows how the oxygen signal (ratioed to the Ni 849 eV peak) varied with oxygen dose and sulphur coverage. Each of the sets of data is taken from one randomly orientated grain of a polycrystalline sample but the overall behaviour is reminiscent of the behaviour observed by Liu et al.[8] on Ni(111) surfaces. A typical three-stage growth process is observed, although the oxygen uptake starts at lower exposures than on single crystal surfaces as might be expected for random (vicinal) surfaces. In the absence of sulphur, no plateau region was observed at low oxygen doses. The data need to be treated with some caution, however, since it was also observed that different grains appeared to behave in slightly different ways, as discussed later in section 3.3.

The variation of sulphur signal with oxygen exposure is also shown in Fig. 1(b). The decrease in sulphur signal with oxygen dose is less than observed by other authors on single crystal surfaces. This may be due to the vicinal nature of the surfaces, although the high primary beam voltage of the Auger instrument will also make a small contribution to this effect. In order to confirm whether any sulphur had been removed during oxidation several calculations of the expected changes in the sulphur signal with different overlayers were carried out using a standard approach.[9] Some of the results are summarised in Fig. 2, where the experimentally measured peak height ratios

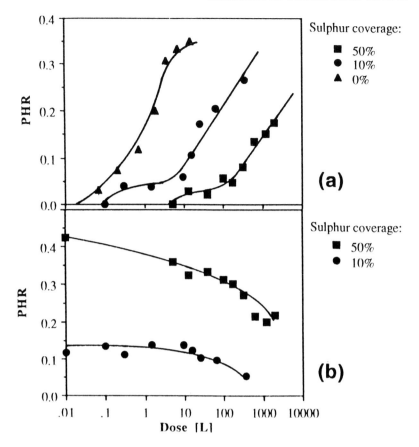

Fig. 1 Variation of (a) oxygen/nickel peak height ratios (PHR) and (b) sulphur/nickel peak height ratios with oxygen exposure for different sulphur coverages on polycrystalline nickel.

for sulphur/nickel were compared with model calculations, assuming that the sulphur layer is covered with a layer of nickel oxide. The sulphur signal measured experimentally drops off faster than the prediction, which implies that some of the sulphur must be lost from the surface during the experiment. (Various assumptions and approximations have to be made during the calculations but these alone cannot be used to explain the large discrepancy between the theoretical calculations and the experimental results.)

3.3 Variation in oxygen take up from grain to grain

Although all the foregoing data are typical of the results which could be collected by low spatial resolution microanalysis of polycrystalline samples, it was clear with the high spatial resolution available on our instrument that there were marked variations in behaviour from area to area on the samples. This has already been reported for the segregation of sulphur to surfaces via

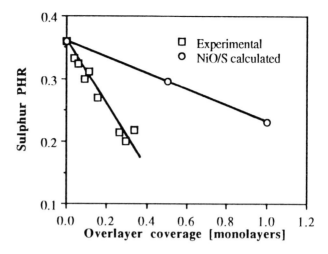

Fig. 2 Comparison of experimental results and model calculations of the sulphur peak height ratio (PHR) to nickel, for a nickel oxide on sulphur overlayer.

grain boundaries and subsequent site competition between sulphur and carbon,[5] but *in situ* oxidation has now added a new dimension to the investigation.

When samples were heated to 625 K for prolonged periods, the coverage of the carbon overlayer was quite uniform from grain to grain and varied little across individual grains. However, as soon as oxygen was introduced, quite dramatic changes were observed in the secondary electron images. One example of this is shown in Fig. 3(a). The sample had been ion wiped, heated to 625 K for 1 h and then exposed to a large dose of oxygen (>300 L). The variation in secondary electron yield from grain to grain could be correlated directly with the oxygen level on the surface. The lightest grains in the image gave a high oxygen signal while the dark grains showed only a small trace of oxygen in the Auger spectra. The carbon was completely removed from the surface in both cases. After prolonged heating with no further oxygen exposure, the oxygen left the surface and was replaced with another uniform layer of carbon. The corresponding secondary electron image is shown in Fig. 3(b) and little contrast difference is now observed from one grain to another. After re-exposure of the sample to oxygen, the contrast effect returns, although the difference between grains is less marked (Fig. 3c). However, grains which previously had a high oxygen level maintain this difference on re-exposure.

When samples were heated to 825 K to promote a sulphur overlayer, small variations in sulphur coverage were observed from grain to grain. 12 h heating led to an average coverage of 0.45 ML, with a variation of ~0.05 ML between different grains. After allowing the samples to cool to room temperature, successively larger doses of oxygen were introduced. Some inverse correlation between sulphur coverage and oxygen uptake was observed, although in view

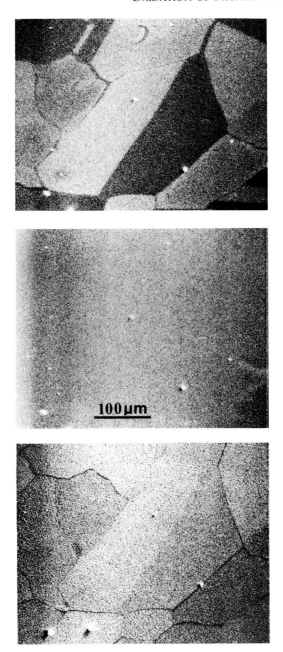

Fig. 3 Secondary electron image of a polycrystalline nickel surface after (a) oxygen exposure, (b) oxygen removal and replacement by carbon and (c) re-exposure to oxygen.

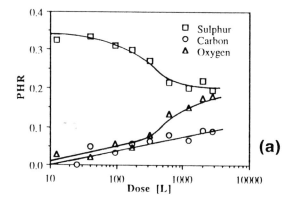

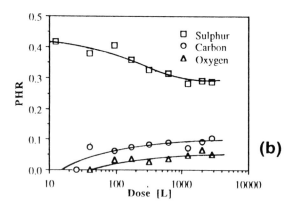

Fig. 4 Peak height ratios (PHR) versus oxygen dose for sulphur, carbon and oxygen for (a) a grain with high oxygen uptake and (b) a grain with low oxygen uptake.

of the low temperature results with carbon overlayers, it was thought that the orientation of the underlying grains was a more dominant influence on the oxygen uptake, than the slight variation in sulphur coverage. Typical plots of the sulphur, carbon and oxygen coverage with oxygen dose are shown in Fig. 4 for two grains – one with a high oxygen uptake (Fig. 4a), and one with a low uptake (Fig. 4b). In both cases the sulphur level appears to remain almost constant to approximately 100 L of oxygen before decreasing. On the grains with a low oxygen uptake the oxygen coverage never exceeds 10% with respect to the nickel signal but reaches 35% on some oxygen rich grains.

Following oxidation, samples were depth profiled with 3.5 keV Ar ions. The results for two different grains with high and low oxygen uptakes are shown in Fig. 5. Both the oxygen and sulphur signals were decreased significantly after 1 min bombardment on grains with a low initial oxygen uptake and no trace remained after 4 min, although the sulphur level decreased faster than the oxygen level. On grains with a high oxygen uptake, the oxygen signal dropped

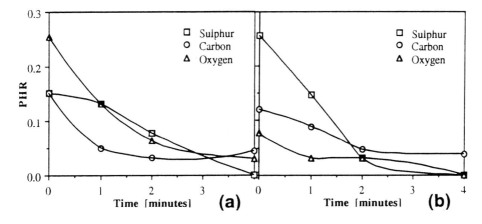

Fig. 5 Variation in peak height ratios (PHR) for sulphur, carbon and oxygen with sputtering time for (a) a grain with high oxygen uptake and (b) a grain with low oxygen uptake.

markedly after 1 min while the sulphur signal remained almost constant. Further sputtering led to a reduction in both signals, which could indicate the presence of an oxygen-rich overlayer on top of a mixed sulphur–oxygen layer. After all the impurities were removed from the surface by ion wiping, reoxygenation with a dose of over 500 L gave variable oxygen uptakes once more. Careful measurements against nickel oxide standards suggested that low oxygen grains were covered with 1 ML of oxide, while high oxygen grains saturated at approximately 1.5 ML. Once again this suggests that the underlying grain orientation dominates the oxidation mechanism.

3.4 Variation in oxidation mechanism with temperature

Samples which were exposed to oxygen above room temperature showed a totally different behaviour from that described so far. Figure 6 shows a sample which had been heated to 875 K until a 25% sulphur coverage had been obtained and then exposed to 70 L of oxygen, while the temperature was maintained at 720 K. The formation of oxide islands can clearly be seen. The size and distribution of the islands varies from grain to grain, while an island-free zone is apparent along each grain boundary. However, this exclusion zone does not appear to apply to twin boundaries as shown in Fig. 6(b). At much higher doses of oxygen (>3000 L), a continuous layer of oxide is formed.

4. Discussion

From the foregoing results it is clear that at room temperature an oxygen overlayer is formed on all samples, although graphitic carbon and heavily

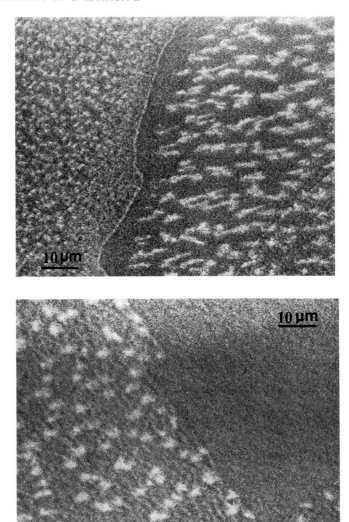

Fig. 6 Secondary electron images of nickel oxide islands formed on polycrystalline nickel samples covered with 0.25 ML of sulphur oxidised at 720 K.

sulphated surfaces take longer to oxidise. When carbon is removed by exposure to oxygen, the amount of oxygen left behind depends on the underlying orientation of the nickel grains. Repeated oxidation and annealing cycles tend to reduce the variation in the oxygen uptake between grains to give a more uniform overlayer, although further work is needed to clarify this effect.

Oxidation at room temperature in the presence of a sulphur overlayer leads to a rather different mechanism. In most cases the sulphur signal appears to drop slightly on exposure to oxygen and calculations would appear to suggest that some of the sulphur is removed at this time. However, most of the sulphur remains on the surface and may form a mixed oxygen sulphur overlayer for a range of sulphur concentrations and oxygen doses. Recent single crystal experiments with an STM appear to confirm these findings. In particular at sulphur coverages from 12 to 25% on Ni(100) a p(2 × 2) structure was shown to be preserved after dosage with oxygen, and both species appeared to coexist in domains on the nickel surface.[10]

When polycrystalline nickel surfaces covered with sulphur have been dosed with oxygen, the ease with which both species can be removed by ion wiping suggests that little, if any, nickel oxide is formed at room temperature and that the sulphur overlayer 'passivates' the nickel surface from oxidation. This is in sharp contrast to exposure of clean nickel surfaces at room temperature when large doses of oxygen (>300 L) lead to the formation of approximately 2 ML of oxide. The presence of nickel oxide was confirmed by XPS analysis of the surface. Oxide formation is also enhanced at elevated temperatures and oxide islands are readily formed at 720 K on samples already overlaid with a 25% coverage of sulphur. However, the sulphur was still present between the islands, and it would therefore appear likely that the oxide islands are nucleated at defect sites in the substrate, which may also affect the local coverage of the sulphur overlayer. The nucleation and growth of the islands is affected by both the substrate orientation and the proximity of emergent grain boundaries, although twin boundaries have much less influence in this respect. This is perhaps not surprising if the grain boundaries are acting as defect sinks. The fact that individual islands remain after prolonged exposure would suggest that the sulphur overlayer is still passivating the remainder of the surface and the oxygen is only very loosely bound in these areas and free to move easily across the surface.

5. Conclusions

Results of oxygen dosing of polycrystalline nickel surfaces at room temperature have some similarities with previous work on nickel single crystals. However, the dosage required to give oxygen coverage is lower on these vicinal surfaces and sulphur appears to be more adherent to the polycrystalline nickel substrate when exposed to oxygen. Marked changes in secondary electron emission from grain to grain can be correlated with variations in oxygen uptake with grain orientation. This in turn affects the work function of the surface and hence the secondary electron yield. Depth profiles through the surface layers on grains which show a high oxygen uptake suggest the presence of an oxygen-rich overlayer on top of a mixed sulphur–oxygen layer.

When sulphur-covered surfaces are dosed with oxygen at elevated temperatures, oxide islands are nucleated, although their distribution is dominated by the underlying substrate orientation. Sulphur can also be detected between the oxide islands and may still continue to have a 'passivating' role during the development of the oxide.

Clearly the results presented here are of a preliminary nature, but they have major implications for the study of the oxidation of nickel surfaces contaminated with impurities or bulk nickel-based alloys containing mobile impurities.

6. Acknowledgement

The financial support of the Science and Engineering Research Council is gratefully acknowledged.

7. References

1. E. D. Hondros and M. P. Seah, *International Metal Reviews*, **222**, 291–298, 1977.
2. J. L. Smialek, *Microscopy of Oxidation* (eds M. J. Bennett and G. W. Lorimer), Institute of Metals, London, pp. 258–270, 1991.
3. D. Fujita and T. Homma, *Surface and Interface Analysis*, **19**, 430–434, 1992.
4. E. N. Sickafus, *Surface Sci.*, **19**, 181, 1970.
5. A. Partridge and G. J. Tatlock, *Scripta Metall. Mat.*, **26**, 847–852, 1992.
6. P. H. Holloway, *J. Vac. Sci. Technol.*, **18**, 653, 1981.
7. C. R. Brundle and J. Q. Broughton, *The Chemical Physics of Solid Surfaces and Heterogeneous Catalysis* (eds D. A. King and D. P. Woodruff), Elsevier, Amsterdam, 1990.
8. J. Liu, J. P. Lu, P. W. Chu and J. M. Blakely, *J. Vac. Sci. and Technol.*, **A10**, 2355–2360, 1992.
9. C. G. H. Walker, D. C. Peacock, M. Prutton and M. M. El Gomati, *Surface and Interface Analysis*, **2**, 266–278, 1988.
10. A. Partridge and G. J. Tatlock, in preparation.

A Sulphur Segregation Study of PWA 1480, NiCrAl, and NiAl Using X-ray Photoelectron Spectroscopy with *in situ* Sample Heating

D. T. JAYNE and J. L. SMIALEK*

Department of Mechanical and Aerospace Engineering, Case Western Reserve University, Cleveland, OH 44106, USA
NASA LeRC, Cleveland, OH 44135, USA

ABSTRACT

Some nickel-based superalloys show reduced oxidation resistance from the lack of an adherent oxide layer during high-temperature cyclic oxidation. The segregation of sulphur to the oxide–metal interface is believed to effect oxide adhesion, since low-sulphur alloys exhibit enhanced adhesion. X-ray photoelectron spectroscopy (XPS) was combined with an *in situ* sample heater to measure sulphur segregation in NiCrAl, PWA 1480, and NiAl alloys. The polished samples with a 1.5–2.5 nm (native) oxide were heated from 650°C to 1100°C with hold times up to 6 h. The sulphur concentration was plotted as a function of temperature vs time at temperature. One NiCrAl sulphur study was performed on the same casting used by Browning[1] to establish a base line between previous Auger electron spectroscopy (AES) results and the XPS results of this study. Sulphur surface segregation was similar for PWA 1480 and NiCrAl and reached a maximum of 30 at% at 800–850°C. Above 900°C the sulphur surface concentration decreased to about 3 at% at 1100°C. These results are contrasted to the minimal segregation observed for low sulphur hydrogen annealed materials which exhibit improved scale adhesion.

1. Introduction

Interfacial sulphur segregation has been related to the spalling of protective Al_2O_3 scales formed on NiAl and NiCrAl coating alloys as well as those formed on structural nickel-based superalloys.[1-8] Current research is now directed toward relating the bulk sulphur content of purified alloys to segregation potential and cyclic oxidation resistance. In an effort to demonstrate the importance of sulphur removal compared to the gettering effect of reactive element additions, various materials were processed to have less than 1–2 ppm sulphur. The dramatic improvements in cyclic oxidation resistance have been well documented.[2,6,7] However, only a limited amount of segregation information is available for low sulphur alloys which indeed show minimal (1%) segregation for 'high purity' NiCrAl.[9] Furthermore we know of no published sulphur segregation data for the superalloys.

The primary purpose of the present study is to document the sulphur segregation behaviour of a commercially available single crystal PWA 1480 superalloy containing about 10 ppm sulphur. The segregation of the as-received superalloy was determined by hot stage XPS from 650 to 1100°C and for hold times up to 6 h. This was compared to the segregation behaviour of the same material after desulphurising to less than 1 ppm S by hydrogen annealing at 1200°C for 100 h. The results were compared to similar experiments for NiCrAl and NiAl.

2. Experimental Procedure

2.1 Sample preparation

The arc-melted NiAl (88 ppm S), NiCrAl (13 ppm S), and vacuum induction melted polycrystalline PWA 1480 (11 ppm S) samples were fabricated into 1 cm diameter coupons 0.3–2 mm thick. The bulk composition of each sample is shown in Table 1. A second PWA 1480 sample (PWA 1480/H) was H_2 annealed for 100 h at 1200 C in 5% H/95% Ar forming gas. The bulk S concentration of PWA 1480 annealed under the same conditions in 100% H_2 was 0.3 wt ppm.[7] The surface of each sample was polished with 2400 SiC paper in water, rinsed with electronic grade methanol, and dried with nitrogen gas. The samples were mounted on a hot stage sample holder 1 cm ID × 3 mm deep.[11] The samples were inserted in the vacuum chamber within 30 min after polishing. No sputter cleaning was performed, allowing the native oxide formed in room temperature air to remain on the sample. The sample heater has a Mo body with a W filament potted in alumina and is itself capable of 1400°C. The samples were analysed at room temperature and at 650–1100°C in 50°C steps for hold times of 1–6 h.

2.2 Sample analysis

The sample surfaces were characterised using XPS in a VG Scientific ESCALAB MkII with a 5 channeltron hemispherical analyser and Mg kα X-rays. The analyser was operated in the constant analyser energy (CAE) mode with a pass energy of 100 eV for survey spectra and 20 eV for the individual elemental spectra. A spot size of 5 × 2 mm was used.

Survey spectra were taken at the beginning and end of each experiment. Individual spectra were obtained for each element (Table 1) at room temperature and at the end of each heating experiment. The sample surface temperature was measured using an optical pyrometer calibrated with a type K

	S	Ti	Co	W	Ta	C	O	Al	Cr	Ni
NiAl	(88) ppm nom.	–	–	–	–	–	–	50	–	50
NiCrAl	13 ppm	–	–	–	–	–	–	24	14.5	61.4
PWA 1480	11 ppm	1.63	6.12	1.08	3.32	–	–	9.36	12.72	65.76
PWA 1480/H	0.3–1 ppm	1.39	5.96	1.12	3.35	–	–	8.53	11.65	68.00

Table 1 Bulk composition of each alloy at%

thermocouple. During each time series at temperature the S 2p photoelectron peak was acquired in 4–12 min time intervals.

Spectra from individual elements were smoothed,[12] the peak areas measured and normalised, and the binding energies determined. The normalised peak areas for each element were compared at the end of each time series at temperature and converted to an atomic percent. At temperatures before and after the loss of the native oxide, angle resolved XPS was used with 20° takeoff angles.[13] This reduced the effective escape depth of photoelectrons by a factor of three, making them comparable to AES electron escape depths. The sulphur peak areas obtained from sulphur profiles alone were converted to an atomic percent using the alloy compositions obtained at the beginning and end of each time series at temperature. The atomic concentration obtained in this manner assumes a homogeneous elemental distribution within the analysed volume (5×2 mm, 2–3 nm deep). This is clearly not the case for sulphur concentrated at the metal–oxide interface, free metal surface, cavities and voids, or metal and oxide grain boundaries. However, the total concentration and chemical state of sulphur within the analysed volume can be measured and compared for different samples.

3. Results and Discussion

3.1 Sulphur segregation in NiCrAl, PWA 1480, and NiAl

Previous AES sulphur segregation studies were performed on sputter cleaned NiCrAl surfaces[1,4,8–10] with the oxide layer and all surface contamination removed. This was necessary because the AES electron escape depths were typically less than the native oxide thickness, preventing measurement of sulphur below or at the metal–oxide interface. Also, the model used for monolayer coverage of sulphur at a surface assumes that sulphur is present in a single overlayer at the free surface, allowing a determination of the degree of saturation at the free surface. In this study sulphur segregation was measured in the presence of a thin native oxide (1.5–2.5 nm).

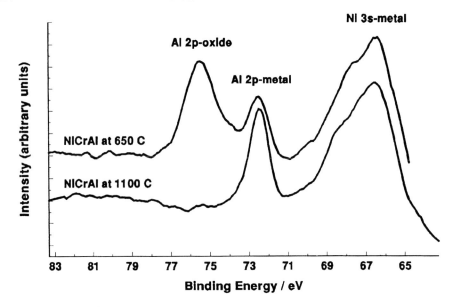

Fig. 1 Al 2p and Ni 3s spectra with and without the native oxide showing the difference between the Al oxide and Al metal.

The XPS depth of analysis is approximately 3 times that of AES allowing the metal–oxide interface to be seen through the native oxide. Figure 1 shows spectra obtained for Al 2p with both Al metal and Al oxide contributions. Since Al metal attenuated by the overlying oxide can be seen, the oxide metal interface is therefore detected. This allows the determination of differences in sulphur segregation behaviour when the surface free energy and near surface stoichiometry have not been altered due to preferential sputtering and when the native oxide and metal–oxide interface are concurrently observed. Preferential sputtering has been shown to modify the normally Al-rich surface of NiAl and Ni_3Al to 90% Ni.[14] All measurements were made here with the native oxide initially present.

The room temperature sulphur 2p spectrum is shown in Fig. 2. The chemical state of the room temperature sulphur (if present) was a sulphate. It was no longer present at 650°C due to evaporation of SO_x species.[2] From 650 to 1100°C the sulphur, found as a metal sulphide, experienced little change in chemistry.[10]

Sulphur segregation on the PWA 1480 alloy was first measured at 650, 800, 1000, and 1100°C for hold times of 4–6 h at temperature and is shown in Fig. 3. Sulphur, which was present at <1 at% at room temperature, segregated to about 1% at 650°C for 2 h, increased to a little over 1% after 5 h at 800°C. The native oxide was no longer seen at 1000°C and sulphur concentrations increased to >8% within 1 h and decreased to a steady state value of 5.5–6% over 2–6 h. At 1100°C sulphur concentrations decreased to 3% within 1 h and

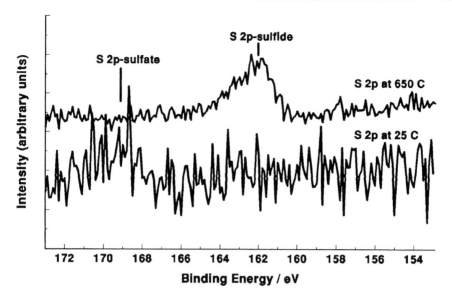

Fig. 2 Sulphur 2p spectra obtained at room temperature and at elevated temperatures. The room temperature sulphate was gone at 650°C. The metal sulfide found at higher temperature decreased in binding energy with increased temperature.

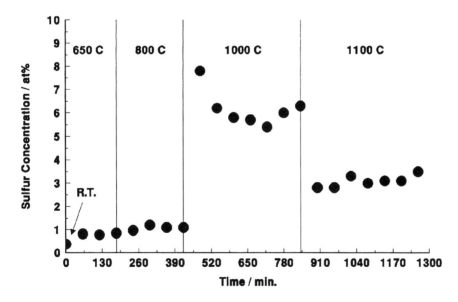

Fig. 3 The sulphur concentration of PWA 1480 vs time at temperature. After 1–2 hours at each temperature the sulphur concentration reached a steady state value. The native oxide was lost between 800–1000°C.

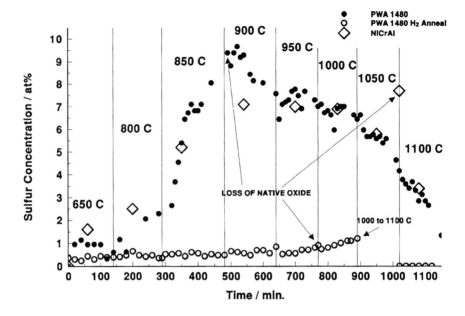

Fig. 4 A comparison of sulphur concentrations at temperature vs time at temperature for NiCrAl, PWA 1480, and H₂ annealed PWA 1480. The H₂ annealed PWA 1480 sulphur concentrations are significantly lower than the as-received PWA 1480.

remained at 3% for 2–6 h. Thus 2 h equilibration times at 50°C increments were used in subsequent experiments.

The main results of this study are shown in Fig. 4. Here sulphur concentrations vs time at temperature are shown for NiCrAl, PWA 1480, and hydrogen annealed PWA 1480 (PWA 1480/H).

3.1.2 NiCrAl sulphur segregation

The NiCrAl sulphur concentration was below XPS detection limits at room temperature. Sulphur segregation began at <650°C and reached concentrations of 1–2 at% at 650°C, 2.5–3 at% at 800°C, 5 at% at 850°C, 7 at% at 900°C, and 7–8 at% at 950°C. At 1000°C sulphur decreased to 6–7 at%. However, at 1050°C the native oxide was lost and the sulphur concentration increased to >8 at%. At 1100°C with no oxide sulphur decreased to 3 at%.

The NiCrAl sulphur concentrations peaked at a value of 8 at% immediately following the loss of the native oxide at 900–1000°C. From these results, a significant increase in sulphur concentration was found after the native oxide had disappeared, indicating, as others have noted,[10,15–17] that sulphur concentrates at the metal–oxide interface. It has previously been shown that sulphur can be present within oxide grain boundaries but not in the oxide lattice itself.[16]

3.1.2 PWA 1480 and PWA 1480/H sulphur segregation

As with NiCrAl, no sulphur was seen at room temperature for the PWA 1480, but <1 at% sulphur was seen for the PWA 1480/H. Sulphur segregation for the PWA 1480 was similar to the NiCrAl, although the sulphur segregation at 850°C rapidly increased to 9 at% concurrent with loss of the native oxide. At 900°C the PWA 1480 sulphur concentration reached a maximum of >10 at%, decreased to 6–7 at% at 1000°C, and reached 3–4 at% at 1100°C. In comparison, the PWA 1480/H, which began with <1 at% at room temperature, showed no sulphur segregation below 900°C. At 900°C sulphur concentrations increased to 0.5–1 at%. At 950°C the native oxide layer was lost and the sulphur concentration increased to 1.0 at%. The PWA 1480/H sulphur concentration reached a maximum of 1.2 at% at 1000°C and decreased to zero at 1100°C. The lack of segregated sulphur at 1100°C for PWA 1480/H indicates that the PWA 1480 and NiCrAl, with 3 at% sulphur at 1100°C, are continuously resupplied with sulphur from the bulk. Otherwise, the sulphur lost on the PWA 1480/H should also be lost for the NiCrAl and PWA 1480.

Angle resolved XPS (ARXPS) measurements were taken before and after the disappearance of the native oxide as shown in Table 4. ARXPS at 70° off normal electron takeoff angles effectively decreases the depth of analysis by a factor of three. The sulphur concentration increased by a factor of three to >24 at% at 70° off normal once the oxide was removed. This value can be compared to previous AES studies which claim saturation values at these temperatures of 15–38 at% on the sputter-cleaned free surface.[1,8,10] At temperatures greater than 900–1000°C, sulphur decreased either from increased bulk solubility or evaporation.[18,19] No evidence was found to conclude that the sputter cleaned surface significantly altered the temperature dependence for segregation. The sulphur decrease seen at higher temperatures is apparently suppressed due to the oxide overlayer. XPS and AES are not precise enough to compare critically absolute sulphur concentrations; however, the kinetic and temperature dependencies were similar.

The NiAl sample with nominally 88 atomic ppm sulphur intentionally added showed surprisingly low segregation concentrations of 1–2 at% at temperatures below 900°C and no sulphur at temperatures of 1100 and 1200°C. This may indicate that segregation in NiAl is much less pronounced than on NiCrAl alloys as suggested by others.[10] In any case, it is important to note that spallation is observed for this alloy with minimal segregation.

A comparison of the sulphur segregation behaviour of PWA 1480 and PWA 1480/H shows that clear differences exist. The PWA 1480/H alloy showed no segregation below 900°C and only reached a maximum sulphur concentration of 1/10, the value reached by unannealed PWA 1480. At 1100°C the PWA 1480/H had no sulphur at the surface. From these results a comparison is now made between sulphur segregation, bulk sulphur content, and cyclic oxidation behaviour.[7]

3.2 Sulphur segregation vs cyclic oxidation behaviour for PWA 1480

In an earlier study on the effect of sulphur on scale adhesion,[7] near zero specific weight changes were observed after 200 h at 1100°C for the PWA 1480

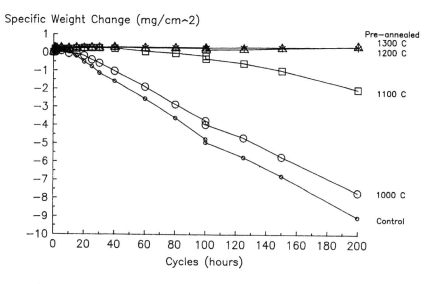

Fig. 5 The effect of H_2 annealing on cyclic oxidation. PWA 1480 polycrystalline.
(1100°C, 1-hr cycles).[7]

samples H_2 annealed at 1200°C and 1300°C (Fig. 5). Also, the cumulative spall areas during the progress of 1100°C cyclic oxidation tests were measured. Figure 6 shows that samples hydrogen annealed at 1000°C showed cumulative spall areas up to 120% while samples annealed at 1200 and 1300°C showed cumulative spall areas of only 1% or less. Thus, Figs 5 and 6 show that H_2 annealing produces a significant improvement in the cyclic oxidation behaviour of PWA 1480.

The improved cyclic oxidation behaviour for hydrogen annealed PWA 1480 has previously been associated with the reduction in the bulk sulphur content.[7] However, it has not been demonstrated whether the reduction of bulk sulphur would correspondingly reduce sulphur segregation. Figure 4 showed no sulphur segregation at temperatures of <950°C and significantly reduced segregation at higher temperatures. A summary of cyclic oxidation, bulk sulphur content, and sulphur segregation results are shown in Table 2. From these results it was concluded that cyclic oxidation behaviour is significantly improved by hydrogen annealing, which has now been shown to reduce both the bulk sulphur content as well as the interfacial segregation potential. Thus reducing the bulk sulphur content reduces the sulphur segregation at the oxide–metal interface, believed to weaken this bond.[20]

3.3 Chemical changes associated with sulphur segregation

In an attempt to correlate sulphur segregation with other changes in the substrate chemistry, all elements in Table 1 along with carbon and oxygen were measured at the end of each heating cycle. Tables 3–5 show the atomic

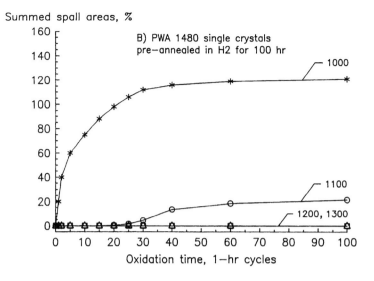

Fig. 6 Cumulative spall areas during 1100°C cyclic oxidation (visual estimate of exposed metal surface).[7]

Sample	Weight change mg/cm²	Cumulative spall area	Bulk sulphur content	Surface segregation concentration
PWA 1480 control	−8	120%	11 ppm	>10% XPS >30% 70° ARXPS
PWA 1480 H₂ annealed	+0.3	1%	0.3–1 ppm	1% XPS

Table 2 Sulphur and oxidation comparison for PWA 1480 and hydrogen annealed PWA 1480

concentration of these species for NiCrAl, PWA 1480, and PWA 1480/H. Angle resolved XPS (ARXPS) before and after removal of the oxide layer are also shown for the PWA 1480 sample.

3.3.1 NiCrAl composition and chemistry

Table 3 shows the change in composition after 1 h at each temperature. The room temperature for the as-polished surface was composed of Ni, Cr, and Al oxides. Carbon was also present at 30 at% composed of adventitious hydrocarbons and C–O species. At 650°C sulphur segregated to the near surface region and was present as a metal sulphide. The binding energy of sulphur 2p was consistent with Cr_2S_3 and NiS but could not be confirmed from the corresponding metal spectra. The binding energy of sulphur decreased from 650–1100°C consistent with a change from Cr_2S_3 to NiS while no chemical change was observed for the other elements in this temperature range. The oxide layer, composed of alumina at >650°C, disappeared between 1 and 2 h at 1050°C. Chromium increased between 800 and 850°C from 4 to 11 at% and then was depleted to 3 at% when the oxide disappeared. The Ni was low between 850 and 1050°C and increased with the decrease in Cr, producing a nearly stoichiometric NiAl phase with 3 at% Cr and S at 1100°C.

3.3.2 PWA 1480 and PWA 1480 H_2 annealed

The effects of heating on the overall PWA 1480 and hydrogen annealed PWA 1480 surface compositions are summarised in Tables 4 and 5, respectively. The room temperature Ni, Cr, and Al composition of PWA 1480 was similar to NiCrAl. The native oxide was lost between 800 and 850°C and was associated with a factor of two increase in Ti, Co, W, and Ta, while Cr remained the same. The Al decreased by a factor of two with the oxygen loss, while Ni, no longer being attenuated by the oxide overlayer, increased by a factor of five. Major compositional changes are shown for both PWA 1480 and hydrogen annealed PWA 1480 in Figs 7–9. These data were obtained after 2 h at temperature.

Figure 7 shows the oxygen and sulphur concentrations vs temperature. The hydrogen annealed alloy lost oxygen at 1000°C compared with 850°C for the unannealed alloy. The sulphur concentration reached a maximum value immediately following the loss of the oxide in both cases. The sulphur began to decrease at 900°C for PWA 1480 with no oxide but did not decrease on the hydrogen annealed alloy until the oxide was also gone. This indicates the possible trapping of sulphides at the metal–oxide interface.

The concentrations of Ni, Cr, Al, and O vs temperature for PWA 1480 are

Test conditions	S 2p	Ni 2p₃	Cr 2p₃	Al 2p	O 1s	C 1s
As polished room temperature	–	14	5.5	16	36	29
650°C 1 hr	1.6	9.5	3.5	42	37	6.5
800°C 1 hr	2.5	9.2	3.8	45	39	–
850°C 1 hr	5.2	14	11	42	28	–
900°C 1 hr	7.1	15.3	12	41	25	–
950°C 1 hr	7.0	16	13	41	23	–
1000°C 1 hr	6.9	17	12	41	23	–
overnight 1000°C/1 hr	6.2	18	13	40	22	–
1050°C/1 hr	5.8	27	8.7	38	20	–
1050°C/2 hr	7.7	40	8.0	45	–	–
1100°C/1 hr	3.4	49	2.8	45	–	–

Table 3 NiCrAl elemental concentration (at%) vs time at temperature

	S 2p	Ti 2p3	Co 2p3	W 4f	Ta 4f	C 1s	O 1s	Al 2p	Cr 2p3	Ni 2p3
R.T.	–	–	–	–	–	33	42	6.9	5.6	13
650°C	0.6	3.2	2.5	.9	3.5	3.2	39	31	5.7	11
800°C	2.1	2.1	3.2	1.7	3.8	2.7	31	33	11	9.3
800°C 70° off	3.4	1.7	2.7	1.8	2.8	3.7	36	33	7.4	7.2
850°C	7.9	4.1	6.8	2.3	5.7	–	–	18	11	45
850°C 70° off	22	6.1	4.8	3.3	4.0	–	–	13	15	32
900°C	7.8	4.2	5.4	2.8	5.8	–	–	18	8.2	48
900°C 70° off	24	6.2	4.3	3.0	3.8	–	–	15	14	30
950°C	7.3	3.1	4.7	2.6	6.8	–	-	19	4.0	53
1000°C	6.6	3.4	6.4	2.7	7.3	–	–	20	–	54
1050°C	4.7	4.6	3.8	2.6	7.8	–	–	21	–	56
1100°C	1.3	4.0	3.0	2.3	8.6	–	–	21	–	56

Table 4 Elemental concentration (at%) vs time at temperature for PWA1480

	S 2p	Ti 2p3	Co 2p3	W 4f	Ta 4f	C 1s	O 1s	Al 2p	Cr 2p3	Ni 2p3
R.T.	0.3	0.2	2.0	0.3	2.3	26	46	7.4	3.3	13
800°C	0.3	2.0	2.4	0.8	3.7	6.2	39	32	4.4	9.3
850°C	0.4	1.5	4.1	1.0	3.5	2.0	33	31	7.9	17
900°C	0.7	1.1	5.3	1.4	3.4	–	29	30	9.0	21
950°C	1.0	1.1	4.5	1.7	4.2	–	21	26	11	31
1000°C	1.2	3.5	7.0	1.8	7.3	–	2.7	19	2.3	55
1100°C	–	2.6	2.4	1.8	10.3	–	–	23	–	60

Table 5 PWA 1480 elemental concentration (at%) after H_2 anneal 100 hrs at 1200°C

shown in Fig. 8. From 850 to 1100°C Ti, W, and Al remained constant. Ta slowly increased with temperature. Cr and Co decreased with no Cr above 1000°C. Ni increased with the decrease in Cr with the Ni : Al ratio consistent with Ni_3Al at 1100°C. No chemical changes were observed once the native oxide was gone. The sulphur chemical state was similar to that found on NiCrAl. The PWA 1480/H shown in Fig. 9 showed no difference in chemistry from the PWA 1480 with the exception of the presence of sulphur as a sulphate at room temperature, the lack of sulphur segregation below 900–1000°C, and the native oxide stable up to 950–1000°C. No explanation is apparent for the increased temperature at which the oxide disappeared.

To summarise, the changes in composition are primarily affected by the presence of the native oxide.

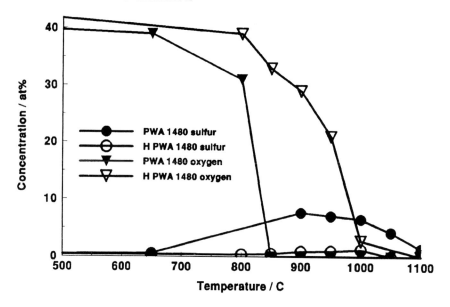

Fig. 7 Sulphur and oxygen concentration after two hours at each temperature for PWA 1480 before and after the H₂ anneal.

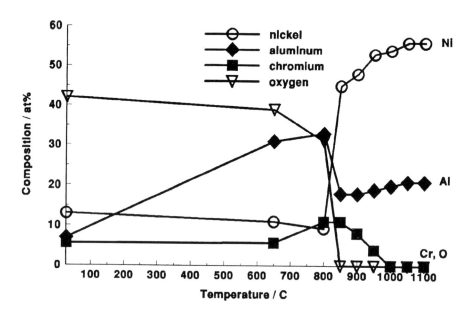

Fig. 8 Ni, Cr, Al, and O concentrations for PWA 1480.

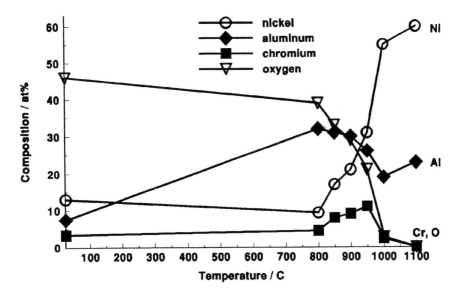

Fig. 9 Ni, Cr, Al, and O concentrations for PWA 1480 after hydrogen annealing at 1200°C for 100 hours. The composition was measured after two hours at each temperature.

4. Conclusion

X-ray photoelectron spectroscopy allows the chemistry and composition of the oxide–metal interfacial region to be determined. Sulphur segregation was present on all alloys studied. Total sulphur concentrations reached values of 8–10 at% from the normal exit data and 25–30 at% at grazing exit angles for the PWA 1480 and NiCrAl samples after loss of the native oxide. This is consistent with previous Auger data in which values of 30 at% were reported for NiCrAl and is consistent with the assumption that the Auger sampling depth is one-third that of XPS. Total sulphur concentrations on the PWA 1480/H initially began at 0.5 at% and never exceeded 1.2 at%. No sulphur was seen at 1100°C. The low concentration of segregated sulphur on the PWA 1480/H sample was correlated with reduced bulk sulphur content and improved cyclic oxidation behaviour.

References

1. J. L. Smialek and R. Browning, *High Temperature Materials Chemistry III* (eds Z. A. Munir and D. Cubicciotti), Electrochemical Society, Pennington, NJ, p. 258, 1986.
2. J. G. Smeggil and G. G. Peterson, *Oxid. of Met.*, **29** (1/2) 103, 1988.

3. A. W. Funkenbusch, J. G. Smeggil, and N. S. Bornstein, *Metall. Trans. A*, **17**, 1164, 1985.

4. H. J. Grabke, D. Wiemer, and H. Viefhaus, *Appl. Surf. Sci.*, **47**, 243, 1991.

5. J. L. Smialek, *Metall. Trans. A*, **18A**, 164, 1987.

6. J. L. Smialek, *Corrosion and Particle Erosion at High Temperatures* (eds V. Srinivasan and K. Vedula), TMS-AIME, Warrendale, PA, p. 425, 1989.

7. B. K. Tubbs and J. L. Smialek, *Corrosion and Particle Erosion at High Temperatures* (eds V. Srinivasan and K. Vedula), TMS-AIME, Warrendale, PA, p. 459, 1989.

8. C. G. H. Walker and M. M. El Gomati, *Appl. Surf. Sci.*, **35**, 164, 1988–1989.

9. C. L. Briant and K. L. Luthra, *Metall. Trans. A.*, **19A**, 2099, 1988.

10. J. G. Smeggil, A. W. Funkenbusch, and N. S. Bornstein, *Metall. Trans. A*, **17A**, 923, 1986.

11. Spectramat Inc., 100 Westgate Dr., Watsonville, CA, USA 95076.

12. M. P. Seah and W. A. Dench, *J. Elect. Spect.*, **48**, 43, 1989.

13. C. S. Fadley, *Prog. Surf. Sci.*, **16**, 275, 1984.

14. M. P. Thomas and B. Ralph, *Surf. Sci.*, **124**, 151, 1983.

15. P. Y. Hou and J. Stringer, *Oxid. of Metals*, **38**, 323, 1992.

16. P. Fox, D. G. Lees, and G. W. Lorimer, *Oxid. of Metals*, **36** (5/6), 491, 1991.

17. M. M. El Gomati, C. Walker, D. C. Peacock, and M. Prutton, *Corr. Sci.*, **25**, 351, 1985.

18. L. B. Mostefa, D. Roptin, and G. Saindrenan, *Mat. Sci. and Tech.*, **6**, 885, 1990.

19. T. Miyahara, K. Stolt, D. A. Reed, and H. K. Birnbaum, *Scripta Metall.*, **19**, 117, 1985.

20. A. B. Anderson, S. P. Mehandru, and J. L. Smialek, *J. Electrochem. Soc.*, **32**, 1695, 1985.

Effect of Sulphur in the Initial Stage of Oxidation of Ni–20Cr Single Crystals

R. DENNERT, H. J. GRABKE and H. VIEFHAUS

Max-Planck-Institut für Eisenforschung GmbH, 40237 Düsseldorf, Germany

ABSTRACT

Ni–20Cr single crystals were investigated having the low-index orientations (100), (110) and (111). The experiments were conducted using Auger electron spectroscopy (AES) and low energy electron diffraction (LEED). Upon heating the specimens an enrichment in chromium on the surface relative to the bulk concentration was observed. This enrichment was increased by cosegregation of chromium and sulphur. The surfaces with a saturated sulphur layer showed ordered LEED superstructures at a coverage of $\Theta \approx 0.5$. Heating the samples to temperatures of 500, 600 and 700°C, the oxidation was made at oxygen pressures between $1*10^{-6}$ and $1*10^{-7}$ mbar. Oxygen exposures up to 500 Langmuirs were offered to the surface. Adsorption of oxygen on clean surfaces induced an enrichment in chromium. On sulphur precovered surfaces the oxidation was retarded.

1. Introduction

In this study the alloy Ni–20Cr is used as a model system for high temperature alloys. This type of Ni-base alloys forms chromia scales for protection against corrosion. The aim of this work is to investigate the influence of non-metal elements such as C, N and S on the initial stages of oxidation. The results of the oxidation behaviour study of single crystal surfaces with and without sulphur are reported.

2. Experimental

A single crystal of Ni-20Cr (fcc-structure) was produced using the Bridgman crystal growth technique and oriented by Laue diffraction. The crystal was spark-cut perpendicular to the crystal directions (100), (110) and (111). Before inserting into the vacuum chamber, the samples were mechanically polished by diamond paste up to $1\,\mu$.

The base pressure in the UHV-chamber was about $2*10^{-10}$ mbar. We used an Auger electron spectroscopy (AES) system equipped with a cylindrical mirror analyser to estimate the chemical composition of the surfaces. A standard four grid LEED (low energy electron diffraction) optics was used to determine the state of order of the surfaces. The samples were cleaned under the vacuum by sputtering with argon ions while heating the samples at a temperature of about 600°C. The samples were heated indirectly using a tantalum coil. This construction enabled reaching temperatures up to 1000°C.

The samples were oxidised in the UHV-chamber at oxygen partial pressures of $1*10^{-6}$ or $1*10^{-7}$ mbar at temperatures of 500, 600 and 700°C, oxygen doses up to 500 Langmuirs (L) were dosed on the surface.

The Auger spectra were recorded before, during and after oxidation. The LEED patterns were obtained after cooling the samples to room temperature.

The surfaces were covered with sulphur by thermal segregation of S from the bulk at temperatures between 900 and 950°C. The bulk concentration of sulphur was about 5 ppm.

2.1 Clean and sulphur covered surfaces

After sputtering of the surfaces with argon ions and heating to 600°C only chromium and nickel were detected as shown in Fig. 1(a). From this Auger spectrum it can be seen that in relation to the peak-to-peak signal of Ni that of Cr was higher than would be expected, taking into account its bulk concentration of 20 wt%. Using sensitivity factors,[1] the ratio between Cr to Ni was calculated to be 1:2. Thus there was an enrichment in Cr at the surface. In Fig. 1(b) the spectrum of a sulphur-covered surface is shown, which indicates co-segregation of Cr and S. This causes an increase of the Cr/Ni ratio of the peak-to-peak signals (Fig. 1a and b). The shoulder to lower energies of the Cr 36 eV peak in Fig. 2(b) is another indication for the interaction of Cr and S.

The LEED patterns of the clean surfaces show well-ordered (1×1) structures; no reconstruction or relaxation was observed.

On the sulphur-covered (100) oriented surface a $c(2 \times 2)$ superstructure was observed, similar to that on the Ni (100) plane reported in the literature.[2] For this superstructure the coverage of S is $\Theta = 0.5$. On the (110) oriented surface a LEED pattern was observed, which cannot be described by a simple superstructure. The LEED pattern of the (111) oriented surface covered with S showed a (3×3) superstructure. This superstructure was not observed on the S-covered Ni (111) surface.

2.2 Oxidation of clean surfaces

In Fig. 2(a) the low energy range (20–80 eV) of the Auger spectra are shown; the first spectrum was recorded before, and the last one after oxidation. Other spectra were recorded successively during the oxidation. The $M_{23}M_{45}M_{45}$ Auger transition of chromium is sensitive to the chemical state of the

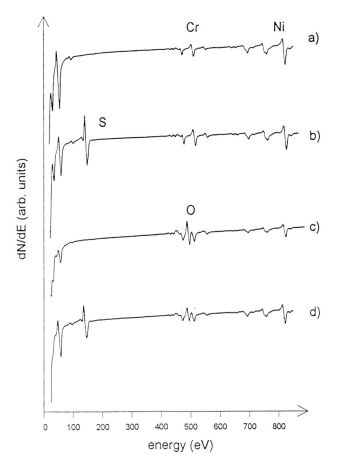

Fig. 1 Auger spectra of the (110) oriented surface at 600°C. (a) Clean surface; (b) sulphur-saturated surface; (c) oxidised surface, $100 \, L$, $p(O_2) = 1*10^{-6} \, mbar$; (d) sulphur saturated and oxidised under the same conditions as in (c).

chromium atoms. For metallic chromium the minimum of this peak in the dN/dE mode is at 36 eV. In the case of oxidised chromium, this peak shifts to 32 eV.[3] In the spectra of Fig. 2(a) this effect can be observed. After an exposure to the 10 L oxygen dose all chromium present on the surface was oxidised, the peak was shifted to 31 eV. The appearance of the Auger peaks at 44 and 47 eV is other evidence for oxidised chromium. With increasing exposure period a small shift of the peak at 31 eV to about 33 eV was observed. The $M_{23}M_{45}M_{45}$ Auger transition of nickel did not show any effects like shifting in energy or changing of the peak shape during oxidation. The only change could be seen in Fig. 2(a), i.e. a decrease of the signal height. An enrichment in chromium on the surface was induced by the oxidation of the surface. This effect can also be observed in Fig. 1 by comparison of the signal ratios of Cr [529] and Ni [848] of the spectra in Fig. 1(a) and (c).

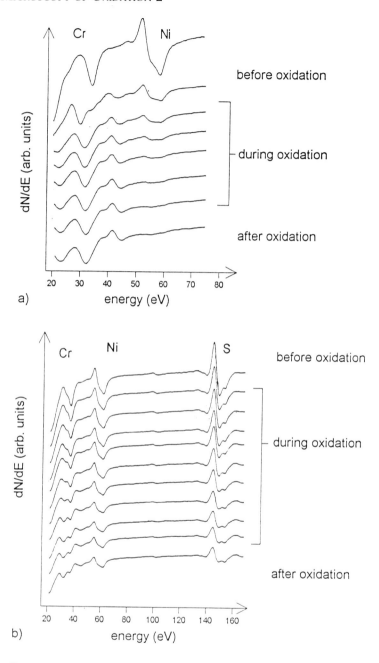

Fig. 2 Low energy range of Auger spectra, (a) (100) Oriented surface before, during and after the oxidation at 500°C, $p(O_2) = 1*10^{-7}$ mbar; (b) (110) oriented surface, sulphur precovered, before, during and after the oxidation at 500°C, $p(O_2) = 1*10^{-7}$ mbar.

A decrease of the signal ratio between O [512] and Ni [848] was observed with increasing temperature for the same oxygen exposure conditions. There is another indication for this effect by the change of the $M_{23}M_{45}M_{45}$ peak of chromium. There was a higher O_2 exposure necessary for a shift of the Auger peak from 36 to 33 eV. This can be explained by a decreasing reaction probability for O_2 with increasing temperature.

During the initial stages of oxidation (exposure ≤ 5 L), the LEED patterns of all three orientations showed weak incommensurate superstructure spots. At higher exposures on all three surfaces ordered superstructures were formed by oxidation of the surface. All these superstructures have a hexagonal symmetry. The LEED pattern of the (100) surface showed the hexagonal $\left(\begin{smallmatrix}2 & 0 \\ 1 & \sqrt{3}\end{smallmatrix}\right)$ superstructure, which consisted of two domains twisted by 90° to each other. This superstructure was also observed on the Ni (100) plane[4] at oxygen coverage above $\Theta = 0.5$. In Fig. 3(a) the LEED pattern of the (111) oriented surface is shown. For better understanding a scheme of this pattern is drawn in Fig. 3(b). The intensity of the supplementary spots (open circles) near the original spots (full circles) shows a different dependence of the electron energy than the supplementary spots rotated 30°C to the direction of the original spots. This superstructure can be interpreted as a superposition of two adlayers. The ratio between the values of the reciprocal vector $\mathbf{b_1^*}$ of the supplementary spots in the direction of the original spots and the reciprocal vector $\mathbf{a_1^*}$ of the original spots is about 7 to 8. Calculation of the distances between the two nearest neighbour atoms in the alloy ($d_{alloy\text{-}alloy} = 2.51$ Å) and oxygen atoms ($d_{O\text{-}O} = 2.86$ Å) from the lattice constants of Ni–20Cr (3.55 Å) and Cr_2O_3 (4.94 Å), result in the ratio ($d_{alloy\text{-}alloy}/d_{O\text{-}O} \approx 7/8$. The most probable explanation is that the oxygen atoms were adsorbed on the Ni-20Cr (111) surface at the same nearest neighbour distances as in Cr_2O_3. The other supplementary spots in Fig. 3(b) are located on the $\sqrt{3}$ sites relative to the supplementary spots near the original spots, described by $\mathbf{b_1^*}$. The structure of Cr_2O_3 is the hcp structure with the oxygen atoms on the lattice sites and the Cr atoms are located on 2/3 of the octahedral interstitial positions in the oxygen lattice. In Fig. 3(c) the first layer of oxygen and chromium of Cr_2O_3 perpendicular to the (0001) direction are shown; the Cr atoms form a $(\sqrt{3} \times \sqrt{3})R30°$ lattice with two atoms in the basis of the unit mesh on the oxygen layer. The interpretation derived from the LEED pattern shown in Fig. 3(a) were two adlayers on the alloy surface, first a monolayer of oxygen and on this a Cr layer as illustrated in Fig. 3(c).

2.3 Oxidation of surfaces precovered with sulphur

In Fig. 2(b) the Auger spectra are shown of the low energy range (20–170 eV) before, during and after the oxidation of a sulphur precovered surface. By increasing the oxygen exposure period, the low energy Cr peak at 36 eV was shifted to 32 eV, but the oxidation was retarded in comparison to that on the surfaces without sulphur. During oxidation only a small decrease of the Ni 61 eV peak signal height was observed. The signal height of the sulphur peak at 152 eV decreased during the oxidation, but after an exposure of about 45 L the amount of sulphur remained constant as observed for different temperatures in

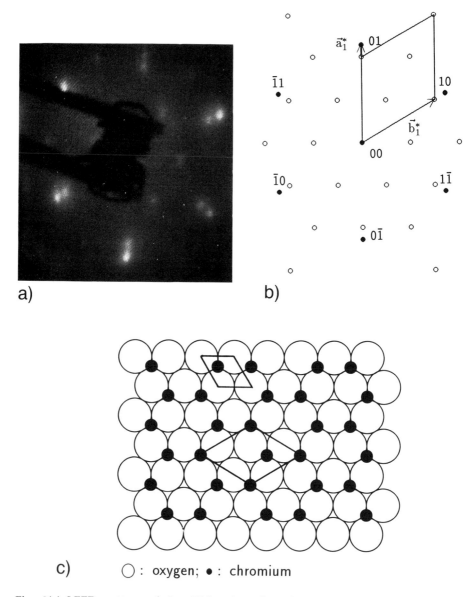

c)   ○ : oxygen; ● : chromium

Fig. 3(a) LEED pattern of the (111) oriented surface after oxidation with an exposure of 100 L at 600°C, E = 102 eV; (b) scheme of this LEED pattern; (c) schematic representation of Cr_2O_3 perpendicular to the (0001) direction.

Fig. 4. This can be explained by an equilibrium of sulphur segregation from the bulk and desorption of volatile sulphur-oxide species (SO_2). Sulphur retarded the oxidation of the surface as indicated by a comparison of Fig. 1(c) and 1(d). In Fig. 5 this effect is shown by plotting the ratio of O_{512} to Ni_{848} as a function of the oxygen exposure for a clean surface, a sulphur-covered surface (no

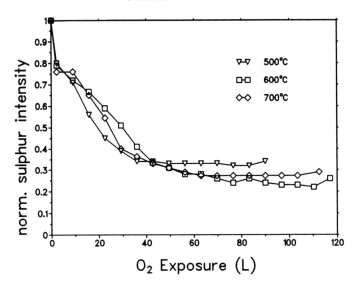

Fig. 4 Sulphur signal height during the oxidation related to the sulphur signal before the oxidation, $p(O_2) = 1*10^{-7}$ mbar.

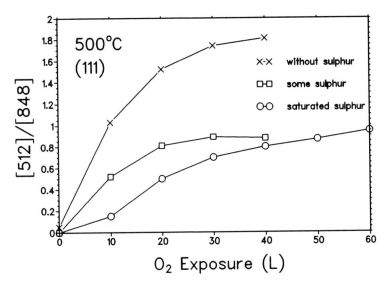

Fig. 5 Ratio between oxygen [512] and nickel [848] Auger signals as a function of oxygen exposure, $p(O_2) = 1*10^{-6}$ mbar.

superstructure is observed) and an initially saturated sulphur-covered surface (S(3 × 3) superstructure). For increasing temperatures the same effect of a decreasing reaction probability of oxygen is observed as on surfaces without the sulphur being precovered.

After adsorption of oxygen on a sulphur-precovered surface no ordered LEED structure of oxide was observed, but at low oxygen exposures the LEED pattern of segregated sulphur was still visible. During the oxidation of the surface the diffuse scattering increased, but the long-range order of the sulphur atoms was not destroyed. Recording sputter profiles indicated that initially the sulphur signal vanished and later on the signal of the oxygen disappeared. These observations lead to the following model: O_2 cannot be adsorbed on the sulphur layer, but only in the vacancies or defects of the sulphur layer. The diffusion of the oxygen atoms is hindered by the sulphur, so the oxygen atoms cannot order. In this way the disordered oxygen can only disturb the order of the sulphur atoms on the surface.

3. Conclusions

No significant differences in oxidation behaviour were observed for the Ni–20Cr single crystals having the following orientations: (100), (110) and (111). However, the LEED patterns indicated differences in the structure. The reaction probability of oxygen decreased with increasing temperature for surfaces with and without sulphur-precovered surfaces. On the (111) oriented surface an initial stage of growth of Cr_2O_3 along the (0001) direction could be deduced from the LEED pattern. The oxidation of a sulphur precovered surface was retarded compared to a clean surface because of a blocking of adsorption sites for the oxygen by sulphur. The long-range order of the sulphur atoms was disturbed by oxygen adsorption, but not destroyed.

References

1. P. W. Palmberg, G. E. Riach, R. E. Weber and N. C. MacDonald. *Handbook of Auger Electron Spectroscopy*, Physical Electronics Industries, 2nd edition 1976.
2. W. Daum, *Surface Science*, **182**, 521–529, 1987.
3. C. Palacio, H. J. Mathieu and D. Landolt, *Surface Science*, **182**, 41–55, 1987.
4. G. Dalmei-Imelik, J. C. Bertolini and J. Rousseau, *Surface Science*, **65**, 67–78, 1977.

An Investigation of the Effect of Sulphur on the Adhesion and Growth-Mechanism of Chromia Formed on Low-Sulphur Chromium

W. DONG*, D. G. LEES*, G. W. LORIMER*
and D. JOHNSON†

*Manchester Materials Science Centre, University of Manchester and UMIST,
Manchester, M1 7HS, UK
†Centre for Surface and Materials Analysis Ltd., Armstrong House, Oxford Road,
Manchester, M1 7ED, UK

ABSTRACT

The oxidation behaviour of low sulphur (<5 ppm) chromium in 0.1 atm oxygen at 900°C has been investigated by means of TEM and imaging SIMS on cross-sectioned specimens. Most of the scale was very adherent even after oxidation for 282 h, and it was possible to prepare cross-sectioned TEM specimens which included adherent oxide and the oxide/metal interface. The microstructure of the scale varied continually from large columnar grains near the oxide/gas interface to small equiaxed grains near the oxide/metal interface. No sulphur was detected at the oxide/metal interface or at the oxide grain boundaries. SIMS linescans on a specimen oxidised sequentially in natural oxygen and oxygen enriched in oxygen-18 showed that the scale had grown at least primarily by oxygen diffusion. The results support the sulphur effect theory which states that sulphur has a deleterious effect upon the adhesion of chromia scales and promotes cation transport, and that reduction in the sulphur content improves adhesion and promotes oxygen transport.

1. Introduction

The high-temperature oxidation behaviour of chromium has been studied for many years. Lees[1] suggested that sulphur promotes cation transport in the scale by segregating to oxide grain boundaries and its removal changes the growth mechanism to one in which oxygen transport predominates. Combined with the suggestion that poor scale adhesion is caused by presence of sulphur

205

at the oxide/metal interface, proposed independently by Ikeda et al.[2] and Funkenbusch et al.,[3] the theory is called the *sulphur-effect*. This theory also implies that the beneficial effect of *reactive elements* on the oxide scale adhesion is due to these elements reacting with sulphur, thus preventing its segregation to the oxide/metal interface and the oxide grain boundaries.[1]

Previous work by Fox et al.[4] on the oxidation of normal *pure* chromium (Goodfellow, 150 ppm sulphur content) showed that the sulphur was present at the Cr_2O_3 scale/Cr interface and the oxide grain boundaries but for the low-sulphur Ducropur chromium no sulphur was found using the same technique.[7] The presence of sulphur at the oxide/metal interface of both Al_2O_3 and Cr_2O_3-forming alloys was also reported by Hou and Stringer.[5] The TEM observations by Fox[6] of cross-sectioned specimens of the low-sulphur Ducropur chromium indicated that the oxide grain structure varied from large columnar near the scale/gas interface to small equiaxed at the oxide/metal interface. For the Goodfellow chromium[6] the grains had similar, equiaxed shape over the entire section.

The growth mechanism of the oxide scale on chromium is believed to be dominated by short-circuit transportation,[8] either Cr chromium cation outward or oxygen anion inward. Oxygen isotope tracer investigations of the normal sulphur content Goodfellow chromium showed that outward short-circuit diffusion of chromium cations predominated.[9]

The present research is entirely focused on further investigation of the high-temperature oxidation of the low-sulphur content Ducropur chromium.

2. Experimental

The chromium used in the present investigation was obtained from Metallwerk Plansee with the trade name Ducropur. The impurities (in ppm wt) were C 15, N 6, O 9 F <0.02, Na 0.03, Mg <0.004, Al 45, Si 14, P 0.2, S 1, Cl 0.08, K <0.05, Ca 10, Sc 0.01, Ti 37, V 14, Mn 0.2 Fe 46, Co 8, Ni 6, Cu 0.09, Zn 0.4 As <0.01, Sn 0.08, Sb 0.3,Pb <0.02.[7] As-received Ducropur coupons approximately $7 \times 10 \times 0.9$ mm were mechanically ground to 1200 SiC and then polished with $3 \mu m$ alumina. The surface of the coupons was cleaned ultrasonically in ethanol and acetone before being oxidised. The oxidation was carried out isothermally in manometric apparatus which was developed at Manchester.[9] The apparatus consists of two quartz glass tubes, both with one end inserted into a furnace and the other end connected to the gas supply and evacuation system. One tube is for accommodating the specimen and the other serves as a pressure reference. After the tubes were filled with oxygen to the desired pressure, they were isolated and a computer-controlled servo system moved the specimen from the cold zone to the hot zone. The pressure difference between specimen and reference tubes was continuously monitored and converted into the oxide weight gain. The oxidation conditions used in the present investigation were 900°C, 0.1 atm pure oxygen, and the oxidation times ranged from 0.5 to 282 h. The surface morphology of the oxidised coupons was examined using a Philips

525 SEM equipped with EDAX energy-dispersive X-ray analyser before the specimens were cut into small pieces for examination with TEM.

Cross-sectioned specimens of the oxide/metal interface for TEM examination were made by a slightly modified method developed by Fox.[6] Firstly, the specimens were mechanically ground to a thickness of about 30 μm, then ion beam thinned to electron transparency. A parallel section through the scale was made by first jet electrode polishing the sample to remove the chromium metal then Ar ion beam thinning the remaining oxide. Transmission electron microscopy was carried out with either Philips EM430 or EM400 analytical electron microscopes operating at 300 KV and 120 KV, respectively. Energy dispersive X-ray (EDX) microanalysis was carried out with an ISI/ABT EM002B analytical electron microscope operating in the 2.5 nm probe mode.

A specimen was oxidised at 900°C in natural oxygen for about 2.5 h and for a further 6.5 h in gas enriched in $^{18}O_2$; the pressure in both cases was 0.1 atm. The specimen was cross-sectioned and the oxygen-16 and oxygen-18 distributions in the scale were determined by means of imaging secondary ion mass spectroscopy (SIMS) at the Centre for Surface and Materials Analysis Ltd. The equipment was manufactured by VG Ionex Ltd. A 30 KeV, 0.5 nA beam of Ga^+ ions with a minimum diameter of less than 0.5 μm was used.

3. Results

The chromium oxidation kinetics (Fig. 1) show a slight deviation from the parabolic rate law with $k_p = 3 \times 10^{-12}$ g^2/cm^4/s. Scanning electron microscopy of the scale showed it was mainly flat and adherent but with buckled areas near the edge of the specimen, possibly caused by surface contamination (see Fig. 2).

The X-ray diffraction (XRD) pattern from the surface of a specimen oxidised for 100 h contained only peaks corresponding to Cr_2O_3 (Fig. 3). There was a significant difference in the intensities of the peaks from the sample and the JCPDS standard. This deviation in intensities (Fig. 3) indicated a texture in the oxide layer; however, significant deviation from random texture was not detected in pole figures (Fig. 4).

TEM micrographs of the scale (Fig. 5a) from the section parallel to the oxide/metal interface showed that the oxide grains were equiaxed with a diameter of about 0.3 μm. Cross-sections of the scale from the specimens oxidised for 1 h (Fig. 5b,c) and 80 h (Fig. 5d) showed similar oxide grain microstructures. The outer part of the scale consisted of relatively large columnar grains (about 0.3×0.8 μm) with the long axis normal to the oxide/metal interface, and small equiaxed grains, <0.2 μm in diameter, next to the oxide/metal interface. No clear boundary could be distinguished between the two kinds of grains.

As expected, only chromium was found in both parallel and cross-sectioned specimens using normal EDX microanalysis techniques on the TEM. The 2.5 nm probe windowless EDX microanalysis of the oxide/metal interface and

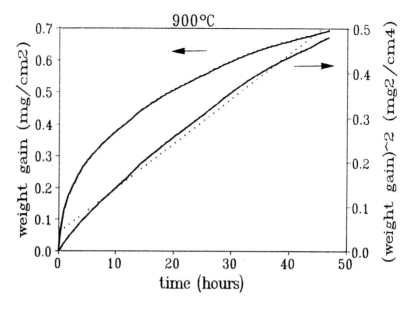

Fig. 1 Oxidation kinetics of Ducropur chromium, oxidised at 900°C, 0.1 atm O_2; weight gain vs. time of oxidation, and square of the weight gain vs. time.

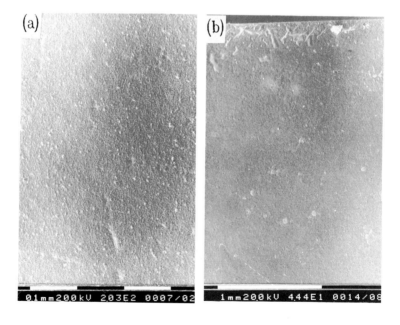

Fig. 2 SEM of surface morphology of Ducropur chromium oxidised at 900°C, 0.1 atm O_2, 282 hours. (a) The flat and adherent region; (b) buckles near the edge of the coupon.

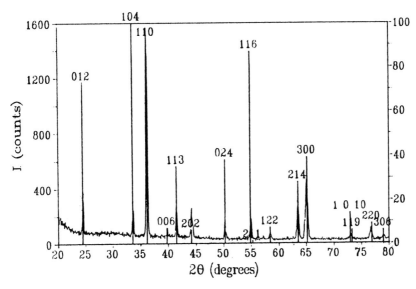

Fig. 3 XRD of the surface of a Ducropur chromium coupon, oxidised at 900°C, 0.1 atm O_2, 100 hours, CuKα radiation, showed the deviation of the intensities of reflections between the specimen and the standard. Standard data are from JCPDS file No. 38-1749.

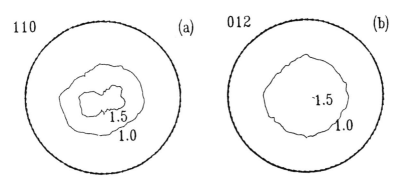

Fig. 4 X-ray pole figures of the surface of a Ducropur chromium coupon, oxidised at 900°C, 0.1 atm O_2, 100 h, 110(a) and 012(b) reflections; CoKα radiation. No significant deviation from random texture was detected.

the oxide grain boundaries did not reveal any evidence of the existence of sulphur.

The distribution of oxygen-16 and oxygen-18 in the cross-sectioned scale of a sequentially oxidised specimen was determined by SIMS line scans and mapping in three areas, each about 20 μm square. The line scans have two types of ^{18}O profiles, typical examples of which are shown in Fig. 6(a,b); the average of the seven lines in one area is shown in Fig. 6(c).

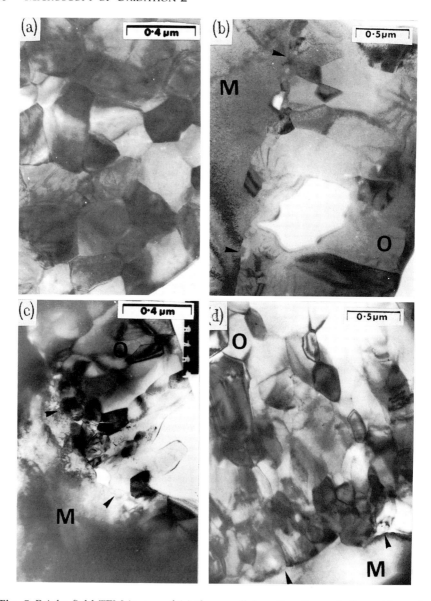

Fig. 5 Bright field TEM image of (a) the parallel section through the oxide scale; (b), (c) and (d) the cross-section through scale and the oxide/metal interface. Ducropur chromium specimens oxidised at 900°C, 0.1 atm O_2, for 1 h (a, b and c) and 80 h (d). M, chromium metal; O, oxide. The oxide/metal interfaces are indicated by arrows.

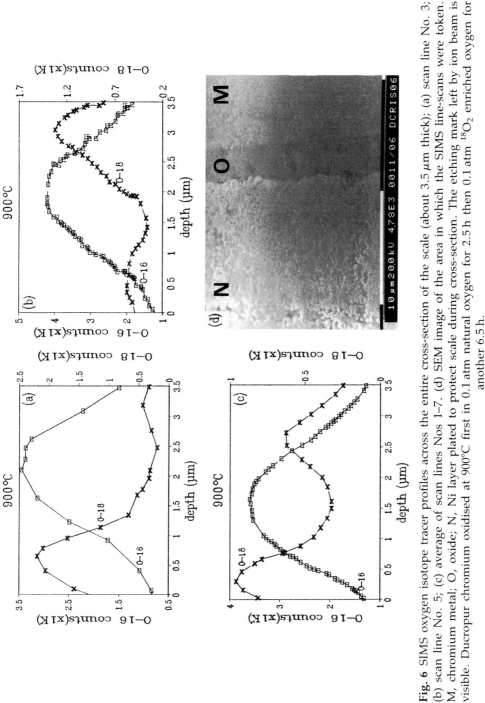

Fig. 6 SIMS oxygen isotope tracer profiles across the entire cross-section of the scale (about 3.5 μm thick); (a) scan line No. 3; (b) scan line No. 5; (c) average of scan lines Nos 1–7. (d) SEM image of the area in which the SIMS line-scans were token. M, chromium metal; O, oxide; N, Ni layer plated to protect scale during cross-section. The etching mark left by ion beam is visible. Ducropur chromium oxidised at 900°C first in 0.1 atm natural oxygen for 2.5 h then 0.1 atm $^{18}O_2$ enriched oxygen for another 6.5 h.

4. Discussion

4.1 Oxygen Isotope Tracer

The distribution profiles of ^{18}O in the cross-sectioned specimen sequentially oxidised revealed that the ^{18}O appeared at both of the oxide/gas and the oxide/metal interfaces. Compared with the models proposed by Basu and Halloran,[11] the present averaged ^{18}O distribution profile was very similar to that produced predominantly by oxygen inward short circuit (such as grain boundaries) transport plus some isotope exchange. According to their calculation based on this model, the changes in the lattice and grain boundary diffusion coefficient and grain size will alter the distribution of oxygen tracer inside the oxide scale. The technique used to trace the oxygen isotopes has superior spatial resolution to a sputtering technique, and no obvious micro-cracks were found by examining the same areas by SEM (Fig. 6d). It is proposed that the local variation of the ^{18}O distribution profiles reveals a true non-uniformity in the rate of the inward diffusion of oxygen.

4.2 Grain Structure

Compared with the oxidation of the relatively high sulphur Goodfellow chromium,[9] the oxidation of Ducropur chromium was slow and the oxide scale was flat in most areas with good adherence after 282 h oxidation at 900°C (see Fig. 2a). It is worth noting that this result was achieved without the presence of any reactive elements.

   In the case of Goodfellow chromium (in which the new oxide is believed to be formed first at the oxide/gas interface and then inside the scale by predominant chromium cation transportation[9]) the oxide grain size was similar throughout the oxide scale.[10] In the present investigation of as-received Ducropur chromium there was an uneven distribution of grain size with large columnar oxide grains at the oxide/gas interface and fine equiaxed grains near the oxide/metal interface. This was also observed in the hydrogen-annealed Ducropur chromium.[10]

   According to Rapp and Pieraggi[12] the chromia formed at the metal/scale interface is expected to be fine-grained and tightly adherent due to inward oxygen diffusion. This is in agreement with the SEM of scale morphology and TEM observations of the scale microstructure in the present investigation. However, the grain size in the layer next to the metal/scale interface was different at different parts of the same specimen (see Fig. 5b,c). It is believed that this is the result of the non-uniform growth or recrystallisation of oxide grains. The Pilling–Bedworth ratio is 2.07[13] so that the growing oxide scale will be under severe compressive stress which will be a maximum in the plane of the specimen and a minimum perpendicular to the surface. The stress-assisted growth or recrystallisation of the oxide grains will tend to occur preferentially perpendicular to the specimen surface, and this will, eventually, result in the formation of columnar grains, even though growth is not occurring at the oxide/gas interface. The transition from equiaxed grains at the metal/scale interface to columnar grains at the oxide/gas interface is not abrupt but there is a transition through the thickness of the scale; this observation supports the proposal of a change from equiaxed to columnar grains during scale growth.

5. Conclusion

Most of the scale on this low-sulphur chromium was adherent even after oxidation for 282 h at 900°C. No sulphur was detected either at the oxide/metal interface or at the oxide grain boundaries. The scale grew either entirely or primarily by oxygen transport, and the microstructure of the scale was consistent with this. The results support the predictions of the sulphur effect theory for scale adhesion and growth-mechanism.

Acknowledgments

This work was supported by a grant from SERC and the authors thank Metallwerk Plansee for the provision of the high-purity chromium, and the Electron Microscope Centre of Glasgow University for the use of the ISI/ABT EM002B electron microscope.

References

1. D. G. Lees, *Oxid. Met.*, **27**, 75, 1987.
2. Y. Ikeda, K. Nii and K. Yoshihara, *Trans. Japan Inst. Met. Suppl.*, **24**, 207, 1983.
3. A. W. Funkenbusch, J. G. Smeggil and N. S. Bornstein, *Metall. Trans.*, **16A**, 1164, 1985.
4. P. Fox, D. G. Lees and G. W. Lorimer, *Oxid. Met.*, **36**, 491, 1991.
5. P. Y. Hou and J. Stringer, *Oxid. Met.*, **38**, 323, 1993.
6. P. Fox, *Microscopy Research and Technique*, **21**, 369, 1992.
7. G. W. Lorimer and D. G. Lees, *SERC report*, 1991.
8. A. Atkinson and R. I. Taylor, *AERE-R 11314*, June 1984.
9. M. Skeldon, J. M. Calvert and D. G. Lees, *Oxid. Met.*, **28**, 109, 1987.
10. P. Fox, D. G. Lees and G. W. Lorimer, *Microscopy of Oxidation* (Ed. M. J. Bennet and G. W. Lorimer), The Institute of Metals, 1991.
11. S. N. Basu and J. W. Halloran, *Oxid. Met.*, **27**, 143, 1987.
12. R. A. Rapp and B. Pieraggi, private communication, 1992.
13. Per Kofstad, *High Temperature Corrosion*, Elsevier Applied Science, 1988.

Contribution of the Water Vapour-Induced Effects to SIMS-Spectra from Scales Growing on Chromium During Isotopic Exposures

A. GIL,* J. JEDLIŃSKI,[†1] J. SŁOWIK,[†]
G. BORCHARDT[†] and S. MROWEC*

*Faculty of Materials Science and Ceramics, Academy of Mining and Metallurgy,
al. Mickiewicza 30, PL-30-059 Kraków, Poland
[†]Institut für Allgemeine Metallurgie und SFB 180, Technische Universitat
Clausthal, Robert-Koch-Str. 42, D-3392 Clausthal-Zellerfeld, Germany
[1]Now at the Faculty of Materials Science and Ceramics, Academy of Mining and
Metallurgy, al. Mickiewicza 30, PL-30-059 Kraków, Poland

ABSTRACT

Monitoring of distribution of hydroxide ions in addition to those related to oxygen isotopes is shown to be helpful in interpreting the tracer experiments results for scales formed on pure chromium. The simple procedure is proposed which allows the corrected SIMS profiles of both oxygen isotopes, ^{16}O and ^{18}O, to be obtained from raw data.

1. Introduction

Two-stage oxidation and marker methods have been widely used in studying corrosion mechanisms, providing information on matter transport in the growing scale layer. In the case of very thin scales, thickness of few micrometres or less, the determination of oxygen isotopes distribution or inert marker location calls for more sophisticated analytical techniques, such as SIMS, SNMS, RBS or AES, the most frequently used being SIMS. It has been found lately that the marker method in certain cases may give misleading results.[1,2] Therefore it is believed that the two-stage oxidation will remain the most reliable method for studying the transport processes in thin scale layers. The principles of the tracer/SIMS(SNMS) approach as well as the main problems related to its application were recently discussed in more detail elsewhere.[3–5] The conventional version of this approach gives reliable information in those cases only where homogeneous scales of uniform thickness are formed, without any physical defects, out- or ingrowths. It does not seem correct to use this method for multiphase materials, in particular when the scale is formed by selective oxidation of one element. The additional difficulties are related to a number of possible interfering phenomena such as initial instabilities of ion beam, increased number of counts due to sputtering of clusters formed from water molecules adsorbed on scale surface, or different etching rates for scale and alloy.

In the present paper, chromia scale formed on pure chromium substrate in a two-stage oxidation experiment has been taken as an example to illustrate that additional information on the concentration profiles of ^{16}O and ^{18}O in the scale can be obtained by simultaneous monitoring of spectra for ions having a mass-to-charge ratio of 17 and 19, corresponding to hydroxide ions.

2. Materials and Experimental

The two-stage oxidation experiments have been carried out using a spectrally pure chromium sample (GoodFellow) at 900°C for 5 min in $^{16}O_2$ and, subsequently, for 15 min in $^{18}O_2$. No cooling of the samples occurred between both the oxidation stages. The scale formed in these conditions fulfilled most of the necessary requirements. It was homogeneous, flat, without cracks or other discontinuities. The SIMS analysis has been carried out with the aid of Cameca SMI 300 spectrometer using a neutral argon primary beam (NPB-SIMS). The principles of this method are described elsewhere.[6] The oxygen- and water vapour-related ions have mass-to-charge (denoted m/z in the following text) ratios of 16, 17, 18 and 19 were analysed.

3. Results and Discussion

The row profiles obtained for m/z=16 (corresponding to $^{16}O_2$) and m/z=18 (corresponding to $^{18}O_2$) are shown in Fig. 1. Except for the increased ^{16}O-intensity in the outermost part of the scale, such profiles are consistent with previous results of other authors who observed predominant outward metal diffusion.[7–9] Some authors reported also a very limited contribution (about 1%) of the oxygen inward diffusion.[7] The presence of $^{16}O_2$ peak at the external surface of the scale would be in agreement with this finding. However, as shown in Fig. 2, the total oxygen SIMS intensity, being a sum of both these oxygen isotopes, distinctly increased at this region, contrary to the expected constant intensity across the whole scale.

Figure 3 shows the profiles for ions having m/z=16 and 17. The latter corresponds to the hydroxide ion $^{16}OH^-$. A similar relationship between the shapes of profiles was obtained for m/z=18 and 19 (not shown here). In Fig. 4, in turn, the total intensity of the water vapour-related hydroxide ions (m/z=17 and 19) is shown. Comparison of profiles in Figs 1, 2 and 3 indicates that the increased oxygen SIMS-intensity observed at the outermost part of the scale

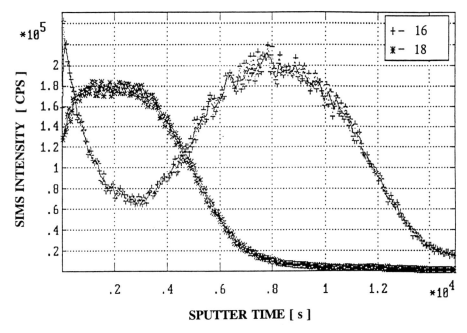

Fig. 1 Row SIMS profiles of mass-to-charge ratios 16 and 18 attributed to oxygen isotopes ^{16}O and ^{18}O in scales formed on chromium at 900°C (5 min in $^{16}O_2$ and 15 min in $^{18}O_2$).

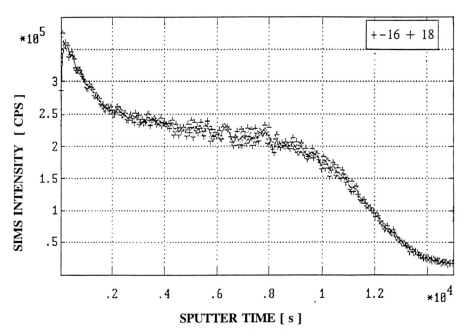

Fig. 2 Profile being a sum of the row profiles of both oxygen isotopes shown in Fig. 1, attributed to the total oxygen.

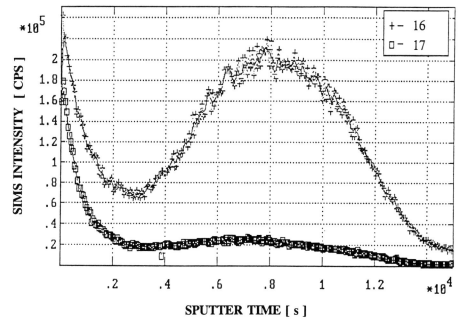

Fig. 3 Row SIMS profiles of mass-to-charge ratios 16 and 17 attributed to oxygen isotope ^{16}O and hydroxide ion $^{17}OH^-$ in scales formed on chromium at 900°C (5 min in $^{16}O_2$ and 15 min in $^{18}O_2$).

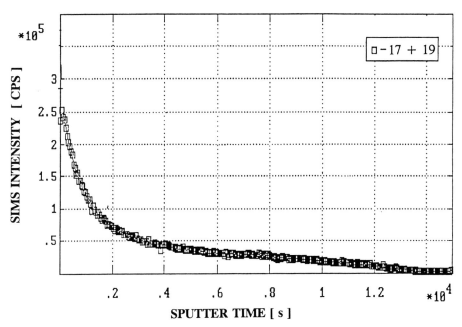

Fig. 4 SIMS profile corresponding to sum of the mass-to-charge ratio 17 and 19 related to hydroxide ions ($^{16}OH)^-$ and ($^{18}OH)^-$ measured in scales formed on chromium at 900°C (5 min in $^{16}O_2$ and 15 min in $^{18}O_2$).

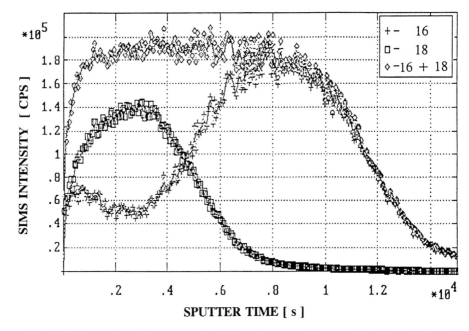

Fig. 5 SIMS profiles of oxygen-related ions (mass-to-charge ratios 16 and 18) and their sum in scales formed on chromium at 900°C (5 min in $^{16}O_2$ and 15 min in $^{18}O_2$) after correction for water vapour-related hydroxide ions.

can be mainly related to the presence of water molecules on the scale surface. However, no additional analyses have been performed in order to confirm this hypothesis.

Assuming that the emission of oxide and hydroxide ions resulting from decomposition of the water molecule is equally probable, it is possible to make corrections for the profiles obtained for oxygen isotopes (m/z=16 and 18) by subtraction of those for hydroxide ions (m/z=17 and 19). The corrected profiles of both oxygen isotopes and their sum are presented in Fig. 5. As can be seen, the corrected profile of the $^{16}O^-$ exhibits only a small increase of SIMS-intensity at the outermost part of the scale. Moreover, in contrast to the row profiles, the intensity of the total oxygen signal remains practically constant across the whole scale, as expected. Thus, the correction procedure results in a more reliable picture of the ions 16 and 18 distribution in the scale compared to uncorrected profiles (Figs 1 and 2).

4. Conclusions

An example is given which indicates that for any conclusions inferred concerning the scale transport properties using the tracer/SIMS approach it is not sufficient to determine the in-depth profiles of ^{16}O and ^{18}O isotopes only. It is shown that simultaneous analysis of the distribution of water vapour-related ions, having mass-to-charge ratios 17 and 19, may be helpful in estimating the real distributions of both oxygen isotopes in the scale. Furthermore, as pointed out elsewhere,[3,7,10], the profiles of oxide-related polyatomic species, in this case of $[Cr(^{16}O)$ and $[Cr(^{18}O)]^-$, should be monitored and correlated with those of oxygen isotopes.

Acknowledgments

Financial support of the Polish Committee for Scientific Research and Development (Project No. 18.160.92) and Deutsche Forschungsgemeinschaft are gratefully acknowledged. Two of the authors (A.G. and S.M.) are indebted to the Deutsche Akademischer Austauschdienst for support of their short-term stay in Germany.

References

1. E. W. A. Young and J. H. W. de Wit, *Solid State Ionics*, **14**, 39, 1985.
2. A. Brückman and A. Gil, *Proc. 2nd Workshop of German and Polish Research on High Temperature Corrosion of Metals* (Ed. W. J. Quadakkers, H. Schuster, and P. J. Ennis) Reports of KFA Jülich, No. Jül-Conf-76, Jülich, p. 44, 1988.
3. J. Jedliński, *Oxid. Met.*, **39**, 61, 1993.
4. J. Jedliński, M. J. Bennett and G. Borchardt. These proceedings.
5. H. Bishop. These proceedings.
6. G. Borchardt, S. Scherrer and S. Weber, *Microchimica Acta*, **II**, 421, 1981.
7. M. J. Graham, J. I. Eldridge, D. F. Mitchell and R. J. Hussey, *Materials Science Forum*, **43**, 207, 1989.
8. C. M. Cottell, G. J. Yurek, R. J. Hussey, D. F. Mitchell and M. J. Graham, *Oxid. Met.*, **34**, 173, 1990.
9. M. J. Bennett, A. T. Tuson, D. P. Moon, J. M. Titchmarsh, P. Gould and H. M. Flower, *Surface and Coatings Technology*, **51**, 65, 1992.
10. J. Jedliński, J. Slowik and G. Borchardt, *Applied Surface Science*, to be submitted.

A Study of the Oxidation Behaviour of Sputtered Oxygen-enriched 304 Stainless Steel

L. ZHOU,*‡ D. G. LEES* and R. D. ARNELL†

*Manchester Materials Science Centre, University of Manchester/UMIST,
Manchester M1 7HS, UK
†Department of Aeronautical and Mechanical Engineering, University of Salford,
Salford, M5 4WT, UK
‡K. C. Wong Royal Society Fellow, on leave of absence from University of Science
and Technology, Beijing

ABSTRACT

Oxygen-enriched 304 stainless steel has been prepared by reactive sputtering, and the same steel without oxygen addition has also been sputter-deposited for comparison. Isothermal oxidation kinetics have been measured continuously at 900°C. The oxygen-enriched steel oxidised more slowly, and reached a constant value of the instantaneous parabolic rate constant earlier. It is believed that there may be a silica layer at the scale/metal interface, which formed earlier on the oxygen-enriched steel. For both materials the scale consisted of an outer layer of spinel oxide and an inner layer of chromia. The grain size of the outer-layer spinel on the oxygen-enriched steel was noticeably smaller. The adhesion of the outer layer of spinel on the oxygen-enriched steel was better, while no significant difference in the adhesion of the inner layer of chromia was found.

1. Introduction

Dispersions of rare earth oxides for chromia-forming and alumina-forming alloys,[1–3] ZrO2, HfO2 and TiO2 for a chromia-forming alloy,[4] and even of chromia for a chromia-forming alloy[5] and alumina for alumina-forming alloys[6–8] and chromia-forming alloys[9,10] have been found to benefit the oxidation resistance of the alloys in two aspects: improving the oxide adhesion and reducing the oxide growth rate. An important feature in these cases is that the growth mode of the chromia scale has been changed from an outward cation diffusion dominant to an inward anion diffusion dominant, although for the alumina scale this still needs to be confirmed.

Many mechanisms have been suggested to explain these observations; these have been reviewed in refs 1–3. Among them the Sulphur Effect theory[11–14] seems to be most general and therefore fundamental because it is applicable to most of the experimental facts. It proposed a combined effect of sulphur: (1)

220

segregating to the scale/metal interface inducing a poor adhesion; (2) segregating to oxide grain boundaries promoting the short-circuit diffusion of cations along the boundaries and hence leading to a cation diffusion-dominated growth mode. This has recently been confirmed in pure chromium.[13,14] According to this theory, dispersions of any stable oxides would give beneficial effects. Previous work with a chromia-dispersed Co–Cr alloy produced by reactive sputtering has confirmed this deduction.[5] The aim of the present work was to investigate this further in the case of a commercial alloy. An oxygen-enriched alloy was produced by the same method as that used in ref. 5, but at a much lower deposition rate and without annealing afterwards. At the present stage we have to call it oxygen-enriched instead of oxide-dispersed, because we have not yet been able to find particles of dispersed oxide in the highly defective as-sputtered alloy, although we believe that the enriched oxygen will precipitate as an oxide dispersion, at least after the heating at the beginning of the oxidation test.

Impurity oxygen commonly exists in various metallic materials and can be closely controlled in the metallurgical industry. Therefore the present investigation should also be commercially important in optimising the oxidation resistance of metallic materials.

2. Materials and Methods

2.1 The Sputtering and the Sputter-deposited Materials

A d.c. magnetron sputtering device with a facility for controlled inlet of oxygen was employed for the deposition. The target for sputtering was made from commercial 304 stainless steel. The oxygen-enriched steel was produced by conventional reactive sputtering with a controlled inlet of oxygen into the sputtering chamber, while a similarly sputtered reference steel without oxygen addition was prepared for comparison. The deposition rate was about 2 μm/h. The sputtered 304 stainless steels were deposited on substrate of cold rolled sheet of 304 stainless steel with as-rolled surfaces. The compositions of both the sputtered steels are very close and are typical of a commercial 304 stainless steel, as listed in Table 1. The topography of both as-sputtered steels is shown in Fig. 1(a,b), and a TEM image of the as-sputtered oxygen-enriched steel is shown in Fig. 1(c). As can be seen from Fig. 1, the sputtered steels were

	Cr	Ni	Mn	Si	Fe
Oxygen-enriched steel	19.33	9.74	1.62	0.41	bal
Reference steel	19.28	8.83	1.67	0.45	bal

Table 1 Chemical composition of the sputter-deposited 304 stainless steels (%wt, by EDAX)

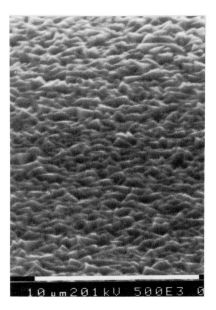

10 μm 29.9 kV   500E3   0

10 μm 201 kV   500E3   0

0.5 μm

Fig. 1 Microstructure of the as-sputtered 304 stainless steels. (a) Surface morphology, the reference steel; (b) surface morphology, the oxygen-enriched steel; (c) TEM image, the oxygen-enriched steel.

fine-grained and highly defective. The diffraction pattern of the oxygen-enriched steel in TEM indicated a bcc structure. The reference steel is expected to have the same structure because, like the oxygen-enriched steel, it was ferromagnetic.

2.2 The Isothermal Oxidation Tests

The isothermal oxidation kinetics were measured manometrically. Details of this apparatus have been described previously.[15] The oxidation tests were conducted at 900°C in pure oxygen at 0.1 atm pressure.

3. Results

3.1 Kinetics

The sputtered oxygen-enriched 304 stainless steel was found to oxidise more slowly than the same sputtered steel without oxygen addition (reference steel), as can be seen in Fig. 2(a). Further analysis of the instantaneous parabolic rate constant, as illustrated in Fig. 2(b), revealed that a constant value of the instantaneous parabolic rate constant was reached on the oxygen-enriched steel within the first 5 min of exposure, while on the reference steel it was not reached until after 3 h of exposure. This strongly suggests an earlier build-up of some protective layer on the oxygen-enriched steel.

3.2 Oxidation Products and Morphology

X-ray diffractometry on both the materials oxidised for 100 h showed that Cr_2O_3 and $Mn_{1.5}Cr_{1.5}O_4$ (spinel) were formed. Figure 3 shows the surface morphology and chemical composition of the oxidised specimens. A layer of oxide with obviously smaller grain size was formed on the oxygen-enriched steel, as shown in Fig. 3. On the reference steel, a kind of spot-like spallation occurred, as shown in Fig. 4, while no spallation occurred on the oxygen-enriched steel. The exposed area was a sublayer of chromia, as indicated by Fig. 4(c). It is therefore reasonable to deduce that, on both materials, the oxide scale consisted of an outer layer of spinel and an inner layer of chromia. The results shown in Figs 3 and 4 indicate that the outer layer of spinel on the oxygen-enriched steel was finer-grained with better adhesion, while no difference in adhesion was found for the inner layer of chromia.

In order to examine the oxide scale formed in the early stage of oxidation, specimens of both materials oxidised together at 900°C for 30 min were analysed. No obvious difference between them in surface morphology was found with SEM. EDAX analyses of the oxidised surface were carried out. The EDAX spectra were detected from an area of about $0.12 \times 0.10 \text{ mm}^2$. In order to be aware of the extent of the contribution of the substrate to the analysed result, the EDAX analyses were conducted systematically at decreasing accelerating voltage of the incident electron beam (corresponding to decreasing

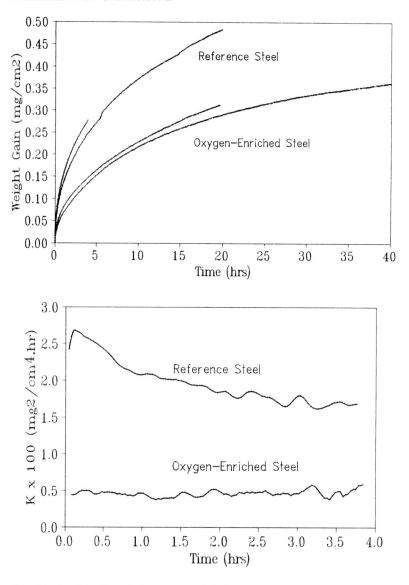

Fig. 2 Oxidation kinetics of the sputtered 304 stainless steels at 900°C. (a) Weight gain versus time; (b) instantaneous parabolic rate constant versus time.

depth of detection). These analyses could also give some information about the depth profile of the elements detected, which is almost unlikely to be obtained through the cross-section of the specimens because the oxide scales at this stage (900°C, 30 min) are very thin. The results are shown in Fig. 5. It is important to note that all the values of concentrations in Fig. 5 include contributions from surface oxides. As can be seen, the results changed with the accelerating voltage, showing that there were contributions from substrate,

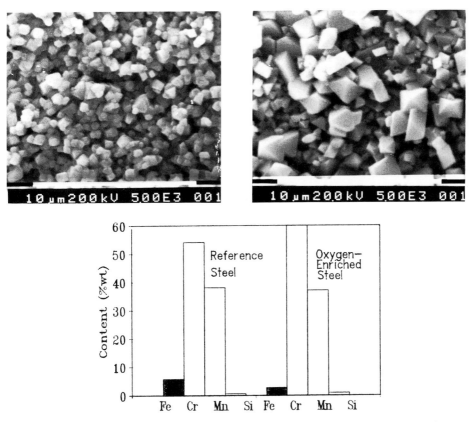

Fig. 3 Surface morphology and EDAX analyses of the sputtered 304 stainless steels oxidised at 900°C for 100 h. (a) The oxygen-enriched steel; (b) the reference steel; (c) the results of EDAX analyses.

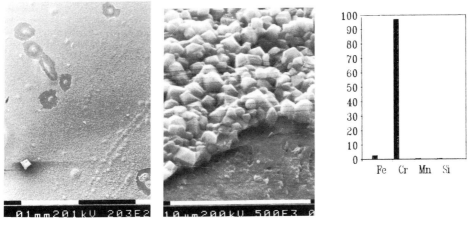

Fig. 4 Spot-like spallation of the reference steel oxidised at 900°C for 100 h. (a) The spallation spots; (b) a closer observation of the edge of a spallation spot; (c) the result of EDAX analysis of the exposed area.

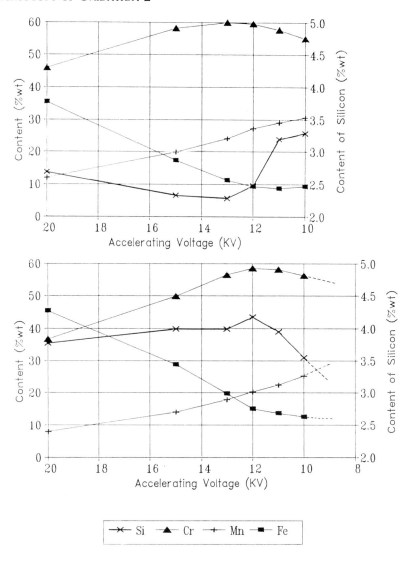

Fig. 5 The results of EDAX analyses at decreasing accelerating voltage for the sputtered 304 stainless steels oxidised at 900°C for 30 min. (a) The reference steel; (b) the oxygen-enriched steel.

which decreased with decreasing depth of detection, i.e. decreasing accelerating voltage. The curve of Fe concentration makes it possible to see the contribution of the substrate to the analysed results, because the Fe content in the substrate is much higher than in the scale. As indicated by the curve, the Fe concentration decreases more and more slowly with the decreasing accelerating voltage, and tends to reach a minimum at some point of low voltage. At voltages lower than this point, where the gradient of the Fe concentration curve reaches zero, there should be no contribution from the substrate. For the reference steel, such a condition was reached at about 11 KV, while for the oxygen-enriched steel, this condition could be reached by extrapolating the profiles to about 9 KV. The difference between the Fe concentration curves indicates that the oxide scale on the oxygen-enriched steel was thinner, which agrees with the kinetic results. The concentration of Cr, Mn and Fe in the scales and the distributions of them in the underlying alloy were quite similar for the two materials, as can be seen from Fig. 5. The concentrations of Si in both the scales were also similar, as the value for Si at accelerating voltage of 11 KV for the reference steel and 9 KV for the oxygen-enriched steel are approximately the same. As the accelerating voltage increased, the curve of Si for the reference steel goes down, whereas that for the oxygen-enriched steel goes up (see Fig. 5). As the concentrations of Si in the sputtered substrate were the same for each case, it would appear that the only explanation for this result is that a Si-enriched layer was present at the oxide/scale interface of the oxygen-enriched steel but not at that of the reference steel.

4. Discussion

4.1 The Effect of Oxygen Enrichment on the Early Stage of Oxidation

In Fe–Cr–Ni-based alloys, it has been repeatedly demonstrated that silicon plays an important role in enhancing oxidation resistance by forming an inner thin layer of silica between the oxide scale and the alloys, which acts as a diffusion barrier.[16-18] At the present stage it should be reasonable to assume that the Si-enriched layer suggested by Fig. 5 for the oxygen-enriched steel is just a silica barrier. The formability and effectiveness of the silica layer depends on the silicon content.[17] The silicon contents of both the present sputtered steels are the same because they were sputtered from the same target (see Table 1). Therefore, the results plotted in Fig. 5 would suggest that the oxygen enrichment has promoted the formation of the inner silica barrier. Such a barrier may form later on the reference steel as happens in other silicon-containing Fe–Cr–Ni alloys, but on the oxygen-enriched steel it formed much earlier. This agrees well with the kinetic features revealed in Fig. 2b.

4.2 The Effects of Oxygen Enrichment on the Later Oxidation Rate

With the same content of silicon, the reference steel will eventually build up an inner silica barrier layer probably as effective as that in the oxygen-enriched steel. However, as also shown in Fig. 2(b), the oxidation rate of the reference steel was still significantly higher after almost only 4 h. This may be because, for some reason, the silica layer did not grow as thick on the reference steel. However, there may be another explanation: if there are internal oxide particles in the oxygen-enriched steel, as expected, impurity sulphur could segregate to the interfaces between these particles and alloy matrix, thus reducing the amount of sulphur incorporated into the scale; Lees[12] has suggested that this will reduce the cation diffusion through the scale and hence the growth rate of the scale. More investigations are to be carried out to see if this is the explanation of the lower scale growth rate in the oxygen-enriched steel.

5. Conclusion

The sputtered oxygen-enriched 304 stainless steel oxidised more slowly, and reached a constant value of the instantaneous parabolic rate constant earlier at 900°C than the same steel sputtered without oxygen addition. It is suggested that a silicon-enriched layer was present at the oxide/metal interface of the oxygen-enriched steel in the early stage of oxidation, and that no such layer was present in the case of the reference steel oxidised for the same time. The scales on both materials consisted of an inner layer of chromia and an outer layer of spinel. The spinel layer on the oxygen-enriched steel had a smaller grain size and better adhesion, while no significant difference between the adhesion of the inner chromia layers was found for the two materials oxidised at 900°C for up to 100 h.

Acknowledgement

We wish to thank the K. C. Wong Education Foundation for a grant via The Royal Society to one of us (L.Z). We greatly appreciate the assistance of Mr Robin Bates of the Department of Aeronautical and Mechanical Engineering, University of Salford in preparation of the sputter-deposited specimens.

References

1. D. P. Whittle and J. Stringer, *Phil. Trans. R. Soc. Lond.*, **A295**, 309, 1980.
2. J. Stringer, *Mat. Sci. Eng.*, **A120**, 129, 1989.
3. H. Hindham and D. P. Whittle, *Oxid. Met.*, **18**, 245, 1982.
4. D. P. Whittle, M. E. El Dashan and J. Stringer, *Corr. Sci.*, **17**, 17, 1977.
5. Zhou Lang, Ye Ruizeng, Zhang Shouhua and Gao Liang, *Corr. Sci.*, **32**, 337, 1991.
6. V. Srinivasam, L. A. Harris, R. A. Padgett, Jr and D. R. Baer, *Oxid. Met.*, **195**, 36, 1991.
7. L. M. Kingsley and J. Stringer, *Oxid. Met.*, **32**, 371, 1989.
8. J. K. Tien and F. S. Pettit, *Met. Trans.*, **3**, 1587, 1972.
9. I. G. Wright, B. A. Wilcox and R. I. Jaffe, *Oxid. Met.*, **9**, 275, 1975.
10. H. T. Michels, *Metall. Trans.*, **A7**, 379, 1976.
11. D. G. Lees, I. S. Grant and G. W. Lorimer, *An Investigation of The Mechanism by Which Reactive Elements Improve Scale-Metal Adhesion*, Science and Engineering Research Council Proposal, 1982.

12. D. G. Lees, *Oxid. Met.*, **27**, 7, 1987.
13. P. Fox, D. G. Lees and G. W. Lorimer, *Oxid. Met.*, **36**, 491, 1991.
14. W. Dong, D. G. Lees, G. W. Lorimer and A. Paul, An investigation of the role of sulphur in the adhesion and growth mechanism of chromia formed on low-sulphur chromium (these proceedings).
15. M. S. Skeldon, J. M. Calvert and D. G. Lees, *Oxid. Met.*, **28**, 109, 1987.
16. M. J. Bennett, J. A. Desport and P. A. Labun, *Proc. R. Soc. Lond.*, **A412**, 223, 1987.
17. R. C. Lobb, J. A. Sasse and H. E. Evans, *Mater. Sci. Technol.*, **5**, 828, 1989.
18. R. C. Lobb and H. E. Evans, *Proc. Int. Conf. on Microscopy of Oxidation*, March, 1990, Cambridge (Ed. M. J. Bennett and G. W. Lorimer), p. 119, 1991.

4

OXIDATION OF ALUMINA-FORMING FERRITIC STEELS

Alumina and Chromia Scales Formed at High Temperature

P. FOX, D. G. LEES,* G. W. LORIMER* and G. J. TATLOCK

The Department of Materials Science & Engineering, The University of Liverpool, P.O. Box 147, Liverpool L69 3BX, UK

Manchester Materials Science Centre, University of Manchester and UMIST, Grosvenor Street, Manchester M1 7HS, UK

ABSTRACT

The morphology and behaviour of scales formed on Fe–5%Al and chromium during high-temperature oxidation have been studied. A wide range of physical techniques was used to examine the respective alumina and chromia scales produced and to identify any morphological differences and similarities. This research provides further evidence that implicates sulphur as the major protagonist in the formation of convoluted and poorly adhered scales and identifies a mechanism by which plastic deformation may occur during scale convolution.

1. Introduction

The mechanism(s) by which the addition of reactive elements alters the adhesion of chromia and alumina scales has been disputed for a long time.[1,2] Early work on sintered alumina showed that yttrium segregated to the grain boundaries.[3] More recent work considering adherent chromia scales formed on alloys containing a reactive element reported the segregation of the reactive element to the oxide grain boundaries. The segregation of impurity elements to the oxide grain boundaries was not detected.[4] It has been proposed[5–7] that the presence of the reactive elements segregated to the oxide grain boundaries changes the diffusion of species through the scale, but later it was proposed that it was the absence of sulphur which had this effect.[8] The change in elemental diffusion was thought to alter the oxide morphology.

Recent work has shown that flat scales can be formed on chromium that does not contain reactive elements.[9] This showed that if the level of non-metallic impurities within chromium was reduced by annealing the sample in hydrogen at a high temperature, the scale formed on oxidation became flat and more adherent. Further work using a commercially available chromium, with very low sulphur content (<5 ppm), also produced scales that were flat

233

and adherent.[10] It was, however, possible to produce convoluted and poorly adherent scales by contaminating the surface of these alloys with sulphur or chlorine before oxidation.[10] With both the hydrogen-annealed chromium and the high-purity chromium there were no reactive elements present within the metal. In this case, therefore, the growth morphology could not be altered by segregation to the oxide grain boundaries and alteration of the grain boundary structure.

STEM techniques similar to those used to detect yttrium segregation to oxide grain boundaries[4] were used to examine the convoluted scales formed on chromium. In those experiments STEM analysis of the scale and the metal/oxide interface was carried out using cross-section TEM samples. This analysis showed conclusively that with the chromium that formed convoluted scales, sulphur could be detected both within the oxide grain boundaries and at the metal/oxide interface. Sulphur was not detected within the scale except at the grain boundaries and in the boundaries it was not present as sulphur-rich particles but was present all along the boundaries. However, as will be shown later in this paper, the scales that were flat and adherent did not contain any detectable sulphur.

This work, therefore, suggests a link between the presence of sulphur and the formation of a convoluted scale, since sulphur segregation was associated with convoluted scales and a lack of sulphur occurred with flat scales. It was not possible during this work to quantify the distribution of sulphur on the oxide grain boundaries. It was, however, possible to approximate the concentration and to determine that sulphur was present on the boundary. The sulphur may be present in several possible distributions. It may occur as a distribution about the boundary or may be concentrated in the core of the boundary. Another possibility is that the sulphur may lead to the formation of thin amorphous bands at the oxide grain boundaries. This effect is reported to occur during the sintering of alumina powder with impurities of silicon and calcium.[11]

The presence of sulphur at the oxide/substrate interface of oxidised chromium has been confirmed recently using Auger and XPS techniques, where the oxide was spalled from the surface using an indenter under UHV conditions and then the surface analysed.[12] However, this technique could not be used to analyse the grain boundaries of the oxide, nor did it identify the chemical form of the sulphur or its distribution around the interface. Experiments with alumina-forming alloys have suggested that impurities are also responsible for the convolution of alumina scales.[13-16] However, other workers consider that another mechanism is operating.[17]

To date, there is no satisfactory explanation of the mechanism by which sulphur segregation affects the scale morphology, and therefore, until the mechanism is determined, it is impossible to say categorically whether the sulphur is causal or just covariant. In this paper, the results obtained for chromia scales are compared with preliminary work on alumina scales formed by the oxidation of an iron–aluminium alloy at 1000°C for short oxidation times. The paper considers scale morphologies and possible deformation mechanisms. It has not yet been possible to carry out STEM analysis of the alumina scales, but this should be possible in the near future.

2. Experimental

The iron-based material had a nominal composition of Fe–5%Al and was supplied by the UKAEA at Harwell as a 1 mm thick sheet. This material was cut to 1 cm × 1 cm samples and then ground to 1200 grit silicon carbide. The samples were oxidised in air at 1000°C for a range of times between 1 and 20 min.

The oxide scales formed were examined in a number of ways:

(a) The oxide was stripped from the surface, using adhesive tape, and then examined using TEM techniques. Oxides that were too thick to be electron transparent were thinned using atom beam milling.
(b) Cross-sections of the metal oxide interface and scale were produced for TEM analysis.[18]
(c) TEM cross-sections of the stripped oxide were also used to allow examination of the oxide scales without the added complication of examining a ferritic sample.

The chromium samples with which these results are compared were prepared in a similar way. The composition of the chromium that formed convoluted scales is given in reference 19, while the low-sulphur chromium was supplied by Metall. Plansee under the trade name Ducropur.

Results and Discussion

Convolution of Chromia and Alumina Scales

In many ways the behaviour of chromia and alumina scales is similar, although they are reported to grow by different mechanisms. They are both affected in the same way by the addition of reactive elements and in both cases the morphology of the convoluted scales changes in the same way with oxidation time and temperature. With both chromia and alumina scales, where the scale does not remain flat, it is, however, adherent at low temperatures (Fig. 1) or at short times (<2 min) at high temperatures. After longer times at high temperatures, the scales become poorly adhered. The scales convolute to form large ridges and the oxide tends to spall on cooling (Fig. 1). In both cases, the oxides tend to lose contact with the base metal with the formation of voids, but without the loss of protection. In other regions, the metal stays attached to the oxide and also develops a somewhat convoluted metal/oxide interface.

The formation of these fine convolutions can be suppressed by initially oxidising the sample at a temperature where it normally forms a flat scale (600°C for Fe–5%Al) and then heating to 1000°C. This oxidising treatment reduces the tendency for the scale to convolute, although it does not prevent loss of adhesion (Fig. 2). Instead of convoluting, the oxide bows out like a dome above the surface of the metal. This produces a scale that cracks and spalls easily and has little contact with the metal substrate. The reduced tendency to convolute is probably due to the increased thickness of the oxide scale, which is thickened during the low-temperature oxidation, and then is

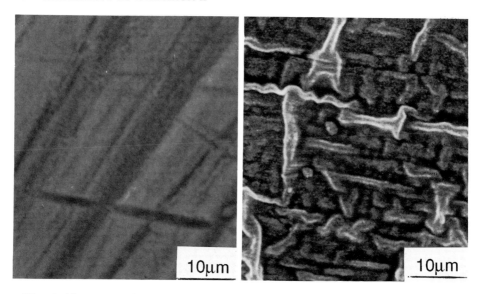

Fig. 1 Alumina scales formed on Fe–5%Al at 600°C and 1000°C after 24 h oxidation.

Fig. 2 Fe–5%Al oxidised for 24 h at 600°C and then 6 h at 1000°C.

subjected to a compressive stress when the temperature is raised which cannot be relieved by convolution, and so the scale bows out instead. Thus this shows that if the ability of a scale to convolute is controlled by scale thickness then only thin scales can convolute.

The mechanism by which the scale convolutes has been reviewed.[20] For normal plastic deformation to occur dislocations must glide through the material. It was shown by Von Mises that this is only possible for a polycrystalline material if at least five independent slip modes exist. With alumina and chromia, glide at temperatures below 1600°C can only occur on the basal plane (alumina: $\langle 11\bar{2}0 \rangle$ $\{0001\}$[21] and chromia: assumed to be the same although the slip systems have not been studied).[22] This only corresponds to two slip systems. It is assumed, therefore, that polycrystalline chromia and alumina scales cannot deform by slip alone. Instead, deformation may occur either by creep, or by grain boundary slip. Both processes are sensitive to the structure of the grain boundaries and may be affected by impurity segregation. Since with grain boundary slip the accommodation of the sliding of triple points is also difficult, creep is assumed to be the most likely process.

SEM analysis of a surface oxidised for a short time reveals that the convolution of the scale is associated with any surface scratches. The convolutions are initially fairly fine and occur over only small sections of the surface. At longer times they spread to cover the whole surface and coarse convolutions can develop.

To understand the processes that are occurring within the scale, it is necessary to use TEM techniques. With oxides that have just started to convolute, some regions of the scale are still attached to the substrate while others have become detached. Where chromia scales become detached, the bending of the oxide away from the substrate is concentrated into a small region of the scale (Fig. 3). The rest of the oxide remains fairly undeformed. Closer examination of these 'hinge regions' reveals the presence of dislocations within the grains (Fig. 4). Similar dislocations were found within alumina scales (Fig. 5). It must be appreciated that dislocations are rare in alumina and chromia scales that have been growing for any length of time. This suggests that at least some convolutions in alumina and chromia scales are formed by a plastic deformation mechanism rather than by creep.

Some dislocations within the alumina grains have been analysed and are of the form $\langle 11\bar{2}0 \rangle$ $\{0001\}$. This is consistent with dislocations formed in bulk alumina. The dislocations present in the chromia samples still have to be analysed, but are likely to be similar to those in alumina. Further analysis of similar regions is required to confirm that these are the only dislocation types that occur.

Although plastic deformation has been considered unlikely at normal oxidation temperatures,[20] the evidence suggests that it occurs at these short oxidation times. It is reasonable to assume that bulk polycrystalline alumina and chromia cannot deform plastically without five independent slip systems. However, it may be possible with thin scales because they are only one or two grains thick. The scales are not constrained in three dimensions, as are bulk polycrystalline materials, and thus plastic deformation may occur with only two slip systems.

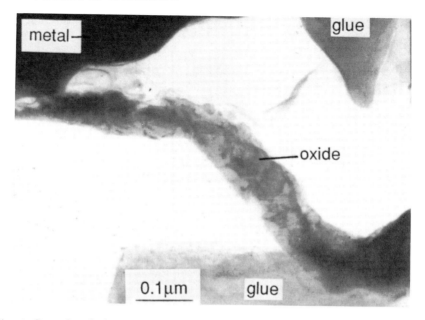

Fig. 3 Convoluted chromia scale formed on chromium oxidised for 15 min at 950°C.

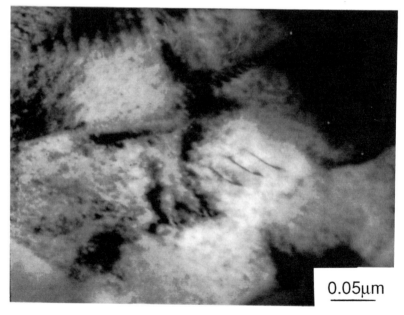

Fig. 4 Dislocations within the chromia scale formed on chromium oxidised for 15 min at 950°C.

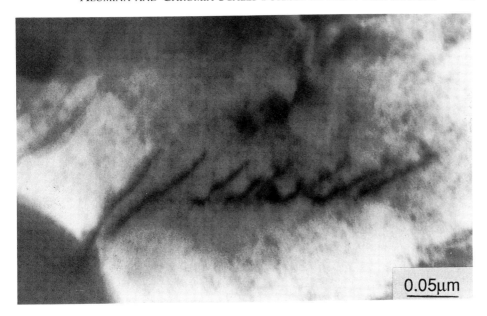

Fig. 5 Dislocations within an alumina scale formed on Fe–5%Al after 10 min at 1000°C.

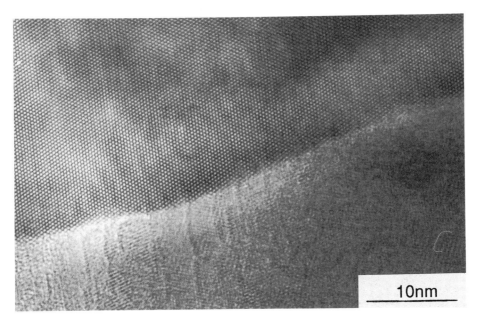

Fig. 6 HREM micrograph of alumina grain boundaries formed on Fe–5%Al oxidised at 1000°C for 5 min.

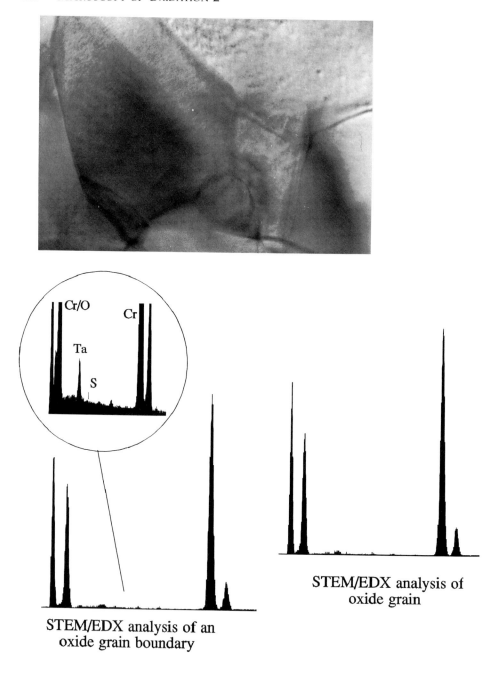

STEM/EDX analysis of an
oxide grain boundary

STEM/EDX analysis of
oxide grain

Fig. 7 STEM/EDX analysis of a flat adherent chromia formed on Ducropur chromium cyclically oxidised at 950°C.

3. HREM Analysis of the Grain Boundaries within the Alumina Scale

Stripped alumina scales formed on Fe–5%Al oxidised for 5 min at 1000°C were examined (Fig. 6). No amorphous layer was detected at the oxide grain boundaries, which is consistent with the deformation of the scale occurring by plastic deformation. Any amorphous layer at the grain boundaries would increase the likelihood of grain boundary sliding and would also increase creep. It has not yet been possible to determine the segregation within these scales.

4. STEM Analysis of Flat Adherent Chromia Scales Formed on High Purity Chromium

STEM analysis was carried out on the oxide scales formed on the Ducropur chromium cyclically oxidised for eighty 6-min cycles at 950°C. Examination of the scale revealed that there was no detectable sulphur segregation to the oxide boundaries (Fig. 7). This is in contrast to the strong segregation of sulphur found within convoluted chromia scales. It has not yet been possible to analyse the alumina scales using this technique.

5. Conclusions

1. The presence of sulphur at oxide grain boundaries is associated with the formation of convoluted chromia scales, while sulphur is not detected at the grain boundaries of flat adherent scales.
2. Thin alumina and chromia scales may deform plastically by a dislocation movement mechanism during the initial stages of oxidation, producing convolution. If the oxide scale is thicker, this mechanism is less likely.
3. Amorphous layers are not found at the grain boundaries of alumina scales formed on Fe–5%Al oxidised for 5 min at 1000°C. This is consistent with deformation being by a plastic process rather than by grain boundary sliding or creep.

Acknowledgements

The authors wish to thank the Science and Engineering Research Council for the financial support of this work.

References

1. D. P. Whittle and J. Stringer, *Phil. Trans. Roy. Soc.*, A **295** (1980), 309–329.
2. D. P. Moon, *Mat. Sci. Tech.*, **5** (8) (1989), 754–764.

3. P. Nanni, C. T. H. Stoddart and E. D. Hondros, *J. Mat. Chem.*, **1** (1976), 297–320.
4. C. M. Cotell, G. J. Yurek, R. J. Hussey, D. F. Mitchell and M. J. Graham, *Oxid. Met.*, **34** (3/4) (1990), 173–216.
5. G. M. Eccer and G. H. Meirer, *Oxid. Met.*, **13** (2) (1979), 159–180.
6. P. Skeldon, J. M. Calvert and D. G. Lees, *Phil. Trans. R. Soc.*, **296** (1980), 557–565.
7. J. L. Smialek, *J. Electrochem. Soc.*, **126** (12) (1979), 2275–2276.
8. D. G. Lees, *Oxid. Met.*, **27** (1/2) (1987), 75–81.
9. I. Melas and D. G. Lees, *Mat. Sci. Tech.*, **4** (5) (1988), 455–456.
10. P. Fox, D. G. Lees and G. W. Lorimer, *Microscopy of Oxidation*. The Institute of Metals, London (1991), pp. 92–97.
11. D. W. Susnitzky and C. B. Carter, *J. Am. Ceram. Soc.*, **73** (8) (1990), 2485–2493.
12. P. Y. Hou and J. Stringer, *Oxid. Met.*, **38** (5/6) (1992), 323–345.
13. J. G. Smeggil, A. W. Funkenbusch and N. S. Bornstein, *Metall. Trans. A*, **17A** (6) (1986), 923–932.
14. J. G. Smeggil, *Mat. Sci. Eng.*, **87** (1987), 261–265.
15. D. R. Sigler, *Oxid. Met.*, **29** (1/2) (1988), 23–43.
16. A. W. Funkenbusch, J. G. Smeggil and N. S. Bornstein, *Metall. Trans. A*, **16A** (6) (1985), 1164–1166.
17. R. Prescott and M. J. Graham, *Oxid. Met.*, **38** (3/4) (1992), 233–254.
18. P. Fox, *Micros. Res. Tech.*, **21** (4) (1992), 369–370.
19. P. Fox, D. G. Lees and G. W. Lorimer, *Oxid. Met.*, **36** (5/6) (1991), 491–503.
20. J. Stringer, *Corr. Sci.*, **10** (1970), 513–543.
21. M. L. Kronberg, *Acta Met.*, **5** (1957), 507–524.
22. H. J. Frost and M. F. Ashby, *Deformation Mechanisms Maps: the plasticity and creep of metals and ceramics*, Pergamon Press, Oxford (1982), pp. 98–104.

Oxidation of an Alumina-forming Alloy: Morphological and Structural Study

M. BOUALAM, G. BERANGER and M. LAMBERTIN

*LG2mS (URA 1505 CNRS), Université de Technologie de Compiègne,
BP 649–60206 Compiègne Cedex, France*

ABSTRACT

A cerium-doped ferritic steel (Fe–22Cr–6Al) was oxidised in air in the temperature range 850–1100°C. In the entire domain, alumina is formed on the alloy, but its structure depends on several parameters. After oxidation for 24 h, a transition alumina (identified as θ-Al_2O_3) is obtained at low temperature (<900°C); at higher temperature (>1000°C) α-Al_2O_3 is formed. That latter phase is always observed after long oxidation times. These phases are not observed at the beginning of the reaction. During the first few minutes, a transition alumina (γ-Al_2O_3) is obtained. Its transformation into α- or θ-alumina occurs quite rapidly and is accompanied by an acceleration of the oxidation. The morphologies of these alumina phases are different. At high temperature, the superficial oxide is in the form of equiaxial small crystals. At lower temperature, there are platelets (which are sometimes described as whiskers). We have also tested the influence of humidity in the oxidising atmosphere. At 900°C, the addition of water vapour first increases the oxidation rate then decreases it. The maximum rate is observed when the dew point in the atmosphere is approximately 40°C. A change of the alumina structure occurs. θ-Al_2O_3 is obtained in dry air and with low water contents. When the humidity increases, the oxide is changed to α-Al_2O_3.

1. Introduction

Among the alumina-forming alloys, the ferritic steels have been and continue to be extensively studied. The reason is that protective scales (α-Al_2O_3) are generally developed on their surface during high-temperature oxidation.

A literature review was recently published by Prescott and Graham.[1] Several results should be highlighted. The first one is that several phases are observed on the surface after oxidation: these were transition (γ, δ or θ) aluminas and α-alumina. Their formation depends mainly on the reaction temperature,[2–7] but transition aluminas are also observed during a transient stage[8–12] at the

243

beginning of the oxidation. Microscopical studies show that the phase transformations are accompanied by morphological changes: δ- or θ-alumina is generally in the form of whiskers; α-alumina is in the form of small equiaxial grains. In addition, the effect of active elements has also been examined.[13–17]

In a series of preliminary experiments, we have compared the high-temperature oxidation resistance of several ferritic steels.[18,19] From the results which were obtained, we have selected one of these alloys to study the morphology and structure of the oxide which forms and to examine the influence of experimental conditions such as temperature and humidity.

2. Experimental

The alloy which was studied during this work is a ferritic steel of composition given in Table 1. It contains chromium (22%) and aluminium (5%). It is doped with mischmetal (Ce+La) to increase its oxidation behaviour at high tempera-ture. It also contains zirconium (and titanium) to trap carbon and avoid chromium carbide formation. Samples are thin foils (50 μm). They were obtained by cold rolling then annealed for a few minutes at 950°C in a protective atmosphere.

Oxidation tests were generally performed in room air, in the temperature range 850–1100°C. Some tests were carried out in synthetic air, dry or wet (with controlled water vapour pressures). The oxidation time was generally 24 h but some tests were longer (up to 120 h) or very short (a few minutes).

After oxidation, specimen morphology was characterised using a scanning electron microscope. Analyses were performed by X-ray diffraction and X-ray spectrometry. Oxide scales which were formed on samples oxidised for very short times were examined by electron diffraction. In some cases, oxidation kinetics was recorded using a microbalance.

3. Results

The first part of this work concerns the oxidation of the steel in room air. Observations of the surface were made after 24 h of oxidation. They show clearly that the oxide morphology strongly depends on the temperature. At

Elmt	C	Si	S	P	Mn	Cr	Ni	Ti
wo%	0.025	0.56	0.001	0.0014	0.29	22.38	0.17	0.004
Elmt	Al	Cu	Co	Ce	La	N	Zr	
wo%	4.71	0.01	0.017	0.021	0.011	0.0032	0.23	

Table 1: Actual composition of the alloy

850°C	900°C	950°C	1000°C	1100°C
		3.49(α)	3.49(α)	3.49 (α)
2.732(θ)	2.732(θ)	2.736(θ)	–	–
–	–	2.55(α)	2.55(α)	2.55 (α)
2.393(θ)	2.393(θ)	2.37(θ, α)	2.37(α)	2.37 (α)
–	–	2.09(α)	2.09(α)	2.09 (α)
–	–	1.74(α)	1.74(α)	1.74 (α)
–	–	1.604(α)	1.60(α)	1.607(α)
1.405(θ)	1.405(θ)	1.407(θ, α)	1.40(α)	1.407(α)
–	–	1.377(α)	1.37(α)	1.375(α)

Table 2 X-ray diffraction lines determined after 24 h of oxidation

low temperature (850 and 900°C), the sample surface is covered with oxide platelets (whiskers). These platelets are very few at 850°C (Fig. 1), but are much more developed at 900°C. At high temperature (1000–1100°C), the morphology is completely different: the surface is entirely covered with small equiaxial crystals. At the intermediate temperature (950°C), both morphologies are simultaneously observed.

By X-ray diffraction, similar modifications are observed. The results are reported in Table 2. At low temperature, a transition oxide is formed (θ-Al$_2$O$_3$), whereas at high temperature the α-Al$_2$O$_3$ (corundum) is produced. At 950°C, both oxides are formed.

As it was established that the formation of α-Al$_2$O$_3$ is favoured by increasing the temperature, we tried to examine the influence of long-term oxidation. Several tests were performed at 900°C, the oxidation length increasing up to 120 h. Morphological observations show that the oxide platelets (whiskers) disappear progressively (Fig. 2) and α-Al$_2$O$_3$ grows and replaces the transition alumina (Table 3).

4. Short Oxidation Tests

In the previous results it has been shown that the alumina phase which formed was strongly dependent on two parameters: the temperature and the oxidation time. However, 24 h oxidation tests were too long at 1000 and 1100°C to observe θ-Al$_2$O$_3$. Therefore we performed a series of experiments (in the entire temperature domain) to determine which phases are formed during the first stages of oxidation.

Kinetic results are reported in Fig. 3. Different curve shapes are observed. At 850°C, the curve is approximately parabolic. Morphological observations show that, after 100 min, only very small crystals are formed on the surface. Electron diffraction analysis reveals that γ-Al$_2$O$_3$ is formed. At 900 and 950°C, an acceleration in oxidation rate is observed. Morphological observations show that this corresponds to the formation of whiskers on the alloy surface (Fig. 4).

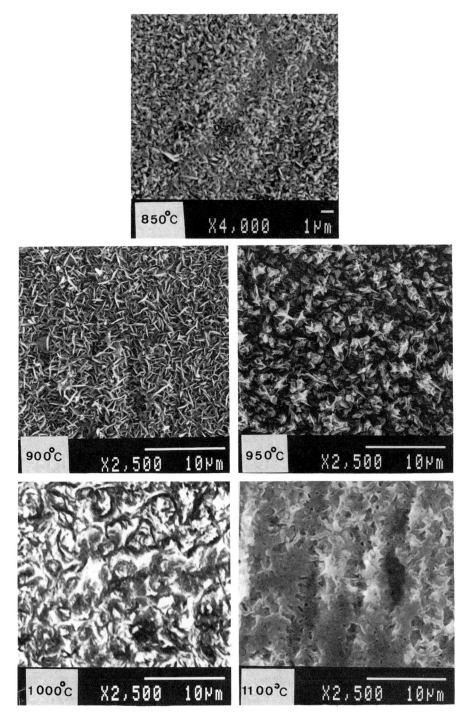

Fig. 1 Influence of temperature on the oxide morphology (24 h of oxidation).

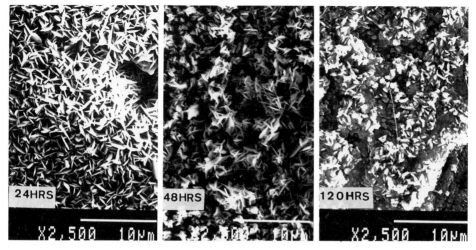

Fig. 2 Morphology of oxide grown at 900°C after different times.

24 h	48 h	120 h
	3.50(α)	3.50(α)
2.74(θ)	2.73(θ)	2.73(θ)
	2.54(α)	2.55(a)
2.40(θ)	2.39(θ, α)	2.39(θ, α)
	2.09(α)	2.09(α)
	1.74(α)	1.74(α)
	1.604(α)	1.605(α)
1.407(θ)	1.406(θ, α)	1.406(θ, α)
	1.376(α)	1.376(α)

Table 3 X-ray diffraction lines determined after different oxidation times

At 1000 and 1050°C, oxide platelets are observed only for a very short time. They disappear completely after less than 1 h of oxidation. At 1100°C it was never possible to observe oxide whiskers. Electron diffraction analyses, after very short oxidation runs (5 min), show that γ-Al$_2$O$_3$ is always formed: the alloy surface is covered with very small 'cubic' crystals (similar to those observed after longer oxidation runs at 850°C).

5. Discussion

From the above results it can be seen that the oxide morphology and the crystalline structure of the alumina which forms are strongly related. Two parameters must be considered: the reaction time and the temperature. After a very short oxidation time (5 min) a transition phase (γ-Al$_2$O$_3$) is always formed.

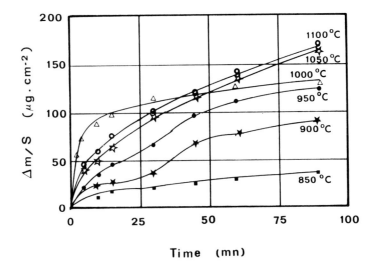

Fig. 3 Oxidation kinetics (short runs) at different temperatures.

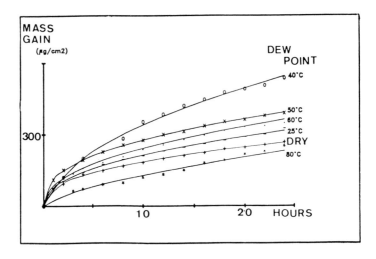

Fig. 5 Oxidation kinetics at 900°C in air with various humidity contents.

The oxide is in the form of small 'cubic' crystals. It transforms quite rapidly into another transition oxide (θ-Al_2O_3) and platelets (whiskers) are observed on the surface. Finally a last transformation occurs, from θ to α-Al_2O_3, which is in the form of small equiaxial crystals. The transformation rates are dependent on the temperature. At 1100°C, α-Al_2O_3 is observed after less than 30 min; it appears after less than 48 h at 900°C and is never detected at 850°C.

It may be concluded that α-Al_2O_3 is the only stable phase but it does not grow immediately on the alloy. There is a transient stage during which transition phases (γ and θ-Al_2O_3) are successively obtained.

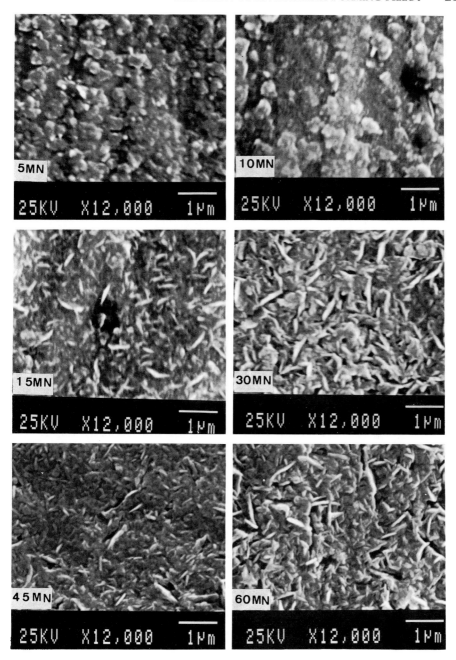

Fig. 4 Morphology of oxide formed after short oxidation runs at 950°C.

Room air	Dew point (25°C)	Dew point (40°C)	Dew point (50°C)	Dew point (60°C)	Dew point (80°C)
		3.50(α)	3.50(α)	3.51(α)	3.475(α)
2.74(θ)	2.74(θ)	2.74(θ)			
			2.569(α)	2.56(α)	2.55(α)
2.40(θ)	2.41(θ)	2.40 (θ, α)	2.398(θ, α)	2.398(θ, α)	2.398(θ, α)
		2.10(α)	2.088(α)	2.09(α)	2.09(α)
			1.75(α)	1.75(α)	1.74(α)
			1.605(α)	1.607(α)	1.603(α)
1.407(θ)	1.406(θ)	1.408(θ, α)	1.405(θ, α)	1.409(θ, α)	1.402(θ, α
			1.380(α)	1.377(α)	1.375(α)

Table 4 Influence of water vapour on the oxide phase

6. Influence of water vapour

In order to test the influence of the humidity content in the oxidising atmosphere, a series of experiments was performed at 900°C in dry and wet air (with controlled water vapour pressures). Kinetic results are shown in Fig. 5. It may be observed that, when the humidity content increases in the atmosphere, the reaction rate first increases then decreases. The highest oxidation rate is observed when the dew point is approximately 40°C. Morphological observations show that whiskers are grown in low-humidity contents whereas equiaxial crystals are formed when vapour pressures are sufficiently high (Fig. 6).

At the same time X-ray diffraction analyses allow the conclusion that, for low dew points, the oxide which forms is θ-Al$_2$O$_3$, whereas for high dew points α-Al$_2$O$_3$ is produced (Table 4).

From these results, kinetic measurements may be explained as follows. When the vapour pressure in the atmosphere is sufficiently low, the oxidation rate increases with the humidity content. Similar results were previously mentioned in the literature.[20–22] This is probably due to an increase of defect concentration in the scale. When the vapour pressure is increased, it was observed that α-Al$_2$O$_3$ forms. That phase is known to be more protective than transition aluminas, which reflects the decrease in oxidation rate.

7. Summary

In this work we have examined the high-temperature behaviour of a ferritic steel. When the reaction is effected in air, it is observed that the morphology of oxide which grows depends on the temperature and is strongly related to

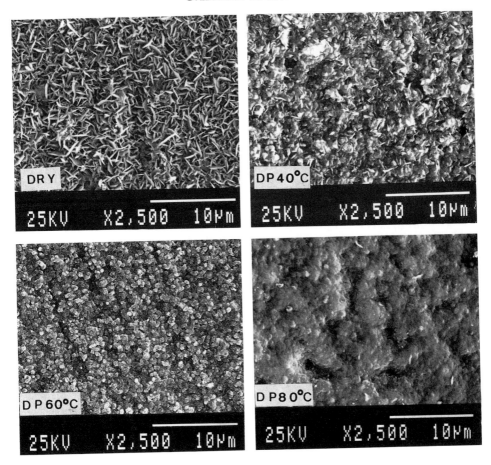

Fig. 6 Influence of the humidity content in the atmosphere on the oxide morphology at 900°C.

the crystalline phase. At low temperature (<950°C) a transition alumina is observed which is in the form of small platelets (whiskers). At higher temperature α-Al$_2$O$_3$ is produced; that phase is in the form of small equiaxial grains.

Even at low temperature (900°C), long-term oxidation (>48 h) leads to the formation of the corundum phase.

The addition of water vapour in the atmosphere at low pressure increases the oxidation rate. When the humidity content increases the oxidation rate is lowered. This result is due to the formation of α-alumina which replaces the transition phase progressively.

Observations and analyses which were made after very short oxidations show that, whatever the temperature, another transition phase (γ-alumina) always grows during the first transient stage.

References

1. R. Prescott and M. J. Graham, *Oxid. Met.*, **38** (3/4) (1992), 233.
2. G. C. Wood and B. Chattopadhay, *Corros. Sci.*, **10** (1970), 471.
3. G. C. Wood and B. Chattopadhay, *Oxid. Met.*, **2** (1970), 373.
4. R. E. Grace and A. E. Seybolt, *J. Electrochem. Soc.*, **105** (1958), 582.
5. T. Nakagama and K. Kaneko, *Corrosion*, **26** (7) (1970), 187.
6. P. A. van Manen, E. W. A. Young, D. Schalkoard, C. J. Van der Wekken and J. H. W. de Wit, *Surf. Interface Anal.*, **12** (1988), 391.
7. P. T. Moseley, K. R. Hyde, B. A. Bellamy and G. Tappin, *Corros. Sci.*, **24** (6) (1984), 547.
8. J. L. Smialek and R. Gibala, *Met. Trans.*, **14** (A10) (1983), 2143.
9. C. J. P. Steiner, D. P. H. Hasselman and R. M. Spriggs, *J. Amer. Ceram. Soc.*, **54** (8) (1971), 412.
10. J. Peters and H. J. Grabke, *Werkst. Korros.*, **35** (1984), 385.
11. T. A. Ramanarayanan, M. Raghaven and R. Petkovic-Luton, JIMIS-3, *Trans. Japan Inst. Metals, Suppl.* (1983), 199.
12. E. J. Felton and F. S. Pettit, *Oxid. Met.*, **10** (3) (1976), 184.
13. F. H. Stott and G. C. Wood, *Mater. Sci. Eng.*, **87** (1987), 267.
14. A. M. Huntz, *Mater. Sci. Forum*, **43** (1989), 131.
15. A. M. Huntz, *Mater. Sci. Eng.*, **87** (1987), 251.
16. J. Jedlinski, *Proc. 11th Int. Corr. Cong.*, **4** (1990), 21.
17. A. M. Huntz, *The Role of Active Elements in the Oxidation Behaviour of High Temperature Metals and Alloys*, Elsevier Applied Science, London (1988), p. 81.
18. B. Widyanto, C. Bracho-Troconis and M. Lambertin, *Mater. Sci. Eng.*, A **120/121** (1989), 207.
19. B. Widyanto, Thesis, Université de Technologie de Compiègne (1990).
20. H. Pfeiffer and H. Thomas, *Zunderfeste Legierungen*, Springer Verlag, Berlin (1964).
21. D. Caplan, *Corros. Sci.*, **6** (1966), 509.
22. F. Armanet, Thesis, Université de Technologie de Compiègne (1984).

Microstructural and Diffusional Studies in α-Aluminas and Growth Mechanism of Alumina Scales

J. PHILIBERT and A. M. HUNTZ

Laboratoire de Métallurgie Structurale, CNRS UA 1107, bat.413,
Université Paris XI, 91405 Orsay, France

ABSTRACT

In order to clarify how yttrium modifies the microstructure of α-alumina and to determine the nature of the point defects created by this doping element, TEM, STEM and EXAFS experiments were performed on yttria-doped α-alumina polycrystalline samples. For the sake of completeness, self-diffusion was studied in undoped and Y-doped α-alumina samples.

The apparent solubility of yttrium in polycrystalline alumina at 1550°C is about 300 ppm while, due to *yttrium segregation in the grain boundaries*, the yttrium solubility in the *bulk* of a single crystal is about 6 ppm. When the amount of yttrium ≥ 300 ppm Y_2O_3, yttrium is precipitated as $Y_3Al_5O_{12}$. For yttrium ≤ 300 ppm Y_2O_3, yttrium ions in solid solution are localised on aluminium sites and induce *point defect complexes* made of *oxygen vacancies and oxygen interstitials* which act as 'donors'.

Oxygen and aluminium diffusion coefficients in undoped α-alumina are of the same order of magnitude. Yttrium doping slightly increases oxygen bulk diffusion, but decreases oxygen grain boundary diffusion on account of yttrium grain boundary segregation.

These results confirm the role of yttrium on the transport properties of alumina and justify why, according to the substrate purity and composition, literature data indicate that alumina scales can grow either by predominant cationic or anionic diffusion.

1. Introduction

The beneficial effect of yttrium on the oxidation resistance of alumina-forming alloys has been known for many years[1,2] and has been the object of a great deal of experimental work.[3-5] Yttrium is known for slowing down the oxidation kinetics in many cases, and in particular it strongly increases the scale adherence. Though many attempts have been made to elucidate the mechanisms by which yttrium acts on alumina scales, various and sometimes opposite mechanisms have been suggested in order to explain the yttrium effect, particularly on transport properties.[3–5] According to the preponderant diffusion process (cationic or anionic diffusion) which ensures the alumina scale growth, it has been proposed that yttrium decreases either aluminium or oxygen diffusion.

Simultaneously, it is well known that yttrium incorporated in alumina acts as a 'donor', but this observation is not clearly explained. Yttrium is isoelectronic with aluminium and the 'donor' effect is attributed to the larger size of yttrium compared to aluminium without any explanation about the effect of yttrium on the point defect nature. Some studies also point out that yttrium has a strong tendency to segregate either along grain boundaries or at the surface of massive aluminas,[6,7] but the incidence of such a phenomenon on the behaviour of alumina scales is not discussed. There is also a lack of experimental data about self-diffusion in alumina.[8–14]

The main problem with alumina is that it is one of the most stoichiometric oxides and, consequently, its transport properties are dominated by the impurities. Up to now, it was not possible to compare cationic and anionic diffusion coefficients. Thus, it appeared necessary to perform new aluminium and oxygen self-diffusion experiments in alumina,[15,16] on the same materials treated in the same conditions and with the same procedure in order to perform a correct comparison of oxygen and aluminium diffusivities in the bulk and in the grain boundaries of undoped or Y-doped alumina.

This paper will relate some of the results obtained on the effect of yttrium on the microstructure of aluminas doped with variable amounts of Y_2O_3, on the grain boundary segregation phenomena, on the influence of yttrium on the nature and the amount of point defects in alumina and on the self-diffusion coefficients in undoped or Y-doped aluminas.

2. Materials

Massive undoped and Y_2O_3-doped (0.1 and 0.03 mol.%) α-aluminas for microstructural, microanalytical and EXAFS analyses have been obtained by sintering performed by P. Carry (Ecole Polytechnique de Lausanne), at 1550°C under a uniaxial load of 45 MPa.[7,17] For the diffusion experiments, undoped and Y_2O_3-doped (300 wt.ppm) α-Al_2O_3 single crystals were grown by the 'Verneuil technique' by Baikowski Chimie Company.[16] The diffusion surface is parallel to the (0001) plane. Dense polycrystals were obtained by hot pressing (P. Carry, Ecole Polytechnique de Lausanne),[16] using a 500 wt.ppm Y_2O_3-doped alumina powder provided by Baikowski Chimie. All these procedures induced a silicon contamination.

Diffusion experiments were performed using either $^{18}O^2$–$^{16}O^2$ isotopic exchange or thin film deposition of ^{26}Al.[16] Depth profiling for oxygen diffusion was made by SIMS (CNRS Bellevue), and for Al diffusion by γ counting.[16]

3. Effect of Yttrium on the Alumina Microstructure

In case of undoped alumina, the samples as sintered or heat-treated in air at 1500°C have a grain size of ~2–5 μm, with some grains lengthened in a direction perpendicular to the uniaxial load (Fig. 1a). The grain boundaries are particularly planar. After heat treatment at higher temperature, some grains become larger (~10 μm) (Fig. 1b). Pores were never observed.

In the 0.1 mol.% Y_2O_3-doped alumina samples, curved grain boundaries are associated with fairly equiaxed grains. The grain size is homogeneous and

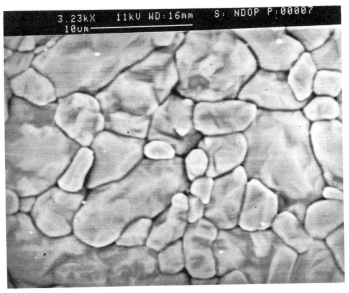

Fig. 1 Microstructure of (a) as sintered undoped alumina, (b) then heat-treated for 72 h at 1650°C in air.

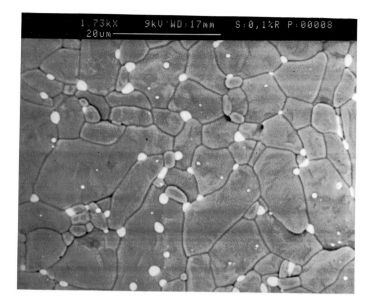

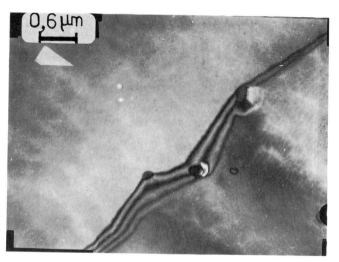

Fig. 2 Microstructure of 0.1 mol.% Y_2O_3-doped alumina, (a) after a heat treatment for 72 h at 1650°C in air, (b) and (c) after a heat treatment for 24 h at 1350°C in air.

somewhat smaller than for undoped alumina (Fig. 2a) A precipitated phase made of $Y_3Al_5O_{12}$ appears along the grain boundaries and in the bulk (Fig. 2a). After heat treatment at 1350°C, some grain boundaries are locked by the intergranular precipitates (Fig. 2b), and the intragranular precipitates are surrounded by a dislocation network (Fig. 2c). For this amount of Y_2O_3, most of the doping element is precipitated as $Y_3Al_5O_{12}$ in the bulk and along grain boundaries. These precipitates stabilise the alumina grain size. The main action of this doping element will consist in modifying the 'short circuit' diffusion

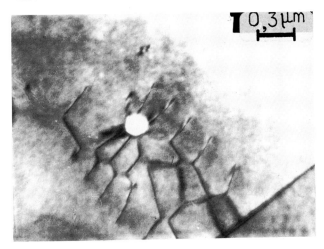

Fig. 2(c)

phenomena on account of the creation of new and different interfaces. In addition, the doping element will act on the plastic properties of alumina scales as it decreases the grain size.

For 0.03 mol.% Y_2O_3-doped alumina samples, this second phase is not observed, whatever the heat treatment. It means that the solubility limit of yttrium in a polycrystalline sample is ≥ 300 ppm. The grain size is homogeneous and equal to ~3.5 μm. Statistical analyses of the Lα ray of yttrium have been performed along and perpendicularly to the grain boundaries of thin foils. The apparent yttrium concentrations are C_b~0.5 a.u. and C_{gb}~0.8 a.u. in the bulk and in grain boundaries respectively. C_{gb} is an apparent concentration since it takes into account contributions from the boundary and from a part of the bulk due to the probe size. From these data, the real ratio of the yttrium concentration in the grain boundary to that in the bulk was calculated equal to C_{gb}/C_b~50.[7] If it is assumed that the 0.03 mol.% Y_2O_3-doped alumina is homogeneous at the grain scale, the real solubility limit of yttrium in the bulk of a single crystal is equal to C_b~6 ppm mol.Y_2O_3. Nevertheless, in our samples, due to the grain boundary density, 96% of yttrium is localised in the lattice, and only 4% is segregated in the grain boundaries.[7]

4. Effect of Yttrium on Point Defects

Figures 3 and 4 are relative to the Fourier transformed spectrum obtained by EXAFS on as-sintered 0.1 and 0.03 mol.% Y_2O_3-doped aluminas respectively.[7,17] The comparison with the spectra obtained with $Y_3Al_5O_{12}$ and Al_2O_3 (with Y^{3+} in substitution) shows that:

- in the sample doped with 0.1 mol.% Y_2O_3, most yttrium is precipitated as $Y_3Al_5O_{12}$;

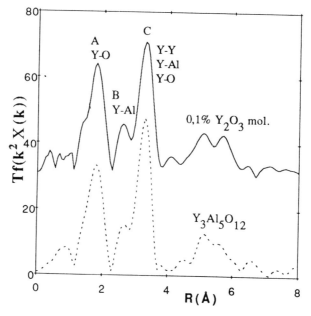

Fig. 3 Fourier-transformed spectrum of 0.1 mol.% Y_2O_3-doped alumina and of $Y_3Al_5O_{12}$.

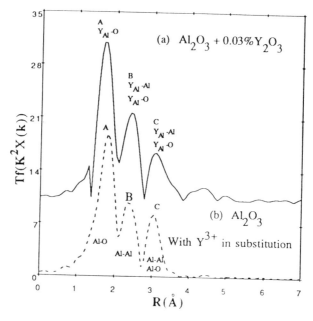

Fig. 4 Fourier-transformed spectrum of 0.03 mol.% Y_2O_3-doped alumina and of Al_2O_3 calculated by considering that large yttrium ions are in substitution on aluminium ion sites.

- in the sample doped with 0.03 mol.% Y_2O_3, yttrium ions substitute on aluminium sites.

Calculations of the distances R_j between an yttrium ion and its neighbouring atoms and of the average number N_j of first, second, etc., neighbours indicate that, due to the great size of the yttrium ion as compared to aluminium ($R_{Y3+} = 0.092$ nm and $R_{Al3+} = 0.054$ nm), oxygen vacancies are created in the alumina lattice around Y^{3+} ions and induce a displacement of other oxygen atoms in interstitial positions. So, from these results, it appears that yttrium ions localised on aluminium sites induce the creation of oxygen vacancies $V_{\ddot{O}}$ (on the first shell) and of oxygen interstitials O_i'' on a second shell. This means that yttrium induces localised point defect complexes according to:

$$2Y_{Al}^x + 3O_O^x \leftrightarrow 2Y_{Al}^x + 3V_{\ddot{O}} + 3O_i''$$

According to the literature,[21] yttrium acts as a donor for alumina. Due to the stable electronic configuration of Y^{3+} ions and to the great ionisation energy of the Y^{3+} ions compared to that of O^{2-} ions,[7] the species responsible for the donor effect must be oxygen on interstitial sites (for steric reasons). So the donor effect would be related to the following reaction:

$$2Y_{Al}^x + 3O_O^x \leftrightarrow 2Y_{Al}^x + 3V_{\ddot{O}} + 3O_i' + 3e'$$

It should be pointed out that EXAFS analyses are relative to yttrium located in the bulk and that the contribution of yttrium segregated in the grain boundaries (4%) to the EXAFS signal is negligible.

5. Self-diffusion Coefficients

The diffusion parameters obtained for oxygen and aluminium diffusion[15,16] in single crystals and polycrystals of undoped or Y-doped aluminas are gathered in Table 1 and in Fig. 5 (with some literature data comparison[18]) and Figs 7–9.
 Aluminium and oxygen lattice diffusion coefficients determined in the same undoped alumina with the same experimental conditions are of the same order of magnitude (Fig. 5). Aluminium diffusion is somewhat greater than oxygen diffusion but the difference is very small. Aluminium diffusion coefficients determined in our single crystals[16] are three to four orders of magnitude smaller than those determined by Paladino and Kingery[8] in polycrystalline samples. The activation energy for oxygen lattice diffusion[15] in alumina is of the same order of magnitude as most results of the literature shown in Fig. 5 for oxygen diffusion.
 For the determination of the oxygen and aluminium sub-boundary diffusion coefficients in undoped α-alumina,[15,16] it was verified by MET that the penetration curve tails were correlated with the presence of sub-boundaries (i.e. aligned dislocations) in the alumina single crystals (Fig. 6). The Arrhenius plot is given in Fig. 7. Aluminium sub-boundary diffusion is faster than oxygen

		Do (cm².s⁻¹)	AE (kJ.mol⁻¹)
Undoped alumina	Oxygen Diffusion		
	Lattice	206	630
	Sub-boundaries	$3.1\ 10^{14}$	896
	Grain boundaries	$1.6\ 10^{16}$	921
	Aluminium diffusion		
	Lattice	*0.16*	*510*
	Sub-boundaries	*$1.3\ 10^{14}$*	*850*
Y-doped alumina	Oxygen Diffusion		
	Lattice	67	590
	Sub-boundaries	10^{17}	980
	Grain boundaries	$7\ 10^{10}$	800

Values in italics: in the case of aluminium diffusion, in spite of the very few experimental points, an attempt has been made to determine the diffusion constants.

Table 1 Diffusion laws (δ_{sb} has been arbitrarily taken as $3\ 10^{-8}$cm)

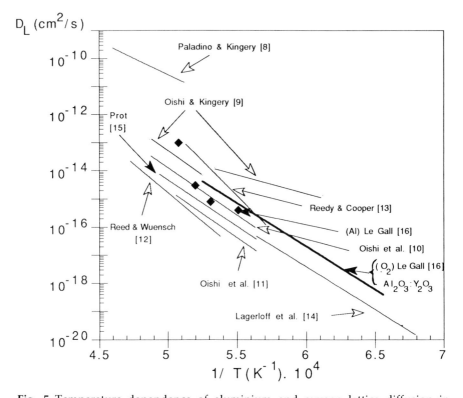

Fig. 5 Temperature dependence of aluminium and oxygen lattice diffusion in α-alumina. Our results in single crystals and literature data.

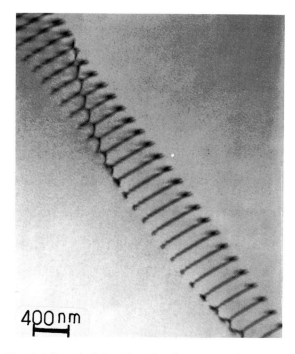

Fig. 6 Aligned dislocations in alumina single crystals.

sub-boundary diffusion. The difference is more pronounced than in the case of lattice diffusion (see Fig. 5). In both cases (Al and O diffusion), it is observed that the activation energy of sub-boundary diffusion is greater than the activation energy of lattice diffusion.

In single crystals of yttrium-doped α-alumina, only oxygen diffusion has been studied.[16] Oxygen diffusion in sub-boundaries only occurs for temperatures higher than 1400°C and it was also verified that the curve tails were correlated with aligned dislocations. The comparison of the results obtained in undoped[15] and Y-doped[16] alumina is given in Fig. 8. It again appears for Y-doped alumina that the activation energy of sub-boundary diffusion is greater than the activation energy of lattice diffusion. It can be considered, due to the experimental uncertainties, that the yttrium doping does not modify the activation energy of lattice diffusion and sub-boundary diffusion but, in both the lattice and the sub-boundaries, it slightly enhances the diffusion coefficients.

In the grain boundaries, oxygen diffusion was studied in undoped[15] and Y-doped[16] polycrystalline aluminas. The results are reported in Fig. 9. The oxygen grain boundary diffusion coefficients in Y-doped alumina are smaller than in undoped alumina by about two orders of magnitude, but the activation energy can be considered of the same order of magnitude. The activation energy of grain boundary diffusion (as well as that of sub-boundary diffusion) is greater than the activation energy of lattice diffusion.

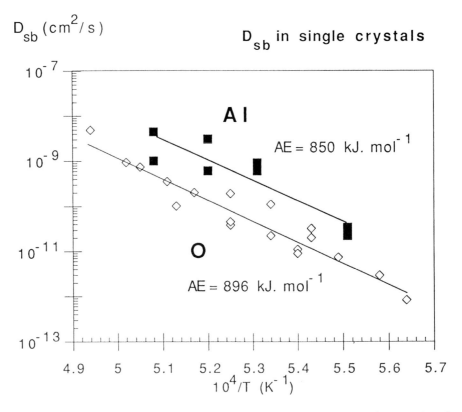

Fig. 7 Arrhenius plot of the sub-boundary self-diffusion in single crystals of undoped α-alumina.

For the lattice self-diffusion coefficients determined in undoped alumina single crystals, considering the non-negligible amount of Si and the low values of the diffusion activation energy, it is suggested that diffusion occurs by an extrinsic mechanism, via oxygen interstitials O_i'' for oxygen diffusion and V_{Al}''' for aluminium diffusion. The fact that the activation energy of aluminium diffusion is smaller than that of oxygen diffusion (according to theoretical calculations[19]) can justify the greater diffusivity of aluminium.

The larger diffusivity of aluminium, compared with that of oxygen, in the sub-boundaries of undoped alumina, can be due to the size differences between aluminium and oxygen ions ($R_{Al^{3+}}$ = 0.054 nm and $R_{O^{2-}}$ = 0.138 nm) and to the fact that oxygen ions have a greater affinity for cations segregated along the sub-boundaries. In both cases, the activation energy of sub-boundary diffusion would be equal to the sum of the defect migration enthalpy and of an interaction (or trapping) enthalpy term (between the diffusing species and the segregated impurities). Both terms would be greater in case of oxygen diffusion. Similar considerations also justify why the activation energy of oxygen grain boundary diffusion in undoped alumina is greater than the activation energy for lattice diffusion.

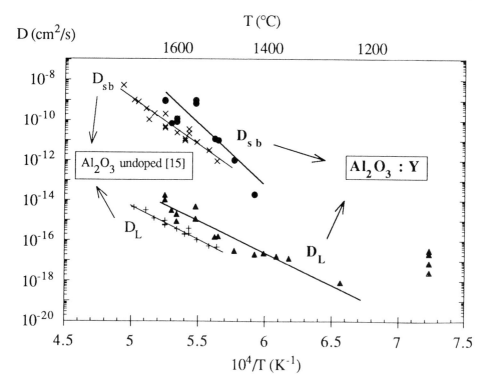

Fig. 8 Arrhenius plot of the lattice and sub-boundary diffusion of oxygen in single crystals of undoped and Y-doped α-alumina (300 wt.ppm Y_2O_3).

The yttrium doping of alumina induces a slight increase of oxygen lattice and sub-boundary diffusion, but a substantial decrease in the oxygen grain boundary diffusion. In case of lattice diffusion the differences are explained by considering that the extrinsic lattice diffusion mechanism is now controlled by silicon and yttrium impurities, which both act as 'donors'. In the case of sub-boundary diffusion, it is suggested that, at temperatures lower than 1400°C, yttrium is precipitated on sub-boundaries and notably decreases the oxygen sub-boundary diffusion (no curve tails were observed in these cases). At temperatures higher than 1400°C, the dissolution of yttrium precipitates induces a greater density of dislocations than in undoped alumina. These dislocations are free for diffusion and consequently increase oxygen sub-boundary diffusion in doped alumina.

In the case of grain boundary diffusion, yttrium decreases the oxygen diffusivity probably because of its segregation or precipitation in the grain boundaries. The activation energy of sub-boundary and grain boundary diffusion is always greater than the activation energy of lattice diffusion, since $(A.E.)_L = \Delta H_m$ when $(A.E.)_{sb \text{ or } gb} \sim \Delta H_m + \Delta H_{int}$.

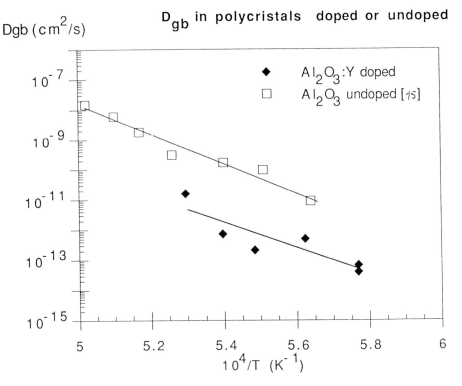

Fig. 9 Arrhenius plot of the grain boundary diffusion of oxygen in undoped and Y-doped α-alumina (500 wt.ppm Y₂O₃).

6. Application to the Alumina Growth Mechanism

6.1 Diffusion Coefficients and Scale Growth

Though all the results concerning self-diffusion are not available for a single alumina, it is interesting to calculate the parabolic oxidation constants k_c from the various possible diffusion mechanisms and to compare them with the literature data on k_c values given by oxidation experiments. From the following equation:

$$k_c = \int_{pO_2(i)}^{pO_2(e)} (1{,}5D_{\mathrm{eff}}^{\mathrm{Al}} + D_{\mathrm{eff}}^{\mathrm{O}})d \ln pO_2$$

with D_{eff}^Z the effective diffusion coefficient: $D_{\mathrm{eff}}^Z = (1-f-\lambda)D_L + fD_{\mathrm{gb}} + \lambda D_{\mathrm{sb}}$ where f is the volumic fraction of grain boundaries: $f = 3\delta/\Phi$, δ is the grain boundary width (taken as 1 nm), Φ is the diameter of the grains (taken equal to 1 μm), λ is the volumic fraction of sub-boundaries given by $\lambda = \pi u^2 \rho_d$, u the dislocation radius ($u \sim 1$ nm), ρ_d the dislocation density whose value is estimated to about 10^{10} cm/cm³.[16]

The calculations were carried out at 1100°C, by extrapolating the self-diffusion results. At this temperature, $\ln([pO_2(e)]/[pO_2(i)]) = 75.9$ in equilibrium conditions. The different diffusion constants in 'undoped' alumina are given in Table 1. Due to the lack of data on aluminium grain boundary diffusion, it has been assumed that $(D_{sb}^{Al})/(D_{sb}^{O}) \approx (D_{gb}^{Al})/(D_{gb}^{O})(\approx 24 \text{ at } 1100°C)$ in order to estimate D_{gb}^{Al}.

Aluminium grain boundary is found to be as important as lattice diffusion with:

$$(fD_{gb}^{Al}) = 7.10^{-21} \text{ cm}^2 \text{ s}^{-1} > (\lambda D_{sb}^{Al}) \quad \text{and} \quad D_{eff}^{O} \ll D_{eff}^{Al} = 1.3 \ 10^{-20} \text{ cm}^2 \text{ s}^{-1},$$

which gives

$$k_c = 1.5(D_L^{Al} + fD_{gb}^{Al}) \ln \frac{pO_2(e)}{pO_2(i)} = 1.5 \ 10^{-18} \text{ cm}^2 \text{ s}^{-1}$$

$$k_c \ll 10^{-13} \text{ cm}^2 \text{ s}^{-1} \text{ (experimental } k_c \text{ value)}.$$

It appear that aluminium diffusion should predominate over oxygen diffusion, at least in the case of the growth of alumina scales doped with donor elements. Clearly, literature data indicate that alumina scale growth is often controlled by oxygen diffusion,[3-5] even if cationic diffusion is observed in some cases (for instance NiAl alloys; but in this case Ni is incorporated in the scale and should act as an acceptor).

Lattice diffusion alone cannot justify the growth rate of alumina scales. Moreover, even considering the intergranular diffusion, the calculated oxidation constants are smaller than the experimental ones. Such a difference was already observed by Sabioni et al.[20] in the case of chromia scales. This suggests that grain boundaries and all diffusion short-circuits can chemically and physically differ according to the preparation route of the materials. Differences in mass transport rates could also be due to differences in the nature and amount of impurities in scales and in massive oxides, and modifications of the purity of scales can occur as the growth is going on. Another parameter, which should be considered, is the fact that, in the case of Al_2O_3 scale growth, other Al oxide phases can form before $\alpha\text{-}Al_2O_3$ nucleates and grows.

6.2 Effect of Yttrium

Calculations similar to those made in the previous paragraph can be carried out for Y-doped alumina. In this case, only oxygen diffusion results are available (see Table 1). At 1100°C, the oxygen lattice diffusion contribution predominates. Then:

$$k_c = D_L^{O} \ln \frac{pO_2(e)}{pO_2(i)} = 2.10^{-19} \text{ cm}^2 \text{ s}^{-1} \ll 10^{-13} \text{ cm}^2 \text{ s}^{-1} \text{ (experimental } k_c \text{ value)}$$

This value is smaller than that found in 'undoped' alumina.

Taking into account the microstructural results and considering the influence of yttrium in solid solution, it can be said that if the growth of alumina scales occurs at the outer interface (high oxygen pressure) by predominant cationic diffusion, yttrium will increase the growth rate as far as it is localised in the outer part of the scale. If the alumina scale growth occurs at the inner part of the scale (low oxygen pressure) by preponderant oxygen diffusion, yttrium will decrease the oxygen vacancy concentration and the oxidation rate, as far as it is localised in the inner part of the scale. This point is supported by variation of the ionic transport number of Y-doped alumina with the oxygen pressure.[21] These apparently opposite effects could correspond to the contradictory results reported for the dependence of growth rate on Y doping. This is related to the fact that lattice self-diffusion coefficients of oxygen and aluminium in undoped alumina are not very different, and according to the impurities incorporated in the alumina scale, oxygen or aluminium diffusion becomes preponderant.

Nevertheless, yttrium in solid solution is expected to have a negligible effect on the alumina scale growth on account of the low amount of yttrium in solution and especially if, in agreement with most authors, alumina scales grow mainly by short-circuit diffusion. Whether segregated or precipitated, yttrium would decrease the growth rate of alumina scales since, below 1400°C, no sub-boundary diffusion is observed and at all temperatures grain boundary diffusion is decreased by yttrium segregation or precipitation.

7. Conclusions

The apparent solubility of yttrium in polycrystalline alumina is as high as 300 ppm while, due to yttrium segregation in the grain boundaries, its solubility in the bulk of a single crystal is only about 6 ppm at 1550°C. With $Y_2O_3 \geq 300$ ppm, yttrium is precipitated as $Y_3Al_5O_{12}$. With $Y_2O_3 \leq 300$ ppm, most yttrium (96%) is in solid solution in the grains of the polycrystals. Yttrium ions are localised on aluminium sites and, due to their greater size, induce point defect complexes made up of oxygen vacancies and oxygen interstitials that act as donors due to oxygen interstitial ionisation.

According to this model, yttrium should increase cationic diffusion by aluminium vacancies and decrease anionic diffusion by oxygen vacancies. However, opposite effects will be observed if aluminium interstitials or oxygen interstitials are the preponderant point defects in alumina scale. Moreover, the yttrium influence will depend on its localisation in the scale depth.

On massive α-alumina, aluminium lattice diffusion coefficients are slightly larger than oxygen lattice diffusion coefficients and very much lower than values previously published (for polycrystals). In both cases, lattice diffusion occurs by an extrinsic mechanism controlled by Si impurities. The activation enthalpy of lattice diffusion corresponds to the migration enthalpy of oxygen interstitials in the case of oxygen diffusion, and to aluminium vacancies in the case of aluminium diffusion.

Sub-boundary (Al and O) diffusion and grain boundary oxygen diffusion are characterised by an activation energy greater than the activation energy of lattice diffusion probably because of segregation and trapping phenomena.

Doping of alumina with Y_2O_3 induces an increase of lattice and sub-boundary diffusion but a decrease of grain boundary diffusion. In the lattice, this effect is due to an increase in the amount of 'donor' impurities. In the sub-boundaries, it is due to a local increase in the dislocation density. The decrease of the intergranular diffusion is related to yttrium segregation or precipitation in the grain boundaries.

Comparison of our calculations and experimental oxidation results strongly suggests that the growth of alumina scales by oxidation of alumina-forming alloys is controlled by short-circuit diffusion. These short-circuits must be physically and chemically different from those found in massive alumina: indeed, the k_c values calculated from the fastest diffusion phenomenon are still smaller than the experimental ones determined by oxidation experiments. As short-circuit diffusion is the preponderant process for alumina scale growth, it is expected that yttrium should decrease the growth rate of alumina scales.

Acknowledgements

The following people have contributed to this work: D. Prot, M. K. Loudjani, M. Le Gall, C. Monty, R. Cortès, B. Lesage, J. Bernardini, M. Miloche, N. Brun, C. Haut.

References

1. L. B. Pfeil, UK Patent no. 459848 (1937).
2. D. P. Whittle and J. Stringer, *Phil. Trans. R. Soc. Lond. A.*, **295** (1980), 309–329.
3. F. H. Stott and G. C. Wood, *Mat. Sci. and Eng.*, **87** (1987), 267.
4. A. M. Huntz, in *'The Role of Active Elements in the Oxidation Behaviour of High Temperature Metals and Alloys'* (Ed. by E. Lang), Elsevier Applied Science (1989), p. 81.
5. J. Jedlinski, *Solid State Phenomena*, **21, 22** (1992), 335–390.
6. G. Petot-Ervas, C. Monty, D. Prot, C. Séverac and C. Petot, in 'Structural ceramics – processing, microstructure and properties', *Proceedings of the 11th RISØ Int. Symp. on Metallurgy and Materials Science* (Ed. J. J. Bentzen, J. B. Sorensen *et al.*), RISØ National Laboratory, Denmark (1990), pp. 465–470.
7. M. K. Loudjani, A. M. Huntz, R. Cortès, submitted to *J. Mat. Sci.* (part of M. K. Loudjani, doctoral thesis, University Paris XI, Orsay, France, 1992).
8. A. E. Paladino, W. D. Kingery, *J. Chem. Phys.*, **37** (5) (1962), 957–962.
9. Y. Oishi, W. D. Kingery, *J. Chem. Phys.*, **33** (2) (1960), 480.
10. Y. Oishi, K. Ando, Y. Kubota, *J. Chem. Phys.*, **73** (1980), 1410–1412.
11. Y. Oishi, K. Ando, N. Suga, W. D. Kingery, *J. Amer. Ceram. Soc.*, **66** (1983), C 130–131.

12. D. J. Reed, B. J. Wuench, *J. Am. Ceram. Soc.*, **63** (1980), 88–92.
13. K. P. R. Reddy, R. A. Cooper, *J. Am. Ceram. Soc.*, **65** (1982), 634–638.
14. K. D. D. Lagerlof, B. J. Pletka, T. E. Mitchell, A. H. Heuer, *Radiation Effects*, **74** (1983), 87–107.
15. D. Prot, Thesis, University Paris VI, France (1991).
16. M. Le Gall, Thesis, University Paris XI, Orsay, France (1992).
17. M. K. Loudjani, R. Cortès, submitted to *J. Eur. Cer. Soc.*
18. M. Le Gall, A.M. Huntz, B. Lesage, C. Monty, J. Bernardini, submitted to *J. Mat. Sci.*
19. G. J. Dienes, D. O. Welch, C. F. Fisher, R. D. Hatcher, D. Lazareth, M. Samberg, *Phys. Rev.*, **B11** (8) (1975), 3060.
20. A. C. S. Sabioni, A. M. Huntz, J. Philibert, B. Lesage, C. Monty, *J. Mat. Sci.*, **27** (1992), 4782–4790.
21. M. M. El Aiat, F. A. Kröger, *J. Am. Cer. Soc.*, **65** (6) (1982), 280.

Initial Stage of Oxidation of Fe–20Cr–5Al Single Crystals With and Without Additions of Yttrium

M. SIEGERS, H. J. GRABKE and H. VIEFHAUS

Max-Planck-Institut für Eisenforschung GmbH, Düsseldorf, Germany

ABSTRACT

The initial oxidation behaviour of Fe–20Cr–5Al single crystals (all compositions will be given in weight percent) with and without additions of yttrium (Y) has been investigated to elucidate the favourable influence of Y on the formation and adhesion of the protective Al_2O_3-scale.

By using two surface analytical methods [LEED (low energy electron diffraction) and AES (Auger electron spectroscopy)] it is shown that additions of Y promote the formation and adhesion of an Al–oxide layer and Y suppresses the sulphur (S) segregation to the surface. The Al–oxide that has been formed during the initial stage of oxidation is present as a surface compound whose structure depends on the orientation of the single crystal.

1. Introduction

Fe–20Cr–5Al forms a protective α-Al_2O_3 scale; for this reason the alloy shows a good oxidation resistance even at temperatures above 900°C. Y improves the adhesion of the protective Al_2O_3-scale; this effect is well established as one of the much studied 'reactive element effects'.[1,2,3]

To investigate the influence of Y on the formation and adhesion of the scale it was intended to analyse the oxidation behaviour of the doped material as well as the behaviour of the undoped material. The chemical composition at the surface and the surface structure during the initial stage of oxidation were investigated using AES and LEED.

2. Experimental

Fe–20Cr–5Al single crystals [orientations (100), (110), and (111)] with and without additions of Y were prepared using the Bridgman method. The compositions of the samples are shown in Table 1.

269

	Cr	Al	Y
Undoped	19.5	4.5	–
Doped	24.9	5.46	0.044

Table 1 Chemical composition of undoped and doped Fe–20Cr–5Al (in wt.%)

The oxidation procedure and the measurements were performed in an UHV-chamber equipped with a LEED- and an AES-system. Oxidation of the sample was carried out at different temperatures in the system which limits the oxygen pressure to a range of about 10^{-7} mbar. The sample could be heated indirectly by tantal wires and the temperature was measured pyrometrically.

At the beginning of oxidation the oxygen inlet and the heating current were started simultaneously so that the temperature rises during the oxidation (10–12 min) to its final value. Auger spectra were taken before, during, and after oxidation to show how the chemical composition at the surface changes. To obtain information on the surface structure, LEED-patterns were observed. For obtaining good LEED-patterns it was necessary to anneal the sample; however, the annealing causes an Al-enrichment at the surface. The annealed and unannealed samples show different oxidation behaviour. For this reason oxidation experiments were conducted for annealed (with Al-enrichment at the surface) and for unannealed (without Al-enrichment at the surface) samples.

3. Results and Discussion

The results of the oxidation experiments of Fe–20Cr–5Al (110) (undoped and doped with Y) are summarised in Tables 2 and 3. For reasons of brevity the

Oxidation temperature	Al-enrichment at the surface	Oxide resp. enrichment at the surface after oxidation
$T \leq 580°C$	No	Spinel
$T \leq 740°C$	No	At first formation of spinel; later formation of Al-oxide and reduction of spinel
$T \leq 1020°C$	No	At first formation of Cr-oxide; later formation of Al-oxide and reduction of spinel; Al-segregation to the surface; decreasing Al (68 eV) and Al-oxide peak (55 eV); S-segregation to the surface
$T < 500°C$	Yes	Al-oxide
$T \leq 950°C$	Yes	Al-oxide; metallic Al
$T \leq 900°C$	Yes	Hardly any oxide, but enrichment of Al, C, Cr, and S at the surface (22 min of oxidation)

Table 2 Oxidation experiments with Fe–19.5Cr–4.5Al (110)

Oxidation temperature	Al-enrichment at the surface	Oxide/enrichment at the surface after oxidation
$T < 500°C$	No	Formation of spinel; component of Fe-oxide predominates compared to the undoped alloy
$T \leq 730°C$	No	First formation of spinel; then pronounced formation of Al-oxide (more Al-oxide than on the undoped alloy)
$T \leq 1000°C$	No	First formation of Cr-oxide; then pronounced formation of Al-oxide, later segregation of Al (metal) to the surface (66 eV) and decrease of Al-oxide peak (55 eV); comparison to the undoped material: formation of Al-oxide starts at lower temperatures and is more pronounced on the doped alloy, the segregation of Al (metal) to the surface towards the end of oxidation (when Al-oxide is already present at the surface) starts at higher temperatures and there is no S-segregation being observed on the doped material
$T < 500°C$	Yes	Al-oxide
$T \leq 930°C$	Yes	Exclusively formation of Al-oxide; increasing amount of Al (66 eV) at the surface with rising temperature
$T \leq 900°C$	Yes	Exclusively formation of Al-oxide; with rising temperature decrease of the Al-oxide peak (55 eV) and increase of the Al peak (66 eV); comparison to the undoped material: after oxidation there is still some Al-oxide remaining on the surface whereas on the undoped alloy there is no Al-oxide left on the surface after oxidation; no S-segregation to the surface has been observed on the doped sample (22.5 min of oxidation)

Table 3 Oxidation experiments with Fe–24.9Cr–5.46Al–0.044Y (110)

experimental results of the (100) and (111) oriented surface are not described in detail, but they are roughly the same as for the orientation (110).

If the oxidation of unannealed samples without Y is performed at temperatures around 500°C where no Al-segregation occurs, only spinel was observed (in this case and in the following with the expression 'spinel' an Fe–Cr-spinel is meant). For thermodynamical reasons it can be supposed that Fe-oxide and Cr-oxide (indicated by Auger peaks) are present in the form of spinel.[4] If the temperature rises to a higher value (about 700°C on the (110) oriented material) during oxidation causing Al-diffusion to the surface the previously developed spinel is reduced while Al-oxide is preferentially formed. At temperatures of about 800–1000°C very strong Al-segregation to the surface occurs and the Al-oxide peak (55 eV) is diminishing while the Al peak (66 eV) increases.

Within this temperature range the Al segregates faster to the surface than it can be oxidised at an oxygen pressure of about 10^{-7} mbar. At even higher temperatures (about 900–1000°C) and at oxidation times of about 22 min the Auger peaks of Al-oxide (55 eV) and Al (68 eV) are diminishing and S and Cr segregate to the surface. S-segregation was observed only on the undoped material. On the surface of annealed samples Al-oxide is formed immediately upon oxidation due to Al-enrichment at the surface; in this case spinel was not observed.

On the samples doped with Y the formation of Al-oxide is promoted. On these samples Al-oxide is stable in a wider temperature range compared to the undoped samples. Y seems to promote the Al-segregation at even lower temperatures compared to the undoped material (500°C for (111) orientation and 540°C for (100) orientation) and Y seems to slow down the S-segregation at higher temperatures.

Before and after oxidation LEED-patterns were observed, giving information about the surface structures of the alloy and the oxide. But to obtain LEED-patterns it was necessary to have well ordered surfaces; they were obtained by annealing the samples. LEED-patterns from spinel could not be observed and it was not possible to anneal these samples. Annealing them leads to an Al-enrichment at the surface; the spinel is reduced and Al-oxide is formed previously.

LEED patterns from oxidised samples were only visible from the ones covered with Al-oxide. The Al-oxide is present at the surface in form of a surface compound and its structure depends on the orientation of the sample. On the (110) surface Al-oxide grows non-epitaxially in an hexagonal arrangement ($a \approx 3$ Å (see example, Fig. 4b). The surface compound of Al-oxide grows epitaxially on the (111) oriented surface in a $(\sqrt{3} \times \sqrt{3})$R30°-overlayer structure. On the (100) oriented surface the Al-oxide is present in a (2×1)-overlayer structure (there are two domains twisted 90° against each other).

To illustrate the measurements, the oxidation behaviour ($T \leq 900$°C) of the annealed (110) oriented Fe–20Cr–5Al single crystal (undoped and doped with Y) is described in detail:

On the undoped material no oxide has been formed after the oxidation procedure (Fig. 1), but S-segregation and cosegregation of Cr are shown by the Auger spectrum (Fig. 1b). In contrast, on the doped material, Al-oxide is formed (indicated by an Auger peak at 55 eV and a shoulder at 40 eV, Fig. 3d–f) which remains stable up to a temperature of 900°C and no S-segregation is observed at the surface (Fig. 2b). Towards the end of the oxidation procedure the Auger peak at 66 eV is increasing (Fig. 3-i), indicating metallic Al at the surface.

The LEED-pattern after oxidation shows an hexagonal arrangement of streaklike spots (Fig. 4b) additional to the spots of the annealed clean surface (Fig. 4a). The streaking of the spots is most probably caused by polycrystalline parts of the oxide.

The lattice constant of the hexagonal overlayer structure has been calculated to be 3 Å. Compared to the lattice constant of α-Al$_2$O$_3$ ($a = 4.76$ Å) it can be

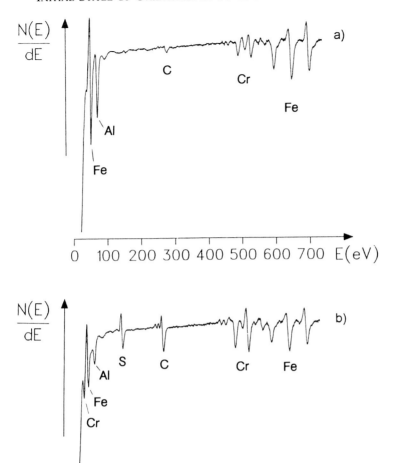

Fig. 1 Auger-spectra of Fe–19.5Cr–4.5Al (110). (a) Before oxidation; (b) after oxidation (22 min, $T \leq 900°C$, $p(O_2) = 1.5 \cdot 10^{-7}$ mbar).

concluded that Al-oxide is present in the form of an hexagonal surface compound and not in the form of α-Al$_2$O$_3$.

4. Conclusions

The processes observed in this study are mainly governed by kinetics, they depend on the surface composition and the temperature at which oxide is being formed. Spinel is formed on unannealed samples if the temperature is not high enough for Al-diffusion and Al-segregation. If after annealing Al has

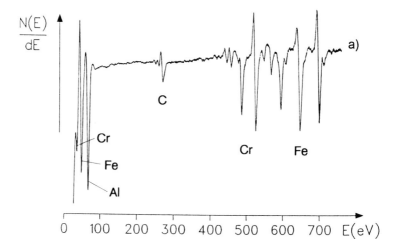

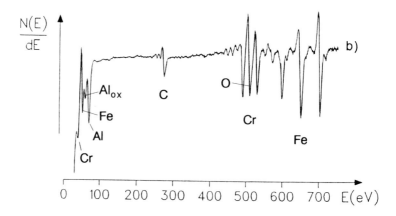

Fig. 2 Auger-spectra of Fe–24.9Cr–5.46Al–0.044Y (110). (a) Before oxidation; (b) after oxidation (22 min, $T \leq 900°C$, $p(O_2) = 1.5 \cdot 10^{-7}$ mbar).

segregated to the surface, Al-oxide is formed upon oxidation previously instead of spinel. If the oxidation is performed in a temperature range were Al-segregation is very rapid, the Auger peak of Al (66 eV) is increasing faster than the Auger peak of Al-oxide at 55 eV. This fact indicates that the oxygen pressure (about 10^{-7} mbar) is not high enough to oxidise the Al as fast as it segregates to the surface.

Additions of Y promote the formation of Al-oxide and the oxide remains on the surface up to a temperature range (about 900–1000°C) where it disappears from undoped material. On the doped samples Al starts segregating to the surface at lower temperatures compared to the undoped material. Y suppresses S-segregation to the surface at temperatures where it disappears from the undoped material during an oxidation (and heating) time of 22 min.

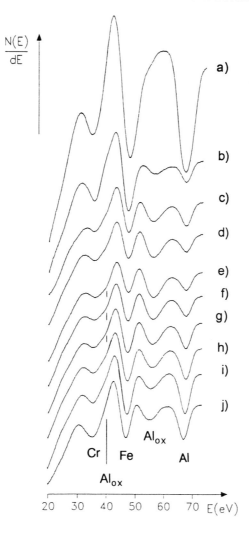

Fig. 3 Auger-spectra of Fe–24.9Cr–5.46Al–0.044Y (110). (a) Before oxidation; (b–i) during oxidation; (j) after oxidation (22 min, $T \le 900°C$, $p(O_2) = 1.5 \cdot 10^{-7}$ mbar).

LEED patterns show that the Al-oxide is present at the surface in form of a surface compound and its structure depends on the orientation of the sample. On the (110) oriented surface a non-epitaxially growth is observed, the Al-oxide is present in the form of an hexagonal arrangement with a lattice constant of 3 Å. On the (100) and the (111) orientations of the sample Al-oxide grows epitaxially; in the first case the Al-oxide is present in a (2×1)-overlayer structure (there are two domains twisted 90° against each other) and in the last case the Al-oxide is present in a ($\sqrt{3} \times \sqrt{3}$)R30°-overlayer structure. One important conclusion can be drawn concerning the reactive element effect:

The presence of Y already affects processes of nucleation and growth of

 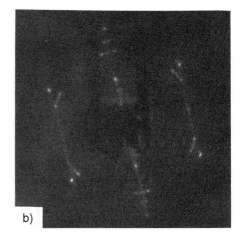

Fig. 4 LEED patterns of annealed Fe–24.9Cr–5.46Al–0.044Y (110). (a) Before oxidation ($E = 55$ eV); (b) after oxidation ($E = 55$ eV).

oxide monolayers, for which processes many mechanisms proposed for this effect cannot be effective, i.e. pegging, change of growth mechanism, enhanced scale plasticity, and the vacancy sink model.

Acknowledgment

The authors are grateful for support by the Deutsche Forschungsgemeinschaft.

References

1. J. Jedlinski, *Solid State Phenomena*, **21**, **22** (1992), 335–390.
2. D. P. Moon, *Mat. Sci. Technol.*, **5** (1989), 754–764.
3. D. P. Whittle, J. Stringer, *Phil. Trans. R. Soc. Lond.*, *A*, **285** (1980), 309–329.
4. I. Barin, *Thermochemical Data of Pure Substances*, VHC, Weinheim (1989), pp. 436, 438, 560, 563, 564.

Methodology Underlying Oxide Scale Growth Mechanism Investigation by Sequential Isotopic Exposures

J. JEDLIŃSKI,[1]* M. J. BENNETT,[2] and G. BORCHARDT[1]

[1]*Institut für Allgemeine Metallurgie and SFB 180, Technische Universität Clausthal, Robert-Koch-Str. 42, D-3392 Clausthal-Zellerfeld, Germany*
[2]*AEA Industrial Technology, Surface Technologies Department, B.393, Harwell Laboratory, Didcot, Oxon OX11 0RA, UK*
Present address: Faculty of Materials Science and Ceramics, Academy of Mining and Metallurgy, al. Mickiewicza 30, Kraków, PL-30-059 Poland

ABSTRACT

More efficient application of modern surface analytical tools in studying the mechanisms of growth of thin oxide layers during high-temperature oxidation requires better understanding of the experimental methodologies employed. It is timely to review the use of the so-called tracer method relying on sequential high-temperature exposures in oxygen with different ^{18}O enrichments. The underlying principles of this approach and current experimental procedures are described. An assessment is made of the surface analytical methods (such as nuclear reaction analysis and secondary ion mass spectrometry) used to analyse the elemental distributions. Examples are given from successful studies using this procedure. Limitations of the method are discussed, while areas for future applications of isotopic exposures are outlined.

1. Introduction

Understanding how protective oxide scales grow and fail on alloys and coatings and how these processes are affected by crucial parameters (such as incorporated active elements) is of fundamental importance in high temperature corrosion research. The growth rate of compact and continuous oxide layers is controlled by transport of the reactants. Since the diffusion rates of the oxygen and the metal frequently differ considerably, it is important to identify the predominant transport modes, which determine the oxidation rate of each material. Three major types of matter transport in scales can be distinguished:

1. Outward metal transport prevails, which may lead to the loss of contact at the scale/substrate interface due to the formation there of voids and/or cavities;
2. Inward oxygen transport predominates, which can induce stresses in the reacting systems resulting from the formation of 'new' oxide at the scale/substrate interface;
3. Simultaneous counter-current transport of both reactants occurs at comparable rates, which may cause substantial stress generation within the scale due to the formation there of a reaction product and result in a lateral scale growth.

For many years two methods have been used to assess the nature and rate of transport processes in oxide scales, namely: (i) the marker method, and (ii) the tracer method.[1,2] The earlier studies using these methods were carried out mainly on relatively thick scales ($\geq 10\,\mu$m) and the examination techniques were autoradiography, scanning electron or even optical microscopy. The subsequent need to study transport processes in $\leq 1\,\mu$m thick protective chromia and alumina oxide scales has led to the development of more sophisticated methods for the marker deposition and to the application of advanced surface analytical methods, such as AES (auger electron spectroscopy), RBS (Rutherford backscattering spectrometry), SIMS (secondary ion mass spectrometry), and SNMS (sputter neutral mass spectrometry) to determine the elemental distributions across the oxide scales. For many reasons,[3] the marker method cannot be employed for studies of the growth mechanisms of thin oxide layers. The only viable experimental approach, the tracer method, is based on oxide scale formation during sequential exposure in oxidants with a widely different enrichment of the isotope ^{18}O followed by analysis of the respective elemental distributions.

Apparently contradictory results obtained during studies of the growth of alumina scales on different substrates by the tracer technique,[4,5] have indicated that a fundamental reappraisal of this approach is needed. The main relevant issues will be discussed, improvements of the experimental procedure will be suggested, while its limitations will be considered. The most spectacular results obtained with the improved procedure will be presented. Finally, potential new applications of this approach will be proposed.

2. Two-Stage-Oxidation Exposure

In the standard experimental procedure reaction in an ^{18}O$_2$-enriched atmosphere is carried out either as the first or second stage, while the content of this isotope is negligible during the other oxidation stage, in essentially ^{16}O$_2$. Subsequent analysis of the respective distributions of both oxygen isotopes provides information on the transport processes in the scale growing during the second oxidation stage only. Since oxidation is a dynamic process and the scale morphology and microstructure are changing continuously and depend on the oxidation time and reaction temperature, it is crucial for the correct interpretation of the results to study additionally both these scale properties. This is particularly important during the early stages of oxidation, which lead to the formation of mature protective scales. Thus, in planning the experiments it has to be recognised that: (i) the longer the duration of the first oxidation stage, the less information which is provided on the early oxidation stages, and (ii) the longer the duration of the second oxidation stage, the more difficult is the interpretation of the results. For a systematic study some sequence of exposure periods should be chosen. One possibility is to set the duration of the first oxidation stage of the next experiment as equal to the total time of the previous two-stage oxidation, while the other is to fix the duration of the first oxidation stage and vary the second exposure period. In the first case it is possible to identify any changes in the oxidation mechanism over an extended period of time, while the second approach enables a more detailed study of any changes occurring during scale growth. Thus, the first procedure should be used in preliminary investigations, while the second should be followed in more detailed studies. It is essential to maintain the temperature during gas exchange between the consecutive oxidation stages in order to avoid cracking and/or spalling of the scale, which may occur, in particular, during cooling the oxidised material from the reaction to ambient temperature.

3. Elemental Distributions: Conventional Approach

The oxygen isotopes can be distinguished analytically either in terms of their different mass numbers or after conversion of the ^{18}O into ^{15}N by the nuclear reaction resulting from irradiation of the scale with protons. In the first case usually the SIMS or SNMS methods are used,[4,6–8] while in the second case the products of the nuclear reaction are analysed by means of the Rutherford backscattering spectrometry (RBS) or nuclear reaction analysis (NRA) methods.[9–11] The first group of methods is destructive, which is not the case for the second group. All the procedures are based on the interaction between the bombarding species (ions or neutrals) with the target material occurring on an atomic scale. This approach offers much better information concerning the transport modes in the scales but only by taking into account all factors which might affect the experimental results derived. Therefore, it is necessary to identify the method-induced disadvantages which need to be considered.

Excellent reviews exist which describe the fundamentals and applications of these examination methods.[12,13] As far as the use of SIMS and SNMS to determine the oxygen isotope distributions is concerned, the following points should be noted:[14]

1. The analysis is *ex situ*, as it is carried out after cooling down samples to room temperature and transportation to the measuring chamber.
2. SIMS/SNMS intensity depends on several factors, not only on the concentration of the element. In particular, effects related to the oxide roughness, porosity, cracking, as well as layered structure may be significant.
3. Water vapour, either derived from an inadequate vacuum in the measuring

chamber or adsorbed at the oxide's surface frequently contributes to the spectra. Monitoring of water vapour-related ions, i.e. (OH), may help evaluation of the extent of these effects. In addition, distributions of polyatomic oxygen-related species ($Me^{16}O$)-type and ($Me^{18}O$)-type should be correlated with the calculated profiles of the oxygen isotopes.

4. Isotopic exchange, which occurs during the scale growth, should be taken into account in interpretation of the results.

5. Significant problems appear when the materials being analysed are insulators. Modifications, such as neutral primary beam (NPB) SIMS or additional bombardment with electrons or ions, have to be introduced to combat charging effects.[15]

6. In most conventional spectrometers the diameter of the analysed area is of the order of 100 μm, which is far larger than the size of important microstructural features such as cracks and pores. It results in the information derived being global rather than the local.

7. A high-energy ion beam is used to reveal the consecutive layers during in-depth SIMS- or SNMS- profiling. An interaction between the bombarding ions and the target causes not only sputtering of the latter but also other effects, mostly undesirable, such as recoil implantation or ion beam mixing. These effects could affect the elemental distributions across the scales. Therefore, a compromise must be found concerning the energy of ion beam: it should be high enough to assure reasonably high sputtering rates yet, as low as possible to reduce the undesirable effects.

No reliable quantitative interpretation of tracer experiment results in terms of diffusion processes in the oxide scale has been undertaken so far, although it is commonly accepted that short-circuit diffusion processes, mostly grain boundary transport, significantly contributes to the growth mechanisms of protective chromia and alumina scales.[16] Attempts have been made to analyse theoretically different possible cases, [17–20] but, despite remarkable progress achieved recently,[18–20] this research must still be considered as ongoing. As yet, the measured distributions of the tracer are being interpreted in terms of some qualitative models and the scheme shown in Fig. 1 can be used for this purpose.

4. Elemental Distributions: Dedicated Methods

The necessity to study the short circuit transport processes was a driving force to improve the analytical spatial resolution. Recent progress in SIMS ion source development has enabled the analysed area to be reduced to ca. 100 nm diameter, which is still above that for revealing grain boundary distributions but does allow cracks and pores to be distinguished.[21,22] This enables more local information to be derived than was possible by conventional analysis (by ca. a factor of 1000).

The utilisation of modern, highly sensitive surface analytical methods, and in particular, realisation of their full potential, is dependent on the ability to

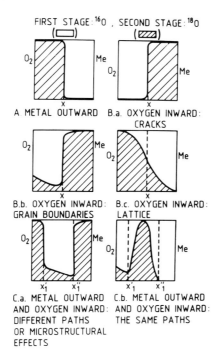

Fig. 1 Scheme illustrating the interpretation of tracer distributions in terms of transport processes in oxide scales [partially after references 4 and 7].

reveal the required features for examination. Thus, the satisfactory preparation of specimens is an essential prerequisite for surface analysis. Recently, the precision of the specimen preparation has been improved significantly by using advanced taper sectioning relying on surface microsurgery.[23] This enables considerable enlargement of a transverse cross-section of a surface layer and magnifications in excess of 1000 times can be achieved (Fig. 2). For each sample the taper angle has to be chosen to match the nature of the oxide scale, especially its thickness and roughness, in the light of the resolution required. In producing the taper angle considerable care has to be exercised to prevent loss of thinned oxide at the metal interface but this can be achieved with the precision polishing procedures described.[23]

Another advantage of examination of polished taper sections is the minimal topographical distortion caused by preferential sputtering on SIMS analysis. Only a moderate depth has to be eroded in order to remove the surface contamination and reveal the elemental images required. It should be noted however that taper sectioning is less useful for studying the elemental distributions in microstructural features having perpendicular orientation with respect to the scale/substrate and/or scale/gas interfaces, as for example through-scale cracks. In the latter case, the in-depth profiling still seems to be a better approach.

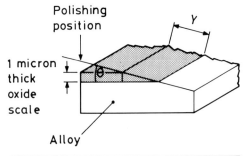

Taper section length (Y) microns	Taper angle θ
10	5°44"
100	34"
1000	3"
2000	1"

Fig. 2 Angle of taper polishing required to magnify exposed through scale section.[23]

5. Application of Improved Procedures: Four Recent Studies

An excellent investigation using isotopic exposures and high resolution imaging SIMS of the oxidation mechanisms of Fe–Al and Ni–Al alloys has indicated the necessity of a local and not a global approach to the transport processes in the growing scales and their complex nature.[22] In particular, the mechanism of growth of ridges, which are formed in cracks developing due to the phase transformation of unstable aluminas to the α-Al$_2$O$_3$, could be distinguished from that of the remaining oxide scale.

Use of the same approach has enabled correlation of the distributions of chromium and natural oxygen isotope (^{16}O$_2$) in local inhomogeneities of the outermost part of the scale on yttrium implanted β-NiAl doped with Cr (Fig. 3).[24] The deployment of the tracer isotope (^{18}O$_2$) in the rest of the scale, being the alumina, strongly indicates different growth mechanisms of Cr-free and Cr-containing regions, in agreement with an inward growth of chromia and outward growth of alumina on yttrium containing materials.

More sophisticated analysis of the transport modes in chromia on chromium was possible due to the combined application of taper sectioning and high resolution imaging SIMS; the most spectacular results of which are shown in Fig. 4.[21] The ^{18}O bearing streaks into the scale formed in ^{16}O$_2$ were interpreted in terms of limited anion transport into and exchange along the grain boundaries, in addition to the prevailing outward cation transport.

A systematic study of the early stages of oxidation of alumina-forming materials and of the effect of yttrium has been carried out using conventional SIMS and carefully planned two-stage oxidation experiments. Additional phase

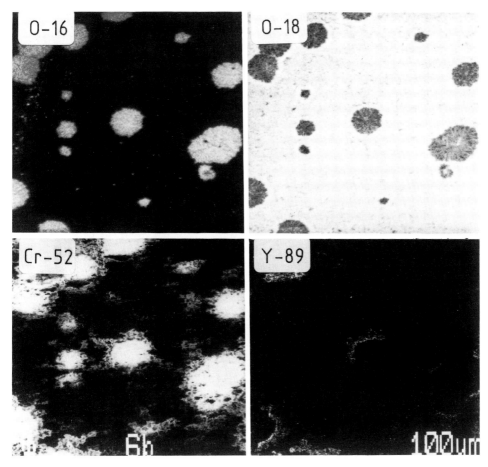

Fig. 3 SIMS-image of the surface of the scale formed on chromium-containing and yttrium implanted β-NiAl during oxidation at 1473 K for 9.5 min in $^{16}O_2$ and then for 15.5 min in $^{18}O_2$.[24] Note: actual width of region shown in each photograph is 100 μm.

composition studies and model considerations provided a consistent model and a comprehensive explanation of the discrepancies previously existing in the literature.[25,26]

6. Potential New Applications of Isotopic Exposures

Hitherto, the isotopic exposure technique has been aimed essentially at defining the relative contribution of various transport modes during growth of oxide scales and the extent of their applicability has reflected the current achievements in analytical techniques used for examining the elemental distributions. Most efforts have been directed towards the early stages of oxidation and elucidation of the formation and growth mechanisms of thin

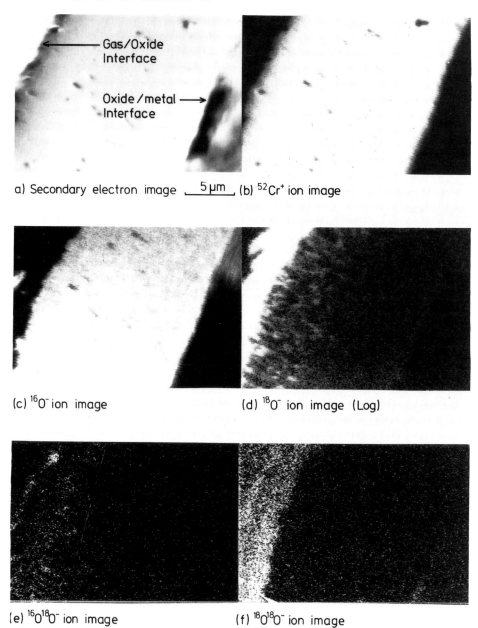

a) Secondary electron image ⌞5 µm⌟ (b) ^{52}Cr$^+$ ion image

(c) ^{16}O$^-$ ion image   (d) ^{18}O$^-$ ion image (Log)

(e) ^{16}O^{18}O$^-$ ion image   (f) ^{18}O^{18}O$^-$ ion image

Fig. 4 Transverse section of the oxide formed on pure chromium during oxidation at 1223 K for 45.75 h in ^{16}O$_2$ and then for 20.25 h in ^{18}O$_2$.[21]

oxide layers. New areas can be studied by taking advantage of increased spatial resolution, application of advanced sample preparation methods, i.e. taper sectioning, and better understanding of the two-stage oxidation procedure. The most attractive field for further study would seem to be investigation of the failure mechanisms of protective oxide scales and the degradation mechanisms of high temperature materials including composites. Appropriate application of isotopic exposures should facilitate studies of scale cracking and spalling. In fact, little is still known concerning the mechanisms of these processes.

Some promising preliminary results were obtained recently using conventional SIMS in studying the reasons for the presence of a thin oxide layer on the substrate after its cooling down from the reaction to ambient temperature and resulting in massive spalling of the scale.[27] Samples were oxidised in $^{16}O_2$ and cooled down in $^{18}O_2$, and vice versa, in order to verify whether this layer developed during cooling, or whether it was formed at high temperature and remained after spalling of the rest of the oxide scale. The first case would suggest an adhesive spalling mode, while a cohesive one would be indicated in the second case. The distributions of oxygen isotopes on Fe–20Cr–5Al alloy oxidised at 1373 K for 46 h and cooled down indicated the adhesive mode of spalling, at least in some areas.

Extension of the isotopic exposure method to more complex atmospheres would facilitate elucidation of the oxidation behaviour of materials in other industrial aggressive environments. Recent studies have showed that nitrogen penetration into the metallic substrate accompanies its degradation.[28,29] In this case overlapping $^{18}O_2/^{15}N_2$ exposures would be highly recommended. A similar procedure could be applied in sulphur/oxygen atmospheres. Moreover, it seems to be worthwhile to apply multi-stage exposures relying on alternation of gas composition more than once.

A need to keep abreast of the design of materials for high-temperature applications must direct attention towards the composites. Their layered, complex structures, with a large fraction of interfacial areas, make them particularly sensitive to local degradation in aggressive environments. It is well known that despite their outstanding good mechanical properties over a wide temperature range, carbon/carbon (C/C) composites cannot be deployed at temperatures exceeding 800 K due to the poor oxidation resistance.[30] Complex, sometimes porous, coatings applied to protect the C/C composite crack during temperature changes. Thus, it is crucial to study the extent and mechanism of the penetration of oxidant into the cracks and pores, which should be possible using isotopic exposures. A similar experimental approach is valid for metal matrix–ceramic fibre composites, as have been recently reported for the β-NiAl-Al$_2$O$_3$[31] and Ni$_3$Al–Al$_2$O$_3$[32] systems.

7. Conclusions

Critical analysis of the methodology underlying investigation of high-temperature oxidation mechanisms by sequential isotopic exposures has indicated that the required progress can be achieved by careful planning and execution of experiments, taking advantage of advances in surface analytical methodology, and the use of some dedicated techniques for sample preparation.

Three potential future applications of isotopic exposures could be: (i) studying the failure mechanisms of protective oxide scales; (ii) overlapping and multi-stages exposure in oxygen-, nitrogen- and/or sulphur-containing atmospheres; and (iii) investigation of degradation mechanisms of composite materials.

Acknowledgements

Helpful discussions with D. F. Mitchell, M. J. Graham (NRC Ottawa), A. Bernasik (ESMN Nancy), J. Słowik (TU Clausthal), and A. Gil (Kraków), as well as the assistance in manuscript preparation of E. Ebeling (TU Clausthal), and financial support of Deutsche Forschungsgemeinschaft are gratefully acknowledged.

References

1. P. Kofstad, *High Temperature Corrosion*, Elsevier Applied Science, London (1988).
2. S. Mrowec, *An Introduction to the Theory of Metals Oxidation*, The National Bureau of Standards and The National Science Foundation, Washington DC (1982).
3. J. Jedliński, G. Borchardt and M. J. Bennett, *Proc. 3rd European Ceramic Society Conference*, Madrid, September 1993, in press.
4. J. Jedliński and G. Borchardt, *Oxid. Met.*, **36** (1991), 317.
5. J. Jedliński, *Solid State Phenomena*, **21**, **22** (1992), 335.
6. J. Jedliński and S. Mrowec, *Mater. Sci. Eng.*, **87** (1987), 281.
7. M. J. Graham, J. I. Eldridge, D. F. Mitchell and R. J. Hussey, *Mater. Sci. Forum*, **43** (1989), 207.
8. W. J. Quadakkers, A. Elschner, W. Speier and H. Nickel, *Applied Surface Science*, **52** (1991), 271.
9. D. G. Lees and J. M. Calvert, *Corros. Sci.*, **16** (1976), 767.
10. K. P. R. Reddy, J. L. Smialek and A. R. Cooper, *Oxid. Met.*, **17** (1982), 429.
11. E. W. A. Young and J. H. W. de Wit, *Solid State Ionics*, **16** (1985), 39.
12. *Practical Surface Analysis*, 2nd edn (Ed. D. Briggs and M. P. Seah). J. Wiley & Sons and Salle+Sauerländer, Chichester (1992).
13. G. C. Smith, *Quantitative Surface Analysis for Materials Science*, The Institute of Metals, London (1991).
14. J. Jedliński, *Oxid. Met.*, **39** (1993), 61.
15. G. Borchardt, S. Scherrer and S. Weber, *Fresenius J. Anal. Chemistry*, **341** (1991), 255.
16. A. Atkinson, *Rev. Mod. Phys.*, **57** (1985), 437.
17. N. S. Basu and J. W. Halloran, *Oxid. Met.*, **27** (1987), 143.

18. W. Wegener and G. Borchardt, *Oxid. Met.*, **36** (1991), 339.
19. K. Bongartz, W. J. Quadakkers, J. P. Pfeifer and J. S. Becker, *Surface Science*, in press.
20. Y. Mishin, G. Borchardt, W. Wegener and J. Jedliński (these Proceedings).
21. M. J. Bennett, A. T. Tuson, D. P. Moon, J. M. Titchmarsh, P. Gould and H. M. Fowler, *Surface and Coatings Technology*, **51** (1992), 65.
22. R. Prescott, D. F. Mitchell, G. I. Sproule and M. J. Graham, *Solid State Ionics*, **53–56** (1992), 229.
23. J. A. Desport and M. J. Bennett, *Proc. 3rd Int. Symp. on High Temperature Corrosion*, Les Embiez, May, 1992, to be published.
24. J. Jedliński, D. F. Mitchell, M. J. Graham and G. Borchardt. To be published.
25. J. Jedliński and G. Borchardt, *Proc. Symp. on Oxide Films on Metals and Alloys*, The Electrochemical Society Inc. (1992), Vol. 92-22, p. 67.
26. J. Jedliński, *Oxid. Met.*, **39** (1993), 55.
27. J. Jedliński and G. Borchardt. To be published.
28. M. J. Bennett, J. A. Desport, C. F. Knights, J. B. Price and L. W. Graham. To be published in *Corrosion Science*.
29. B. A. Pint. Ph.D. thesis, Massachusetts Institute of Technology, Cambridge, 1992.
30. L. D. de Castro and B. McEnaney, *Corros. Sci.*, **33** (1992), 527.
31. J. Doychak, *Oxid. Met.*, **38** (1992), 45.
32. P. F. Tortorelli, J. H. De Van, M. J. Bennett and H. E. Bishop. To be published.

The Effect of Yttria Content on the Oxidation Resistance of ODS Alloys Studied by TEM

A. CZYRSKA-FILEMONOWICZ,[1] R. A. VERSACI,[2] D. CLEMENS,
W. J. QUADAKKERS

*Research Centre Jülich, Institute for Reactor Materials, PO Box 1913, 5170 Jülich,
Germany*
[1] *Academy of Mining and Metallurgy, Krakow, Poland*
[2] *CNEA, Buenos Aires, Argentina*

ABSTRACT

The effect of yttria content on the oxidation behaviour of FeCrAl-based ODS alloys of the type MA 956 and PM 2000 at 1200°C has been investigated. The growth rate of the alumina surface scale increases with yttria content, especially if the yttria content exceeds ca. 0.5%. SEM studies of oxide fracture surfaces and TEM analyses of the oxide microstructure revealed that the decreased oxidation resistance of high-yttria-containing alloys is caused by finer oxide grains resulting in an increased number of oxygen diffusion paths. After longer times formation of cavities and microcracks at the oxide grain boundaries leads to a further increase in growth rate because of additional oxygen transport. The formation of the cavities is probably correlated with the presence of coarse yttrium-containing particles at the oxide grain boundaries. From the viewpoint of oxidation resistance, the optimum yttria content of FeCrAl-based ODS alloys is significantly smaller than the value of 0.5% usually used in commercial alloys.

1. Introduction

The presence of finely dispersed yttria (mostly 0.5 wt.%) in iron–chromium–aluminium-based oxide dispersion strengthened (ODS) alloys not only increases the creep strength[1] of the alloys but also their oxidation resistance.[2,3] In spite of the excellent oxidation resistance, the possible service lives of FeCrAl based ODS alloys are not infinite, especially if the alloys are being envisaged to be used at temperatures of 1200°C or 1300°C.[4] Temperature changes of the components during service can lead to oxide spalling after the scale has reached a critical thickness. The rehealing of the scale consumes aluminium from the bulk alloy. If by this effect the alloy aluminium content decreases beneath a critical level, the protective alumina scales can no longer form and a catastrophic breakaway oxidation occurs.[4] It could be shown by mathematical modelling and confirmed experimentally that the time at which this breakaway occurs strongly decreases with only slight increases of the oxide growth rate.[4,5] One of the factors which appears to affect the growth rate of the alumina scale on iron chromium aluminium-based ODS alloys is the yttria content.[6] For ODS alloys of the type MA 956 it was found that variations of the yttria content between 0.17% and 0.7% lead to an increase in oxide growth rate during oxidation at 1100°C with increasing yttria content; however, the mechanism could not be defined exactly. In the present study the effect of yttria content on the growth rates of alumina scales on iron chromium aluminium ODS alloys is being investigated for the commercial materials MA 956 and PM 2000. The main emphasis will be placed on SEM and TEM analyses with the aim to investigate the effect of alloy yttria content on the alumina scale microstructure.

2. Experimental

The ODS materials used in the present study were four alloys of the type MA 956 (base composition in wt.%: Fe–20Cr–4.5Al–0.3Ti–Y$_2$O$_3$; supplier INCO International, Hereford, UK) and two alloys of the type PM 2000 (Fe–20Cr–5.5Al–0.3Ti–Y$_2$O$_3$; supplier PM Hochtemperaturmetall, Frankfurt, Germany). The four MA 956-type alloys were produced from the same base powder mixtures; however, they contained different yttria contents of 0.17%, 0.3%, 0.5% and 0.7%. Besides the PM 2000 batch with the standard yttria content of 0.5%, a second alloy (PM 2002) was investigated which additionally contained 0.25% of metallic yttrium. During the mechanical alloying process[1] this metallic yttrium is practically completely oxidised to an yttria-based dispersion.

Oxidation studies were carried out at 1200°C with flat specimens, of size 20×10 mm and 2 mm thickness, with an 800 grit surface finish. Oxide growth rates and spalling behaviour were investigated by gravimetrical analysis after intermediate cooling at regular time intervals during the long time exposures up to 3000 h in air. Scale cross-sections and morphology were studied by optical metallography and scanning electron microscopy (SEM) with energy-dispersive X-ray analysis (EDX). Oxide scale microstructure was studied by SEM analysis of fracture surfaces. Selected samples of alloy PM 2000 and PM 2002 were chosen for transmission electron microscopy (TEM) investigations of the oxide microstructure and of the yttrium distribution in the scale as well as in the bulk material. For planar section sample preparation of the oxide a specimen was cut into 3 × 5 × 0.3 mm strips after oxidation using a low-speed diamond saw. After hand grinding to 0.2 mm thickness, 3 mm diameter discs were punched. The discs were mounted in a GATAN apparatus and dimpled from the metal side to about 30 μm in the centre of the discs. Further thinning was accomplished by ion beam milling of the rotating sample from the metal side. Thin foils of bulk specimens were prepared by conventional double-jet polishing using a solution of 10% perchloric acid in ethanol (temperature: −30°C).

Extraction double-replicas were prepared by evaporation of carbon onto both sides of a thin specimen (thickness ca. 0.05 mm) followed by the dissolution of the metallic matrix in a 10% solution of bromine in ethanol.[7] Replicas were used for EDS analysis of the dispersoid composition and for measurements of dispersoid size distribution. Quantitative microstructural analysis of TEM micrographs was carried out using an interactive image analysis system (IBAS) of Kontron Co. for dispersoid size and scale grain size determination. The thinned specimens were investigated using a JEM 200CX electron microscope equipped with an energy-dispersive X-ray spectrometer (EDS) of Tracor Nothern.

3. Results

3.1 Effect of Yttria Content on Oxidation Behaviour

Figure 1 shows double logarithmic plots of weight changes as a function of time during oxidation of MA 956- and PM 2000-type materials with different yttria contents at 1200°C. The results confirm the data which were recently observed for MA 956-type materials at 1100°C:[8] in the range 0.17 to around 1% yttria, the oxidation rate increases with increasing yttria content. This effect appears to be independent of the alloy type used (Fig. 1b). Detailed analysis of the data revealed that the scale growth rates of the alloy with yttria contents of ca. 0.5% or lower obey a rate law, $\Delta m = k \cdot t^n$, where Δm is the weight change, t the exposure time and k the oxidation rate constant.[8] The value of n is approximately 0.35. The alloys with the highest yttria contents, however, show an n which is near to 0.5; the n even increases after exposure times of ca. 500 h.

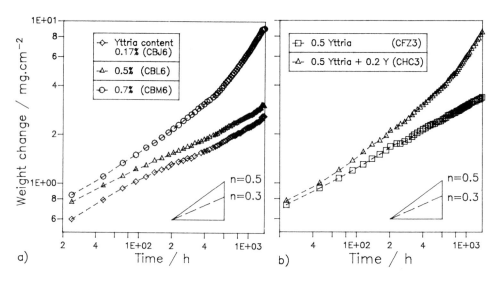

Fig. 1 Effect of yttria content on oxidation at 1200°C in air. (a) MA 956-type alloy; (b) PM 2000-type alloy.

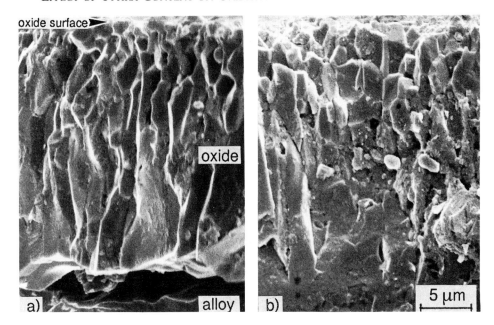

Fig. 2 Fracture surfaces of oxide scales after 200 h oxidation at 1200°C. (a) alloy PM 2000 (0.5% yttria); (b) alloy PM 2002 (0.5% yttria+0.25% yttrium).

A typical example of the effect of yttria content on oxide microstructure is shown in Fig. 2. The low-yttria alloy shows small grains at the scale/gas interface. Beneath these, columnar grains which enlarge in direction of the scale/alloy interface are observed. The scale on the high-yttria alloy does not explicitly show this increasing grain size of the oxide in direction of the interface with the alloy. A second striking difference is that the high-yttria alloy exhibits a large number of pores at the oxide grain boundaries. These effects were observed in all studied cases (1100 and 1200°C for exposure times of 5–1000 h) for the PM 2000-type as well as for MA 956-type materials.

3.2 Microstructure of Bulk Material and Oxide Scale

Previous studies[9,10] have shown that the microstructure of FeCrAl-based ODS alloys consists of mixed Y–Al oxide dispersoids, larger particles of pure Al_2O_3 and titanium carbonitrides Ti(C,N) in a ferritic matrix. The Y–Al oxides which may be present in the ODS alloys are: YAM (*y*ttrium-*a*luminium *m*onoclinic, $Y_4Al_2O_9$), YAH (*h*exagonal, $YAlO_3$), YAP (*p*erovskite, $YAlO_3$), and YAG (*g*arnet, $Y_3Al_5O_{12}$).

Statistical measurements of dispersoid size[9,10] performed on extraction double replicas using TEM and IBAS revealed three size ranges of mixed Y–Al oxides from 3 to 500 nm, the dominant small dispersoids being in the size range 3 to ca. 40 nm (Fig. 3). The comparison of the size distributions of two alloys with different yttria contents exhibited hardly any difference for the

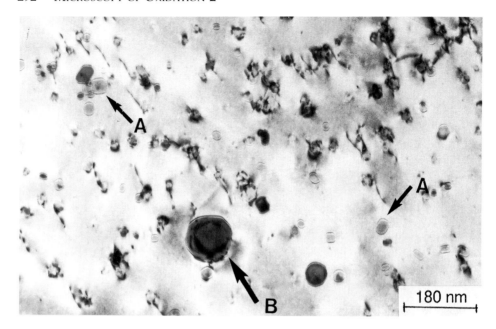

180 nm

Fig. 3 TEM micrograph of alloy PM 2000 exposed for 1000 h at 1200°C showing structure of the bulk material; smaller dispersoids with size up to 40 nm (A) and larger dispersoid about 110 nm in diameter (B).

dispersoids smaller than 50 nm, but alloys with a higher yttria content showed an increased number density of particles larger than 120 nm.[9]

The microstructures of the alumina scales on the alloys investigated after 1000 h oxidation at 1200°C are shown in Figs. 4–8. Figures 4 and 5 present the scale morphologies formed on alloy PM 2002 (0.5% Y_2O_3+0.25% Y) and alloy PM 2000 (0.5% Y_2O_3), respectively. On the grain boundaries, especially in alloy PM 2002, large yttrium-containing particles and cavities were observed. Statistical measurements of scale grain sizes performed on thin foils using TEM and IBAS showed grain sizes of the scales close to the surfaces in both specimens of 1.2–1.4 μm. Quantitative comparison of these oxide grain size measurements is not absolute, because it was not clear whether the planar section examined was at a comparative distance from the oxide/gas interface in both cases (compare Fig. 2).

Figures 6–8 show structural defects within the alumina scales formed on alloy PM 2002: namely a dislocation structure (Figs 6, 7), together with yttria particles and cavities on the grain boundaries (Fig. 7). In a few cases, cavities which were formed homogeneously and on dislocations within the grain (Fig. 8), were observed. The latter were found with the image slightly out of focus and under kinematical diffraction conditions. In underfocused conditions, as presented in Fig. 8, dark Fresnel fringes which surrounded the cavities were observed. An interesting observation was that the cavities in the grain sometimes showed a faceted morphology (Fig. 8).

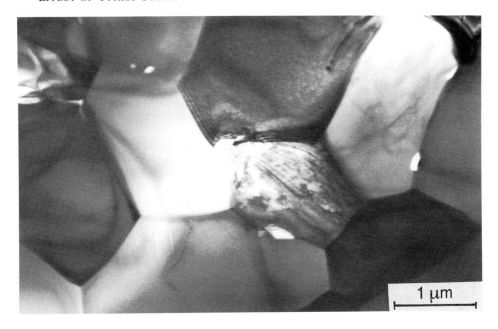

Fig. 4 Morphology of alumina scale (TEM micrograph of planar section) on alloy PM 2000 (0.5% yttria) after 1000 h exposure at 1200°C.

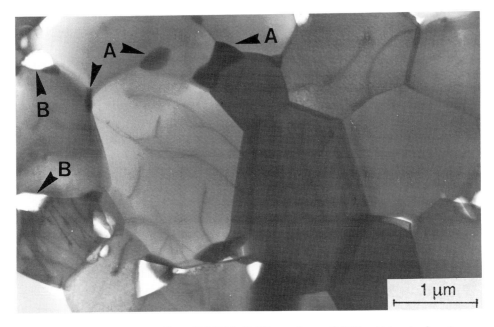

Fig. 5 As Fig. 4, for alloy PM 2002 (0.5% yttria + 0.25% yttrium) showing yttrium-containing particles (A) and cavities (B) on grain boundaries.

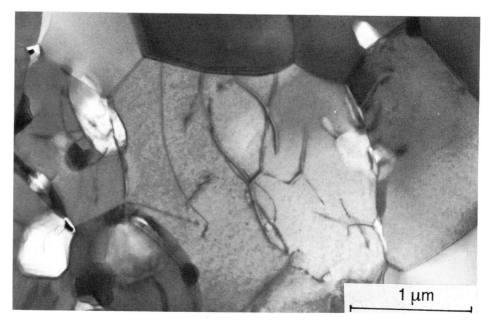

Fig. 6 Dislocation structure within alumina scale on alloy PM 2002 (0.5% yttria+0.25% yttrium) after 1000 h exposure at 1200°C.

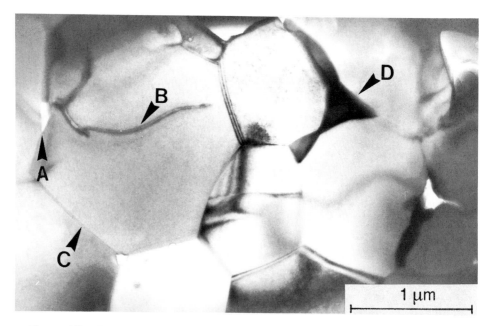

Fig. 7 Alloy PM 2002 (0.5% yttria+0.25% yttrium) exposed for 1000 h at 1200°C. Structural defects within the alumina scale: grain boundary cavities (A), dislocations (B), grain boundaries (C) and yttrium-containing particles (D).

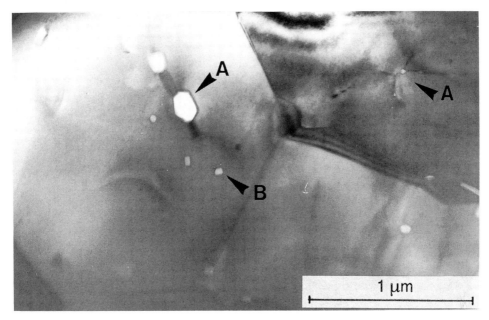

Fig. 8 Alloy PM 2002 (0.5% yttria+0.25% yttrium) exposed for 1000 h at 1200°C. Cavities formed on dislocations (A) and homogeneously (B) within the alumina grain.

4. Discussion and Conclusions

The oxide growth rate of FeCrAl ODS alloys in the temperature range 1000–1200°C increases with increasing yttria content.[8] Yttria levels above ca. 0.5% lead to significantly enhanced growth rates of the alumina scales. For these scales, the growth rate shows a significant deviation from the ideal exponential time dependence $\Delta m = k \cdot t^n$, i.e. the exponent n increases after longer times (Fig. 1). The high growth rate can be explained by the observed scale microstructure. As the scale grows by oxygen grain boundary diffusion,[8] the enlargement of the grains from the scale/gas to the scale/metal interface in the case of the low-yttria-containing alloys leads to a decreased number of oxygen diffusion paths and therefore a sub-parabolic growth rate ($n<0.5$) is observed. This increase of oxide grain size in the scale growth direction is far less pronounced or even absent in the case of alloys with yttria contents higher than 0.5%. This explains why in this case in the early stages of oxidation the exponent n is nearer to the value of 0.5 which would be expected if the grain boundary density would be independent of time, i.e. also of scale thickness.

The significant increase of the exponent n during the oxidation of the high-yttria alloys after longer exposure times, however, cannot be explained by the change in oxide microstructure because it would require that the grain size decreases with time, which was not observed. It can also not be explained by poor scale adherence, because scale spalling would eventually lead to weight loss, which also was not detected (Fig. 1). Recently it was shown that the variations in yttria content of FeCrAl ODS alloys do not lead to a change in oxide growth mechanism, as all the scales grow by oxygen transport, via rapid diffusion paths.[11] The higher growth rates on the alloy with high yttria content were found to be due to enhanced oxygen transport. A possible explanation of this effect could be the porosity and microcracking observed on the grain boundary, which was visible especially in MA 956 and PM 2000 with the highest yttria contents (Fig. 2). As this effect was observed for all alloy types, it is highly unlikely that the voids are induced by specimen preparation. The voids, which are visible in the fracture surfaces, are of the same size as those observed during the TEM analysis (Figs 2, 4, 5). This result and the observation that the voids are often correlated with dislocations in the grains, strongly indicate that the voids which are observed by TEM are not caused by specimen preparation but formed during the oxidation process, although some might be related to preferential etching during the ion beam milling.

These observations suggest that high alloy yttria contents lead to void formation and microcracking at the oxide grain boundaries after extended oxidation times. The formation of these cavities might be related to the counter diffusion processes and/or the formation of yttrium-containing particles at the boundaries. Previous studies have shown that in the alumina scales oxygen diffusion from the gas/oxide interface to the metal and yttrium (and titanium) diffusion from the metal to the oxide/gas interface occur, via grain boundaries.[6,11] If the diffusion coefficient of these oppositely directed fluxes are significantly different, vacancy formation could occur at the interfaces between the yttria particles in the grains[12] and the surrounding alumina. After longer times this might lead to an opening of the oxide grain boundaries finally causing enhanced inward transport of molecular oxygen and consequently an enhanced oxygen uptake, i.e. an accelerated oxide growth rate.

An interesting observation was that the yttrium-containing particles on the boundaries are often correlated with formation of dislocations in the grains (Fig. 8). The reason for this effect is not clear at the moment. It might be related to the generation of stresses which are induced by the scale and/or particle growth at the boundaries or by the thermal cycling because of different thermal expansion coefficients of alumina and the yttria-containing particles.

Acknowledgements

The authors are grateful to Mr Baumanns, KFA/IRW, for carrying out the oxidation experiments and to Mrs Esser for assistance in TEM specimen preparation. Dr Wallura and Mr Els are gratefully acknowledged for carrying out the SEM studies.

References

1. G. Korb and A. Schwaiger, *High Temperatures – High Pressures*, **21** (1989), 475.

2. W. J. Quadakkers, H. Holzbrecher, K. G. Briefs and H. Beske, *Oxid. Met.*, **32** (1989), 67.
3. M. J. Bennett, H. Romary and J. B. Price, Heat resistant materials, *ASM International*, 1991, p. 95.
4. W. J. Quadakkers and M. J. Bennett, *Mat. Sci. Technol.* to be published.
5. W. J. Quadakkers and K. Bongartz, Research Centre Jülich, to be published.
6. W. J. Quadakkers, K. Schmidt, H. Grübmeier and E. Wallura, *Mater. High Temp.*, **10** (1992), 23.
7. A. Czyrska-Filemonowicz, K. Spiradek and S. Gorczyca, *Praktische Metallographie*, **22** (1991), 217.
8. W. J. Quadakkers, *Werkstoffe und Korrosion*, **41** (1990), 659.
9. P. Krautwasser, A. Czyrska-Filemonowicz, M. Widera and F. Carsughi, *Materials Science and Engineering A.*, to be published.
10. P. Krautwasser, M. Widera, D. Esser and B. D. Wirth, *Proc. of 13th Int. Plansee Seminar* 1993, 24–28.05 (1993), Reutte, Tirol, Austria (in print).
11. D. Clemens, K. Bongartz, W. Speier, R. J. Hussey and W. J. Quadakkers, 7. Arbeitstagung 'Angewandte Oberflächenanalytik', 22–25 June, 1992, Jülich, FRG, Proceedings in Fresenius J., *Anal. Chem.*, **346** (1993), 318.
12. K. Przybylski and G. J. Yurek, *Mat. Sci. Forum*, **43** (1989), 1.

Oxidation of Incoloy MA956 After Long-term Exposure to Nitrogen-containing Atmospheres at 1200°C

N. WOOD and F. STARR

British Gas plc, Research and Technology Division, London Research Station, Fulham, UK

ABSTRACT

This paper looks at the effect of nitrogen-containing atmospheres on the long-term corrosion resistance of the ODS alloy, Incoloy MA956. Electron microscopy and X-ray analysis were used to study the effects of different levels of oxygen present in the nitrogen. the depletion of Al within the parent alloy, used in forming the surface film of α-Al_2O_3, and then large AlN cuboids, led to the MA956 exhibiting breakaway corrosion. The onset of breakaway was accelerated when the oxygen levels were low.

1. Introduction

Ferritic iron-based ODS (oxide dispersion strengthened) alloys are strong enough to resist creep at high temperatures (above 1100°C). However, in order to be used in energy conversion devices, they will need to show long-term resistance to oxidising environments.

Most of the research effort put into high-temperature oxidation during the last twenty years has shown that, at such high temperatures, only α-Al_2O_3 films can protect materials from degradation. Hence, the ferritic iron-based ODS alloys tested at British Gas are α-Al_2O_3 formers. One of the energy conversion plants of interest to British Gas will require a working fluid of N_2+2 v/v %O_2. As a consequence, tubes of the ODS alloy Incoloy MA956 have been exposed to nitrogen-containing atmospheres, with and without 2 v/v % oxygen.

The factors which allow the formation of a protective film have been researched and reviewed extensively.[1] Many papers have reviewed the microstructure, adhesion and growth kinetics of protective oxide films. One of the main conclusions was that the O_2 potential in a gas would determine the ability of a metal to form a continuous surface oxide of α-Al_2O_3. However, there has been little published on the effect of nitrogen.

Nitrogen is not generally regarded as a corrosive medium, as most environments contain usually sufficient oxidant impurities to form a protective surface film of α-Al_2O_3. Some work has been carried out on chromic oxide-forming austenitic stainless steels, and Inconels, at up to 700°C, mostly in ammonia-containing atmospheres.[2] However, there is little information on the nitridation of α-Al_2O_3-forming alloys, even though it can lead to significant changes in the mechanical properties of an alloy, as nitrides are hard, and consequently brittle.

A fundamental study of the thermodynamics of high-temperature nitridation and oxidation by Strafford[3] summarised and compared the standard free energies of formation for oxides and nitrides. Nitrides were always found to be less stable than oxides, so much so that internally formed nitrides often oxidise at high temperature.[4]

Thermodynamics show that the free energy of formation of α-Al_2O_3 is high (-849 kJ mol^{-1}) at 1000°C, while it is only -380 kJ mol^{-1} for AlN. The parabolic rate coefficient for AlN is 100 times less than that for α-Al_2O_3 (but data are only available at 600°C). Many other metals, including Ti, Zr, Ta and Nb, nitride in a protective manner, exhibiting parabolic kinetics.[3] The parabolic nitridation rates are 1–2 orders of magnitude less than the parabolic oxidation rate, but an exception to this rule is Cr. Cr is present in Incoloy MA956, and most other ODS alloys, at high levels, and it nitrides rapidly, more so at higher nitrogen pressures.

Strafford[3] also tabulated diffusivity data. The diffusivities of nitrogen in most metal nitrides are similar to those of oxygen in the corresponding metal oxide. This led to the conclusion that the lower weight gains and smaller parabolic rate constants for most nitride systems were only partially due to lower rates of diffusion in nitrides.

In contrast, the solubility of nitrogen in Group IV, V and VI metals was often much higher than for oxygen. The solubility of nitrogen in α-iron is 0.6 at.%, whereas for oxygen at the same temperature, it was only 0.02%.[3] It is reasonable to assume from this that any nitrogen atoms which get through a surface oxide film on a ferritic ODS alloy will be soluble in α-phase FeCrAl. Thermodynamics is against the formation of Fe_4N, and the free energy of formation of Cr_2N is low, compared with the figure of -530 kJ mol^{-1} for Cr_2O_3.[3] Any nitrogen atoms penetrating deep into a ferritic alloy, like MA956, will thus first combine with Ti (high ΔG formation), and then with Al. This still leaves the question as to how the nitrogen penetrates the surface film of α-Al_2O_3. Aydin et al.[5] claim that nitrogen atoms are virtually insoluble in oxides; hence, nitrogen can only transport, by Cr_2O_3 to the substrate, by molecular transport, through pores, cracks and grain boundaries. α-Cr_2O_3 and α-Al_2O_3 have very similar structures and, the latter, a lower defect structure. Hence, nitrogen probably transports through α-Al_2O_3 films in a similar manner. If true, the greater the amount of cracking and spalling, the greater the uptake of nitrogen.

Some work[6] recently carried out on two ODS alloys, in nitrogen, showed differences in performance between Incoloy MA956 and another ODS alloy, ODM331. In these tests, the two alloys were thermally cycled ten times from room temperature to 1200°C and corroded for 7 h. After this exposure, the Incoloy MA956 showed a uniform, adherent film, with no sign of spalling. The weight gain kinetics were parabolic and XRD detected only α-Al_2O_3. No nitrogen was found even using auger electron spectroscopy (AES) and even though precautions were taken to remove all the oxygen in the environment. The α-Al_2O_3 was found to have grown to about 5 μm thickness, with a uniform layer and void-free alloy/oxide interface. The ODM331 alloy also showed a uniform and adherent surface film of α-Al_2O_3, with the kinetics of attack closely following a parabolic law. However, the weight gain for ODM331 was 6 mg cm^{-2}, compared with less than 1 mg cm^{-2} for the Incoloy MA956. The extra weight gain was due to nitrogen pick-up and showed up internally as large cuboids of AlN. These cuboids were not apparent in samples sectioned after 2 or 5 cycles, so had probably formed following failure of the surface α-Al_2O_3 film. The ODM331 contains less Cr and Al than the MA956. Such low levels of both elements allowed significant amounts of nitrogen to diffuse into the substrate. A non-ODS, α-Al_2O_3-forming alloy also showed some internal nitrides. Most of the alloy (Fecralloy) showed no attack, but at its corners, where the nitrides had formed, the surface oxide had exfoliated. This fits in with Aydin et al.'s theory.[5]

In this paper, the results of exposing tubes of Incoloy MA956 to nitrogen-containing atmospheres for long periods at 1200°C will be presented. Light microscopy, electron microscopy and X-ray analysis were used to examine the nature of attack and to provide an understanding of the corrosion mechanism.

2. Experimental

The composition of the alloy tested and experimental conditions are shown in Table 1. In Test A, the stagnant air conditions present on the outer surface of the MA956 tube were induced by the presence of a thick layer of insulation.

	Alloy Composition				
	Fe	Cr	Al	Ti	Y_2O_3
Incoloy MA956	Bal.	16.5	4.5	0.5	0.4

	Operating Conditions	
	Test A	Test B
Test temperature	1200°C	1200°C
Internal atmosphere	N_2–10 ppm O_2	N_2–2 v/v % O_2
External atmosphere	Stagnant air	Natural air
Test time	2500 h	5000 h
Heating rate	3°C min^{-1}	3°C min^{-1}
Cooling rate	3°C min^{-1}	3°C min^{-1}

Table 1 Composition of alloy (wt.%) and operating conditions.

The insulation was wedged tightly at both ends of the furnace – it would have prevented any free flow of air along the outer surface of the tube. The white spot nitrogen (i.e. nitrogen, 10 ppm oxygen) flowed through the tube, at just over atmospheric pressure. In Test B, the white spot nitrogen was changed to nitrogen, 2 v/v % oxygen. Also, the insulation was removed, allowing convective currents to bring some air (and thus oxygen) to the outside of the tube. So, both the inner and the outer surfaces of the tube used in Test B saw higher levels of oxygen.

3. Results

3.1 Incoloy MA956 Tube from Test A

The Incoloy MA956 tube had corroded badly after 2500 h exposure to the stagnant air and nitrogen atmospheres. The hottest part of the tube showed large nodules (Fig. 1). These had effectively coalesced at one point into a nodule several inches long and half a centimetre thick. Closer inspection showed that much of the remainder of the hot zone had fine nodules. Areas a couple of centimetres away from the nodules showed a thick surface film. When examined edge-on using a stereo microscope, and compared with a sheet of aluminium foil 25 μm thick, the film and the foil were of similar thickness. Figure 2 shows that this oxide film had spalled in many places.

The tube was then sectioned across the large nodule, and at an area about 2 cm away. Samples of the tube were examined using X-ray diffraction (XRD). The large nodule was mostly haematite, Fe_2O_3, with very little other metallic ions in the oxide lattice. The inside of the tube was also mostly haematite, but with some Cr substituted, i.e. $(Fe,Cr)_2O_3$. At the area where corrosion was worst, there was no trace of any other phases. Just to the side of the worst corrosion, XRD was able to detect α-Al_2O_3.

Light and electron microscopy of the section taken 2 cm from the large nodule both showed a thick surface oxide film, plus considerable internal attack (Figs 3 and 4). A back-scattered electron image showed the film to consist of two phases (Fig. 5). Analysis of the film, using EDA (energy dispersive X-ray analysis), showed the darker area of the film to be Al-rich, and

Fig. 1 Macrophotograph of the massive corrosion product on the Incoloy MA956 tube tested in nitrogen.

Fig. 2 Surface oxide on the tube, a few centimetres away from the large nodule. Some areas or subscale, where the oxide has exfoliated, are visible.

Fig. 3 Light micrograph showing the non-planar surface film, and considerable internal attack. There is a zone of fine particles ca. 100 μm beneath the film.

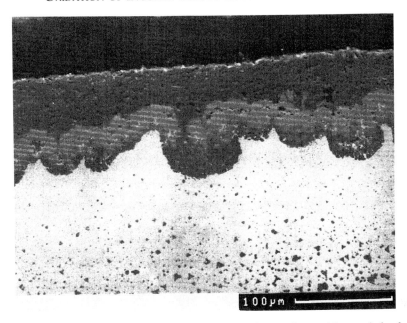

Fig. 4 Micrograph of the tube section, showing the surface oxide, and the large internal cuboids.

the lighter, inner layer to be Cr-rich. The whole film showed considerable voidage.

The small internal particles visible in Figs 3–5 were only present beneath the outer surface of the tube; they were not seen beneath the inner surface. One third of the tube section showed neither the larger nor the smaller internal particles. The surface films here were also thinner. This indicates that the catastrophic oxidation was localised, and that the rapid attack could be associated in some way with the large internal precipitates.

The large internal particles were examined more closely (Fig. 6). The sharp, angular or cuboidal nature of most of them was evident. A few small angular particles appeared to be present within the larger cuboids. They were Ti-rich nitrides. The large cuboids were analysed using wavelength dispersive analysis (WDA). The only elements detected were Al and N. A micrograph of some of the analysed cuboids is shown in Fig. 7, with a N X-ray map of the same area (Fig. 8). The particles were virtually pure AlN. EDA was used to quantify the metallic elements in a number of particles. The results are shown in Table 2.

All the Al-rich regions were analysed for oxygen, using WDA. This confirmed that the surface film was oxide, and the cuboids contained no oxygen, but were nitrides. The matrix of the parent alloy was analysed for any changes in the Al content. The areas surrounding the large particles were almost completely bereft of Al (see Table 2). The Al content in this area ranged from 0.1 to 0.5 wt.%. The matrix *close* (within 1 mm) to the denuded zone also showed low levels of Al, ca. 1.4 wt.%. Further away from the nitrided part of

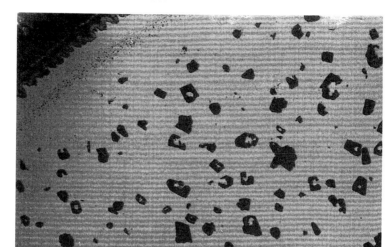

Fig. 5 Back-scattered image of the surface film. The lighter-coloured part of the oxide was rich in chromium.

Fig. 6 Higher magnification image of one of the cuboids, with fine, denser cuboids inside it.

Fig. 7 Back-scattered electron image of several typical cuboids.

Fig. 8 N X-ray map of the cuboids in Fig. 7.

	Element					
Area analysed	Al	Cr	Fe	Si	Ti	Y
Nitride particles	96.7	0.5	1.3	–	0.5*	2.5*
Matrix in nitrided region	0.37	20.3	77.8	0.13*	0.17*	0.88*
Matrix at edge of nitrided region	1.41	20.1	77.6	–	0.51*	1.17*
Matrix 1 mm from nitrided region	1.82	20.2	77.4	–	0.34	–

Note: Figures are weight per cent, averaged.
*Some particles contained Ti or Y, some did not.

Table 2 Energy-dispersive analysis of nitrided area.

the sample, the Al content was 1.8 wt.%. This figure is much less than the 4.5 wt.% present in the original alloy. Some Al is used up in forming the surface film of α-Al$_2$O$_3$, and some would be contained in the YAG particles. If a surface film of α-Al$_2$O$_3$ spalls frequently, the Al in the alloy matrix would be consumed more rapidly. Further Al was lost because its diffusion at 1200°C is quite rapid, and some would have diffused into the heavily attacked part of the tube. It is noteworthy that once nitridation started the Cr level in the alloy was unaffected and seemed to play no real part in protecting the alloy.

The electron micrographs and corresponding series of analyses show that once there is insufficient Al left in the alloy matrix for the surface film of α-Al$_2$O$_3$ to be repaired, nitrogen can pass into the parent alloy, and nitride formation is very rapid. Then the remainder of the Al is tied up in the nitrides, base metal oxides form and rapid corrosion ensues. The change from an area that was nitrided to one that was not was very abrupt.

3.2 Incoloy MA956 Tube from Test B

The tube from Test B was examined after 5000 h. Visual examination suggested that the tube had resisted oxidation well. There was no evidence of enhanced attack at any point. The hottest section of the tube was covered in a thick, white-coloured film which had the appearance of α-Al$_2$O$_3$. Clearly, the better circulation of air, and the 2 v/v % O$_2$ present had allowed the alloy to resist oxidation longer. Further testing (not covered in this report) showed that the sample went into breakaway oxidation after 10000 h.

4. Discussion

After 2500 h exposure to nitrogen at 1200°C, Incoloy MA956 had corroded catastrophically. When 2 v/v % O$_2$ was added to the nitrogen, and natural air currents allowed to reach the outer surface of the tube, the alloy was resistant to enhanced attack for nearly four times that period. Unfortunately, it is not possible to say categorically which is the more important: a very low Po_2 in a nitrogen atmosphere, or oxygen starvation in a stagnant atmosphere. Thus, when oxygen is in abundance, even at such a high temperature, an ODS alloy with 4.5 wt.% Al, such as Incoloy MA956, can resist attack for long periods. It does so by forming a compact, adherent film of α-Al_2O_3. However, if oxygen starvation has occurred, then the alloy does not resist attack by a second oxidant, in this case nitrogen.

A growing film of α-Al_2O_3 is a dynamic system – as the film grows, internal stresses build up, and diffusion occurs. Thus microcracks, or fissures, plus coalesced vacancies, or voids, will always be present. In an atmosphere with a plentiful supply of oxygen, any microcracks or fissures can be repaired by new growth in the film. The oxygen potential at the oxide/metal interface creates the driving force for selective diffusion of Al to the oxide/metal interface. Although Lai[2] found that an alloy will absorb nitrogen during long-term exposure, the amount will be much reduced by the presence of a compact surface barrier layer, such as an α-Al_2O_3 film. Lai admits that more work is needed on the effect of oxygen levels, and the effect of a surface film of α-Al_2O_3.

If oxygen starvation occurs, the microcracks and fissures etc. are no longer repaired. According to Aydin et al.,[5] the film will now allow nitrogen to penetrate it. The solubility of nitrogen in α-Fe is high, so the nitrogen diffuses into the parent alloy and ties up the Al deep within it. When the Al levels are too low, then the surface oxide will not heal at all. In MA956, Ti and Al have the highest affinity for nitrogen. The Al has combined with most of the nitrogen as Ti is present only at low levels. In fact, the small cuboids within the larger ones were TiN, acting as nucleation sites for the growth of the AlN (Fig. 6).

Why was the corrosion so localised? The large internal AlN cuboids were not present within one-third of the tube section. This part of the tube also showed a much thinner surface α-Al_2O_3 film. The breakaway attack only occurred where the Al had been tied up as AlN.

The microscopy and analysis undertaken leads to the conclusion that when the Al level falls to about 1.4 wt.%, a critical level of Al activity, a_{crit}, has been reached. Once the Al level in the MA956 matrix has fallen below a_{crit}, nitridation can occur. It probably results within a few hours at 1200°C, but would take longer at 1100°C, and might not occur at all at lower temperatures.

Another conclusion from this work is that the dividing line between an alloy being able to resist high-temperature oxidation, and undergoing catastrophic attack is very sharp. The Al levels in Table 2 are very significant. The nitrogen has tied up 1.4 wt.% Al. The part of the tube which contained 1.8 wt.% Al behaved like an α-Al_2O_3-forming alloy. Below a_{crit}, however, selective oxidation of Al at the film/metal interface can no longer occur. Oxidation of Cr and, more catastrophically, of Fe will then proceed. The implication here is that even when oxygen is not in short supply, once the Al content of any α-Al_2O_3-forming ODS alloy falls below 1.4 wt.%, the onset of breakaway attack and, hence, catastrophic failure will flow almost immediately. Alloys whose surface film spalls more rapidly will become depleted of Al more rapidly. Hence, the adhesion of the surface film is also crucial, as borne out by both Aydin *et al.*[5] and Karim.[6]

The alloy tested here was Incoloy MA956, but this rapid nitridation, followed by catastrophic corrosion attack, would occur on any α-Al$_2$O$_3$-forming ODS alloy, in a nitrogen-containing atmosphere starved of oxygen. It is also worth noting that work carried out recently by Bennett[7] has shown that the time to breakaway of ODS alloys was *less* in pure oxygen. However, these tests were carried out on polished specimens, whereas the MA956 tubes tested in the present studies were pre-oxidised in air. These results suggest that the 'Bi-oxidant' effect needs further investigation. In particular the effect of having the nitrogen under pressure should be studied.

5. Conclusions

1. In atmospheres with a plentiful supply of oxygen, Incoloy MA956 can resist oxidation at 1200°C for long periods, for at least 5000 h.
2. When oxygen is depleted below a critical level, Incoloy MA956, or any α-Al$_2$O$_3$-forming alloy, can be attacked by nitrogen. The nitrogen diffuses into the alloy to form large, cuboidal-shaped AlN. Once the nitrides have tied up the Al, base metal oxides, principally Fe$_2$O$_3$, form, and catastrophic attack ensues.
3. The formation of the large nitride particles only occurred when the parent alloy was depleted of Al and had fallen below a critical level of about 1.4 wt.%.
4. Even when oxygen depletion is not a problem, α-AL$_2$O$_3$-forming ODS alloys could be attacked by N$_2$ at high temperatures, as nitrogen is soluble in ferrite, and AlN is a very stable nitride. The adhesion of the surface film of α-Al$_2$O$_3$ is thus critical to governing the corrosion resistance of these alloys.

Acknowledgements

The authors are grateful to British Gas plc for permission to publish this paper, to Inco Alloys Limited, for supplying the Incoloy MA956, and to our colleagues, J. Wonsowski and Q. Mabbutt, who helped with the experimental work.

References

1. H. Hindam and D. P. Whittle, *Oxid. Met.*, **18** (1982), 245.
2. G. A. Lai, 'High Temperature Corrosion of Engineering Alloys', *ASM International* (1990).
3. K. N. Strafford, *Corros. Sci.*, **19** (1979), 51.

4. O. T. Goncel, J. Stringer and D. P. Whittle. Paper presented at 18th Corrosion Science Symposium, Univ. of Manchester (1977), ref. by (3).
5. J. Aydin, H. E. Buhler and R. Rahmel, *Workstoffe u. Korrosion*, **31** (1980), 675.
6. N. Karim, The Nitridation of Commercial Alloys at Very High Temperatures. MSc. Thesis, UMIST, Manchester, 1989.
7. M. J. Bennett and R. Perkins, unpublished results.

Nitridation of Ferritic ODS Alloys

M. TURKER and T. A. HUGHES

School of Materials, University of Leeds, Leeds LS2 9JT, UK

ABSTRACT

The nitriding behaviour of two ferritic oxide dispersion strengthened (ODS) alloys, namely MA956 and ODM751, has been examined at temperatures of 1200 and 1300°C in environments of stagnant air and a flowing mixture of 2% oxygen in nitrogen. Breakaway oxidation occurred in both alloys at 1300°C and in MA 956 at 1200°C when the aluminium content of the matrix decreased to a critical level. Subsequently, MA 956 showed evidence of internal nitridation with the formation of coarse AlN particles whereas, after much longer exposures, alloy ODM 751 was found to be both internally oxidised with α-Al$_2$O$_3$ and, at greater depths, internally nitrided with TiN.

1. Introduction

A closed-cycle gas turbine system incorporating a heat exchanger operating above 1000°C has been designed to utilise oxidation-resistant, ferritic ODS alloys produced by mechanical alloying (MA). The working fluid in the turbine will probably be air and will almost certainly contain nitrogen. The long-term oxidation behaviour of these materials at high temperatures has been under investigation for many years and considerable data are now available.[1,2] However, only short-term nitridation tests have been performed[3-5] and a very limited amount of data is available in the published literature.

Short-term ($\leq$100 h) nitridation behaviour of mechanically alloyed iron-based MA 956 was investigated[3,4] in nitrogen-containing atmospheres in the temperature range 1000–1200°C and the alloy was found to be very resistant to nitridation. This behaviour was attributed to the formation of a very adherent alumina scale which acts as a barrier between the nitrogenous atmosphere and the matrix. However, Kane[5] studied nitridation of the same alloy at 1200°C in four different surface conditions and found localised formation of AlN in abrasively cleaned MA 956 after 100 h.

Alloying elements such as Cr, V, Nb, Al, Ti and Zr (increasing order of stability) all form stable nitrides and most superalloys contain at least one of these elements. Aluminium is one of the stronger nitride formers and AlN can form in a range of alloys depending on the temperature and the Al content.[6,8] Low (>1.3 wt.%) Al-containing alloys were found to contain extensive precipitation of AlN whereas high (4.5 wt.% Al-containing alloys formed little AlN.[7,8] These effects were again attributed to the formation of a coherent alumina scale.

2. Materials and Experimental Procedures

Two commercial ferritic ODS alloys were selected for investigation. These were MA 956 (Inco Alloys International Ltd., England) in both sheet and tube form and ODM 751 tube (Dour Metal S.A., Belgium). The compositions of the alloys are given in Table 1.

ODM 751 tube, with an outside diameter of 24 mm and a wall thickness of 2.4 mm, was cut into 10 mm lengths and samples of 1 mm thick MA 956 sheet were cut to give coupons approximately 20×15 mm for long-term isothermal exposure tests in a flowing mixture of 2% oxygen in nitrogen. The sample surfaces of both materials were abraded on 600 grade SiC paper to remove the preformed oxide film and thereby produce a standard surface finish. After ultrasonic cleaning in acetone, the samples were exposed in tube furnaces operating at 1200°C (±3°C) for times of 2.5, 24, 240, 1200, 2400, 4800 and 7368 h. Alumina furnace furniture was used exclusively to support the specimens.

Further exposure tests were carried out at 1300°C in stagnant air on tubular samples of MA 956 and ODM 751 cut to 60 mm lengths and then machined to give three different wall thicknesses, as shown schematically in Fig. 1. The outside surfaces of the tubes were abraded using 600 grade SiC paper while the inner surface was left in the as-received condition. After solvent cleaning, the samples were exposed at 1300°C (±5°C) in a recrystallised alumina tube furnace with sealed ends. The idea of this arrangement was to see how, after depletion of oxygen inside the furnace due to the formation of alumina scale, the remaining nitrogen would interact with the different thicknesses of material. Since the nitridation behaviour is strongly influenced by the Al level within the sample and depletion of Al during exposure is faster in thinner sections, the nitridation kinetics should be thickness dependent. Exposure times up to

Alloy	Fe	Cr	Al	Ti	Mo	Y
MA 956	Bal	20.51	3.77	0.13	–	0.41
ODM 751	Bal	16.70	4.38	0.57	1.56	0.34

Table 1 Compositions of the alloys in wt.%

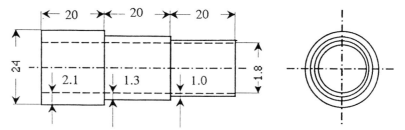

Fig. 1 Geometry of stepped tube samples.

1824 h were used and during exposure the samples were removed from the furnace at intervals to establish whether nitridation had occurred or not.

Exposed samples were mounted in bakelite and then polished using conventional techniques (wet grinding on SiC papers followed by polishing on 'diamond cloths'). Optical examination was performed using a Reichert MeF 3 optical microscope. A Camscan series 4 scanning electron microscope equipped with a windowless energy-dispersive X-ray detector capable of analysing elements down to carbon was used. Quantitative analysis was performed using a CAMECA SX-50 electron microprobe analyser fitted with three wavelength-dispersive spectrometers and a LINK 10/55S energy dispersive system.

3. Results

3.1 Exposure at 1200°C in Flowing Nitrogen–2% Oxygen

Small samples of MA 956 sheet and ODM 751 tube were exposed at 1200°C in a flowing mixture of 2% oxygen in nitrogen for times up to 7368 h. During the initial stages of exposure, both samples developed external scales of alpha alumina which were continuous, adherent and uniform, although the oxide on MA 956 exhibited a significantly higher growth rate than that formed on ODM 751. Further exposure (2400 h) resulted in some spalling of the thicker oxide layer on MA 956 during cooling with relatively little on ODM 751. However, neither material showed any evidence of sub-surface reactions indicative of nitridation.

After 4800 h exposure, the oxide scale on MA 956 was found to be ~80% spalled and beneath the spalled regions there was clear evidence of sub-surface precipitation. Figure 2 shows a longitudinal through-thickness section of this sample in which two distinct regions are clearly evident. Region (A) shows a very irregular thinner scale with a highly convoluted interface between it and the underlying matrix. In this region, the alumina scale has spalled and has been replaced by a chromium-rich oxide with an irregular growth morphology. Immediately below this the matrix contains fine (~4 μm) irregularly shaped alumina particles to a depth of about 200 μm. Beyond this zone, and extending to the centre line of the sample, there are relatively coarse (~10 μm) irregularly shaped AlN particles together with a few similarly sized but more angular TiN

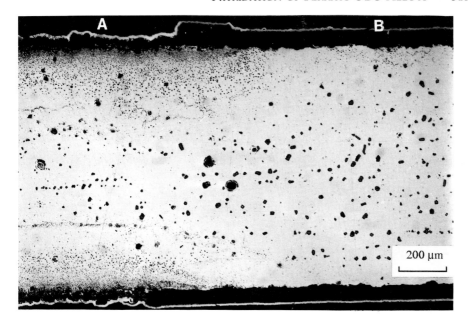

Fig. 2 MA 956 sheet exposed for 4800 h at 1200°C in flowing nitrogen–2% oxygen showing: (A) spalled region; (B) unspalled region.

particles. These latter almost certainly derive from Ti(C,N) particles which were present in both alloys prior to exposure. Electron probe micro-analysis (EPMA) indicates that the Al content of the matrix close to the Cr-rich scale is essentially zero and this increases to about 1 wt.% in the central region of the section.

In areas of non-spalled scale (B), the alumina layer remained adherent and uniform with a thickness of about 50 μm. There was little evidence of alumina formation in the matrix beneath this layer but coarse AlN particles were again present throughout the section. The aluminium content of the matrix was approximately 1.3 wt.% and was uniform across the thickness of the sample.

After the same exposure time (4800 h), the ODM 751 tube had formed an alumina scale about 27 μm thick with relatively little spalling and showed no evidence of sub-surface reactions. The aluminium content was uniform and fairly high (3.2%). Longer exposures of up to 7360 h resulted only in a thickening of the alumina scale.

3.2 Exposure of Stepped Tubes at 1300°C in Stagnant Air

Stepped tube samples of MA 956 and ODM 751 were exposed at 1300°C in stagnant air for times up to 1824 h under effectively cyclic conditions. After 480 h, breakaway oxidation had occurred on the exposed end of the thinnest section of MA 956 (Fig. 3) and extended some 15 mm along the surface. The large friable nodule consisted of a Cr-rich mixed oxide and was centred on the outer corner of the end face. Within the matrix adjacent to the mixed oxide

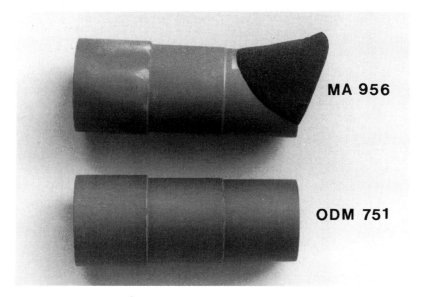

Fig. 3 Stepped tube samples exposed for 480 h at 1300°C in stagnant air.

layer, and extending for a distance of ~1 mm along the length of the tube, there were high densities of coarse (~30 μm) irregular AlN, finer (~4 μm) irregular Al_2O_3 and a few angular TiN particles (Figs 4 and 5). All three types of precipitate appeared to be uniformly distributed throughout the wall thickness of the tube and did not occur in sequential layers. The aluminium content of the matrix in the reaction zone was about 0.5 wt.%. Beyond this region only a slightly coarsened population of TiN particles was visible with a matrix aluminium content of about 1.9 wt.%.

This same MA 956 sample also exhibited small amounts of internal nitridation at the outer corners of the two thicker steps where the alumina layer had spalled locally during cooling. A fine Widmanstatten array of needle/lath shaped AlN particles (~2 μm) was observed in each of the corner regions extending to about 75 μm below the surface (Fig. 6). The aluminium content of the matrix in this region was about 1.7 wt.% compared with approximately 2.6 wt.% towards the central regions of the sample.

Continued exposure of the sample for a further 96 h resulted in rapid growth of the nodule in breakaway oxidation mode and the test was discontinued.

The ODM 751 stepped tube proved much more resistant to both breakaway oxidation and the ensuing sub-surface reactions. After 960 h breakaway oxidation had initiated on the end face of the thinnest step and a 350 μm thick layer of Cr-rich oxide had formed. Adjacent to this oxide scale was a narrow (50 μm) transition region of matrix containing small rounded second phase particles of variable chemistry but involving Al, Cr and oxygen. Beyond this transition region, a significant number of coarse (10 μm) angular TiN particles was observed over a distance of about 1 mm. EPMA results revealed that the Al content within the matrix varied greatly from region to region; it was

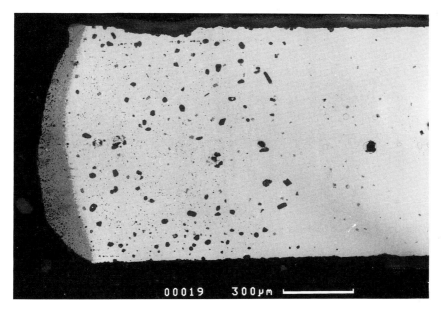

Fig. 4 Longitudinal section of MA 956 stepped tube exposed for 480 h at 1300°C in stagnant air.

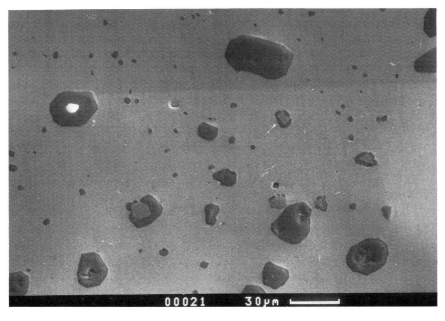

Fig. 5 Distribution of AlN particles in MA 956 exposed for 480 h at 1300°C in stagnant air.

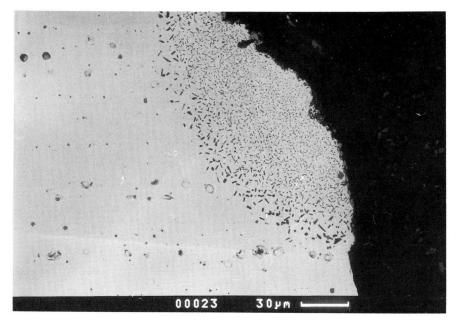

Fig. 6 Internal nitridation beneath spalled corner.

essentially zero in the transition region, very low (~0.5 wt.%) in the nitrided region and significantly higher (~2.1 wt.%) in the bulk.

Continued exposure (1824 h) resulted in an extension of the breakaway oxidation zone and a widening of the accompanying transition region (Fig. 7). Within this latter region the rounded second phase particles were fine and Cr-rich near the surface oxide, becoming coarser and richer in Al at greater depths. At the inner boundary of the transition region, the second phase is almost continuous and consists essentially of pure alumina (Fig. 8). The TiN-containing zone now extended about 10 mm beyond the transition region. The sample showed severe Al depletion (~0.2 wt.%) within this zone close to the transition boundary and remained low (0.5 wt.% even at a depth of 2 mm.

4. Discussion

4.1 Exposure at 1200°C in Flowing Nitrogen–2% Oxygen

Sheet samples of MA 956 showed internal oxidation and internal nitridation after an exposure of 4800 h whereas there was no evidence of sub-surface reactions in ODM 751 tube even after much longer (7360 h) exposure times.

It is well known[8] that alumina surface scale is one of the most protective barriers to nitrogen ingress into metallic substrates. In the presence of macroscopic defects in the scale, nitrogen can penetrate, dissolve in the matrix and then diffuse to react with solute elements for which it has a high affinity.

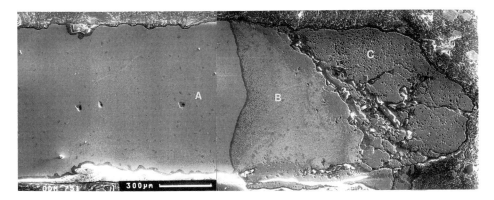

Fig. 7 Longitudinal section of ODM 751 stepped tube exposed for 1824 h at 1300°C in stagnant air. (A) internal nitrided zone; (B) transition region; (C) breakaway oxidation.

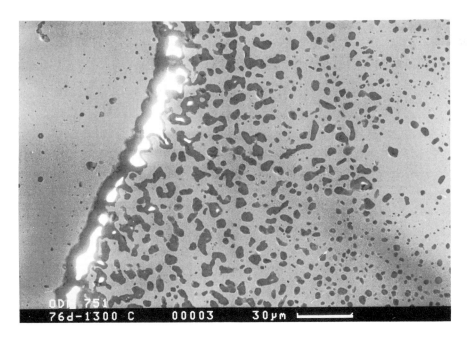

Fig. 8 Internal oxidation zone in ODM 751 tube after exposure for 1824 h at 1300°C.

However in oxidising environments, cracked or spalled layers can be healed if the Al content of the matrix is higher than a critical value. According to Lai,[7] there should be at least 1.3 wt.% Al within the matrix to develop an external alumina scale in nickel-based alloys. There is no published data about the critical level of Al in iron-based alloys but, from previous work, the value appears to lie in the range 1.5 to 1.7 wt.%.

Internal oxidation and internal nitridation on MA 956 sheet (Fig. 2) depend strongly on the Al level within the matrix. Due to the fast oxidation rate[9] and subsequent spallation of the oxide layer, the aluminium content decreased markedly from 3.77 wt.% in the as-received condition to 1.7 wt.% after an exposure of 2400 h. After spalling, the external alumina scale could not reform due to severe aluminium depletion and a porous chromium–iron oxide scale grew in its place. This enabled both nitrogen and oxygen to penetrate and dissolve in the matrix. Due to the slow diffusion of substitutional oxygen in the matrix, internal oxidation occurred in only a very shallow region beneath the surface with the formation of small alumina particles. Since the diffusion of interstitial nitrogen is much faster than oxygen, nitrogen diffused into the matrix to much greater depths and reacted with Ti to form, or coarsen, the TiN. When Ti depletion was complete, the nitridation reaction continued with the formation of less stable AlN. Barnes and Lai[8] found extensive AlN precipitation in Inconel alloy 601 and 617 which have low (1.2 wt.%) Al contents but found much less in higher (4.5 wt.%) Al-containing Haynes alloy 214. The reduction in internal nitridation in the higher aluminium-containing alloy was again attributed to the continuous alumina film on the surface.

In the non-spalled region of MA 956, where the alumina layer acts as a barrier, rapid lateral diffusion of nitrogen occurs from adjacent spalled regions and reacts first with Ti to form TiN then subsequently with Al to form AlN. Only limited lateral penetration of oxygen occurs (essentially as an edge-effect) due to its much lower diffusivity.

ODM 751 tube, on the other hand, did not show any internal oxidation or nitridation even after much longer (7360 h) exposure times. This is principally due to the slow growing, adherent alumina scale on the surface and also the thicker section which results in a much slower depletion of Al within the matrix. The Al level decreased from 4.4 wt.% (as-received) to 3.7 after 4800 h.

4.2 Exposure of Stepped Tubes at 1300°C in Stagnant Air

During exposure, MA 956 oxidised at a faster rate than ODM 751 and this resulted in enhanced Al depletion within the matrix, particularly in the thinnest section of tube. When the Al level dropped to a critical level, breakaway oxidation initiated on the end face of the thinnest section after 480 h (Fig. 3). The resultant Cr/Fe oxide is porous and allows nitrogen to penetrate into the matrix and form TiN and AlN internally. Since these particles were seen only in the regions close to the breakaway oxidation (Fig. 4), it is apparent that nitrogen diffused from this end face rather than from the closer curved surface of the sample where the protective alumina scale remained intact.

A large proportion of the AlN (hexagonal) particles formed during exposure were found to be associated with smaller, more angular TiN (cubic) particles (Fig. 5). This indicates that there is considerable difficulty in nucleating AlN in the matrix due to the very limited number of grain boundaries and the inherently low dislocation density. However, TiN particles which grow *in situ* from smaller pre-existing Ti(C,N) are distributed fairly homogeneously throughout the grains and act as effective heterogeneous nucleation sites.

At the corners of the sample, local concentrations of relatively fine, uniformly distributed, AlN particles were seen beneath spalled oxide regions (Fig. 6). It would appear that this scale spalled during slow cooling (5.5 h) and allowed nitrogen to dissolve unhindered in the matrix whence it diffused and reacted with Al to form AlN. However, since the nitridation depth was small and the precipitate morphology was Widmanstatten, this reaction must have occurred under different circumstances (lower temperature) than those associated with breakaway oxidation.

ODM 751 showed very good resistance to nitridation up to 384 h, after which a little breakaway oxidation, some internal nitridation and a very limited depth of internal oxidation were observed. Within the nitrided region only coarse TiN particles were detected with a marked absence of AlN. This is probably due to the higher Ti content (0.57 wt.%) of ODM 751 compared to MA 956 (0.13 wt.%) when growth again occurs at the larger number of pre-existing Ti(C,N) particles. The higher concentration of Ti within the matrix was not exhausted by the diffusing nitrogen during exposure thus preventing the formation of AlN.

The superior nitridation behaviour of ODM 751 is due to the slower growing alumina scale and lower levels of aluminium depletion in the matrix. The formation of an almost continuous alumina layer between the transition region and the internally nitrided zone (Fig. 8) must also inhibit nitrogen penetration and retard the nitridation kinetics.

5. Conclusions

ODS alloys MA 956 and ODM 751 are both susceptible to internal nitridation in atmospheres containing nitrogen (with or without oxygen). The onset of nitridation is principally dependent on the integrity of the normally protective alumina scale: as the alumina scale thickens and the matrix aluminium content becomes depleted, spalling of the oxide layer occurs and decreased aluminium fluxes result in the formation of a less protective Cr/Fe scale. This allows ingress of nitrogen into the matrix where it diffuses to form internal nitrides. In MA 956, a coarse distribution of AlN is formed, whereas in ODM 751 containing significantly titanium coarse TiN particles grow. However, any process which results in damage to the oxide scale can initiate internal nitridation at least in oxygen depleted environments.

Acknowledgements

This work was undertaken within the COST 501 (work package 4) programme. The authors wish to thank British Gas plc and the Department of Trade and Industry for their financial support and materials used. M. Turker also extends his gratitude to the Turkish Government for sponsoring his studies in the UK.

References

1. W. J. Quadakkers, *Werkstoffe und Korrosion*, **41** (1990), 659–668.
2. M. J. Bennett, H. Romary and J. B. Price, *Proc. Conf. Heat Resistant Materials*, Fortana, Wisconsin, USA, Sept. 1991, pp. 95–103.
3. B. S. Stott, M. Sang and N. Karim, *High Temperature Materials for Power Engineering*, Part 1 (1990), pp. 213–226.
4. G. M. McColvin and G. D. Smith, *High Temperature Alloys* (Ed. J. B. Mariot, M. Merze, J. Nihoul and J. Word). Elsevier Applied Science (1985), pp. 139–153.
5. R. H. Kane, 'Frontiers of high temperature materials II', *Inco Alloy International* (May 1983), pp. 392–418.
6. I. C. Chen and D. L. Douglas, *Oxid. Met.*, **34** (1990), 473–495.
7. Y. G. Lai, 'High temperature corrosion of engineering alloys', *ASM International*, The Metal Information Society (1990), pp. 73–84.
8. J. J. Barnes and G. Y. Lai, *Corrosion and Particle Erosion at High Temperatures* (Ed. V. Sirinivasan and K. Vedula), TSM Meeting, Las Vegas (Feb. 1989), pp. 617–634.
9. M. Turker and T. A. Hughes, to be published.

5

OXIDATION OF ALUMINIUM, TITANIUM, ZIRCONIUM AND MANGANESE AND THEIR ALLOYS

The Influence of Surface Preparation and Pretreatments on the Oxidation of Liquid Aluminium and Aluminium–Magnesium Alloys

S. IMPEY, D. J. STEPHENSON and J. R. NICHOLLS

Cranfield Institute of Technology, Bedford, UK

ABSTRACT

It is shown that the condition and morphology of the first formed oxide on the solid determines the subsequent oxidation rate of molten aluminium alloys. Thus, the oxidation of molten aluminium and aluminium–magnesium alloys at 750°C is influenced by the surface finish and preparation of the solid specimen prior to oxidation.

On pure molten aluminium an initial incubation period prior to breakaway oxidation is extended as the surface finish of the solid specimen is improved. Breakaway oxidation is delayed on polished and grain-refined aluminium–magnesium surfaces oxidised at 750°C in comparison with machined and non grain-refined specimens.

Adsorption of water vapour at the surface stabilises an 'amorphous' oxide film and greatly reduces the rate of oxide crystallisation. Preheating Al–Mg specimens at 550°C in a humid atmosphere sufficiently preconditions the oxide so that on exposure to a dry, oxygen-free environment breakaway oxidation is delayed for over 20 h. Thus dross formation is decreased by melting in the presence of water vapour.

1. Introduction

Dross, a mixture of oxide and entrapped metal, forms on the surface of molten aluminium alloys. Oxidation is particularly severe on molten Al–Mg alloys (beverage can material) and considerable quantities of metal are lost in this form.

To understand the underlying mechanisms by which dross formation occurs studies of the oxidation of molten aluminium alloys were carried out, and have been reported for commercial purity aluminium[1] and aluminium–magnesium.[2]

323

In the secondary aluminium industry, charges may typically comprise low volume, high surface area material. As aluminium and aluminium alloys possess a thin surface oxide film, then the presence of these surface oxides may modify the melting behaviour of aluminium alloys during recycling. Problems within the secondary aluminium melting industry[3] are further compounded by the recovered material carrying an assortment of oxide coatings which may often be heavily contaminated.

The current methodology for studying oxide films on liquid aluminium requires the removal of the old surface film, appreciably thickened by high furnace temperatures. In this way newly grown films are developed for examination. Hence the first formed oxide is removed from the melt surface by skimming. However, this procedure does not reflect industrial practice within the secondary aluminium industry before a charge is melted, although after melting the melt surface is skimmed in order to remove the dross layer prior to casting.

The condition of the original surface oxide film on the recycled alloys is shown to have a fundamental influence on liquid metal oxidation of recycled aluminium alloys and the results of further investigations are presented in this paper.

2. Experimental

Commercially pure aluminium and beverage can aluminium alloys of low and high magnesium content, Al-1% Mg, 5005 alloy and Al-5% Mg, 5182 alloy (with (GR) and without (NGR) titanium diboride grain refiner), were examined.

Specimens were machined, ground with silicon carbide to 600 μm, polished with diamond to 1 μm, or with an alumina 0.3 μm slurry, or electropolished in a 25% nitric acid–75% methanol solution to provide a range of surface finishes. Exposure tests were carried out at temperatures of 550°C and 750°C and materials were exposed to air and argon environments with varying water vapour (50–3000 Pa) and oxygen (60–2120 Pa) content. Oxide samples were removed from the surface of melts after various times using a copper loop technique.[4] Excess metal was removed from the oxide using a 3% bromine–methanol solution, or by jet-electropolishing. These surface films formed during the early stages of the oxidation process were then examined by transmission electron microscopy (TEM).

Development of dross during the oxidation process was studied by monitoring changes in surface morphology using optical and scanning electron microscopy. The structure of the oxide-melt interface was studied optically with cross-sections of solidified melts and in the TEM using microtomed sections. In addition, the types of oxides that formed were characterised using electron and X-ray diffraction.

In parallel, oxidation rates were studied thermogravimetrically and oxidation kinetics were related to changes in oxide structure and morphology as a function of exposure time. These kinetic studies have been reported in detail elsewhere.[5] Examples are presented to support the microscopy studies in this paper.

3. Results and Discussion

3.1 A Review of the Mechanism of Oxidation in Molten Aluminium and Aluminium–Magnesium Alloy

A mechanism for the time-dependent development of dross on liquid aluminium and liquid aluminium magnesium alloys at 750°C has been reported previously.[1,2] This mechanism is pertinent to the understanding of the effects of surface preparation and pretreatments on the oxidation of liquid aluminium and aluminium–magnesium alloys and is therefore summarised below.

An 'amorphous' oxide film instantaneously forms on liquid aluminium and aluminium–magnesium alloys at 750°C. Beneath this film at the 'amorphous' oxide-melt interface crystalline oxides develop (the development of these crystalline oxides is discussed in detail below).

For both aluminium and aluminium–magnesium alloys, the formation and growth of oxide crystals beneath the continuous 'amorphous' film results in considerable stress, primarily due to volume changes. As the degree of crystal growth and stress increases the likelihood of fracture of the surface film increases. Finally the amorphous oxide fails.

Optical and scanning electron microscopy has shown that molten metal exudes through the ruptured oxide to the melt surface, a consequence of the wetting of alumina by liquid aluminium[1] and aluminium-based alloys. The exuded metal structure is enveloped in a rapidly formed 'amorphous' oxide film and is described as an oxide growth. The growth itself is rapidly oxidised, new crystals nucleate and further exudations take place resulting in secondary and tertiary growths. The resulting structure is a network of oxide and trapped metal as illustrated in Fig. 1(a). The growths increase in number and size with time and their appearance (Fig. 1b and c) marks the onset of dross formation and is reflected in oxidation kinetics as breakaway oxidation.

Figures 1(b) and 1(c) demonstrate that oxidation occurs more rapidly for magnesium-containing alloys (5 h) than on pure aluminium (24 h) in air at 750°C. However, the similarity in appearance of both examples supports the theory of one mechanism of dross formation.

3.2 Oxide Development on Molten Aluminium

Oxide films grown directly on the melt after 5 min at 750°C (Fig. 2a), shows oxide crystals (identified as γ-Al_2O_3) developing in an 'amorphous' film. The crystals grow laterally until a monolayer forms. The oxide film thickens and ultimately α-alumina develops. Traces of α-alumina are found after 5 h.

Full crystalline coverage is obtained when oxide films develop from the solid, during rapid heating over 6 min to 750°C. Films typically show advanced oxide development and consist of a matrix of dense regular crystals and randomly

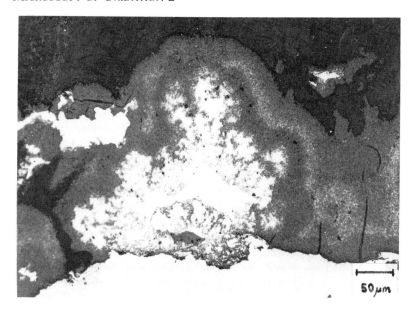

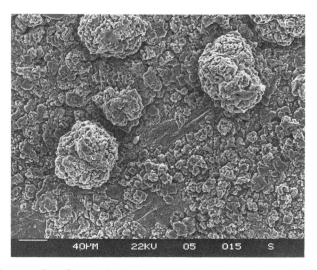

Fig. 1 Oxide growths after oxidation at 750°C. (a) section through oxide growth on Al–5%Mg showing entrapped metal (12 h); (b) Al (24 h); (c) Al–Mg (5 h).

distributed elongated structures. γ-Alumina and traces of α-alumina are identified at early stages by electron diffraction, for example after 5 min. exposure as in Fig. 2(b). With longer exposure times, the oxide crystals grow and become elongated as the oxide film thickens. Microtome sections taken normal to the melt surface show large crystals developing at the oxide-melt interface beneath a crystalline monolayer. According to crystal nucleation theory, the number and size of γ-alumina grains is qualitatively consistent with

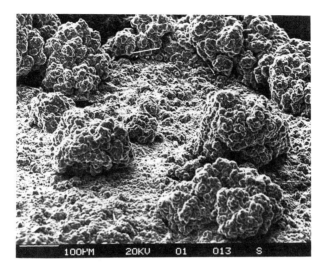

Fig. 1c

slow heating conditions in which a few well ordered crystals nucleate and grow.

By way of comparison during rapid heating conditions, numerous small nuclei form. Such oxide development is observed on skimmed aluminium melts where the oxide films are a mass of fine, regular γ-alumina crystals.

The implication is that a substantial amount of oxide growth occurs on solid aluminium prior to melting during the heating up period. Thus the initial oxide developed on solid aluminium is more advanced than that grown on the melt. The difference is so marked that the oxidation behaviour of skimmed and unskimmed melts is not directly comparable. For example, under the same conditions of dry air at 750°C oxide growths appear on an unskimmed aluminium melt more than 10 h before a single oxide growth protrudes from a skimmed surface.

In industrial practice, and indeed in most gravimetric studies, heating aluminium through the solid state is a necessary prerequisite before melting. Care should be exercised therefore when comparing kinetic data from the literature with reported oxide structures developed on a melt, for the standard practice is to skim melts prior to such investigations in the laboratory, whereas it is not practicable to remove the oxide scales from metal charges during industrial melting processes. This is particularly pertinent for oxides grown on aluminium–magnesium melts.

3.3 Oxide Development on Aluminium–Magnesium Alloy

The oxidation behaviour of molten Al–Mg alloys containing 1 and 5 wt.% of magnesium show striking similarities. The oxide film grown on Al-Mg melts exhibits coarse crystals (200–500 nm) but also fine MgO crystallites (20 nm)

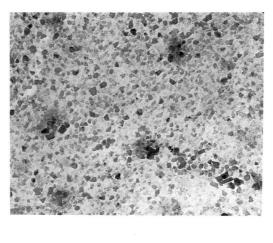

Fig. 2 Oxide development on aluminium in air at 750°C (5 min). (a) skimmed oxide of γ-alumina crystals; (b) unskimmed oxide of γ- and α-alumina.

(Fig. 3b). Microtome sections show coarse primary magnesia crystals form at the oxide–metal interface. The smaller secondary magnesia crystals exist within the oxide film and originate from the secondary reduction reaction of original amorphous γ-Al$_2$O$_3$ surface film:

$$3Mg + \gamma\text{-}Al_2O_3 \rightarrow 3MgO + 2Al$$

The primary MgO crystals coarsen with increasing temperature, and appear as discrete, random nodules at the oxide surface. Early oxide films formed on unskimmed melts are dense and convoluted, indicating that much oxide development occurred on the solid alloy.

Skimming the melt reduces the local magnesium concentration, although magnesium diffuses rapidly to the surface. In regions of localised magnesium depletion or in low magnesium-containing alloys, coarse and discontinuous MgAl$_2$O$_4$ crystals are identified. However, skimming also removes the original

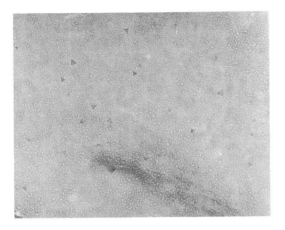

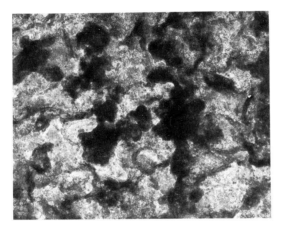

Fig. 3 Oxide development on Al-Mg alloy in air at 750°C (5 min). (a) skimmed oxide of γ-alumina crystals; (b) unskimmed oxide of primary and secondary MgO.

γ-Al$_2$O$_3$ film in which secondary magnesia develops. Figure 3(a) shows an oxide film severely depleted in magnesium in which new γ-Al$_2$O$_3$ crystals are nucleating.

Thus the alloy oxidation is dominated by the preferential oxidation of magnesium. In dry environments growth formation rapidly occurs and does not cease until all the available magnesium is consumed. Kinetic measurements[5] reflect the rapid formation of magnesia. The rate of oxidation reduces until conditions are favourable for transformation of magnesia to magnesium aluminate whereupon rapid oxidation ensues. This is illustrated in Fig. 6. Oxidation may progress further by the formation of α-alumina. The oxide growth in Fig. 1(a) has been identified by X-ray diffraction to contain both MgAl$_2$O$_4$ and α-Al$_2$O$_3$.

Oxide nucleation, growth and the morphology developed in the solid state influences rates of liquid metal oxidation. It is anticipated therefore that parameters influencing the condition of the first formed oxide will affect both the composition and rate of oxidation of molten melts. This is demonstrated by considering the effect of specimen surface finish on molten metal oxidation.

3.4 The Effect of Specimen Preparation and Surface Finish on Molten Aluminium

Aluminium surfaces prepared by machining, grinding, mechanical polishing or electropolishing were examined following oxidation at 750°C for 20 h. Scanning electron micrographs demonstrate variations in the oxide morphology. Figure 4(a) shows a typical morphology on a machine ground surface. This consists of areas of thick oxide in the direction of machining or grinding together with oxide clusters and nodules normally seen in liquid aluminium oxidation. It is likely that these oxide nodules are associated with primary cracking that has occurred in the solid state. These reheal the air formed oxide. Surfaces prepared by mechanical polishing to 1 μm produced extremely convoluted and irregular surfaces after oxidation, as in Fig. 4(b). On smoother electropolished surfaces, clusters of spheroids developed (Fig. 4c). These spheroidal protrusions do not aggregate into clusters typical of the nodular type oxide growths illustrated in Fig. 1(b). In general, as the surface roughness of the starting specimen decreased, the degree of oxide nodule development or surface area increased. This is evident from Figs 4(a)–4(c) where the magnification has been reduced to show the required detail. This behaviour correlates well with the enhanced oxidation rates and weight gains achieved during oxidation in dry and wet air and argon environments at 750°C on electropolished specimens compared to the as-machined surface. This can be seen in Fig. 5.

Data concerning the effect of the initial specimen surface finish on molten metal oxidation have not been reported in the literature. If the only effect of roughening a surface is to increase the surface area available for oxidation the greatest oxidation will be exhibited by machined specimens and the lowest oxidation by electropolished surfaces.[6] This trend is contradictory to observations made in this study for liquid metal oxidation. Starting surface areas were estimated by profilimetry and BET nitrogen adsorption and do not quantitatively or qualitatively account for the observed oxidation rates on molten alloy.[5] This simple interpretation of surface roughness ignores other contributions such as the introduction of cold work, deformation and contamination of surface layers by the roughening process.

The present work demonstrates that the degree of oxidation reflects the ability of the first formed oxide film to provide protection of the melt from further oxidation. Oxidation rates become rapid if this surface oxide is damaged or exhibits a tendency to crack. Hence on highly polished aluminium, rapid oxidation is believed to be a consequence of stressed oxide films which damage and crack easily. The degree of polishing and surface distortion is reflected in the wide scatter in oxidation rates and weight gains obtained with polished specimens. In contrast, machined specimens, less influenced by surface distortion and damage, exhibit high reproducibility during oxidation.

The original condition of the oxide film formed on aluminium–magnesium

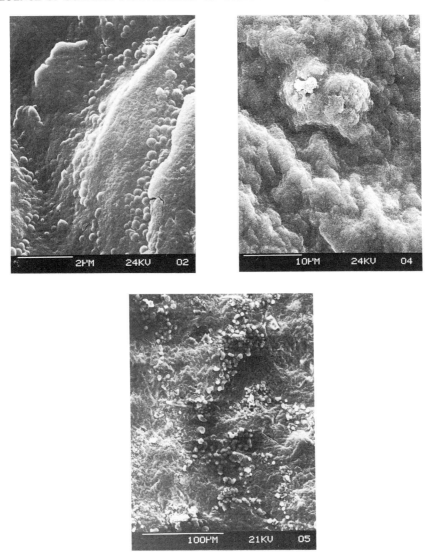

Fig. 4 The effect of surface finish on aluminium oxidation in air at 750°C (12 h). (a) machine ground (600 μm); (b) mechanically polished (1 μm); (c) electropolished.

melts can be inferred by monitoring rates of oxidation and the time to breakaway oxidation. Local stress and defects present in the developing oxide will encourage early breakdown of the protective film, and where mechanical failure results, accelerated oxidation is observed.[7] SEM examination of aluminium surfaces subjected to ultrasonic cleaning from 30–300 s exhibited cavities from gas bubble erosion. For polished aluminium–magnesium 4.5%Mg alloy accelerated rapid oxidation behaviour occurred sooner on ultrasonically treated specimens (Fig. 6).

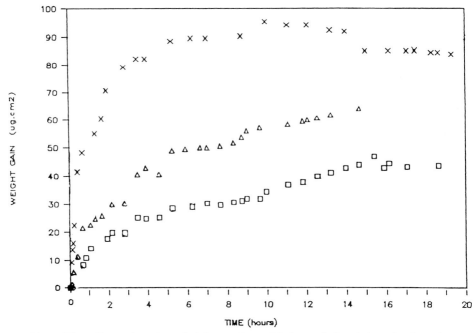

Fig. 5 The effect of surface finish on the oxidation of aluminium in dry air at 750°C. × = electropolished, △ = polished (1 μm), □ = machined.

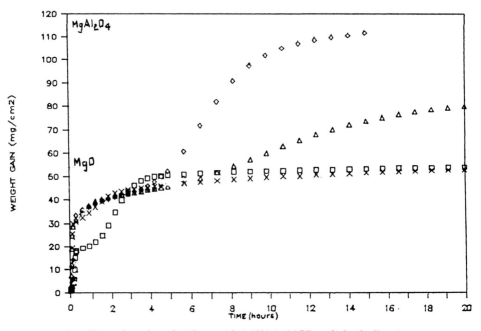

Fig. 6 The effect of surface finish on Al–4.5%Mg NGR polished alloy in argon at 750°C. ◇ = diamond (1 μm) ultrasonic, △ = diamond (1 μm), □ = diamond (1 μm) Al–4.5%Mg, × = alumina (0.3 μm).

Further, specimen edges are areas at which the surface oxide would be characteristically stressed and hence the first to rupture. Figure 1(c) shows that the majority of oxide growths develop initially at specimen edges.

On Al-Mg specimens in dry environments at 750°C, breakaway oxidation occurs faster on machined specimens than on polished surfaces. With pure aluminium under the same conditions this trend is completely reversed. For pure aluminium it was proposed[8] that the level of stress accumulates as the amount of surface preparation increases. This influences the subsequent behaviour of the alumina film and the oxide is less likely to remain intact during the oxidation of polished surfaces than for machined specimens. During the polishing process on small Al–Mg alloy specimens the surface temperature may be raised sufficiently to encourage a build up of magnesium in the outer surface layers. A greater proportion of magnesium at the surface layer would ensure that this region is the first to melt. Stress build-up at the oxide–metal interface is relieved in the polished surfaces before the machined specimens. Thus the oxide growth and onset of rapid oxidation is enhanced for both polished Al–1%Mg and Al–5%Mg alloys oxidised at 750°C in comparison with the non-stress relieved machined specimens.

A fine recrystallised grain structure may be produced during the heating up period leading to melting. Polished surfaces will have more grain boundary diffusion paths than machined specimens and the increased diffusion can lead to a faster rate of surface oxide coverage. By rapidly forming a continuous layer of oxide in the solid state a greater degree of protection is provided for the melt, thereby delaying breakaway oxidation. Indeed grain-refined aluminium alloys with 1% and 4.5% magnesium additions exhibit slower oxidation rates than non-grain refined material.[8] This is illustrated in Fig. 6. Further, if a machined non-grain refined alloy is polished, the resulting oxidation rate is equivalent to that of the grain refined material. This suggests that polishing produces a grain refining effect. Oxidation curves for both polished grain refined and non-grain refined Al–Mg alloys fall in a band of reproducibility reflecting the various degree of polishing and consequent grain refinement.[8]

The effect of surface finish on the oxidation rate of molten aluminium and aluminium magnesium alloys has shown that the rate of liquid metal oxidation is dependent on the initial composition and structure of the oxide formed on the solid metal. In particular, the protective nature of this first formed oxide is critical in determining the subsequent rates of liquid metal oxidation.

3.5 Preconditioning the First Formed Oxide

It has been shown that breakaway oxidation is influenced by mechanical means, for example by surface preparation. Clearly, to limit the onset of rapid oxidation, it is necessary to reduce the chances of scale failure by cracking, subsequent liquid metal exudation and growth development. This section examines the use of environmental treatments to accomplish a delay in dross formation.

To demonstrate this behaviour, the simplest and most effective way of

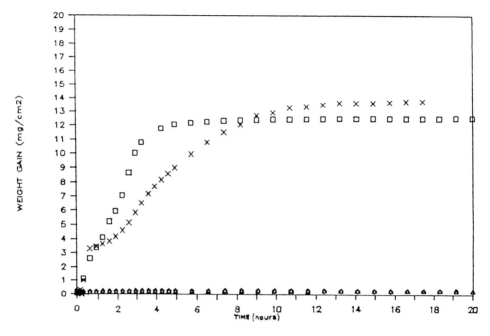

Fig. 7 The effect of preconditioning in humid air on the subsequent oxidation of Al-5%Mg in dry argon at 750°C. □ = no treatment, ◇ = 5 h at 750°C, △ = 15 h at 550°C, × = 15 h at 550°C, cooled.

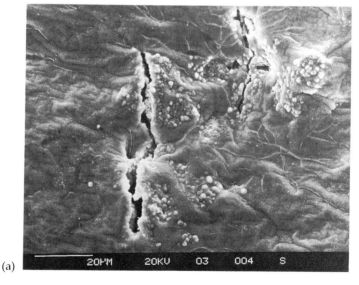

(a)

Fig. 8 The oxide surface of Al–5%Mg after preconditioning at 550°C then exposed to dry argon at 750°C. (a) Preconditioned in humid air; (b) preconditioned in dry argon; (c) enlargement of above (b).

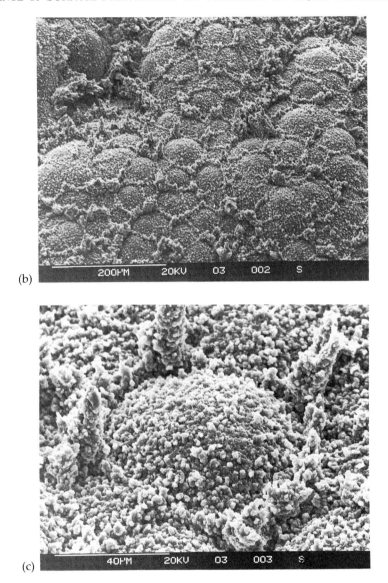

Fig. 8—continued

delaying breakaway oxidation is achieved by melting in an atmosphere containing a high water vapour pressure.[1,2] In the presence of hydroxyl ions the 'amorphous' film is considerably stabilised and the rate of crystal nucleation is reduced. Thus the surface film remains protective for longer periods.

Al–5%Mg specimens were heated in a humid atmosphere to enhance the surface protection period before exposure to a dry environment at 750°C. Figure 7 shows that rapid oxidation is delayed for up to 20 h by preheating in humid air for short times (5 h at 750°C or 15 h at 550°C). With no precondition-ing, breakaway oxidation in dry conditions is instantaneous. If a similarly preconditioned oxide is cooled after preheating, cracks generated in the oxide (from cooling) do not provide any protection from rapid breakaway oxidation on remelting in a dry environment. In this work, the surface oxide of small, highly stressed cylindrical specimens was successfully preconditioned to delay rapid oxidation for up to 20 h by preheating in the presence of water vapour for 5 h at 750°C and for 15 h at 550°C.

The appearance of surfaces following such treatments support the view that rapid oxidation is a consequence of surface oxide damage. The surface oxide is unable to prevent cracking and melt exposure. Figure 8(a) illustrates an Al–5%Mg polished sample preconditioned at 550°C and exposed to gettered argon at 750°C for 2 h. Subsequent cracking on the surface after cooling reveals oxide growth beneath the surface.

As a control experiment, specimens were oxidised for 5 h at 550°C in gettered (dry) argon to encourage non-protective oxide development. Indeed, the Al–Mg surface rapidly becomes non-protective, forming primary magnesia. The temperature was raised to 750°C and the specimen exposed for a further 15 h. The micrographs in Figs. 8(b) and 8(c) show the thick oxide layer which develops, predominantly of magnesium aluminate. The oxide surface de-veloped in gettered argon does not provide an effective barrier to further oxidation. Stacks of crystals are observed developing in a manner consistent with easier diffusion paths at oxide areas of reduced thickness or where local breakdown has occurred.

4. Conclusions

The condition and morphology of the first formed oxide determines the oxidation rate of molten metal.

The oxidation of molten aluminium and aluminium–magnesium alloys at 750°C is influenced by the surface finish and preparation of the specimen prior to oxidation.

To prevent the onset of rapid oxidation and hence dross formation, it is essential to maintain a protective surface oxide on the melt. This can be achieved by preconditioning the alloy in a humid atmosphere to produce a thin stable amorphous γ-alumina film.

References

1. S. A. Impey, D. J. Stephenson and J. R. Nicholls, *Mat. Sci. Tech.*, **4** (1988), 1126.
2. S. A. Impey, D. J. Stephenson and J. R. Nicholls, in *Proc. Conf. on Microscopy of Oxidation* (26–28 March 1990), paper 27.
3. D. V. Neff and R. G. Teller, in *Proc. 2nd Int. Conf. on Molten Aluminium Processing*, Orlando, Florida (6–7 Nov. 1989), pp. 18.1–18.19.

4. G. D. Preston and L. L. Bircumshaw, *Phils. Mag.*, **22** (7) (1936), 654–665.
5. S. A. Impey, D. J. Stephenson and J. R. Nicholls, to be published.
6. C. N. Cochran and W. C. Sleppy, *J. Electrochem. Soc.*, **108** (4) (1961), 322–327.
7. A. S. Nagelberg, *J. Mat. Res.*, **7** (2) (1992), 265–268.
8. S. A. Impey, PhD Thesis. Cranfield Institute of Technology (Dec. 1989).

Photoelectrochemical Studies of the Scales Thermally Formed on Titanium in O_2, H_2O or O_2/H_2O Atmospheres

A. GALERIE,* M. R. DE NICOLA* and J. P. PETIT**

*ENS d'Electrochimie et d'Electrométallurgie de Grenoble, Institut National
Polytechnique de Grenoble, BP 75 Domaine Universitaire,
F-38402 Saint-Martin d'Hères Cedex, France*

Laboratoire Science des Surfaces et Matériaux Carbonés, URA CNRS no. 413
***Laboratoire d'Ionique et d'Electrochimie des Solides, URA CNRS no. 1213*

ABSTRACT

The semiconducting properties of thin TiO_2 scales, thermally grown on titanium in dry oxygen, O_2/H_2O mixtures or pure water vapour, were studied by means of photoelectrochemical techniques. The magnitude of the photocurrents, their wavelength dependence and the shape of the photocurrent versus potential curves were shown to clearly depend on the TiO_2 film thickness and/or the H_2O content of the oxidising atmosphere. The differences in the photoelectrochemical behaviour of the TiO_2 scales could have various causes, such as differences in oxide stoichiometry, microstructural defect density or hydrogen content.

1. Introduction

The thermal oxidation of titanium has been extensively studied[1] and exhibits peculiar features resulting from the important volume change associated with TiO_2 formation, leading, in connection with the inward growing of the scale, to large compressive stresses. After an initial period where the oxidation rate is controlled by oxygen vacancy diffusion through a growing compact scale (parabolic stage), mechanical decohesion occurs leading to an acceleration of the oxidation followed by a constant rate period (linear stage). Kinetic explanation of this stage supposes the existence, in contact with the metal, of a thin compact TiO_2 layer of statistically constant thickness. Long-term oxidised samples therefore exhibit a typical stratified morphology where the thickness of the individual TiO_2 sublayers depends on temperature, in the $1–10\,\mu m$ range.

In presence of water vapour, both kinetics and morphology are greatly modified. The rate law no longer presents the classical acceleration but

338

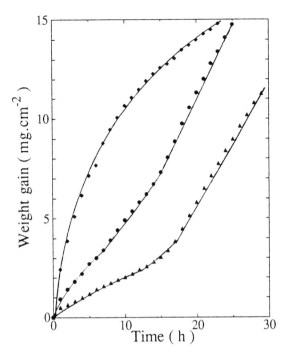

Fig. 1 Influence of the atmosphere on the corrosion kinetics of pure titanium at 850°C. P_{O_2} = 0.133 bar (▲); P_{H_2O} = 0.02 bar (◆); P_{O_2} = 0.133 + P_{H_2O} bar = 0.02 bar (●).

proceeds accordingly to a power law $(\Delta m/A)^n = kt$, with n values between 1 and 2 (Fig. 1). Cross-sections of the oxidised scales exhibit more compact TiO_2 scales, with larger grains and spherical pores. Figure 2 presents the difference in morphology for scales grown in oxygen or water vapour.

A better understanding of the causes of these kinetical and morphological differences requires a better characterisation of the oxide films. For this purpose, we present here first results dealing with the photoelectrochemical characterisation of thin TiO_2 scales grown on titanium in dry oxygen, O_2/H_2O mixtures or pure water vapour. Photoelectrochemistry indeed has been shown in the recent past to be a valuable technique for interestingly characterising semi-conducting materials in terms of electro-optic properties, more or less related to their nature and structure.[2,3,4]

2. Experimental

2.1 Preparation of the scales

Thin TiO_2 films were grown in a microbalance at 850°C on pure, ethanol-cleaned titanium (99.5% supplied by Goodfellow Metals Ltd.) under static atmospheres. Oxygen (99.995%) and water vapour (distilled and outgassed)

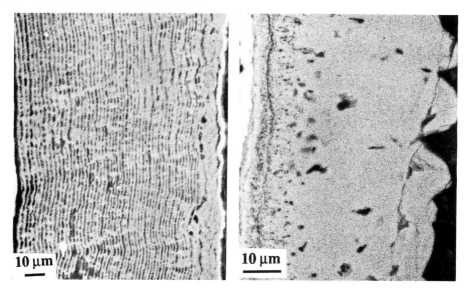

Fig. 2 Scanning electron micrographs of scales formed on titanium specimens oxidised at 850°C in: (a) 0.133 bar O_2 ($\Delta m/A = 12$ mg.cm^{-2}); (b) 0.02 bar H_2O ($\Delta m/A = 10$ mg cm^{-2}).

were used separately or as mixtures prepared using pressure measurements with a deformation sensor and a Pirani gauge.

The films were allowed to grow up to weight gains ranging from $4.25\ 10^{-2}$, to 0.85 mg. cm^{-2}, likely corresponding to thicknesses from 0.05 to 1 μm, much lower than the thickness where scale decohesion occurs in pure oxygen (10 μm).

2.2 Photoelectrochemical measurements

Electrodes were prepared from the oxidised samples by first grinding off the oxide layer from one side, then sticking to a brass electrode holder with silver paste and finally isolating the sides of the samples and the electrode holder with epoxy resin.

All these electrodes were characterised at room temperature in Ar-sparged aqueous 0.5 M H_2SO_4 by using them as the working electrode of a standard three-electrode photoelectrochemical cell with a platinium counter-electrode and a calomel reference electrode. Prior to use, the surface of the working electrode was neither polished nor etched but left in its original state.

The photoelectrochemical experiments were performed using a typical experimental setup, similar to the one presented in reference 3. All optical parts, including the window facing the working electrode, consisted of fused silica, allowing measurements at wavelengths down to 220 nm. The photocurrent spectra, $I_{ph}(\lambda)$ and the photocurrent vs. potential curves, $I_{ph}(E)$, were mostly determined under modulated illumination (205 Hz), using a mechanical chopper. However, several experiments were made under stationary light conditions and showed no significant differences with the modulated ones.

In a typical experiment, the electrode was first allowed to relax in open circuit conditions until a stationary rest potential was reached. Then, after a few voltametric cycles in dark conditions, whose results will not be discussed in this paper, the electrode was illuminated with a monochromatic light spot of constant size ($\sim$0.2 cm^2) and the measurements of the $I_{ph}(\lambda)$ and $I_{ph}(E)$ curves subsequently performed. It is noteworthy that, for each modification of the potential value of the $I_{ph}(E)$ experiments, the photocurrent was measured after a time sufficient enough to allow it to reach a quasi-stationary value. It must be observed also that the potential was kept in the 0–1.5 V.SCE range to avoid electrochemical damage of the oxide film, but allowing measurement to be made of non-zero photocurrents, taking into account that the generally reported value of the flatband potential of the TiO$_2$/0.5 M H$_2$SO$_4$-H$_2$O interface is about -0.2 V/SCE.

All the photocurrent values were light intensity corrected, i.e. normalised to identical light intensity, which was estimated to be 100 μW.cm^{-2}.

3. Results and discussion

Within the investigated potential range, all photocurrents were anodic, confirming the well known n-type semiconducting behaviour of titanium dioxide.

3.1 Oxide films formed in dry oxygen

Typical photocurrent spectra $I_{ph}(\lambda)$ relative to different oxide thicknesses, are shown in Fig. 3. These curves present the same general shape; in particular they exhibit two photocurrent peaks corresponding to energies of about 3.5 eV (peak 1) and 4.5 eV (peak 2) respectively. Before discussing the possible origins of the rather rare occurrence of two peaks, several other points should first be stated.

1. The energy onset value of the photocurrents are located ca. 2.9–3.0 eV. Moreover, when rearranging the data of Fig. 3 near the onset value in the classical $(I_{ph}.h\nu)^{1/2}$ vs. $h\nu$ graphs used for indirect bandgap transitions in semiconductors,[3] it is obvious that these plots are linear and do intercept the energy axis at approximately the same energy value of 2.9(5) eV (Fig. 4), corresponding to the forbidden bandgap energy of the rutile modification of TiO$_2$.
2. The photocurrent magnitudes are clearly decreasing with increasing film thickness. It does not seem reasonable to explain this fact by a higher charge carrier density (i.e. a larger stoichiometry deviation) of the thickest films. It is more likely that the photocurrents are lower in thick oxides due to the increased density of recombination centres for the photogenerated electron/hole pairs. This is indeed in agreement with the results of Fig. 5 showing

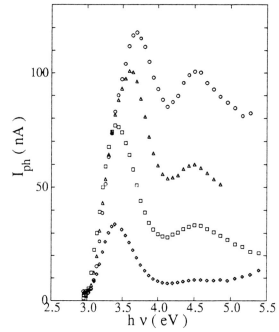

Fig. 3 Photocurrent spectra measured in $0.5\,M\,H_2SO_4$ for TiO_2 thin films formed by thermal oxidation in 0.133 bar O_2 at 850°C. $\Delta m/A = 4.25\ 10^{-2}$ mg cm^{-2} (○); $\Delta m/A = 1.49\ 10^{-1}$ mg cm^{-2} (△); $\Delta m/A = 2.55\ 10^{-1}$ mg cm^{-2} (□); $\Delta m/A = 4.25$ 10^{-1} mg cm^{-2} (◇). Photoelectrochemical conditions: $E = 0.5$ V vs SCE, $f = 205$ Hz.

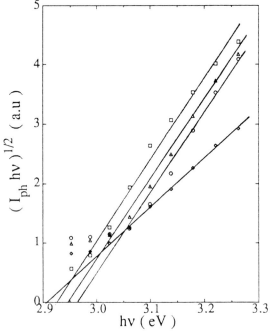

Fig. 4 Evaluation of the bandgap energy of TiO_2 thin films formed by thermal oxidation in 0.133 bar O_2 at 850°C with different weight gains. $\Delta m/A = 4.25\ 10^{-2}$ mg cm^{-2} (○); $\Delta m/A = 1.49\ 10^{-1}$ mg cm^{-2} (△); $\Delta m/A = 2.55\ 10^{-1}$ mg cm^{-2} (□); $\Delta m/A = 4.25\ 10^{-1}$ mg cm^{-2} (◇).

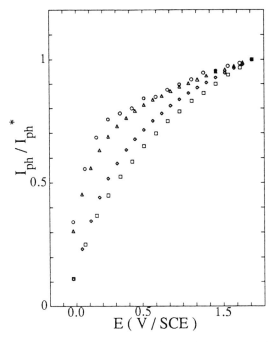

Fig. 5 Photocurrent versus potential curves obtained in 0.5 M H$_2$SO$_4$ for TiO$_2$ thin films formed by thermal oxidation in 0.133 bar O$_2$. Δm/A = 4.25 10^{-2} mg cm^{-2} (○); Δm/A = 1.49 10^{-1} mg cm^{-2} (△); Δm/A = 2.55 10^{-1} mg cm^{-2} (□); Δm/A = 4.25 10^{-1} mg cm^{-2} (◇). The curves are normalised to I_{ph}*, the photocurrent value measured at E = 1.5 vs SCE. Photoelectrochemical conditions: J = 385 nm, f = 205 Hz.

the $I_{ph}(E)$ curves. The shape of these curves is notably and regularly modified when increasing the oxide thickness, gosing from the $I_{ph}(E)$ photocharacteristic of a nearly perfect semiconductor with few recombination sites to the one of a poorer material containing more traps. The question of the nature of the recombination centres, which could be for instance grain boundaries, dislocations, microcracks, etc., is not yet clear but their increasing density with the thickening of the oxide film should be connected with the appearance of cracking for thicker oxides and could possibly reveal precracking phenomena.

As seen above, the general shape of the photocurrent spectra presents two maxima. Figure 6 shows that the ratio between the height of peak $I(h_1)$ and peak 2 (h_2) decreases drastically with increasing oxide thickness. Considering these results and similar ones presented and/or discussed previously by other authors,[4,5,6] a possible model for the oxide scale could include two more or less distinct sublayers, differing either by their stoichiometry or nature. This model is interestingly supported by the results obtained for films grown in H$_2$O-containing atmosphere which are presented now.

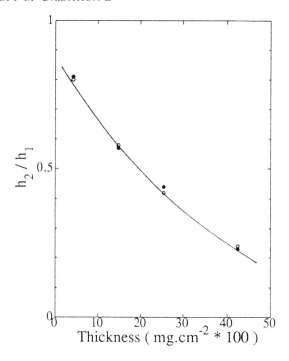

Fig. 6 Dependence of the h_2/h_1 ratio on the film thickness for TiO_2 layers of Fig. 3. Values derived from photocurrent spectra measured at $E = 0.5$ V vs SCE (○) and $E = 1$ V vs SCE (●).

4. Oxide films grown in wet atmospheres

Figure 7 presents a representative set of the photocurrent spectra obtained with oxide films of same thickness (0.255 mg cm^{-2}) but grown in eight different atmospheres ranging from 0 to 100% H_2O. It is obvious from these plots that photocurrent magnitudes are higher than those obtained with films grown in dry oxygen and that an increase of the water content of the oxidising atmosphere results in an increase of the photocurrent.

These experimental observations may originate from two different phenomena consecutive to the increase of the H_2O/O_2 ratio:

1. the decrease of the stoichiometry deviation;
2. the decrease of the number of the recombination centres of the photo-generated electron/hole pairs.

As the deviation from stoichiometry of TiO_2 is well known (colour, mechanical properties) to increase in the presence of H_2O, it could be concluded that the second hypothesis holds. This, in addition, is in good agreement with the evolution of the shape of the $I_{ph}(E)$ curves which were recorded for each atmosphere composition (but are not presented here), and

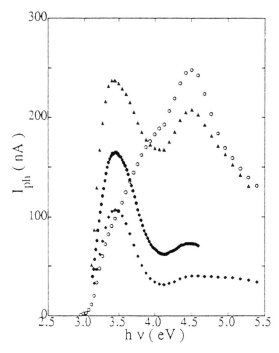

Fig. 7 Photocurrent spectra measured in 0.5 M H_2SO_4 for TiO_2 thin films ($\Delta m/A = 2.55 \ 10^{-1}$ mg cm^{-2}) formed by thermal oxidation at 850°C in O_2/H_2O atmospheres with different H_2O percentages. 100% (○), 95% (▲), 50% (●), 5% (◆). Photoelectrochemical conditions: $E = 0.5$ V vs SCE, $f = 205$ Hz.

could be connected with a decrease of the number of mechanical defects in the oxide and/or a filling of traps by hydrogen species contained in the film.

The two photocurrent peaks observed for oxides grown in dry oxygen are still present here. Furthermore, the h_2/h_1 ratio increases with increasing the H_2O/O_2 ratio as shown in Fig. 8, leading, in the case of the sample grown in pure H_2O, to the quasi disappearance of peak 1.

If one accepts the above mentioned model, such an evolution should involve an increasing influence of the external part of the scale with increased H_2O content. This assumption is not in contradiction with a possible non-equilibrium state of the gas–oxide interface, increasing with the H_2O/O_2 ratio and becoming maximum in pure H_2O.

5. Concluding remarks

Photoelectrochemical experiments applied to thermally grown TiO_2 scales were shown to give results depending largely on the conditions of preparation of the samples. The technique is therefore very suitable for studying the evolution of

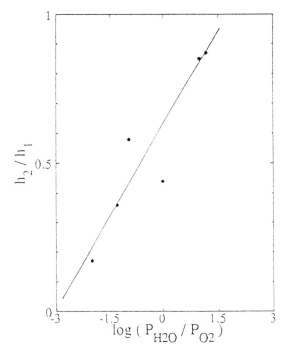

Fig. 8 Dependence of the h_2/h_1 ratio on the P_{H_2O}/P_{O_2} ratio for film formed in water vapour containing atmospheres.

properties in a series of samples relating to temperature, pressure, stress application or doping.

The results presented here have shown, in particular, the possibility of observing the evolution of the amount of mechanical defects but more information is required from other techniques to obtain further facts on the nature of these defects.

References

1. Per Kofstad, *High Temperature Corrosion* (1988), Elsevier Applied Science, London.
2. Yu. V. Pleskov and Yu. Ya. Gurevich, *Semiconductor Photoelectrochemistry* (1986), Consultants Bureau, New York.
3. U. Stimming, *Electrochim. Acta*, **31** (1986), 415.
4. H. Gerischer, *Corr. Sci.*, **29** (2/3) (1989), 257.
5. S. E. Lindquist, A. Lindgren and Z. Yan-Ning, *J. Electrochem. Soc.*, **132** (3) (1985), 623.
6. S. E. Lindquist, A. Lindgren and C. Leygraf, *Solar Energy Materials*, **15** (1987), 367.

A Study of the Role of Alloying Additions During the High Temperature Oxidation of IMI 834

T. J. JOHNSON, M. H. LORETTO, C. M. YOUNES* and
M. W. KEARNS[†]

*I.R.C. in Materials for High Performance Applications, The University of
Birmingham, Edgbaston, Birmingham B15 2TT*

*Interface Analysis Centre, University of Bristol BS2 8BS
†IMI Titanium Ltd, PO box 704, Witton, Birmingham B6 7UR*

ABSTRACT

Analytical transmission electron microscopy (ATEM), scanning electron micro-scopy (SEM), Auger electron spectroscopy (AES) and secondary ion mass spectroscopy (SIMS) have been used in order to investigate the influence of alloying additions on the rate of oxidation of a Ti-based alloy, IMI 834. The results obtained show that the rate of oxidation of the alloy is a complex function of temperature. Diffraction and X-ray analysis of thinned cross-section samples of IMI 834 show unexpectedly high concentrations of alloying elements throughout the oxide layers and these observations have been correlated with analyses of stripped oxides in the TEM and of fractured oxides using AES and SIMS. These results, taken together with the morphological observations made in the SEM, are discussed in terms of the expected activities of the individual alloying additions.

1. Introduction

Highly complex titanium alloy systems have been developed for improved high temperature properties by compositional control, thermomechanical processing and microstructural development.[1] IMI 834 represents the most recent conventional high temperature alloy. It has a number of alloying additions, which although present in only small concentrations do, neverthe-less, apparently contribute to its overall oxidation behaviour. This alloy is employed as both a disc and a blade material and is capable of operating for short times at temperatures up to 650°C although it actually operates in the jet engine at temperatures below this. The development of this alloy has resulted

347

in a marked increase in the working limit of titanium alloys in terms of inherent oxidation resistance. It would seem, however, that whilst continued improvement in the high temperature mechanical properties of such alloys may still be possible, the level of oxidation resistance of the conventional titanium alloys already attained through alloy development is unlikely to be greatly advanced through compositional change alone, at least not without some improvement in our knowledge of the fundamental processes of oxidation in these materials. Therefore, in a continuing effort to increase further the high temperature capabilities of titanium alloys, it is becoming increasingly important to investigate the factors that influence the rate of oxidation of titanium and its alloys.

2. Experimental Techniques

Oxidation trials have been carried out on material supplied by IMI Titanium Ltd. The composition of IMI 834 in atomic percent is Ti–9.8Al–1.6Sn–2.1Zr–0.5Nb–0.15 Mo–0.86Si–0.06C. Specimens were exposed in a thermogravimetric balance at test temperatures above 600°C. Exposure at temperatures above 600°C permit the assessment of oxidation kinetic data and the mechanisms of oxidation under more severe conditions than experienced in service. All exposures were made in static air for a duration of 100 h, unless otherwise stated.

2.1 Cross-Sectional TEM

Transmission electron microscopy was carried out on cross-sectional specimens of oxidised material produced by glueing together two oxidised 10 mm disc specimens of 1.5 mm thickness using an epoxy resin. These were then sectioned to produce rectangular specimens of 3 mm×3 mm×10 mm, which was encapsulated in a thin walled brass tube of 3 mm outside diameter. Thin sections cut from this composite were polished and dimpled to less than 50 μm. The thin discs were finally ion milled for TEM examination, which was carried out on either a JEOL 4000FX or a Phillips CM20.

2.2 Stripped Oxide

Many of the properties of surface oxides may be best studied following the removal of the oxide film from the metal substrate. This technique has been used here to remove the thin scales from specimens of IMI 834 after exposure for 45 min at 600°C. A 10% solution of bromine in ethyl acetate was heated to its boiling point (75°C) and a sample of oxidised material of approximately 5×5×0.5 mm was immersed in the solution. Dissolution of the metal substrate was complete after about 1 h. The oxide was rinsed in ethyl acetate and ethanol before the thin film was carefully collected on to folding copper grids for TEM examination.

2.3 AES and SIMS

Auger analysis was carried out on oxidised specimens which were fractured *in situ*, using a Physical Electronics (PE) 595 Model Scanning Auger Microscope (SAM). Argon has been used as the primary ion source with an incident beam energy of 3 keV and a beam current of 10–20 nA. These conditions yield a standard sputter etch rate of approximately 350 Å per minute over an area of 1 mm^2, calibrated with various oxides of known thickness.

Secondary ion mass spectroscopy (SIMS) has been carried out using either a PE 595 Model SAM with PHI 3500 SIM II with argon ion beam energy of 3 keV, or a scanning ion microscope SIM/SIMS with the gallium ion gun operating at 10 kV and with sample currents of up to 100 nA.

All AES/SIMS specimens were oxidised for 100 h at either 600°C or 900°C to permit the examination of both thin and thick scales.

3. Results and Discussion

3.1 Scale Morphology

After exposure at 900°C the oxide scale (Fig. 1) comprises a coarse grained surface layer, a compact but stratified oxide layer and a thin porous layer at the metal/oxide interface. The oxide remains reasonably adherent up to 800°C, but above this the outer oxide readily spalls. Oxide thicknesses, measured by SEM

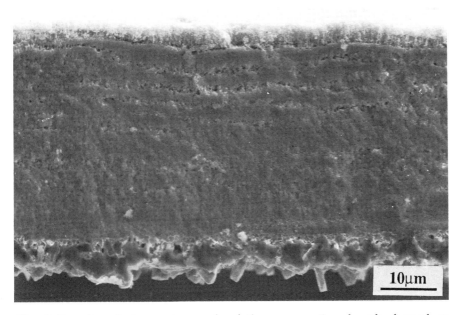

Fig. 1 Scanning electron micrograph of the cross-sectioned scale formed on IMI 834, after exposure for 100 h at 900°C.

Composition (Atomic %)								
Analysis No.	Ti	Al	Zr	Sn	Nb	Mo	Si	Comments
1	93.8	1.6	1.9	1.1	0.5	0.2	0.9	Surface oxide grains
2	11.2	87.9	0.2	0.5	0.7	0.1	–	Sub-surface scale
3	65.9	323.7	–	–	–	0.2	–	Mixed compact scale
4	93.1	1.3	3.1	0.4	0.6	0.4	1.3	Interface scale
5	60.4	4.2	1.8	31.3	0.5	0.3	0.8	Metallic interface layer
6	69.1	18.8	4.2	3.6	0.6	0.5	3.2	Substrate near layer
7	84.8	9.8	2.1	2.0	0.5	0.3	0.5	Bulk substrate

Table 1 EDX line scan analysis through a bulk cross-section after oxidation for 100 h at 900°C, showing the distinct Ti-rich and Al-rich oxide layers and the high concentration of Sn and Al at the interface.

examination of cross-section specimens, show that scales formed on the alloy are not only considerably thinner than those formed under equivalent conditions on pure titanium, but are also more compact and stratification is less distinct.

3.2 Bulk Energy Dispersive X-ray Analysis

EDX has been carried out on bulk cross-sections of oxidised samples which have been exposed at 900°C. Line scan analyses have been performed and successive analyses have been collected from a distance well within the matrix to the oxide surface. The results are presented in Table 1 in the form of AZF-corrected atomic percents. Analyses taken from well within the substrate are close to the nominal composition. Towards the metal/oxide interface, however, there is an increased aluminium concentration and immediately below the metal/oxide interface there is a distinct layer of high tin concentration. The composition of the scale is also found to vary through its cross-section. The surface scale is clearly titanium-rich but with significant concentrations of the other elements. Below this an aluminium-rich scale is found which also has a relatively high dopant concentration. The scale adjacent to the substrate is a mixed scale high in both titanium and aluminium.

3.3 Transmission Electron Microscopy

The alloy scale is found to consist of a compact layer which is found next to the metal substrate and a thin surface layer as shown in Fig. 2. Diffraction ring patterns taken from various positions within the scales have been indexed as TiO_2 (rutile). Micro-diffraction has also been carried out on this scale but gave no further information; the scale was again found to be TiO_2.

EDX line scan type surveys have been performed on the TEM cross-sections. Analyses from various depths within the matrix material are compared with

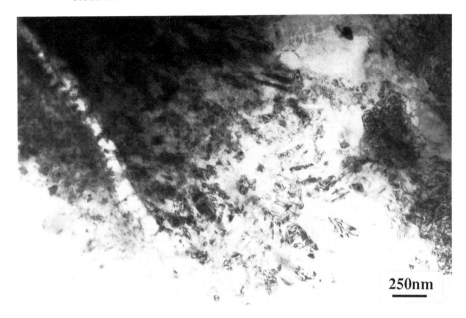

Fig. 2 Transmission electron micrograph of the cross-sectioned oxide formed on IMI 834 after exposure for 100 h at 600°C, with corresponding diffraction ring patterns.

Material	Ti	Al	Sn	Zr	Nb	Mo	Si
Cross-sectioned matrix	86.0	8.7	2.0	2.1	0.4	0.3	1.8
Standard Error (95%)	±1.9	±1.2	±0.1	±0.3	±0.1	±0.1	±0.4
Cross-sectioned oxide	83.4	8.6	1.6	2.7	0.4	1.1	2.1
Standard Error (95%)	±0.9	±0.5	±0.6	±0.4	±0.1	±0.3	±0.3
Stripped oxide	84.9	10.9	1.55	1.7	0.4	0.3	2.28
Standard Error (95%)	±2.0	±1.7	±0.9	±0.3	±0.2	±0.2	±0.85

Table 2 EDX line scan analysis through cross-sectioned TEM specimen, after oxidation of 45 min at 600°C.

that of the oxide in Table 2. For the purpose of statistical integrity, numerous repeat analyses have been taken from both the matrix and the scale. The results are presented as a statistical mean composition for which the standard error has been calculated, and is also shown. Clearly the matrix composition detected is close to the nominal, with the exception of silicon which is higher. This situation does not appear to change in the region of the metal/oxide interface and no concentration of any one particular element is observed. It appears that in general the composition of the oxide reflects that of the matrix.

In view of the high concentration of additional elements detected within the cross-sectional scale and the possibility that this may result from spurious

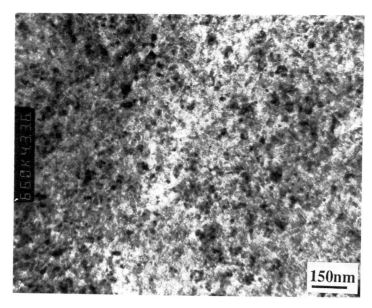

Fig. 3 Transmission electron micrograph of the stripped oxide formed on IMI 834 after exposure for 45 min at 600°C, with corresponding diffraction ring pattern.

contributions and stray X-rays in the TEM, the stripped scale has been similarly analysed. These results are also shown in Table 2. The proportion of alloying elements present in this oxide would appear to be very similar to that of the matrix material. In addition to this general analysis, there are also some regions of high aluminium concentration. A TEM micrograph of the stripped scale is shown in Fig. 3. It is difficult to differentiate between possible TiO_2 (rutile) and α-Al_2O_3 reflections at d-spacings smaller than the first few. It seems likely that the stripped scale is a predominantly TiO_2 scale doped with relatively large concentrations of solute elements in which there are pockets of Al_2O_3.

3.4 AES and SIMS

Analysis of the scale formed after oxidation at 600°C shows that both elemental aluminium and silicon are present in the 'TiO_2' scale. There appears to be very little change in their concentrations across the fractured specimen from the scale surface into the substrate. The scale formed at 900°C is found to be a predominantly TiO_2 oxide in which both Al_2O_3 and SiO_2 are detected. Al_2O_3 and SiO_2 are found in small concentrations throughout the scale. There are, however, some regions of the fractured oxide film which have the appearance of non-metallic intergranular fracture. X-ray mapping of one such area (Fig. 4a and 4b) suggests that there is some concentration of Al_2O_3 along the cracks. Auger analysis of the scale adjacent to the metal/oxide interface finds SiO_2 in greater concentration than that detected within the surface oxide. An AES depth profile showed the layer of SiO_2 is removed after etching for approx-

 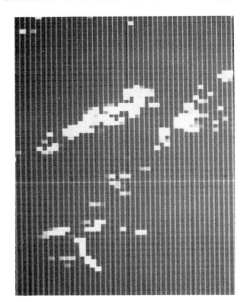

Fig. 4 (a) SE micrograph of the fractured surface scale after exposure at 900°C. (b) X-ray map of the region illustrated in (a), showing the concentration of aluminium along the cracks in the scale.

imately 4 mins. This suggests that the thickness of the layer is about 0.15 μm. An X-ray map for SiO_2 (Fig. 5a and 5b) shows its concentration at the interface with respect to the surface scale.

SIMS has been applied in this case in the analysis of the scale formed after oxidation for 100 h at 600°C. A SIMS survey of the surface scale is shown in Fig. 6. It can be seen that all of the elemental additions are found in the survey and, significantly, all are found in the oxidised condition.

4. General Discussion

Interpretation of the behaviour of individual alloying elements is complicated by their different relative concentrations and effects on the properties of the scale. Their combined effect either as dopants in the rutile scale, or in the substrate material however, is clearly beneficial. The thin scales produced after exposure at 600°C have a high degree of doping with all additional elements, all of which are present in both ionic states and 'in association with oxygen'. Aluminium is thought to play a key role during oxidation and has been detected in both states of the scale, and in relatively high concentrations in the substrate, below the metal/oxide interface.

The behaviour of aluminium in rutile may be examined on the basis of the Wagner-Hauffe theory of mass transport in oxides. The trivalent Al^{3+} cation dopant, dissolved substitutionally in the anion-deficient TiO_{2-x} lattice, should

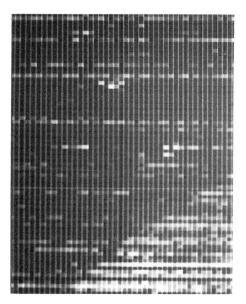

Fig. 5 (a) SE micrograph of both the fractured surface scale and the adherent interface scale after exposure at 900°C, and (b) X-ray map of the region illustrated in (a), showing the concentration of silicon at the interface.

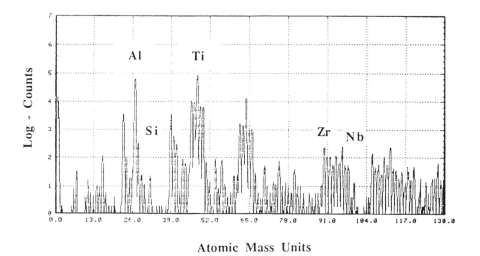

Atomic Mass Units

Fig. 6 SIMs survey of the surface scale after exposure for 100 h at 600°C.

increase the number of oxygen vacancies and thus cause an increase in the rate of oxygen transport through the oxide. This is not observed experimentally, however, and it is noted that the presence of pentavalent cations in the alloy may readily compensate for this effect.

Hence other mechanisms must be operative. If the trivalent cations dissolve interstitially in the TiO_2 then the oxygen vacancy concentration will be reduced, the electron concentration increased and the number of titanium interstitials reduced. In this case the solubility of Al_2O_3 will increase with decreasing oxygen partial pressure.[2] The solubility of Al_2O_3 in TiO_2 at atmospheric pressure is very small and hence, if there were an oxygen activity gradient across the scale, Al_2O_3 in solution in TiO_2 would tend to precipitate out towards the scale surface where the oxygen activity is high. Evidence of an oxygen activity gradient across the scale is seen in the result of Auger analysis of the oxide scales. A distinct shift from the characteristic Al_2O_3 to the Al^{3+} Auger energy was observed as a function of distance within the scale. Further evidence of an oxygen activity gradient is in the observation of more compact oxide layers close to the metal/oxide interface. According to Bertrand,[3] increasing oxygen partial pressure has the effect of reducing oxide crystallite size and producing a finer structure.

The observation of an aluminium-enriched zone in the metal, below the metal/oxide interface and the fact that there is no depletion of aluminium in the metal immediately below this, suggests that there is rapid diffusion of aluminium away from the advancing titanium-rich oxide surface. This results in an increased aluminium activity which may eventually permit the formation of Al_2O_3; hence the observation of Al_2O_3 layers in the scale. Once this aluminium has been removed from the metallic matrix the process is effectively repeated with the subsequent formation of a second highly doped titanium-rich scale. This complex sequence of scale formation is evidently protective at the lower exposure temperatures, although its effectiveness is questionable after higher temperature exposure due to its tendency to spall. A probable reason for this is the formation of a thin layer of SiO_2 at the metal/oxide interface.

The silicon content of IMI 834 is too low for the equilibrium formation of SiO_2, hence its presence in small concentrations throughout the scale and at the metal/oxide interface again suggests that some mechanism of cation diffusion or trapping is in operation at the higher exposure temperatures. The stability of SiO_2 is less than that of TiO_2 and Al_2O_3; it would seem that a significant relative increase in the concentration of silicon in the titanium and aluminium-depleted region of the interface would be required for silicon to react. Silicon is though to play an important role in the prevention of oxygen dissolution into the metallic substrate,[4] however, it would appear that its presence at the interface has a detrimental effect on the adherence of the scale at the higher temperatures. This is in agreement with the findings of Chaze and Coddet[5] who report that small concentrations (less than 0.5 wt.%), reduce scale adhesion to below that of pure titanium. This might be considered in terms of mechanical stresses at the interface. The Pilling–Bedworth ratio of SiO_2 (1.88–2.15) is significantly larger than that of TiO_2 (1.73). This difference in molecular volume results in the superimposition of differential contraction stresses between the two oxides and the metal. This can cause considerable scale cracking and spalling resulting in its loss of adhesion to the substrate.

Initial scale formation at the high temperatures is similar to that observed after exposure at the lower temperatures, that is, a titanium-rich scale is formed with a high concentration of additional elements that appear to have been trapped during the initial rapid formation of the 'rutile' layer. It seems likely that these elements act as sinks or traps for the oxygen diffusing inwards through the scale. Whilst there is a significant concentration of most dopants in the surface layer, the aluminium concentration is relatively low, although there are discrete regions of higher aluminium concentration which appear to be islands of Al_2O_3. This may suggest either that there haa been some aluminium diffusion or that the exposure of trapped aluminium to an oxygen-enriched environment 'forces' its association with the oxygen.

The diffusion of alloying elements, particularly during exposure to the higher temperatures, and hence their increased concentration in the surface layers of the matrix may well contribute to the blocking or trapping of diffusing oxygen either by the occupation of interstitial sites or the closing of interstitial sites in the presence of large substitutional solute atoms. Simple trapping of the diffusing oxygen atoms in the aluminium-rich layer may result in a reduction in the rate of oxidation diffusion. According to the theory of diffusion in very dilute solutions,[6] large solute atoms tend to diffuse more quickly than the solvent. Hence it is perhaps not surprising that tin, being a relatively large atom ($r_{Sn} = 1.58$ Å) of equivalent valence to titanium, is able to diffuse away from the advancing oxygen. This oversized atom may displace solvent atoms outwardly from their equilibrium positions, thus creating a lattice strain and an increased migration energy. It is possible that a concentration of the larger tin atoms has the effect of reducing the size of the interstices available for oxygen diffusion and hence inhibit the process. One might equally argue, however, that lattice strains induced by the presence of a large solute may be relieved by the location of a vacancy in its proximity. Enhanced vacancy concentrations may well compensate for reduced interstitial diffusion.

References

1. P. A. Blenkinsop, *Titanium Science and Technology* (Ed. G. Lutering), **4** (1984), 2323.
2. P. Kofstad, *Nonstoichiometry, Diffusion, and Electrical Conductivity in Binary Metal Oxides*, Krieger & Sons (1983).
3. G. Bertrand, K. Jarraya, and J. M. Chaix, *Oxidation of Metals*, **21** (1/2) (1983), 1–19.
4. A. M. Chaze and C. Coddet, *Oxidation of Metals*, **28** (1/2) (1987), 61–71.
5. A. M. Chaze and C. Coddet, *Journal of Materials Science*, **22** (1987), 1206–1214.
6. P. Shewmon, *Diffusion in Solids*. Warrendale, PA: The Minerals, Metals and Materials Society (1989).

Oxidation of a Near γ Ti-48.8a/o Al-2.2a/oV Alloy

A. I. P. NWOBU, H. M. FLOWER, and D. R. F. WEST

Dept of Materials, Imperial College, London SW7 2BP, England

ABSTRACT

The chemical reactions of a near γ Ti-48.8a/oAl-2.2a/oV alloy with 0.2–1 atm air, oxygen, impure argon and nitrogen gases containing O_2 at 800–1000°C have been investigated Ti$_3$Al, Al$_2$O$_3$, and TiO$_2$ are typically the final reaction products in all the different environments even in impure N_2 gas containing 7 ppm of O_2. The nitrides, TiN and Ti$_2$AlN, are formed as intermediate products in the N_2-containing environments including 'pure' air. The rate of their conversion to oxides increases with the oxidation temperature, total gas pressure, and O_2 to N_2 concentration.

1. Introduction

TiAl γ-based alloys are currently receiving considerable attention as potential high temperature materials. Alloying Al lean γ compositions with small (<3a/o%) Cr, Mn and/or V additions has been shown to improve the otherwise poor room temperature ductility but may degrade oxidation resistance while elements which improve oxidation behaviour (Nb, Ta, and/or Mo) tend to reduce ductility.[1,2]

There have been few studies[3–6] which try to provide a clear understanding of the mechanisms and kinetics of oxidation of the near γ TiAl based alloys in air/oxygen. The alloys tend to form a more protective alumina scale in oxygen than in air.[6] Choudhury et al.[4] suggested that the presence of nitrogen (in air) is responsible for this observation although they reported that the same oxidation products (Ti$_3$Al, Al$_2$O$_3$, and TiO$_2$) are formed on the alloys exposed in oxygen, air, and even in nitrogen containing just about 1 ppm (by volume) of oxygen. Meier and Pettit[5] observed that the rate of oxidation increases continually as increasing amounts of nitrogen are added to pure oxygen but then drops again in pure nitrogen. McKee and Huang[3] studied the effects of

357

small solute additions (Cr, Mn, Nb, Si, Ta, V and/or W) on the oxidation of the near gamma alloys and suggested that thermodynamic factors such as the effect of the elements on the activity of Ti and Al in the alloys, are possibly responsible for the observed differences in oxidation resistance.

In this work, reactions not previously observed[3–6] during the oxidation of near γ TiAl based alloys are reported. A near γ based alloy, Ti-48.8a/oAl-2.2a/oV, has been exposed to a range of oxygen–nitrogen gas mixtures, and nitrogen/argon containing less than 7 ppm of oxygen under controlled conditions of pressure, the amount of gas and gas flow. The observations provide insight into the differences in oxidation behaviour of near γ TiAl based alloys in air and oxygen. The effects of the 2.2a/oV addition and microstructure on oxidation in air are also reported.

2. Experimental Methods

The alloy mainly used in the study is a cast Ti-48.8a/oAl2.2a/oV alloy pancake forged to about a 5 cm thick × 35 cm diameter disc by TIMET USA. Small 30 g ingots of cast Ti-48a/oAl and Ti-48a/oAl-2.2a/oV alloys prepared by electric arc melting were also used. Oxidation specimens were either 2 mm thick 10 mm diameter discs, or about $10 \times 10 \times 2$ mm plates both given 1200 grit SiC finish.

Specimens were weighed to a sensitivity of 0.1 mg, ultrasonically cleaned in acetone, dried before reacting in static air, static or flowing N_2, O_2, argon gases and/or mixtures of the gases at 800–1000°C. They were reweighed after the reactions and cooling to room temperature. Weight changes during reactions in static air, flowing N_2, O_2 and argon gases were continuously monitored using a CI microforce balance with a Mark 2B vacuum head and sensitivity of up to 1 μg. The furnace temperature where the specimen is located in the system was monitored with a Pt/Pt-10%Rh thermocouple controlled to within ±4°C. Before the reaction rig was lowered quickly into a furnace N_2, O_2, or argon gas flowing under pressure at a rate of about 300/1000 ml/min was introduced. Reactions in static N_2, O_2, argon gas/mixtures were also carried out in sealed silica tubes of fixed volume and at total gas pressures of 0.2–0.8 atmospheres, the silica tube initially having been evacuated twice after flushing with the reacting gas.

Microstructural characterisation and microanalysis were carried out using scanning electron microscopy (SEM) and X-ray energy/wavelength dispersive microanalysis. A Philips PW1300 X-ray diffractometer was used for structural analysis of the different oxides and nitrides phases.

3. Results and Discussion

3.1 Oxidation of Cast and Forged Alloy in Air

Weight gain versus time graphs (Fig. 2(a)) of the pancake forged Ti-48.8a/oAl-2.2a/oV alloy (microstructure shown in Fig. 1(a)) oxidised in static air between

 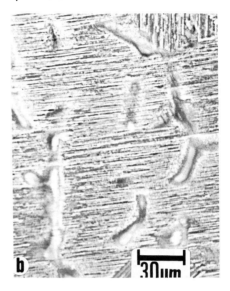

Fig. 1 SEM micrographs of (a) forged Ti-48.8a/oAl-2.2a/oV alloy showing grains containing deformed lamellar γ/α_2 phases, and (b) cast Ti-48a/oAl–2.2a/oV alloy showing lamellar γ/α_2 structures and γ segregates.

800 and 1000°C indicate a mainly parabolic relationship at 800–950°C and mixed parabolic/linear modes at 1000°C. The oxidation rate constants k_p (parabolic rate) varied from $1–2.4 \times 10^{-5}$ (g/cm²)²/h at 1000°C to 7.6×10^{-9} (g/cm²)²/h at 800°C (Fig. 4). Using the relationship

$$(\delta m)^2/t = k_p = \text{const.} \times \exp(-Q/RT) \tag{1}$$

where m mass/unit area (g/cm²), t time (h), Q activation energy, R gas constant, and T temperature, activation energies (Q) were calculated. The apparent activation energy, Q, for oxidation at 850–1000°C has a value of about 274 kJ/mol similar to the value of 264 kJ/mol observed for the oxidation of Ti-32.4w/o(47a/o)Al alloy in air at 600–1300°C.[7] The oxidation products of the forged Ti-48.8a/oAl–2.2a/oV alloy exposed in air are $\alpha_2 + Al_2O_3 + TiO_2$ and this is in general agreement with previous observations.[3–6]

The cast Ti-48a/oAl-2.2a/oV alloy (microstructure shown in Fig. 1b) oxidised in air at 900 and 1000°C was found to have a similar Q value (276 kJ/mol), but higher k_p values (Fig. 4) and weight gains compared to those observed for the forged alloy (e.g. cast alloy oxidised over 24 h at 900 and 1000°C showed weight gains of 119.8 and 267.5×10^{-4} g/cm² respectively compared to the values of 43.5 and 168.9×10^{-4} g/cm² respectively observed for the forged alloy in similar oxidation treatments). The cast binary Ti-48a/oAl alloy (microstructure similar to that of cast Ti-48a/oAl-2.2a/oV alloy, Fig. 1b) oxidised at 900 and 1000°C showed lower weight gains/k_p values but a similar Q value (298 kJ/mol) compared to the ternary alloys (cast Ti-48a/oAl-2.2a/oV and forged Ti-48.8a/

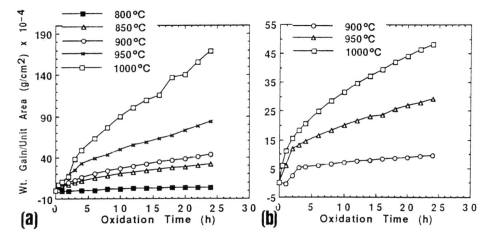

Fig. 2 Weight gain-time graphs for forged alloy exposed to (a) air and (b) oxygen gas flowing at the rate of 300 ml/min at 800–1000°C.

oAl-2.2a/oV alloys, Fig. 4). The 2.2a/oV addition to TiAl thus reduces oxidation resistance.

3.2 Oxidation of Forged Alloy in Oxygen and Impure Argon

Figure 2(b) shows the weight gain graphs for the forged alloy oxidised in dry flowing (at rate of 300 ml/min) oxygen gas at 900–1000°C. Similar results were observed when the forged alloy was oxidised in flowing impure argon gas containing a total oxygen impurity of about 6 ppm and also flowing at a rate of 300 ml/min. The oxidation behaviour is parabolic up to 1000°C but the rate is less than that observed for oxidation in air (e.g. after 24 h oxidation at 1000°C weight gain 48.23×10^{-4} g/cm^2 in oxygen/argon compared to 168.9×10^{-4} g/cm^2 for oxidation in air; and lower k_p values, Fig. 4). The activation energy of 437 kJ/mol obtained for the oxidation in oxygen is larger than that (274 kJ/mol) observed in air.

3.3 Reactions with Impure Nitrogen Gas

3.3.1 Flowing Impure Nitrogen Gas

The weight gain versus time graphs observed for the forged alloy exposed at 1000°C and above over a period of 24 h in dry flowing nitrogen gas containing a total of 7 ppm of O impurity changed from a parabolic behaviour to a linear one as the oxidation time increased (Fig. 3(a)). The time after which this transition occurs increases with the decrease in oxidation temperature (Fig. 3(a)) and is sensitive to other factors such as the impure N$_2$ gas flow rate, and the manner in which the reaction is started (e.g. by a 'cold start', i.e. introducing N$_2$ gas into the reaction chamber at room temperature and slowly heating the

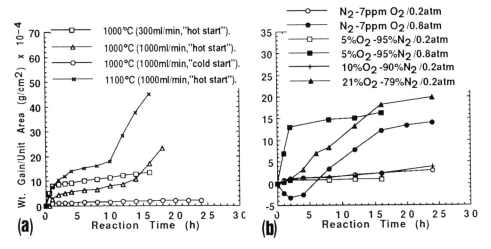

Fig. 3 Weight gain-time graphs of forged alloy reacted with (a) N_2-7 ppm O_2, (300/1000 ml/min) at 1000–1100°C, and (b) a fixed volume of static N_2-7 ppm to 21% O_2 gases at 0.2–1 atm and 1000°C.

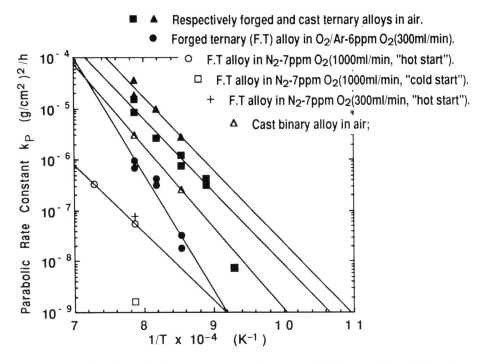

Fig. 4 Graphs of parabolic rate constants k_p versus 1/T for reaction of forged Ti-48.8a/oAl-2.2V, cast Ti-48a/oAl-2.2a/oV and Ti-48a/oAl alloys with air, flowing O_2, and N_2-7 ppm O_2 gases at 800–1100°C.

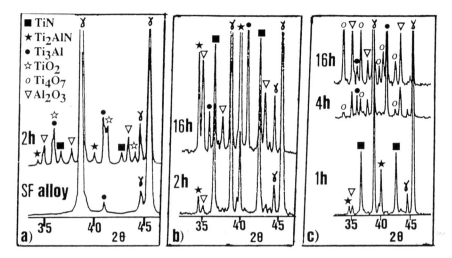

Fig. 5 XRD patterns showing phases observed on surface of the forged alloy reacted at 1000°C with (a) flowing (1000 ml/min) N_2-7 ppm O_2 gas (note the composition of the scale free (SF) alloy) and a fixed volume of (b) 10% O_2-90%N_2 gas at 0.2 atm, and (c) 21% O_2-79% N_2 gas at 0.2 atm for various times.

chamber to the reaction temperature, or by a 'hot start', inserting the sample rapidly into an already heated reaction chamber containing the N_2 gas or moving the reaction chamber containing the sample/gas quickly into already heated furnace). The weight gain/k_p values within the parabolic region of the reaction were usually less than those observed for the parabolic oxidation in oxygen/impure argon (compare Figs. 2(b), 3, 4). XRD analysis of the oxidation products indicates the formation of nitrides (TiN and Ti_2AlN), α_2, Al_2O_3, and TiO_2 (Fig. 5(a)) during the period of low/parabolic rate of reaction; while only the α_2, Al_2O_3, and TiO_2 phases remain during the later linear behaviour. Formation of α_2 phase during the period of slow growth tends to be suppressed using the 'cold start' rather than the 'hot start'. Reactions of the alloy with static impure N_2 at 1000°C (described below) provided further information about the complex reactions processes.

3.3.2 Static Impure Nitrogen Gas of Fixed Volume/Variable Total Gas Pressures

The weight gain versus time graphs for the forged alloy exposed to a fixed volume of impure N_2 gases containing 7 ppm to 21% (by volume) O_2 at total gas pressures 0.2 and 0.8 atm at 1000°C up to 24 h period are shown in Fig. 3(b). At a total gas pressure of 0.2 atm the alloy shows very little weight gain (less than 3×10^{-4} g/cm^2 and k_p less than 10^{-8} (g/cm^2)^2/h over a reaction time of 24 h at 1000°C in the impure N_2 gas containing 10% and less O_2 gas. However, in the 21% O_2/79% N_2 gas mixture at 0.2 atm rapid weight gains above 3×10^{-4} g/cm^2 are observed after just 2 h, and this rises to about 20×10^{-4} g/cm^2 after 24 h (Fig. 3(b)). Detailed XRD revealed that the formation of nitrides, small amounts of Al_2O_3 and α_2 (later) is responsible for the low weight gain of alloy in impure N_2 containing 10% and less O_2 at 0.2 atm (Fig. 5(b)). In contrast,

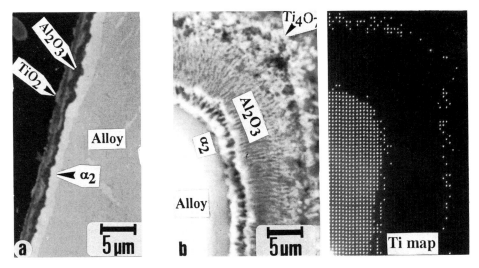

Fig. 6 SEM micrographs/Ti map showing (a) protective Al_2O_3 scale observed on the forged alloy reacted with O_2 flowing at the rate of 300 ml/min at 900°C for 24 h and (b) unprotective porous Al_2O_3 layer formed in a fixed volume of 5%O_2-95%N_2 at 0.2 atm and 1000°C for 24 h.

rapid and larger weight gains in the 21% O_2/79% N_2 atmosphere at 0.2 atm is associated with the transition of the nitrides/Al_2O_3 structure to large quantities of α_2, Al_2O_3 and Ti oxides (Ti_4O_7 and/or Ti_8O_{15}, Fig. 5(c)). When the total gas pressure is raised to 0.8 atm the alloy showed weight loss in the N_2 gas containing 7 ppm O_2 up to 4 h of reaction time after which it rapidly gained weight up to 14×10^{-4} g/cm^2 after 24 h of reaction at 1000°C (Fig. 3(b)). Nitrides and small amounts of Al_2O_3 were mainly observed during the period in which the alloy lost weight in the 0.8 atm N_2-7 ppm O_2 environment, while large quantities of α_2, Al_2O_3, and Ti oxides are observed when the alloy weight gains were above 3×10^{-4} g/cm^2. The weight loss is due to Al evaporation prior to the formation of a protective nitride/Al_2O_3 scale (Fig. 7(b)). However, rapid formation of the α_2, Al_2O_3 and Ti oxides within 30 min of exposure in 0.8 atm gas mixtures containing 5% and more O_2 does not allow Al loss but rather rapid weight gains (Fig. 3(b)).

3.4 Microstructures

A protective continuous thin (less than 5 μm thick) Al_2O_3 scale was observed (between the oxidation induced α_2 layer at the oxide–metal interface and a TiO_2 scale at the surface (Fig. 6(a)) to form on the forged alloy exposed to flowing oxygen/argon gases at 900°C and to air at 800°C. At higher temperatures (950 and 850°C respectively in oxygen and air) the thin continuous Al_2O_3 film layer is replaced by a multi-layered scale of mixed $Al_2O_3 + TiO_2$ oxides usually observed in the oxidation of near γ alloys.[3–6] Microstructural changes during the isothermal reactions: Nitrides + Al_2O_3 → α_2 + nitrides + Al_2O_3 → α_2 +

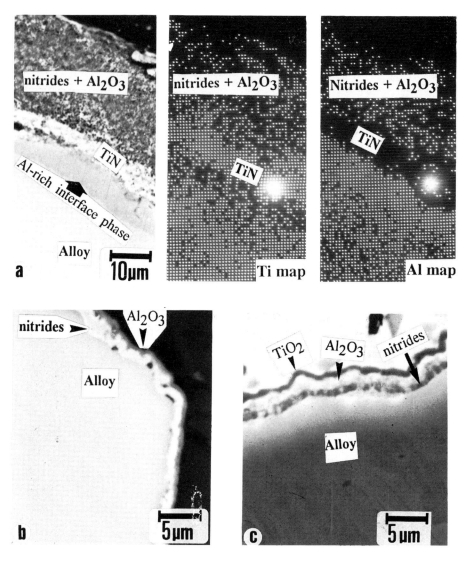

Fig. 7 SEM micrographs, and Ti/Al maps showing (a) an Al-rich interface phase and Al$_2$O$_3$ mixed with surface nitride layer on the forged alloy reacted with a fixed volume of N$_2$-7 ppm O$_2$ at 0.2 atm and 1000°C for 16 h, (b) Al$_2$O$_3$ as surface film layer above the nitrides in a reaction at same temperature for 4 h with the gas filled to the higher gas pressure 0.8 atm, and (c) surface TiO$_2$ grown on the Al$_2$O$_3$ layer in a reaction with the flowing gas (1000 ml/min) at 1000°C for 10 h.

nitrides + Al$_2$O$_3$ + Ti oxide(s) → α_2 + Al$_2$O$_3$ + Ti oxide(s) showed that the initial small quantity of Al$_2$O$_3$ formed is mixed with the nitride layer at the surface, while an inner nitride layer (TiN mainly) is separated from the alloy by an Al-rich interface phase (e.g. composition Ti-57.9a/oAl-2.8a/oV compared to Ti-47.6a/oAl-2.0V of alloy, Fig. 7(a)). Al$_2$O$_3$ tends to grow more at the surface of the scale (Fig. 7(b)) formed at the second stage of the reaction when α_2 begins to form and replaces the Al rich interface phase between the nitrides and metal alloy. Ti oxide(s) subsequently form mainly at the scale surface (Fig. 7(c)). The complete elimination of the nitrides usually leaves a microstructure similar to that formed in air/oxygen or a very porous thick Al_2O_3 layer between the α_2 interface phase and surface Ti oxides layer (Fig. 6(b)).

There is little knowledge on the type of reactions taking place when near γ base alloys oxidise in air and O_2. Thermodynamic considerations[8,9] using data on the activities of Ti and Al in the Ti-Al system reveal that Ti oxide (TiO) is the more stable oxide in the reaction

$$2Al(l) + 3TiO(s) = Al_2O_3(s) + 3Ti \qquad (1)$$

in near γ TiAl alloys. However, a recent study by Li et al.[10] of reactions in the Ti-Al-O system showed that Al_2O_3 is stabilised by the initial absorption of oxygen into α_2 and γ phases during oxidation, which results in an increase in the Al activity. The present results showed that under 0.2 atm of static N_2/O_2 gas mixtures at 1000°C only Al_2O_3 (and no Ti oxide) forms with the nitrides (nitrides forming more readily than Al_2O_3) and support the work of Li et al.,[10] i.e. Al_2O_3 is the more stable oxide phase.

Reaction (1) is apparently not the only one taking place when the alloys are exposed in air. The present results indicate also that nitride formation and their oxidation are taking place. For the oxidation of nitrides, e.g. the reaction:

$$TiN + 1/2O_2 = TiO + 1/2N_2 \qquad (2)$$

the $O_2:N_2$ gas pressure ratio at 1000°C is about 10^{-15} using free energy data.[11] The reaction is thus possible in the $O_2:N_2$ gas pressure ratio 7×10^{-6} (a BOC oxygen free N_2 gas, OFN containing a total of 7 ppm O_2 based impurity) used in the study. This does not suggest that nitriding of Ti and Ti-Al based alloys in BOC OFN N_2 gas is impossible. No transition of nitrides to Ti oxides is observed when Ti-33a/oAl-2.2a/oV alloy (containing only α_2) was exposed at 1000°C to 5%O_2 95%N_2 gas mixture at total gas pressure 0.8 atm. Also TiN formed with Al_2O_3 and TiO_2 oxides on α_2 based Ti-24a/oAl-10a/oNb-3a/oV-1a/oMo alloy exposed to 1 atm high purity air at 800–1000°C over 100 h was not converted to Ti oxides.[12] Other reactions involving the formation of Ti_2AlN from TiN (TiN being unstable in the presence of Al),[13] and the conversion of Ti_2AlN to Al_2O_3/Ti oxides and α_2 may have to be deduced so as to understand why nitrides formed by near γ based alloys in air are more unstable than those formed by α_2 based alloys. This requires thermodynamic data on Ti_2AlN formation which is not yet available.

4. Conclusions

The oxidation of the near γ Ti-48.8a/oAl–2.2a/oV alloy is reduced if the cast structure is forged and if it occurs in pure O_2 gas rather than in air. The oxidation reaction in air involves intermediate nitride (TiN and Ti_2AlN) formation before the formation of the oxidation products α_2, Al_2O_3 and Ti oxides. The sequence of the reactions observed at 1000°C in a fixed volume of air at 0.2 atm and in O_2/N_2 gas mixtures less than air 21%:79% ratio is nitrides + $Al_2O_3 \rightarrow \alpha_2$ + nitrides – $Al_2O_3 \rightarrow \alpha_2$ + nitrides + Al_2O_3 + Ti oxides $\rightarrow \alpha_2$ + Al_2O_3 + Ti oxides. The rate of transition to the final products increased as oxidation temperature, total gas pressure, and the O_2 to N_2 content is increased.

Acknowledgement

This work has been carried out with the support of Procurement Executive Ministry of Defence.

References

1. Young-Won Kim and D. M. Dimiduk, *J. Met.*, **43** (1991), 40–47.
2. D. M. Dimiduk, Y.-W. Kim, D. B. Miracle and M. Mendiratta, *ISIJ Int.*, **31** (1991), 1223–1234.
3. D. W. McKee and S. C. Huang, *Corrosion Sc.*, **33**(12) (1992), 1899–1914.
4. R. N. S. Choudhury, H. C. Graham and J. W. Hinze, *Proc. Symp. on Properties of High Temp. Alloy*, Princeton, Electrochem. Soc. (1977), pp. 668–680.
5. G. H. Meier and F. S. Pettit, *High Temperature Intermetallics*, April–May 1991, London, preprint, Institute of Metals (1991), pp. 66–80.
6. G. H. Meier, D. Appoalonia, R. A. Perkins and K. T. Chiang, *Oxidation of High Temperature Intermetallics*, (ed. T. Grobstein and J. Doychak), TMS, Warrendale, PA (1988) pp. 185–193.
7. E. U. Lee and J. Waldman, *Scripta Metall.*, **22** (1988), 1389–1394.
8. A. Rahmel and P. J. Spencer, *Oxid. of Metals*, **35**(1/2) (1991), 53–68.
9. K. L. Luthra, *Oxid. of Metals*, **36**(5/6) (1991), 475–490.
10. X. L. Li, R. Hillel, F. Teyssandier, S. K. Choi, and F. J. H. Van Loo, *Acta Metall. Mater.*, **40**(11) (1992), 3149–3157.
11. O. Kubaschewski and C. B. Alcock, *Metallurgical Themochemistry*, 5th edn, Pergamon Press, Oxford (1979).
12. T. A. Wallace, R. K. Clark, K. E. Wiedemann and S. N. Sankaran; *Oxid. of Metals*, **37**(3/4) (1992), 111–124.
13. R. Beyers, R. Sinclair and M. E. Thomas, *J. Vac. Sci. Technol.*, **B2**(4) (1984), 781–784.

Microstructural Investigation of Sulphide Scale Formation

ILONA TUREK, MAREK DANIELEWSKI, ALEKSANDER GILL

University of Mining and Metallurgy, Department of Solid State Chemistry,
al. Mickiewicza 30, 30–059 Kraków, Poland

and

JERZY MORGIEL

Institute for Metals Research, Polish Academy of Sciences, 25 Reymonta st.,
30-059 Kraków, Poland

ABSTRACT

The microstructure of managanese sulphide prepared by exposing stripes of manganese in an H_2/H_2S atmosphere (at $ps_2 = 1.10^{6.75}$ Pa, $T = 800$ K, $t = 48$ h) has been examined in a plan-view in the transmission electron microscope. Thin foils were prepared by dimpling and ion milling in an argon atmosphere. The sulphide layer, as identified by electron diffraction analysis, consisted of polycrystalline α-MnS. Two types of grain boundaries were observed: those which were ordered along certain crystallographic planes, and those which were 'jagged'. Comparatively high densities of dislocations of mixed character were observed within the sulphide grains. The influence of these plane and line defects on the diffusion of manganese is discussed.

1. Introduction

Manganese is considered as relatively resistant to sulphidation due to the fact that its sulphide exhibits low concentrations of defects.[1] Manganese can be introduced as an alloying addition for alloys designed to operate in sulphur-containing environments.

The sulphidation behaviour of manganese has been investigated in a very wide temperature (700–1400 K) and sulphur pressure range.[2,3,4] It has been stated that the sulphide scale morphology is very sensitive to temperature. At 800 K, the α-MnS scale is fine-grained, with grain sizes of less than 1 μm, and it grows predominantly by grain-boundary diffusion of Mn^{+2} cations. At temperatures exceeding 1000 K, the scale is coarse grained with a grain size of dozens of micrometers and grows by the outward lattice diffusion of cations. At intermediate temperatures (800–1000 K) both types of scale growth are operating. At the initial, usually non-parabolic, stage a fine-grained scale is formed (typical of intergranular diffusion) which, however, transforms into a coarse-grained one, typical of volume diffusion, as the exposure time increases.

The promising results concerning the resistance of amorphous Mo–Al alloys to oxidation and sulphidation[5] encourage further investigation of the transport properties of pure and doped refractory metal sulphides. For such detailed studies of scale formation on the amorphous layers in oxidising/sulphidising environments, it is necessary to acquire experience in sample preparation procedures for TEM observation. The most useful are 'sandwich' specimens which enable the microstructural observation of both the scale and metal as well as of the interfacial region. These, however, are extremely rare in the literature because of serious experimental difficulties.

The work described here concerns our TEM observations of grain boundaries in αMnS and is a starting point for further investigations concerning the corrosion products formed on selected manganese-based alloys and amorphous Mn–Al type films.

2. Experimental methods

The experiments have been carried out with spectrally pure manganese with impurity levels not exceeding 5 ppm. The samples of irregular shape and total surface area not greater than $2\,cm^2$ were ground on emery papers down to 800 grit, polished with $1\,\mu m$ diamond paste and subsequently sulphidised in H_2/H_2S at $ps_2 = 1.10^{-6.75}$ Pa and $T = 800\,K$ 48 h.

The first step in the preparation of thin foils for TEM consisted of mechanical thinning specimens to less than $30\,\mu m$ and this was carried out with the aid of a SBT inc. 515 dimpler. The specimen, which had a thickness of 1 mm, was washed in acetone and stuck on top of a rotating roller with wax. A brass disk, 3 mm thick and 16 mm in diameter, was fixed to another roller and pressed against the specimen with an applied load of 15 g. At 30 minute intervals the surface of the specimen was treated with an abrasive paste. When a thickness of $100\,\mu m$ had been reached, dimpling of the specimen was continued with the aid of a smaller disk, 1.5 mm thick and 13 mm diameter. The final thickness of the specimen in the thinned zone was $28\,\mu m$. Thin foil was subsequently prepared by ion milling.

Microstructural observations were performed on a Philips CM20TWIN (200 kV) transmission electron microscope equipped with a Link eXL energy dispersive spectroscopy system.

3. Results and discussion

Transmission electron microscope examination of the manganese sulphide scale was restricted to plan-view observations of their surfaces. The scale consisted of grains a few microns in size with a high proportion of characteristically jagged boundaries, as shown in Fig. 1(a), as well as smooth ones (Fig. 1b). This difference is probably related to changes from low to high angle boundary type respectively. The interior of the grains contained a large

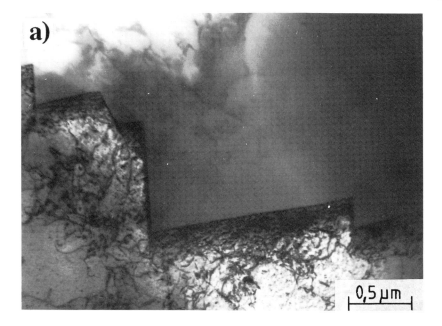

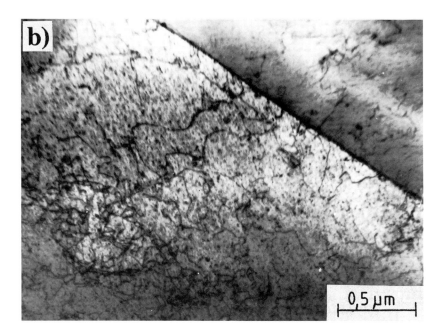

Fig. 1 Two types of grain boundaries observed in α-MnS: (a) jagged; (b) smooth.

0,5 μm

Fig. 2 The structure of dislocations in α-MnS.

number of dislocations which tended to form entanglements and pileups (Fig. 2).

Electron diffraction patterns from the oxide, two of which are shown in Figs 3(a) and 3(b), confirmed that the scale was single phase α-MnS with the NaCl-type structure ($a = 0.5224$ nm).

Microanalysis performed with a probe size of 10 nm generally showed an equal distribution of both Mn and S. However, on the smooth boundaries (Figs. 4a and 4b) a small but reproducible sulphur enrichment in the range of 1 to 2% was detected, as presented in Table 1. In this type of boundary the Mn/S atomic ratio is slightly smaller than within the grains.

	Label	Start keV	End keV	Net integral	Eff. factor	% age, total
Matrix 1	S Kα	2.21	2.41	5822	1.00	35.07
	Mn Kα	5.75	6.03	10777	1.00	64.93
Matrix 2	S Kα	2.21	2.41	6862	1.00	36.67
	Mn Kα	5.75	6.03	11850	1.00	63.33
Matrix 3	S Kα	2.21	2.41	8016	1.00	38.30
	Mn Kα	5.75	6.03	12914	1.00	61.70
Matrix 4	S Kα	2.21	2.41	8255	1.00	37.26
	Mn Kα	5.75	6.03	13899	1.00	62.74
Boundary	S Kα	2.21	2.41	7954	1.00	38.37
	Mn Kα	5.75	6.03	12775	1.00	61.63

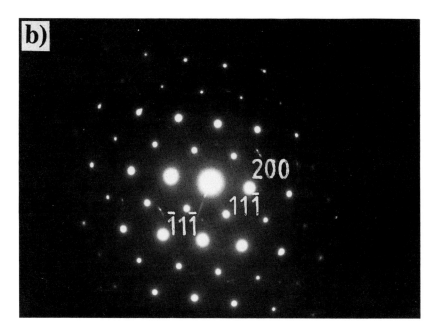

Fig. 3 The electron diffraction patterns with the (a) [001] and (b) [011] zone axis.

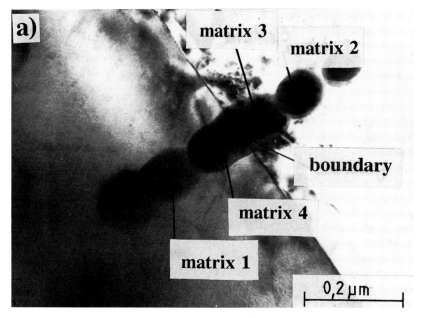

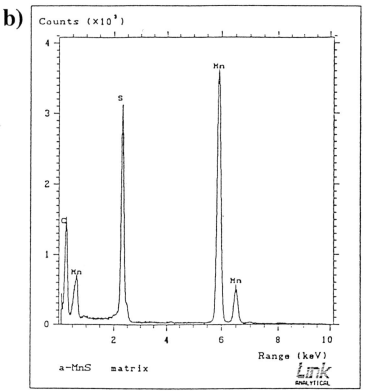

Fig. 4 (a) The post-analysis image of α-MnS grain boundaries; (b) and example of X-ray spectrum.

The experimental results allow us to formulate some interesting hypotheses concerning the mechanism of mass transport in the polycrystalline α-MnS scales. From among a number of the observed features two seem the most important: (1) the high density of dislocations inside the MnS grains and (2) the lower Mn/S ratio in the grain boundary region of the scale by comparison with the interior of the sulphide grains.

The dislocations observed inside the sulphide grains can be formed during sulphidation at high temperature and can be regarded as short-circuit diffusion paths for manganese ions. It has to be verified, however, that the dislocations are not formed in the dimpling and ion milling sample preparation procedures. At this stage it is difficult to give an explanation for the observed differences in the Mn/S atomic ratio. This ratio reflects the relative concentrations of both elements and it does not enable recognition of whether the grain boundaries are actually enriched in sulphur or depleted in manganese by comparison with the interior of the sulphide grains. Taking into account that the doubly ionised cation vacancies (V_{mn}'') are the predominant defects in α-MnS in the chosen experimental conditions,[3] sulphur enrichment at the grain boundaries might be an effect of vacancy segregation. The vacancy segregation can take place during sample quenching to room temperature.

The investigations which are now being performed concern the microstructure of sulphide scales formed on pure manganese at different temperatures and sulphur pressures and it is anticipated that a series of experiments will lead to interesting results.

Acknowledgements

Support of the Polish State Committee for Scientific Research in the course of this work is acknowledged with gratitude, project Stat. AGH No. 11.160.169.

References

1. S. Mrowec and K. Przybylski, *High Temp. Mat. and Proc.*, **6**, 1 (1984).
2. M. Danielewski, *Oxid. Met.*, **25**, 51 (1986).
3. M. Danielewski, S. Mrowec and H. J. Grabke, *Corros. Sci.*, **28**, 1107 (1988).
4. S. Mrowec, M. Danielewski and H. J. Grabke, *J. Mat. Sci.*, **25**, 537 (1990).
5. H. Habazaki, J. Dabek, K. Hashimoto, S. Mrowec and M. Danielewski, *Corros. Sci.*, **34**, 1983 (1993).

Instabilities in the Oxidation Behaviour of Zircaloy-4

D. T. FOORD and S. B. NEWCOMB

Department of Materials Science and Metallurgy, University of Cambridge,
Pembroke Street, Cambridge CB2 3QZ, UK

ABSTRACT

We have examined the microstructural roles played by the developing oxides and sub-scale metal of Zircaloy-4 in the propagation of either healed or accelerated types of oxidation. Our TEM results show that the morphology of the scale formed in oxygen at the metal–oxide interface is controlled by the local activity of oxygen there and there the alloy microstructure can promote instabilities in the oxidation behaviour.

1. Introduction

Zirconium alloys are used extensively in the nuclear industry because they have a small neutron absorption cross-section. The formation of a highly protective non-breakaway oxide scale, which can also act as a permeation barrier to hydrogen ingress on most zirconium alloys,[1,2] has added to their attractiveness. The degree to which these favourable oxidation properties can be exploited even further to optimise fuel burn up times, however, will only come when their mechanisms of oxidation and corrosion are known. Instabilities in the weight gain of Zircaloy oxidation, with the short-term repeated formation of healed scales followed by long-term rapid oxide growth, is an example of a behaviour which has been observed in both aqueous and non-aqueous environments,[3] but is not fully understood. We have examined the microstructures of the oxide scales formed in oxygen at critical times prior to the apparent onset of changes in the weight gain behaviour for Zircaloy-4, and here describe the microstructures of the different oxides and sub-scale alloys in relation to the observed changes in the weight gain behaviour.

2. Experimental Methods

We have investigated the oxidation behaviour of fuel cladding tube of Zircaloy-4, which has a nominal composition (wt%) of Zr-1.50Sn-0.19Fe-0.10Cr-1280 ppmO-30 ppmH. The alloy was supplied as 22 gauge tube which was given a stress relief α-phase anneal prior to oxidation. The resultant alloy typically consists of ~1 μm grains, which are interspersed with (Fe,Cr) Laves phase intermetallics, and is highly textured, the α-Zr c-axis lying predominantly perpendicular to the walls of the tube. Oxidation was carried out at 500°C in pure oxygen and the kinetics were followed by making continuous weight gain measurements for 275 h. Coupons of the same tube, as oxidised under identical conditions, were also removed for microstructural evaluation after oxidation periods of 28 and 68 h. Polished discs containing the oxides formed on both the inner and outer diameters of the Zircaloy-4 tube were examined in cross-section in the SEM, and cross-sectional TEM specimens of the outer diameter oxides were prepared using methods which have been described previously.[4]

3. Experimental Results

3.1 Weight Gain Data

Figure 1(a) shows schematically the first 120 h of the weight gain curve for the Zircaloy-4 tube oxidised in pure oxygen at 500°C. The positions for the two oxides formed after 28 and 68 h oxidation and examined microscopically are marked at P1 and P2 respectively. Three distinct regimes of oxidation, as marked I, II and III in Fig. 1(a) may be seen to have occurred. In the first, there has been an initial period of rapid oxidation followed by a reduction in the rate of scaling with the onset of near parabolic kinetics, as is observed for pure zirconium.[5] A transition to a second shorter weight gain regime occurs after about 43 h oxidation in which the kinetics mirror those of the first by involving an initially fast period of oxidation followed by a slower one. A second transition in the weight gain behaviour then occurs after about 75 h oxidation but this time the approximately linear rates of oxidation which characteristically follow such a transition are sustained over the remaining 200 h period over which the alloy weight gain was followed.

3.2 Oxide Microstructures

3.2.1 Scanning Electron Microscopy

SEM micrographs of the inner and outer diameters of the Zircaloy-4 tubes and the oxides formed on them after 28 and 68 h are shown in Figs 1(b) and 1(c) respectively. The thicknesses of the different scales are irregular, the outer diameter oxide formed after 28 h oxidation, for example, typically varying from 2 to 3.5 μm but in some regions being as thin as 1 μm and as thick as 4 μm. While the scales formed after 68 h are generally more developed (4–7 μm) than

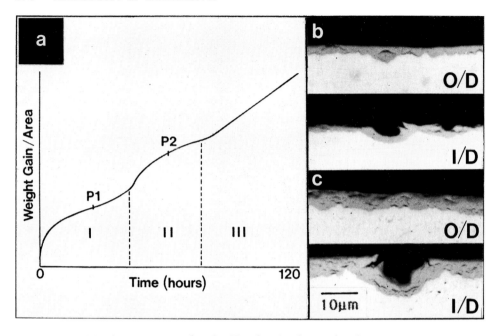

Fig. 1 (a) Weight gain curve for the Zircaloy-4 tube oxidised in pure oxygen at 500°C, and SEM micrographs of cross-sections of the oxides formed on the outer (O/D) and inner (I/D) diameters of the tubes after (b) 28 h and (c) 68 h oxidation.

those seen after the shorter period of oxidation, differences are also apparent between the oxides formed on the two diameters of the Zircaloy-4 tubes, much coarser irregularities being apparent on the inner diameter oxides which may also have partially spalled. The variable nature of both the oxide thicknesses on any one surface as well as the differences between those formed on the different diameters of the same tube demonstrates that the weight gain data described above are necessarily an average of the different rates of oxidation which must be occurring in different regions of the alloy at any one time. While the thickness variabilities are generally consistent with the rather smooth transitions to the higher rates of oxidation which are seen in Fig. 1(a), our TEM results will demonstrate that the oxidation inhomogeneities can be on an even finer scale than the SEM micrographs would suggest because both locally passivated and accelerated processes have been found to occur side by side. It should also be noted that the oxides show a tendency to exfoliate so that cracks, which are typically extended in a direction parallel to the metal–oxide interface, were apparent particularly in the oxide formed after 68 h (Fig. 1c). The cracking may have taken place during the metallographic preparation of the specimens,[6] but its occurrence highlights the stressed nature of the oxide scales and indicates that residual stresses are higher in the oxide formed after 68 h than after 28 h.

3.2.2 Transmission Electron Microscopy

We now describe the microstructures of the different oxide scales formed on the outer diameters of the Zircaloy-4 tube prior to the onset of both the first and second transitions seen in the weight gain behaviour. As an aid to the descriptions, the microstructures observed after 28 and 68 h oxidation are shown schematically in Figs 2(a) and 2(b) respectively, in which different zones of both the oxides and sub-scale alloys have been marked.

Pre-First Transition: 28 hours. The oxide examined in the TEM, as shown schematically in Fig. 2(a), was 2.4 μm in thickness beneath which the sub-scale alloy was modified to a depth of a further 1 μm. The morphology of the oxide scale (zone A) is shown in Fig. 3. Figure 3(a) is a dark field micrograph, where the upper surface of the scale (as at T) may also be seen, which shows the columnar morphology of the oxide. The columnar grains, as seen throughout zone A, are typically 300–500 nm in length and 10–30 nm in width but are also intermixed with equiaxed grains, as arrowed, which are much smaller (~25 nm). The textured appearance of the columnar grains was confirmed by the form of diffraction patterns taken from the scale (see inset, Fig. 3a) which also showed that in the main the oxide was monoclinic ZrO_2. Diffraction patterns further indicated that the scale contained low volume fractions of tetragonal ZrO_2 although, in zone A, this phase was seen almost only in the upper 750 nm of the scale, and it is this region which was formed at the beginning of the oxidation period. Dark field images taken with tetragonal oxide reflections indicated that this phase was fine grained although it should be emphasised that the fine grained oxide seen in Fig. 3(a) is monoclinic. The orientation of the textured grains was found to change through the depth of zone A so that there were 'pools' of the textured oxide in one orientation or another but typically separated by low angles. The 'pools' containing any one orientation of oxide were found to extend over depths of ~1 μm, not dissimilar from the grain size of the unoxidised α-Zr, so that it is likely that we are seeing an oxide morphology which is controlled by the local crystallography of the metal which has been oxidised. A second region of the scale which is typical of zone A, ~0.5 μm from its upper surface, is shown in Fig. 3(b) where Fresnel contrast has been used to enhance the visibility of the pores which were seen. The pores are bimodal tending to be laterally extended (as at P) as well as fine. Examples of the fine, typically vertically aligned, pores are arrowed in Fig. 3(b), from which it is also apparent that they are seen only in the vicinity of the extended pores which were themselves banded through the zone A. Thin window EDX analysis through the depth of zone A showed a constant Zr:O ratio and did not indicate that any regions of the oxide were significantly enriched in either Sn, Fe or Cr.

Differences in both the structure and chemistry of the oxide were observed in the lower 0.5–0.75 μm of the scale, marked as zones B and C in Fig. 2(a), and these regions are now described. Zone B of the scale, which was typically some 0.5 μm in width, is shown in dark field in Fig. 4(a) and again consists of columnar grains of oxide. These grains, however, are now both more laterally developed (60–75 nm) and textured than those seen in zone A, the inset

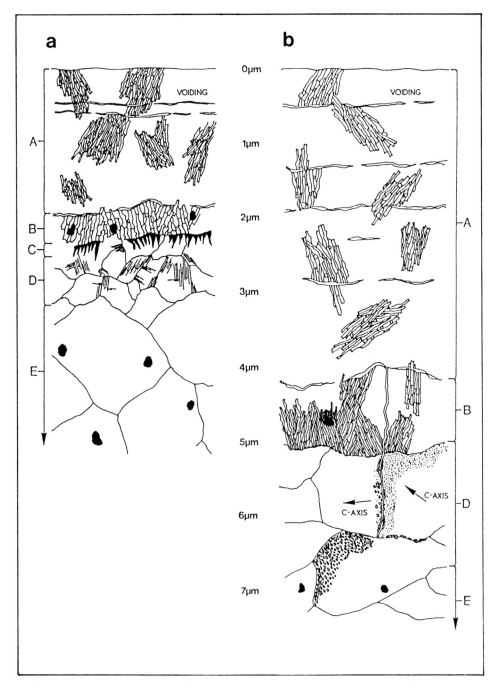

Fig. 2 Schematic summary diagrams showing the microstructures of the oxides and sub-scale alloy formed after (a) 28 h and (b) 68 h oxidation. Different zones within each scale and underlying alloy have been marked.

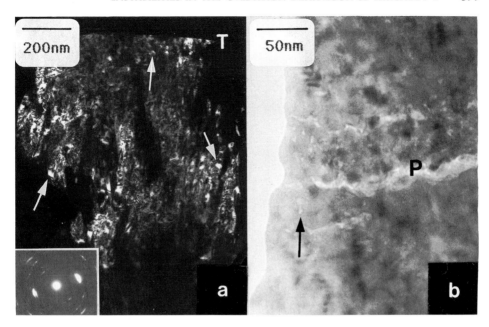

Fig. 3 TEM micrographs of the oxide seen in zone A (as defined in Fig. 2a) and formed after 28 h oxidation showing (a) a dark field micrograph of an upper region (the top surface is marked at T) of the columnar grained oxide and (b) a bright field micrograph where a bimodal distribution of pores may be seen.

diffraction pattern in Fig. 4(a) demonstrating the highly textured nature of the monoclinic oxide. Furthermore, more of the tetragonal ZrO_2 phase was seen in zone B than in Zone A and this too was coarser grained ($\sim$50 nm) than above. Further differences between the zones A and B were apparent with the observation of an (Fe,Cr) intermetallic particle here, as shown at L in Fig. 4b, where the visibility of the pores (as arrowed), which may be seen to have developed in the vicinity of the spherical intermetallic, has been enhanced by their Fresnel contrast. One of the lattice parameters of this hexagonal Laves phase precipitate was found to be much smaller ($a_0 = 0.432$ nm) than that seen in the bulk of the alloy, the implication being that we are seeing partial loss of either Fe or Cr from the intermetallic as they are locally incorporated into the oxide scale. The (Fe,Cr) intermetallics will be found to have been modified even further at the base of the scale formed after 68 h oxidation, and their relationship with the oxide surrounding them acts as a fingerprint for the activity of oxygen in such regions. Here the ratio of Zr:O in zone B was found to be higher than in the bulk of the scale formed above it indicating that this zone may contain zirconium metal.[2] The way in which we are seeing an apparent reduction in the activity of oxygen was emphasied by the appearance of Zone C. This typically extended over depths of $\sim$0.2 μm and, as shown in Fig. 4(c), had a characteristic 'sawtooth' morphology so that the region consists of interlocking oxide and coarser, but modified, α-Zr grains beneath it (zone

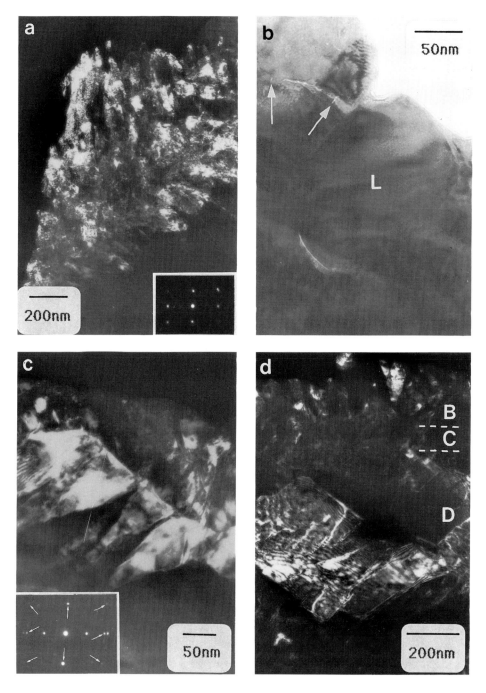

Fig. 4 TEM micrographs showing the microstructures of the oxide and sub-scale alloy after 28 h oxidation for zone B in (a) dark field and (b) bright field, where an intermetallic is marked at L, (c) zone C and (d) zone D showing the recrystallised morphology of the subscale α-Zr.

D). Unlike the scale formed above it, however, the oxide here was the rhombohedral Zr_3O_x phase ($a = 0.563$ nm, $c_o = 1.559$ nm), the inset diffraction pattern in Fig. 4(c) showing a $(\bar{1}010)$ Zr_3O_x normal. The lattice parameter variations of different grains of the 'sawtooth' have shown that the sub-oxide phase described has been formed with a variable stoichivmetry and this was confirmed by the variable Zr:O ratios seen for zone C. It should also be emphasised that, in the regions examined, zone C was not everywhere continuous and this shows that the local variabilities seen in the oxygen profile can occur on a coarser scale. The zone D metal beneath the oxide was found to be both structurally and chemically modified over a depth of approximately 0.75 μm. Here the α-Zr was equiaxed as well as both finer grained (500 nm) and more dislocated than the bulk alloy, the dark field micrograph in Fig. 4(d) showing the typical morphology of zone D. The microstructure is indicative of recrystallisation as has been observed in a number of oxidation processes[7] in which solute atoms are injected into the metal.[8] This is consistent with our thin window EDX analyses which have shown that zone D contains a low, but apparently ungraded, concentration of oxygen unlike the coarser grained alloy in zone E.

Pre-Second Transition: 68 h. The microstructures of the oxide and sub-scale alloy formed on the outer diameter of the Zircaloy-4 tube after 68 h oxidation are shown schematically in Fig. 2(b), where different zones are again marked. While the thickness of the oxide scale was slightly thicker ($\sim$5 μm) than that of the average (4.6 μm) measured in the SEM, the alloy immediately beneath the oxide was also modified over locally variable distances, but to a maximum of some 3 μm from the base of the scale. Given the historical development of the scale, which involves the progressive formation of oxide at the metal–oxide interface, it is perhaps not unsurprising that differences between the scales formed after the two oxidation periods, in relation to the changes seen in the kinetics, were most apparent at the base of the scale. The bulk of the oxide (zone A) formed after 68 h was thus almost indistinguishable from that seen after the shorter oxidation period and again consisted of columnar grains of monoclinic ZrO_2, which were highly textured and interspersed with the minority tetragonal ZrO_2 phase. We noted, however, that there tended to be less of the tetragonal oxide at the surface of the scale than was observed after 28 h oxidation.

Differences in both the morphology and chemistry of the scaling process were apparent in zones B and D (there is now no zone C) and these differences reflect both the averaged low rate of oxidation seen in Fig. 1(a) as well as the ensuing near-linear kinetics. A low magnification dark field micrograph of the metal–oxide region is shown in Fig. 5(a), in which zones B and D are also marked, and it is here that extensive regions of internally oxidised alloy were found to have under-cut areas which were less so and where the diffusional pathway for the oxygen ingress has been determined by the local alloy crystallography. The oxide seen in zone B was also found to be modified by comparison with the same zone formed after less oxidation. While the zone is again found to contain monoclinic and tetragonal ZrO_2, which are both highly

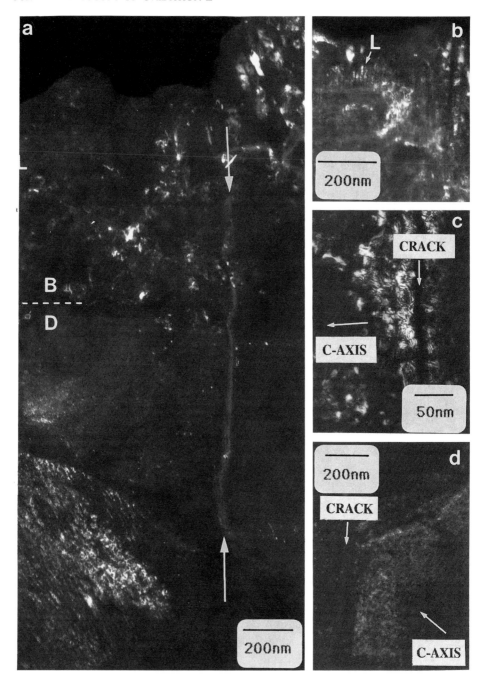

Fig. 5 Dark field TEM micrographs of zones B and D formed on the alloy after 68 h oxidation showing (a) both zones at low magnification, (b) the variable grain size of zone B and (c,d) the internally oxidised alloy grains on either side of crack arrowed in (a).

textured, the widths of the columnar grains shown in Fig. 4(a) are typically finer (20–30 nm) than those described earlier (60–75 nm). The zone B oxide (as at L in Fig. 5a) formed after 68 h is shown in the dark field in Fig. 5(b) where even finer columnar grains (~7.5 nm) may also be seen, as arrowed. The latter part of the zone B oxide was found to contain high concentrations of Fe and Cr and has a morphology (as at L) which suggests that we are seeing the remnants of an intermetallic which has been oxidised leading to the incorporation of Fe and Cr into the scale. That this in itself has led to a localised change in the grain size of the oxide is apparent but also reflects the generally higher oxygen activity in this zone by comparison with the same zone formed after 28 h. Furthermore, we now no longer find a transitional region between the oxide and underlying metal where sub-oxide (Zr_3O_x) formation has taken place, and the zone C seen after 28 h was not observed in the scale described here.

Returning now to the sub-scale of the alloy shown in Fig. 5(a), we now describe the microstructure of zone D. The microstructure is significantly different from that described earlier, which consisted of a thin band of recrystallised α-Zr, and is apparently dominated by a crack which runs through zone B and into the sub-scale alloy for approximately 1 μm, as arrowed in Fig. 5(a). The crack was found to have propagated at an uncharacteristically high angle alloy grain boundary while the alloy grains on both sides of it as well as at its base were modified differently as a function of their different crystallographies, as observed in salt corroded Ti based alloys.[9] The two dark field micrographs shown in Figs 5(c) and 5(d) and taken from regions immediately adjacent to either side of the crack demonstrate that internal oxide formation has been promoted in the alloy grain which has its c-axis lying perpendicular to the crack. The crystallography of the grain in Fig. 5(c), however, has apparently inhibited such oxidation immediately beneath the main body of the zone B scale (even where the oxide was very fine grained), unlike the other side of the crack where internal oxide is seen to a depth of some 0.3 μm. At the base of the crack, the α-Zr was found to be extensively internally oxidised although here any alloy crystallography effects may be more complex.

4. Discussion

In examining the microstructures of the oxides and sub-scale regions formed in Zircaloy-4 after different oxidation periods, our TEM results have pointed to the ways in which instabilities in the observed kinetic behaviour can be either healed or propagated. In order to understand these instabilities, it is essential to see whether the scale microstructures formed prior to changes in the kinetics are, on the one hand, related to one another and, on the other, can explain the passivated or accelerated weight gain behaviours respectively seen after the first and second transitions in the kinetics.

The microstructure of the scale formed on the outer diameter of the Zircaloy-4 tube is consistent with the low rate of oxidation observed after 28 h. The morphology of the oxide at the base of the scale, where most of the passivated growth takes place, as well as of the modified alloy immediately beneath it is indicative of the low oxygen partial pressure there. Evidence that zones B and C have been formed at a lower rate than most of the scale above comes from the fact that the oxide in zone B has both a coarser grain size and is more textured than that in zone A. While the relatively coarse grains seen in zone B will themselves provide fewer diffusion paths for the inward flux of oxygen, both the appearance of the partially modified (Fe,Cr) intermetallic in zone B and the discontinuous formation of the Zr_3O_x sub-oxide[10] in zone C are consistent with a low activity of oxygen. Unlike the higher oxygen activity environment of zone A, where intermetallics have not been seen, the activity here is significantly lower, and very little oxidation of the intermetallic has apparently taken place. Equally, the activity of oxygen in zone D is sufficiently low that oxide has not been formed internally within the metal although oxygen has diffused into it and this has resulted in alloy recrystallisation.

The morphology of the oxide at the base of the scale formed after 68 h was rather different from that seen after 28 h. Here too it must be remembered that breakaway weight gains follow the second transition in the kinetics compared with the healed period which follows the first transition. While the microstructure of the passivated oxide formed after 28 h is apparently controlled by a limited activity of oxygen, comparable regions of the passivated oxide observed prior to the breakaway are again supply controlled but have been formed at higher oxygen partial pressures, and this too results in a faster rate of oxidation at the end of the second passivated regime (see Fig. 1a). Evidence for a higher interfacial oxygen activity comes from both the relatively fine grained morphology of the zone B oxide and the fact that (Fe,Cr) intermetallics within it have been oxidised (resulting in further local modifications to the oxide morphology). The high oxygen activity of the region is further evidenced by the absence of the sub-oxide layer (C) seen earlier and the observation of internal oxide within the sub-scale alloy.

It would seem inescapable that the transitions seen in the weight gains have been initiated by changes in the flux of oxygen to the metal–oxide interface region of the Zircaloy-4. Compressive stresses will develop in the scale as it grows into the alloy and these stresses will increase with the thickness of the oxide though tend to be relieved and this, with local formation of cracks within the oxide, may initiate changes in the scaling behaviour.[3] While the flux of oxygen to the metal–oxide interfaces of the two scales, as then determined by the differences in their compressive stresses, is clearly higher after longer times, the ensuing changes which take place in the sub-scale alloy play a critical role in determining whether healing or breakaway kinetics follow the weight gain transitions. After 28 h the apparently low oxygen activity was sufficient to have resulted in alloy recrystallisation, involving the formation of some high angle boundaries, over a depth of $\sim$0.5 μm. Once oxide stresses have changed the oxygen flux, however, oxidation will occur at an accelerated rate through such a modified region, diffusion being promoted by both its high dislocation content and the high angle boundaries within it. After the longer period of oxidation, where we have already seen higher interfacial activities of oxygen, the same process should itself be accelerated because the sub-scale alloy modifications will be coarser although the crystallographies of such regions have also been shown to favour locally slower rates of oxidation. Parallel changes in the stress distribution of such regions will similarly promote the development of vertical cracks into the scale above so that the oxide is no longer a diffusional barrier. Once initiated, the self-propagating process described should remain catastrophic.

5. Conclusions

Our TEM results have emphasised the apparent rarity with which breakaway processes can be driven solely by 'mechanical' effects in the scale that has been formed. While changes in the flux of oxygen through the oxide formed on Zircaloy-4 have of course dominated the scaling processes seen here, we have shown that the alloy too plays a critical role in determining how accelerated oxidation rates can be maintained. This in itself points to ways in which either structural or compositional modifications could be made to the alloy which would promote 'chemically controlled' protective oxidation.

Acknowledgements

We are grateful to Professor C. J. Humphreys for the provision of laboratory facilities and to both the SERC and Nuclear Electric plc for financial support. We acknowledge useful discussions with Drs H. E. Evans, R. Lobb, K. A. Simpson and W. M. Stobbs.

References

1. G. P. Sabol, S. G. McDonald and G. P. Airey, in Zirconium in nuclear applications, *ASTM STP 551, American Society for Testing Materials* (1974), 435–448.
2. S. B. Newcomb, B. D. Warr and W. M. Stobbs, *Inst. Phys. Conf. Ser. No. 119* (Sept. 1991), Institute of Physics, pp. 221–224.
3. J. K. Dawson, G. Long, W. E. Seddon and J. F. White, *J. Nucl. Mat.*, **25** (1968), 179–200.
4. S. B. Newcomb, C. B. Boothroyd and W. M. Stobbs, *J. Microsc.* **140** (1985), 195–207.
5. G. R. Wallwork, W. W. Smeltzer and C. J. Rosa, *Acta Met.*, **12**, (1964), 409–415.
6. B. Cox, N. Ramasubramanian and V. C. Ling, 'Zircaloy corrosion properties under LWR coolant conditions' (Part 1), E.P.R.I. NP 6979-D, Project X101-2 (October 1990).
7. S. B. Newcomb, W. M. Stobbs and E. Metcalfe, *Phil. Trans. Roy. Soc. (Lond.)*, **A319** (1986), 191–218.

8. K. Lucke and H. P. Stuwe, *Acta Met.*, **19** (1971), 1087–1099.
9. K. Seebaruth, D. Özkaya, S. B. Newcomb and W. M. Stobbs, in *Inst. Phys. Conf. Ser. No. 119.* (Sept. 1991), Institute of Physics, 241–244.
10. T. Arai and M. Hirabayashi, *J. Less Comm. Met,* **44** (1976), 291–300.

Microstructure and Electron Microprobe Study of Oxide Layers Obtained on Zircaloy-4 by Oxidation at High Temperature

M. ABRUDEANU, I. VUCU and V. DUMITRESCU

University of Pitesti, Piata Vasile Milea, 1, 0300 Pitesti, Romania

C. PETOT and G. PETOT-ERVAS

CNRS-LPCM/UPR 211, University of Paris Nord, 93430 Villetaneuse, France

P. ARCHAMBAULT

EM, LSG2M, University of Nancy I. 54042, Nancy, France

M. PETRESCU and N. PETRESCU

Polytechnical Institute of Bucarest, Romania

ABSTRACT

The paper presents the results of analyses carried out for oxide films formed during the isothermal oxidation of Zy-4. The oxidation process was done in dry air at 600–1050°C having a constant duration of 60 min. According to the oxidising temperature, the structure, compactness and adherence of the oxide layers vary from the black, compact, adhering oxide to white-yellowish oxide having pores and fissures. The oxide films were analysed by optical microscopy, scanning electronic microscopy, microprobe and mass spectroscopy of secondary ions. The variation profiles of the concentration was held by the main elements in the black oxide layer as well as white oxide layers, and their deterioration process was analysed in keeping wih the oxidation temperature.

1. Introduction

The oxidation of zirconium alloys at high temperatures in air results in the formation of two types of oxides: a black, compact, adherent oxide and a white oxide showing a defective structure that deteriorates even more as the temperature increased.[1,4,5]

387

The process of oxygen diffusion in the metal under the oxide layer at temperatures above the phase transformation points of the alloy leads to the formation of a stabilised solid solution enriched in oxygen.

A study by scanning electron microscopy, by microprobe analysis and by SIMS has been carried out in this paper on the oxide films obtained at the surface of Zy-4 by isothermal oxidation in a temperature range of 600–1050°C.

2. Materials and Methods

Zircaloy-4 samples, having the chemical composition 1.58%Sn, 0.22%Fe, 0.10%Cr, 50 ppmCu, 1200 ppmO$_2$, and a microstructure consisting of equiaxed grains, have been oxidised isothermally for 60 min in dry air at a series of temperatures stepped up by 50°C in the temperature range 600–1050°C. The oxide layers have been studied by optical microscopy, scanning electron microscopy (SEM), SIMS and by quantitative energy dispersive microprobe analysis (EDAX).

3. Experimental Results and Discussion

3.1 Microstructure of the oxide layers

Optical and scanning electron micrographs show a different structure for the samples oxidised below 750°C and above this temperature. Oxidation below 750°C ends with formation of a black, compact, adherent film (Fig. 1a). At 750°C a new layer of white oxide begins to appear. As one can see in the micrograph in Fig. 1(b), this white oxide layer comprises two phases: a greyish-white phase and a yellow phase located predominantly in the vicinity of the oxide–metal interface.

By comparison with the black oxide layer, the white oxide layer is less compact: pores and cracks exist in the oxide white layer (Fig. 2), the cracks being mostly parallel to the oxide/gas interface but also perpendicular in the vicinity of the oxide/metal interface (Fig. 3). The number of pores and cracks proved to be temperature dependent. It was observed that increasing temperature resulted not only in the increase of the white oxide layer thickness but also in a further deterioration of the compactness of this layer. In order to estimate this deterioration process we have carried out a quantitative metallographic study by using an image analyser. Figure 4 shows an increase of the area fraction of pores and cracks in the white oxide layer from a few percent at 800°C to 20% of 1050°C. Also shown in Fig. 4 is an approximately constant amount of yellow phase (~10%) in the white oxide layer.

The structure of the metal is equiaxed up to transition level (Fig. 1a) after which an oxide layer of a solid stabilised solution α is formed by oxygen dissolution (Fig. 1b), while the core has a needle-like structure formed on the initial β grains. In both situations the increase of oxidation temperature causes

Fig. 1 Micrograph for a zircaloy-4 sample isothermally oxidised for 60 min (a) At 600°C, black oxide; (b) at 1050°C, white oxide.

an increase of grain size. The deterioration process of the grain boundary by oxygen diffusion was observed in the metal.

3.2 Microchemistry of the oxide layers

To confirm the distribution of different component elements existing in the oxide layers, a profiling was carried out by SIMS and by X-ray microprobe analysis. The SIMS analysis allowed determination of concentration profiles only for thin oxide layers whose thickness was less than the penetration depth.

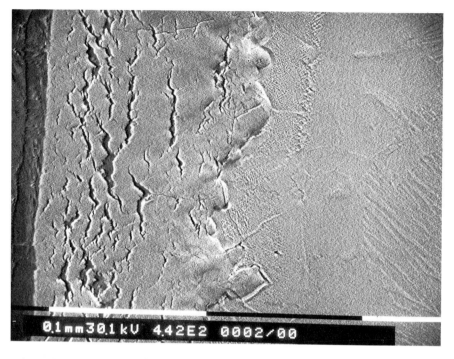

Fig. 2 SEM micrograph for a zircaloy-4 sample isothermally oxidised for 60 min at 1050°C in dry air.

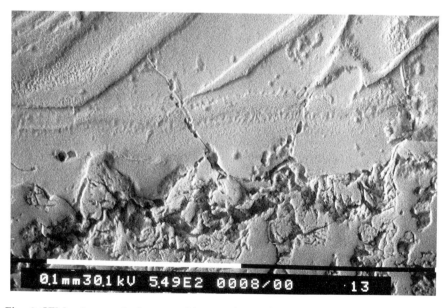

Fig. 3 SEM micrograph for an oxide–metal interface for the sample oxidised at 1050°C.

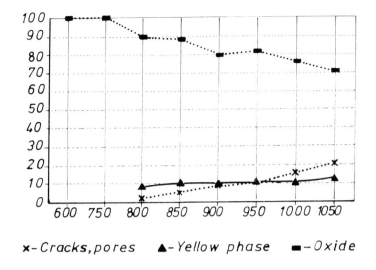

x- Cracks, pores   ▲- Yellow phase   ■- Oxide

Fig. 4 Relative amount of pores, cracks and yellow phase in the white oxide layer
at different oxidation temperatures.

So we have succeeded in carrying out this profiling by SIMS only for the black
oxide layer. Figures 5 and 6 for the sample oxidised at 600°C comprised only
the black oxide layer. For samples oxidised at higher temperature with white
oxide layers, we have not succeeded in obtaining information by SIMS on the
microchemistry of the white layer because this layer was thick and too deeply
located from the samples surface.

Interesting information on the white oxide layer microchemistry and the
solid solution underlying layer were obtained by profiling the distribution of
the component elements by X-ray microprobe analysis. Figure 8 shows the
concentration profiles for the main metallic elements (Zr and Sn) and for two
non-metallic elements (O_2 and N_o) assumed to have penetrated by diffusion in
the sample oxidised at the highest temperature (1050°C).

The oxygen content in the white oxide layer seems to be constant, while a
decrease is noticed in the underlying metal layer. This decrease is typical for a
diffusion layer and so we can infer that a layer of stabilised solid solution has
been formed by the dissolution of oxygen in metallic zirconium.

A very interesting effect was noticed concerning the penetration of nitrogen
from the heating atmosphere into the oxide layer. The nitrogen profile in Fig. 8
shows an increased nitrogen content in the vicinity of the oxide–metal
interface, more specifically in the region where the oxide layer has a yellow
colour and where a new phase is sometimes seen. We consider that the
penetration of nitrogen by diffusion in the depth of white oxide layer is
favoured by the defective microstructure of this layer that comprises a
temperature increasing area fraction of pores and cracks (Fig. 4). The nitrogen
enriched region in Fig. 8 may correspond to a reaction diffusion process that
results in the formation of the stable compound ZrN. It is interesting to

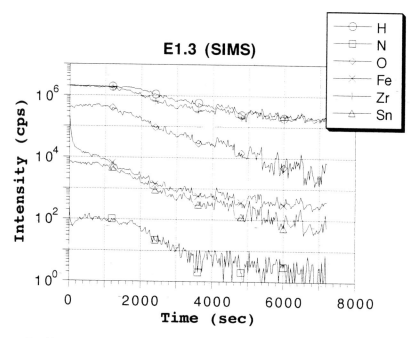

Fig. 5 Profiles for several component elements for the sample oxidised at 600°C (black oxide layer 8 μm thick).

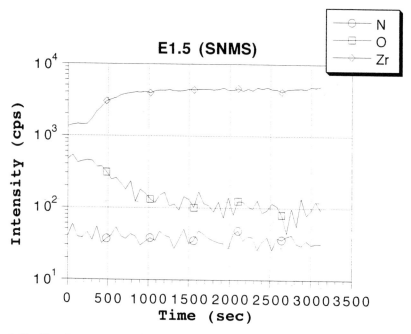

Fig. 6 Profiles for oxygen and zirconium for the sample oxidised at 600°C (black oxide layer 8 μm thick).

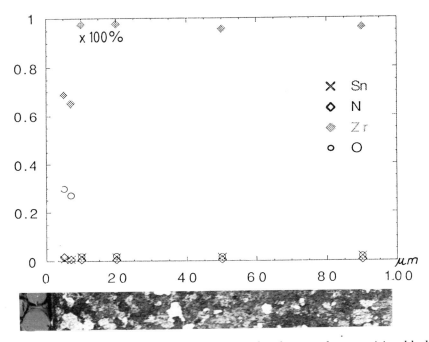

Fig. 7 X-ray microprobe concentration profiles for the sample comprising black oxide formed at 600°C.

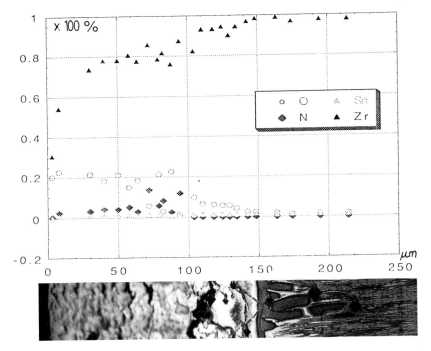

Fig. 8 X-ray microprobe concentration profiles for the sample comprising the white oxide formed at 1050°C.

mention that the micrograph in Fig. 1b for a sample oxidised under the same conditions with the sample in Fig. 8 shows clearly the existence of a thin layer of a new hard yellow phase located just in the region of nitrogen enrichment, namely at the interface between α Zr solid solution and the white oxide layer. The sudden drop of the nitrogen constant in the α solid solution (clearly seen in Fig. 8) may be a consequence of the diffusion barrier provided by the ZrN compound located at the oxide–metal interface and also as a consequence of the slow diffusion rate of nitrogen in the metallic phase.

4. Conclusions

The optical and electron microscopy study and the microprobe profiling carried out in this paper have shown that the oxide layers formed on Zy-4 by isothermal oxidation in dry air in the temperature range 600–1050°C show the following characteristic features:

1. up to 750°C a single layer of black oxide is formed that looks compact and adherent;
2. at 750°C a new inner layer of white oxide begins to appear;
3. the white oxide layer is less compact than the black oxide layer, comprising an area fraction of pores and cracks that increases with the oxidation temperature;
4. the white oxide layer seems to comprise two phases: a greyish-white phase and a yellow phase located predominantly in the vicinity of the oxide–metal interface;
5. the oxygen content in the white oxide layer is approximately constant, but a typical diffusion decrease is manifested in the underlying α Zr solid solution layer;
6. a penetration by diffusion of nitrogen from the air heating atmosphere was observed during the oxidation of Zy-4;
7. a nitrogen enrichment was put in evidence in the white oxide layer in the vicinity of the oxide–metal interface, just in the region where the oxide layer has a yellow colour and seems to comprise a new phase. This nitrogen-rich phase was presumed to be the stable ZrN compound, and its presence at the oxide–metal interface proved to be a barrier for the further diffusion of nitrogen into the underlying metal.

References

1. D. L. Douglas, The metallurgy of zirconium. *Atomic Energy Review*, supp. 1971.
2. C. B. Alcook, K. T. Jacob, S. Zador, O. Kubaschewski von Goldbek, H. Nowotny, S. Seifert and O. Kubaschewski, Zirconium. Physicochemical properties of its compounds and alloys, *Atomic Energy Review*, Vienna, Special Issue no. 6, 1976.

3. M. Coster and J. L. Chermant, *Precis d'analyse d'images*, Edition CNRES, Paris, 1985.
4. A. T. Donaldson and H. E. Evans: Oxidation induced creep in zircaloy-2, *J. Nucl. Mat.*, **99** (1981), 46–57.
5. A. T. Donaldson and H. E. Evans: Oxidation-induced creep in zircaloy-2, *J. Nucl. Mat.*, **99**, (1981), 46–57.
6. R. A. Ploc, Breakaway oxidation of zirconium at 573 K, *J. Nucl. Mat.*, **82**, (1979), 265–271.

6

OXIDATION OF NICKEL, TITANIUM AND NIOBIUM BASED INTERMETALLICS

TEM Observations of the Oxidation of γ'-Ni$_3$Al at Different Oxygen Partial Pressures

E. SCHUMANN and M. RÜHLE

*Max-Planck-Institut für Metallforschung, Institut für Werkstoffwissenschaft,
Seestrasse 92, 7000 Stuttgart 1, Germany*

ABSTRACT

The oxidation of (001) oriented single-crystal γ'-Ni$_3$Al has been investigated using cross-sectional TEM. At first experiments were conducted under low oxygen partial pressure (LOP) at 1223 K. Al$_2$O$_3$ is the only thermodynamically possible oxide under these conditions. Small Ni particles were observed to form at the oxidised surfaces. A phase transformation from γ-Al$_2$O$_3$ to α-Al$_2$O$_3$ occurred between 5 and 50 h oxidation. The transformation started at the metal–oxide interface.

Oxidation of single-crystal γ'-Ni$_3$Al was also performed in air at 1223 K. Initially internal γ(Al$_2$O$_3$ precipitates were observed. Between the internal γ-Al$_2$O$_3$ particles and the Ni matrix a cube-on-cube orientation relationship was detected. A continuous γ-Al$_2$O$_3$ layer was found after 6 min oxidation. The Ni in the internal oxidation zone (IOZ) was oxidised to NiO. During longer term oxidation the γ-Al$_2$O$_3$ in the inner Al$_2$O$_3$ layer transforms to the thermodynamically stable α-Al$_2$O$_3$.

1. Introduction

γ'-Ni$_3$Al is an important intermetallic phase in the Ni–Al system. The understanding of the basic microstructural process in scale formation is important for the control of high temperature oxidation behaviour.

Most work concerning the oxidation of Ni–Al alloys has been carried out on β-NiAl in single- and polycrystalline form.[1-4] The investigations have been conducted using scanning electron microscopy (SEM), transmission electron microscopy (TEM), thermogravimetric analysis (TGA) and secondary ion mass spectroscopy (SIMS). The initial formation of epitaxially oriented γ-Al$_2$O$_3$, δ-Al$_2$O$_3$ and θ-Al$_2$O$_3$ scales has been found.

Several studies are known for the oxidation of γ'-Ni$_3$Al[5,6] but only a few detailed TEM investigations have been conducted.[7,8] In constrast to β-NiAl, which forms Al_2O_3 scales exclusively, oxidation of γ'-Ni_3Al also produces NiO and $NiAl_2O_4$. In the present paper the oxidation behaviour of γ'-Ni_3Al under different oxygen partial pressures will be compared.

2. Experimental procedure

Single crystals of γ'-Ni_3Al were prepared by the Bridgman technique, resulting in an Al concentration of 24 at%. The as-grown single crystals were oriented using the Laue X-ray technique. (001)-Oriented discs were cut by spark erosion. The surfaces were polished with $1\,\mu$ diamond paste.

Oxidation experiments were carried out under low oxygen partial pressure (LOP) and high oxygen partial pressure (HOP). The LOP conditions were selected so that Al_2O_3 is the only thermodynamically stable oxide. The LOP oxidation was conducted in a H_2/H_2O atmosphere at 1223 K. The corresponding oxygen partial pressure was calculated as $p_{O_2} = 4 \times 10^{-14}$ Pa. The HOP oxidation was conducted in air at 1223 K. Oxidation times ranged between 1 min and 50 h.

The surfaces of the oxidised specimens were first characterised by SEM. A special specimen preparation technique for TEM cross-sections was applied, including grinding, dimpling and ion thinning. TEM cross-sections allowed the analysis of the oxide phases and the morphology of the metal–oxide interface.

The TEM work was conducted with a JEOL 2000 FX instrument, using conventional imaging (bright field and dark field), selected area electron diffraction (SAD) and energy dispersive X-ray spectroscopy (Tracor Northern EDS).

3. Experimental Observations

The microstructural development during oxidation under different oxygen partial pressures (LOP, HOP) has been analysed by TEM. In this section the observations during different oxidation stages will be described.

3.1 Observations during oxidation at low oxygen partial pressure

SEM study of the specimen after 1 min oxidation revealed the existence of small Ni particles on top of the oxidised surface (Fig. 1). The particles were faceted with low indexed (100) planes parallel to the particle surfaces, larger particles have been formed by coalescence.

TEM cross-sections after 1 min oxidation (Fig. 2) show that the Ni particles are connected with the underlying γ'-Ni_3Al single crystal by a metallic channel. A cube-on-cube orientation relationship between the particles and the bulk metal could be detected with electron difraction (SAD). Chemical analysis in the TEM revealed that the particles consisted of pure Ni within the resolution limit of 0.5 at%. The oxide scale was identified as γ-Al_2O_3. Protrusions of

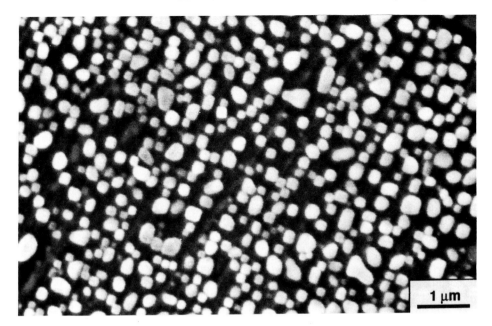

Fig. 1 SEM showing Ni particles after 1 min oxidation at LOP.

γ-Al$_2$O$_3$ into the metal could be found. The typical wavelength of the protrusions was about 30 nm.

During subsequent oxidation, a transition from a rough metal–oxide interface to a planar interface was observed. Figure 3(a) shows the γ-Al$_2$O$_3$ scale and the metal–oxide interface after 15 min oxidation. The γ-Al$_2$O$_3$ layer has thickened, but the interface is still rough on a larger scale.

Figure 3(b) shows a superlattice dark field image after 5 h oxidation. The oxide scale consisted entirely of fine grained (~20 nm) γ-Al$_2$O$_3$, and the metal–oxide interface is essentially flat. A disordered zone consisting of a solid solution of Ni and ~10% Al and γ'-Ni$_3$Al precipitates exists under the oxide scale.

Figure 4 shows a TEM cross-section after 20 h oxidation. Randomly oriented α-Al$_2$O$_3$ has formed between the γ-Al$_2$O$_3$ and the metal. Due to the growth of α-Al$_2$O$_3$, the metal–oxide interface has become rough again. The grain size of the α-Al$_2$O$_3$ is in the range of 200 nm. Pores have been found between the γ-Al$_2$O$_3$ and the α-Al$_2$O$_3$ as well as within the α-Al$_2$O$_3$ grains. The disordered zone between the Al$_2$O$_3$ scale and the γ'-Ni$_3$Al single-crystal has thickened. Between this disordered zone and the γ'-Ni$_3$Al a distinct interface can be seen. After 50 h oxidation, a continuous α-Al$_2$O$_3$ layer was observed.

3.2 Observations during oxidation at high oxygen partial pressure

After 1 min oxidation, a NiO scale was observed by SEM. TEM cross-sections after 1 min oxidation (Fig. 5) show a continuous NiO layer in contact with the

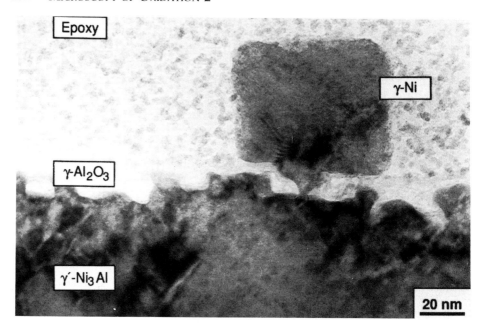

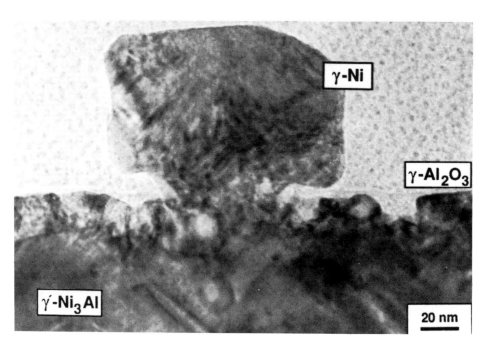

Fig. 2 TEM cross-sections of different Ni particles (a, b) after 1 min oxidation at LOP.

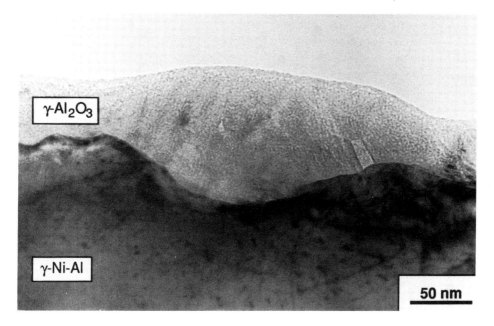

Fig. 3 TEM cross-sections showing γ-Al$_2$O$_3$ and metal after 15 min (a) and 5 h (b) oxidation at LOP.

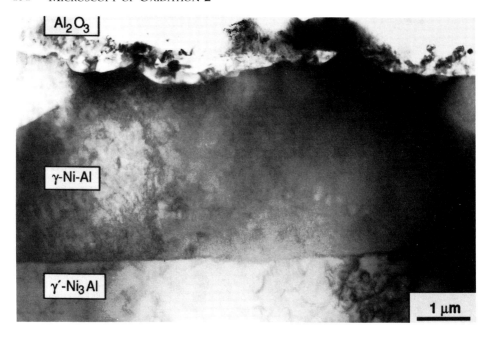

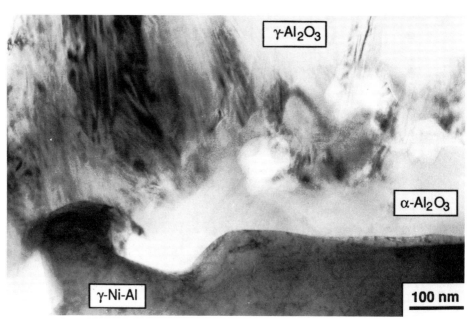

Fig. 4 TEM cross-sections showing disordered zone, γ-Al$_2$O$_3$ and α-Al$_2$O$_3$ after 20 h oxidation at LOP (a) and larger magnification (b).

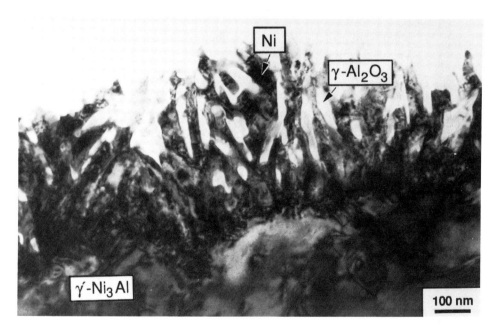

Fig. 5 TEM cross-sections showing NiO scale (a) and internal oxidation zone (b) after 1 min oxidation at HOP.

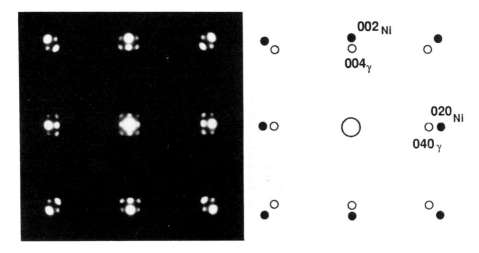

Fig. 6 SAD pattern (a) and indexed schematic (b) from internal oxidation zone after 1 min oxidation at HOP.

internal oxidation zone (IOZ). The interface between NiO and the IOZ is flat. The IOZ consisted of γ-Al$_2$O$_3$ precipitates within a Ni matrix. The length of the internal oxide particles ranged between 50 and 200 nm. The precipitates were cube-on-cube oriented with respect to the Ni matrix (Fig. 6):

$$(001)[110]_{\gamma\text{-Al}_2\text{O}_3}||(001)[110]_{Ni}$$

Oxidation for 6 min resulted in the formation of a continuous γ-Al$_2$O$_3$ layer (Fig. 7a). The orientation of the continuous γ-Al$_2$O$_3$ layer was cube-on-cube as verified by electron diffraction. Between the γ-Al$_2$O$_3$ layer and the γ'-Ni$_3$Al single crystal a disordered zone exists. The metal–oxide interface is very rough.

After 30 min oxidation the microstructure was similar to that after 6 min, but the metallic Ni in the IOZ was oxidised to NiO (Fig. 7b). Between the NiO and the γ-Al$_2$O$_3$ there exists a cube-on-cube orientation relationship.

A TEM cross-section after 50 h oxidation revealed that the oxide scale consisted of three different layers: an outer NiO layer, an intermediate NiAl$_2$O$_4$ layer and an inner Al$_2$O$_3$ layer. The NiAl$_2$O$_4$ layer exhibited a cube-on-cube orientation to the γ'-Ni$_3$Al (Fig. 8a). The Ni:Al ratio within the spinel was mesured with EDS as 14.0:30.4 at% ($\pm$1.5 at%). Within the inner Al$_2$O$_3$ layer, α-Al$_2$O$_3$ has formed. A special orientation relationship was found:

$$(0001)[1\bar{1}00]_{\alpha\text{-Al}_2\text{O}_3}||(111)[1\bar{1}0]_{\gamma\text{-Al}_2\text{O}_3}$$

Dark field imaging (Fig. 8b) shows that large α-Al$_2$O$_3$ grains have formed at the metal–oxide interface.

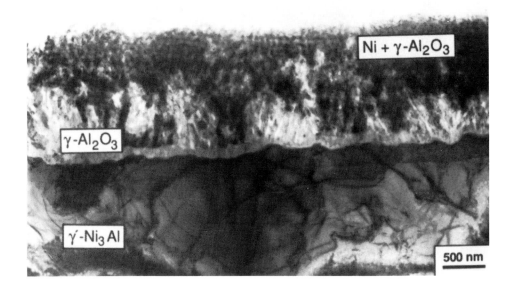

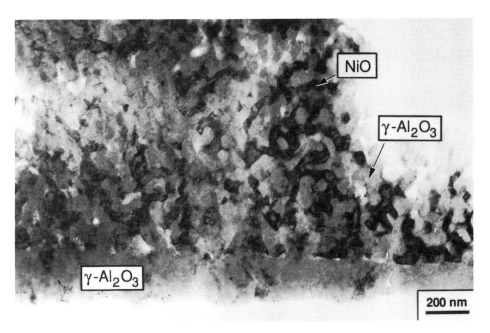

Fig. 7 TEM cross-sections showing continuous γ-Al$_2$O$_3$ layer after 6 min (a) and γ-Al$_2$O$_3$/NiO two phase zone after 30 min (b) oxidation at HOP.

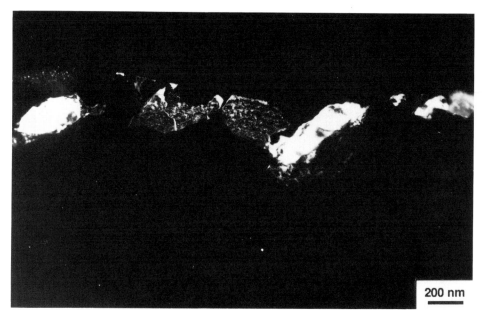

Fig. 8 Dark field image showing NiAl$_2$O$_4$ (a) and α-Al$_2$O$_3$ (b) after 50 h oxidation at HOP.

4. Discussion

In this section, the experimental observations will be discussed and an overall model for the microstructural development during oxidation of γ'-Ni$_3$Al will be presented.

4.1 Oxidation at low oxygen partial pressure

Metastable Al$_2$O$_3$ phases have been observed in different oxidation experiments.[1,2] Their formation is caused by the lower free energy of nucleation in the initial stage compared with the thermodynamically stable modification α-Al$_2$O$_3$. The interfacial energy during the nucleation of γ-Al$_2$O$_3$ is reduced through the formation of epitaxial orientation relationships.

In the early stage of oxidation, γ-Al$_2$O$_3$ protrusions and Ni particles on top of the oxidised surfaces have been observed. The development of oxide protrusions into the metal can be explained by different mechanisms:

1. Random formation of γ-Al$_2$O$_3$ nuclei beneath the amorphous Al$_2$O$_3$ layer.
2. Localized oxide growth due to defects in the initial oxide layer.
3. A morphological instability due to non-linearities of the growth mechanism.

During longer term oxidation, a transition to a planar metal–oxide interface has been observed. Theoretical calculations[9] suggest that the smoothening of the metal–oxide interface can be interpreted as a transition from oxidation rate control by alloy interdiffusion to rate control by diffusion in the oxide. The growth of γ-Al$_2$O$_3$ protrusions into the metal is accompanied by a 28% volume increase, which results in large internal stresses. Micromechanical modelling[10] shows that internal stresses up to the 1 GPa range can be reached during the growth of the oxide protrusions. Such high stresses probably set up the driving force for the outward diffusion of Ni. This corresponds to the observation of small Ni particles on top of the oxide scale.

A transformation from metastable Al$_2$O$_3$ to the thermodynamically α-Al$_2$O$_3$ has been observed between 5 h and 20 h oxidation. The formation of α-Al$_2$O$_3$ starts at the metal–oxide interface. The transformation is very sluggish. The volume decrease (14%) resulting in large stresses seems to be a high nucleation barrier. A transformation to α-Al$_2$O$_3$ has also been observed by other authors.[2,3] It has been shown that the transformation is accompanied by a two orders of magnitude decrease of the parabolic rate constant k_p.

4.2 Oxidation at high oxygen partial pressure

The early oxidation stage of γ'-Ni$_3$Al at high oxygen partial pressure is characterised by internal oxidation and simultaneous NiO scale growth. The growth of internal oxide particles is accompanied by a large volume increase (28%), as is the growth of γ-Al$_2$O$_3$ protrusions under LOP conditions. The mechanical stresses can be relieved by Ni diffusion toward the surface. Because the ambient partial pressure during HOP oxidation is higher than the dissociation partial pressure of NiO, the Ni is oxidised, forming the outer NiO scale.

The classical Wagner criterion[11] predicts a transition from internal to external oxidation under the following condition:

$$N_{Al} > \left(\frac{\pi g^*}{2\nu} \cdot N_O^{(S)} \cdot \frac{D_O}{D_{Al}} \cdot \frac{\Omega_{Ni}}{\Omega_{\gamma\text{-}Al_2O_3}} \right)^{1/2}$$

where g^* is the critical volume fraction of oxide in the IOZ, ν a stoichiometric factor for the oxide, $N_O^{(S)}$ the solubility of oxygen in Ni, D_O the diffusivity of oxygen in Ni, D_{Al} the diffusivity of Al in Ni, Ω_{Ni} the atomic volume of Ni and $\Omega_{\gamma\text{-}Al_2O_3}$ the molecular volume of γ-Al_2O_3.

With $\nu = 1.5$, $g^* \approx 0.3^{12}$, $D_{Al} = 6.3 \times 10^{-16}\,m^2\,s^{-1},^{13}$ $N_O^{(S)} = 3.7 \times 10^{-4},^{14}$ $D_O = 4.9 \times 10^{-13}\,m^2\,s^{-1},^{14}$ $\Omega_{Ni} = 1.1 \times 10^{-29}\,m^3$ and $\Omega_{\gamma\text{-}Al_2O_3} = 4.8 \times 10^{-29}\,m^3$ the minimum concentration of Al in the alloy can be appproximated as $N_{Al} = 0.14$. The concentration of Al in the bulk alloy, $N_{Al}^{(0)} = 0.24$, is higher than this value; therefore external scale formation is expected in agreement with the observation of a continuous γ-Al_2O_3 scale after 6 min oxidation.

During subsequent oxidation the Ni in the internal oxidation zone was oxidised to NiO resulting in the two phase zone shown in Fig. 6 after 30 min oxidation. From this observation it can be concluded that the spinel layer, which was found after 50 h oxidation between the outer NiO layer and the inner Al_2O_3 layer, must have formed by a solid state reaction within the intermediate two-phase zone. This can be explained by the following reaction:

$$NiO + \gamma\text{-}Al_2O_3 = NiAl_2O_4$$

Because NiO and γ-Al_2O_3 are in a cube-on-cube orientation, the solid state reaction probably occurs by the classical mechanism of cation-counter diffusion.[15] The oxygen sublattice will essentially be fixed, and the reaction takes place only by interdiffusion of cations.

As under LOP conditions, γ-Al_2O_3 transforms to α-Al_2O_3 during longer term oxidation. In contrast to LOP oxidation, an orientation relationship has been found whereby close-packed planes of α-Al_2O_3 are parallel to close-packed planes in γ-Al_2O_3 and close-packed directions are also parallel. This topotactic orientation relationship is caused by the higher degree of texture of the γ-Al_2O_3 scale under HOP conditions. The orientation between both lattices is related to the atomistic transformation mechanism from the fcc arrangement of the oxygen sublattice in γ-Al_2O_3 to the hcp arrangement of oxygen ions in α-Al_2O_3. This requires a change in the stacking sequence of the oxygen layers. Different possible mechanisms for such a transformation have been discussed in the literature.[16]

5. Summary

Oxidation behaviour of γ'-Ni_3Al under high (HOP) and low oxygen partial pressure (LOP) has been studied. In both cases during the early oxidation stage an instability occurred; γ-Al$_2$O$_3$ protrusions (LOP) and internal γ-Al$_2$O$_3$ precipitates have been found (HOP). The volume change due to oxidation resulted in growth stresses, which can be relieved by out-diffusion of Ni. During the transient stage a planar metal–oxide interface developed, corresponding to rate control by diffusion in the oxide. Under LOP conditions as well as under HOP conditions, a phase transformation from metastable γ-Al$_2$O$_3$ to the thermodynamically stable modification α-Al$_2$O$_3$ was observed.

Acknowledgements

The authors thank M. Bobeth and W. Pompe for helpful discussions and U. Salzberger for valuable support during TEM preparation. The work was funded by the Deutsche Forschungsgemeinschaft (DFG).

References

1. J. Doychak, J. L. Smialek and T. E. Mitchell, *Met. Trans. A*, **20A** (1989), 499.
2. G. C. Rybicki and J. L. Smialek, *Oxid Met.*, **31**, (1989), 275.
3. M. W. Brumm and H. J. Grabke, *Corros. Sci.*, **33** (1992), 1677.
4. J. Jedlinski, *Sol. State Phenom.*, **21 & 22** (1992), 335.
5. J. D. Kuenzly and D. L. Douglas, *Oxid. Met.*, **8** (1974), 139.
6. J. H. DeVan, J. A. Desport and H. E. Bishop, *Proc. Conf. on Microscopy of Oxidation, Cambridge* (1990).
7. J. Doychak and M. Rühle, *Oxid. Met.*, **31** (1989), 431.
8. E. Schumann, G. Schnotz, K. P. Trumble and M. Rühle, *Acta metall. mater.*, **40**, (1992), 1311.
9. M. Bobeth, W. Pompe and M. Rockstroh (these Proceedings) (1993).
10. M. Bobeth, W. Pompe, E. Schumann and M. Rühle, *Acta metall. mater.*, **40** (1992), 2669.
11. C. Wagner, *Z. für Elektrochemie*, **63** (1959), 772.
12. R. A. Rapp, *Acta metall.*, **9**, (1961), 730.
13. W. Gust, M. B. Hintz, A. Lodding, H. Odelius and B. Predel, *Phys. Stat. Sol. (a)*, **64** (1981), 187
14. J. W. Park and J. Altstetter, *Met. Trans. A*, **18A** (1987), 43.
15. C. Wagner, *Z. Physik. Chem. (B)*, **34** (1936), 309.
16. C. B. Carter and H. Schmalzried, *Phil. Mag.*, **52** (1985), 207.

Modelling of the Structural Development of Oxide Scales on Ni₃Al Single Crystals During the Initial Stage of Oxidation

M. BOBETH, W. POMPE and M. ROCKSTROH

*Max-Planck-Gesellschaft, Arbeitsgruppe Mechanik heterogener Festkörper,
Hallwachsstr. 3, D-01069 Dresden, Germany*

ABSTRACT

Recent experiments on the oxidation of γ'-Ni₃Al at low oxygen partial pressure revealed the formation of a rugged oxide–metal interface (OMI) in the early stage of oxidation (1 min). In the course of further oxidation (>15 min) a flattening of the interface occurred. In the present study, the emergence of a rough OMI due to a morphological instability is investigated by means of a linear stability analysis and a two-dimensional simulation. The characteristic wavelength of interface roughness is evaluated taking into account interface smoothing due to the capillary effect. The observed flattening of the OMI during oxidation is interpreted as a transition from scale growth controlled by the interdiffusion in the metal to scale growth controlled by transport processes through the oxide.

1. Problem

In recent experiments on the oxidation of γ'-Ni₃Al single crystals at 1223 K by Schumann *et al.*[1] the formation of a γ-Al₂O₃ scale with pronounced protrusions of the oxide into the metal was observed after 1 min oxidation time. Oxidation was performed in a H₂/H₂O atmosphere at $p(O_2) = 4 \times 10^{-14}$ Pa where NiO could not form. TEM cross-sections of the scale for longer oxidation times (>15 min) showed a flattening of the OMI. Furthermore, a planar phase boundary in the metal between γ-Ni–Al solid solution and γ'-Ni₃Al was identified.

The initial formation of oxide protrusions in these experiments has not been explained previously. The following mechanisms are under consideration: (i) Beneath an initially amorphous oxide layer, crystalline nuclei are formed by heterogeneous nucleation (Fig. 1a). (ii) Defects in the scale (grain boundaries, cracks) result in short-circuit diffusion paths which can lead to localised oxide growth (Fig. 1b). (iii) Non-linearities of the diffusion–reaction system affect a morphological instability with respect to small perturbations of a planar OMI.

412

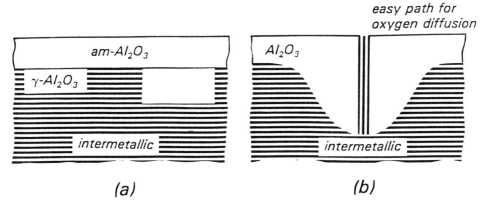

Fig. 1 Possible mechanisms initiating protrusion formation: Heterogeneous nucleation of γ-Al$_2$O$_3$ crystals underneath an amorphous oxide layer (a) and short-circuit diffusion along defects in the scale accompanied by localised oxide growth (b).

The mechanism (i) combined with short-circuit diffusion has been proposed in connection with the observed formation of γ-Al$_2$O$_3$ crystals during the initial oxidation of pure aluminium near a temperature of 800 K.[2,3] In the present context, this mechanism could affect the formation of γ-Al$_2$O$_3$ nuclei during heating of the sample.

The stability of a planar OMI under the influence of small perturbations was initially investigated by Wagner[4] and later by Whittle *et al.*[5] According to their analysis, a morphological instability can occur if interdiffusion in the alloy is slow compared to transport processes through the scale. The mechanism of the instability is illustrated in Fig. 2. The enhanced slope of the Al-profile in front of a perturbation leads to an enhanced Al-flux into the perturbation area, and therefore, to a higher recession rate of the metal near the perturbation.

Regardless of the special mechanism initiating protrusion growth, the further development of protrusions is governed by a competition between an unstable roughening of the interface in the presence of a morphological instability and a flattening of the interface due to the capillary effect. In the following, these processes are investigated theoretically, where major emphasis is on the questions whether the observed interface roughening in the early stage of oxidation can be understood as a morphological instability, and what is the mechanism which affects the interface flattening in the later oxidation stage.

2. Oxidation Model

A general difficulty in modelling the growth of alumina scales is the poor knowledge of the defect properties and relevant transport processes in alumina, in particular in γ-Al$_2$O$_3$ (cf. review[6]). In the present case, the observations suggest that in the early oxidation stage the scale grows by

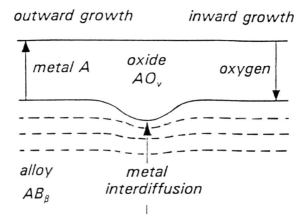

Fig. 2 Unstable scale growth affected by an increased concentration gradient in front of the protrusion; dashed curves represent isoconcentration lines.

inward diffusion of oxygen. Particularly, the observation of Ni-particles on top of the oxide scale formed by the out-diffusion of Ni due to internal growth stresses supports this suggestion.[7]

A schematic of the relevant transport processes for the selective oxidation of Ni₃Al by inward oxide growth is shown in Fig. 3. At moderate temperatures grain boundary diffusion of oxygen through the scale seems to be predominant. Correspondingly, we consider in the following an interstitial mechanism and introduce the mean density of oxygen interstitials n_O. The migration of oxygen through the scale and the interdiffusion of Al in the alloy are described by the diffusion equations

$$\partial_t n_O = D_O \Delta n_O, \qquad \partial_t n_A = D_A \Delta n_A \tag{1}$$

where n_A represents the Al-concentration in the alloy, D_O is the defect mobility, and D_A the interdiffusion coefficient. As a peculiarity of the system (Ni, Al), a miscibility gap between 13 and 24 at% Al has to be considered (cf. Fig. 3).

Boundary conditions. The particle balance at the OMI leads to

$$\frac{\nu}{1 - \Omega_A n_A} D_A \nabla_n n_A = -D_O \nabla_n n_O, \qquad v_i = \frac{\Omega_A D_A}{1 - \Omega_A n_A} \nabla_n n_A, \ z = z_i \tag{2}$$

where $\nu = 3/2$ is a stoichiometric factor, Ω_A the volume of an Al atom in the alloy, v_i denotes the normal velocity of the OMI, and ∇_n the normal derivative. The velocity of the phase boundary in the metal is given by

$$v_p = \frac{1}{n_{Af} - n_{Ae}} [(D_A \nabla_n n_A)_{z=z_p-0} - (D_A \nabla_n n_A)_{z=z_p+0}] \tag{3}$$

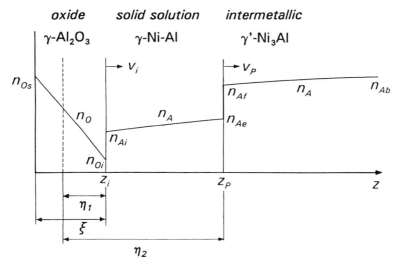

Fig. 3 Schematic of the diffusion-reaction model for the selective oxidation of γ'-Ni₃Al by inward oxide growth taking into account the formation of a solid solution zone in the metal. The dashed line corresponds to the original surface of the metal prior to oxidation.

A second boundary condition follows from chemical equilibrium at the interface. By assuming Henry's Law in the alloy and the relation $n_O \sim p(O_2)^{1/2\zeta}$ ($\zeta = 1, 2, 3$ refers to the oxygen charge O, O^{1-}, O^{2-}), one obtains

$$n_A \cdot n_O^{\nu\zeta} = K_n(1 + \kappa L), \qquad z = z_i \tag{4}$$

where K_n represents the reaction constant. The term κL considers the capillary effect in the case of a curved OMI where κ denotes the interfacial curvature and L the capillary length. If one takes into account additionally interfacial diffusion, the term $v_i^+ \sim \partial^2\kappa/\partial s^2$, proportional to the second derivative of the curvature, must be added to v_i in equation (2).[8]

3. Early Oxidation Stage (1 min)

3.1 Linear Stability Analysis

The present analysis follows essentially the approach by Wagner[4] with the primary difference being that the case of inward oxide growth is considered here and the capillary effect is taken into account because of the small size of observed oxide protrusions.

The stability analysis is performed by calculating the evolution of an infinitesimal interface perturbation appearing at some time t_0

$$z_i(y, t) = z_i^{(1)} \exp[\omega(t - t_0)] \sin(qy) \tag{5}$$

to the first order in $z_i^{(1)}$ from the system of equations (1), (2) and (4). As a result, the relative growth velocity of the perturbation, $\omega(\lambda) = \dot{z}_i/z_i$, is obtained where $\lambda = 2\pi/q$ is the wavelength of the perturbation. For $\omega > 0$, an unstable growth of the perturbation results, i.e. a planar interface is morphologically unstable in this case.

Since the rigorous derivation of the expression for ω is rather lengthy (cf. for example Ref. 9), a quasi-stationary approximation is presented here which typically describes the physically relevant cases with sufficient accuracy. The simplification consists of the omission of time derivatives and drift terms in the corresponding diffusion equations (1) in a moving frame. The range of validity of the approximation is given by the relations

$$|\omega| \ll D_\alpha q^2, \qquad v_i \ll D_\alpha q \qquad \text{with} \qquad \alpha = O, A \qquad (6)$$

The solutions of the diffusion equations (1) are written as the sum of the solutions for the planar interface, $n_O^{(0)}$ and $n_A^{(0)}$, and first order corrections

$$n_\alpha = n_\alpha^{(0)} + n_\alpha^{(1)} \exp[\omega(t - t_0) + \sigma_\alpha qz] \sin(qy), \quad \alpha = O, A \qquad (7)$$

with $\sigma_O = 1$ and $\sigma_A = -1$. The perturbations of the corresponding concentration profiles decay exponentially with distance from the interface. The planar interface is described by $z_i^{(0)} = 0$. The unperturbed solutions are now expanded near the planar interface

$$n_\alpha^{(0)} = n_{\alpha i} + G_\alpha z \qquad (8)$$

where $n_{\alpha i}$ are the interfacial concentrations and G_α are the gradients of the concentration profiles ($G_O < 0$). For a planar interface, equations (2) and (4) can be written as

$$\tilde{\nu}\tilde{\Omega}D_A G_A = -\Omega_A D_O G_O, \qquad v_i^{(0)} = \tilde{\Omega}D_A G_A, \qquad n_{Ai} \cdot n_{Oi}^{\nu_\zeta} = K_n \qquad (9)$$

where the variable $\tilde{\Omega} = \Omega_A/(1 - \Omega_A n_{Ai})$ has been introduced. By inserting equation (8) into equation (7), one obtains at the interface to the first order in z_i

$$n_\alpha = n_{\alpha i} + (G_\alpha z_i^{(1)} + n_\alpha^{(1)}) \exp[\omega(t - t_0)] \sin(qy), \qquad (10)$$

$$\nabla_n n_\alpha = G_\alpha + \sigma_\alpha q n_\alpha^{(1)} \exp[\omega(t - t_0)] \sin(qy) \qquad (11)$$

Insertion of equations (10) and (11) into equation (2b), together with

$$v_i = v_i^{(0)} + v_i^{(1)}, \qquad v_i^{(1)} = \dot{z}_i = \omega z_i \qquad (12)$$

and equation (9b), leads to

$$\omega z_i^{(1)} = \tilde{\Omega}D_A(\tilde{\Omega}G_A^2 z_i^{(1)} - q n_A^{(1)}) \qquad (13)$$

According to relations (6b) and (9b), the term $\tilde{\Omega}G_A$ in equation (13) and in the following is neglected compared to q. Furthermore, insertion of equations (10) and (11) into equation (2a) together with equation (9a) yields

$$D_O q \Omega_A n_O^{(1)} - \nu D_A q \tilde{\Omega} n_A^{(1)} = -\nu D_A \tilde{\Omega}^2 G_A^2 z_i^{(1)} \tag{14}$$

and insertion of equation (10) into equation (4) together with equation (9c) yields

$$\nu \zeta (n_{Ai}/n_{Oi}) n_O^{(1)} + n_A^{(1)} = [n_{Ai} q^2 L - G_A - \nu \zeta (n_{Ai}/n_{Oi}) G_O] z_i^{(1)} \tag{15}$$

The system of equations (14), (15) is solved for $n_A^{(1)}$ and the result is inserted into equation (13). Thus, one obtains the approximate relation

$$\omega = \frac{1}{1+Q} \tilde{\Omega} D_A q [(1-Q) G_A - n_{Ai} L q^2] \tag{16}$$

with

$$Q = \frac{\nu^2 \zeta}{1 - \Omega_A n_{Ai}} \frac{D_A n_{Ai}}{D_O n_{Oi}} \tag{17}$$

From equation (16), we find $Q < 1$ as a necessary condition for the appearance of an instability. Otherwise, ω is negative for all wavelengths of the perturbation.

In the discussion above, the effect of interface diffusion has been neglected. However, the spatial variation of the chemical potential along the rough OMI gives rise to an interfacial flux of atoms which leads to an additional interface flattening. This additional contribution to the interface velocity is approximately given by[8]

$$\dot{z}_i^{(add)} = -a_0 L D_i q^4 z_i \tag{18}$$

where a_0 is the lattice constant and D_i represents the effective interfacial diffusivity of the oxide components. The capillary length $L = \gamma_i \Omega/kT$ is determined by the interface energy of the OMI, γ_i, and the atomic volume of the oxide components Ω.

In the limiting case of unstable scale growth with $Q \ll 1$ and $\Omega_A n_{Ai} \ll 1$, the sum of equation (16) and $\dot{z}_i^{(add)}/z_i$ yields finally

$$\omega(\lambda) = \Omega_A G_A D_A (2\pi/\lambda) - \Omega_A n_{Ai} L D_A (2\pi/\lambda)^3 - a_0 L D_i (2\pi/\lambda)^4 \tag{19}$$

The first term in equation (19) affects the unstable growth whereas the other terms lead to an interface flattening via interdiffusion of Al in the metal ($\sim \lambda^{-3}$) and interface diffusion of the oxide components ($\sim \lambda^{-4}$). The effect of volume Al-diffusion in the oxide has been neglected because of the presumably low volume diffusivity of Al in the oxide (for a general analysis, cf. Ref. 8).

Interface perturbations with wavelengths $\lambda > \lambda_0$, where $\omega(\lambda_0) = 0$, grow

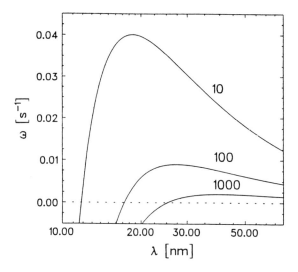

Fig. 4 Relative growth rate of an interface perturbation, ω, as a function of wavelength λ at different times $t_0 = 10,\ 100$ and $1000\,\mathrm{s}$ $(D_A = 10^{-17}\,\mathrm{m}^2/\mathrm{s},$ $D_i = 3 \times 10^{-18}\,\mathrm{m}^2/\mathrm{s},\ L = 1.5\,\mathrm{nm}).$

unstably. The wavelength λ_m of the fastest growing mode is determined by the maximum of $\omega(\lambda) = \omega(\lambda_m) = \omega_m$ (cf. Fig. 4). The most unstable wavelength is expected to correlate with the lateral size of oxide protrusions. The wavelength λ_m as well as the maximum growth rate ω_m depend on time owing to the temporarily decreasing gradient G_A of the Al-profile. Assuming parabolic scale growth with $G_A \sim t^{-1/2}$, one finds $\lambda_m \sim t^{1/4}$ if volume diffusion is the predominant mechanism of interface flattening and $\lambda_m \sim t^{1/6}$ for interface diffusion.

3.2 Two-dimensional Simulation of Unstable Growth

In the case of unstable scale growth, the development of the OMI is controlled by interdiffusion in the metal. The evolution of the interface shape has been calculated from equations (1b) and (2b) by means of a finite element model. In the first approach, the presence of a second interface in the metal has been omitted. The initial perturbation of the interface has been represented by introducing a small bump (cf. Fig. 5). The boundary condition for the Al-concentration at the interface is approximately given by $n_{Ai} = n_{Ai}^{(0)} \exp{(\kappa L)}$ when $n_{Ai}^{(0)}$ corresponds to a planar interface. The example of the evolution of a localised interface perturbation in Fig. 5 shows in principle how oxide protrusions are formed due to a morphological instability. For simplicity, the Pilling–Bedworth ratio in this simulation has been chosen equal to one. In general, metal oxidation is accompanied by a volume expansion and, consequently, the formation of internal stresses. These stresses and the possibility of plastic deformation in the metal have to be taken into account. Furthermore, the effect of interface diffusion has been omitted in the example in Fig. 5.

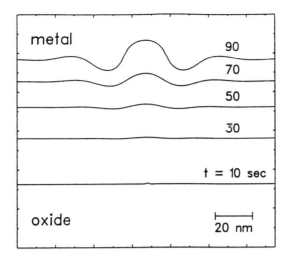

Fig. 5 Evolution of a localised interface perturbation after the appearance of the perturbation (t = 10, 30, 50, 70 and 90 s, D_A = 10^{-15} m²/s).

3.3 Experimental Comparison

The question regarding the appearance of an initial instability cannot be directly answered from the criterion $Q < 1$ because of the lack of known input parameters. An analysis of the present model shows however that unstable growth occurs for $n_{Ai} \ll n_{Ae}$. Evidence for an instability is thus given by the measurement of small Al-concentrations in the metal between the protrusions up to only 1 at%.[1]

Assuming an instability is present, the model parameters D_A, n_{Ai} and D_i can be fitted to the observed mean scale thickness (≈ 10 nm), the lateral protrusion size (≈ 15 nm) as well as their amplitude (≈ 7 nm) after 1 min oxidation. From the mean thickness growth, the rough estimate D_A = 10^{-17} m²/s follows, which is considerably smaller than the reported data $D_A \approx 10^{-15}$ m²/s.[10] Because of the high binding energy of alumina a small equilibrium concentration n_{Ai} is expected. As a consequence, the results of the stability analysis suggest that interface flattening occurs mainly by interface diffusion. The value of the corresponding interfacial diffusivity, $D_i \approx 10^{-18}$ to 10^{-17} m²/s, fitted to the observed wavelength of interface roughness is compatible with data of the surface diffusivity of alumina D_s = 2×10^{-18} m²/s.

The TEM observations revealed a rapid initial growth of the amplitude of oxide protrusions compared to the mean thickness growth which was not obtained by our preliminary simulations where a spatially homogeneous diffusivity, D_A, had been assumed (cf. Fig. 5). This result suggests that an additional mechanism strongly decreases the Al-diffusion between the protrusions. A possible mechanism could be a vacancy depletion near the interface, where the interface acts as a vacancy sink. The absorption of vacancies at the interface leads to the relief of oxide growth stresses. In this way one could also understand the relatively small value of D_A estimated from the mean thickness growth.

Although it has been derived that the lateral extension of protrusions increases with time, the observed flattening of the OMI in the course of oxidation cannot be explained in this way since the ratio of the normal and lateral protrusion growth velocities, $\omega_m/(\lambda_m/\lambda_m)$, increases with time. Thus, the observed rapid interface flattening suggests a transition from metal- to oxide-diffusion controlled scale growth in the course of oxidation.

4. Later Oxidation Stage (15 min $< t <$ 20 h)

The development of a planar OMI at longer oxidation times demonstrates the presence of stable scale growth controlled by transport processes through the oxide. The observed thicknesses of the oxide scale and the solid solution zone in the metal reveal an approximate parabolic scale growth: $\xi = \Pi\eta_1 = \Pi\sqrt{2k_{p1}t}$ and $\eta_2 = \sqrt{2k_{p2}t}$ where Π denotes the Pilling–Bedworth ratio (cf. Fig. 3 and Table 1). For parabolic growth, the current model with a concentration-independent interdiffusion coefficient has an analytic solution. In the experiments under consideration,[1] $n_{Ab} \approx n_{Af}$ was realised, i.e. $n_A \approx n_{Ab}$ for $z > z_p$. In this special case, insertion of the analytic solution into equations (2b) and (3) yields

$$\sqrt{\pi}\,\mathrm{erf}(u_2) + \frac{1/\Omega_A - n_{Ae}}{n_{Ab} - n_{Ae}}\frac{1}{u_2}\exp(-u_2^2) = \sqrt{\pi}\,\mathrm{erf}(u_1) + \frac{1}{u_1}\exp(-u_1^2), \quad (20)$$

$$n_{Ai} = 1/\Omega_A - (n_{Ab} - n_{Ae})\sqrt{k_{p2}/k_{p1}}\exp[(k_{p2} - k_{p1})/2D_A] \quad (21)$$

where $u_\alpha^2 = k_{p\alpha}/2D_A$ with $\alpha = 1, 2$. Equation (20) determines the interdiffusion coefficient D_A for known parabolic rate constants k_{p2} and k_{p1}, and equation (21) connects the interfacial concentration, n_{Ai}, with D_A. The rate constants may be evaluated by measuring the thicknesses ξ and $\eta_2 - \eta_1$ from TEM cross-sections. Unfortunately, according to the data in Table 1, the rate constants are not well

t [s]	ξ [nm]	$\eta_2 - \eta_1$ [nm]	k_{p1}^a [nm^2/s]	k_{p2}^a [nm^2/s]	k_{p2}/k_{p1}
60	$\approx 10^b$	—	≈ 0.23	—	—
900	100	—	1.54	—	—
3600	210	650	1.70	80.3	47.2
18 000	500	1700	1.92	107	55.7
72 000	1000	3100	1.92	91.3	47.6

[a]$\xi = \Pi\eta_1$ with $\Pi = 1.9$, $\eta_1^2 = 2k_{p1}t$ and $\eta_2^2 = 2k_{p2}t$
[b]mean value of a strongly varying scale thickness

Table 1 Oxide scale thickness, ξ, and thickness of solid solution zone, $\eta_2 - \eta_1$, at different oxidation times, t, measured from TEM micrographs and the resulting parabolic rate constants.[1]

defined. Especially, the data at 5 h oxidation lead to unreasonable results. From the other data, we find the estimate $D_A \approx 2 \times 10^{-15}$ m²/s and an interfacial concentration, n_{Ai}, of about 12.6 at% compared to the value n_{Ae} of 13 at% ($1/\Omega_A = 79.4$ nm^{-3}, $n_{Ab} = 21.2$ nm^{-3}, $n_{Ae} = 11.5$ nm^{-3}). Note however that slight changes of the parameters Ω_A, n_{Ab} and n_{Ae} affect strong changes of the result for D_A so that the present estimate is only accurate for the order of magnitude of D_A. The estimated interdiffusion coefficient is compatible with the value $D_A = 2.4 \cdot 10^{-15}$ m²/s at an Al-concentration of 10 at% as reported by Shankar and Seigle.[10] The presence of a flat Al concentration profile in the solid solution zone ($z_i < z < z_p$) is confirmed by the observation of γ'-Ni₃Al precipitates in this zone.[1] Precipitates are formed during cooling of the sample because the Al concentration exceeds the solubility limit of Al in Ni which is roughly 11 at% at 1050 K.

5. Conclusions

The initial formation of a rough OMI can be explained in principle by a morphological instability initiated by a fast initial oxygen penetration through the scale. However, other mechanisms of protrusion formation cannot be excluded. Within the framework of the present model, a decrease of the interdiffusion near the interface is suggested for the case of unstable scale growth. This is possibly affected by atomic ordering processes during the formation of a second phase boundary in the metal or by a vacancy depletion near the interface due to internal growth stresses.

The flattening of the OMI in the subsequent stage of oxidation is attributed to a transition from scale growth controlled by interdiffusion in the metal to growth controlled by transport processes through the scale. The reason for this transition is presumably the disappearance of easy diffusion paths in the scale due to defect elimination. For the later stage of oxidation the theoretical predictions are in satisfactory agreement with experimental observations and confirm the presence of stable scale growth.

Acknowledgements

This work was supported in part by the Deutsche Forschungsgemeinschaft. The authors would like to thank M. Rühle and E. Schumann for helpful discussions.

References

1. E. Schumann, G. Schnotz, K. P. Trumble and M. Rühle, *Acta Metall. Mater.*, **40** (1992), 1311.
2. M. J. Graham, J. I. Eldrige, D. F. Mitchell and R. J. Hussey, *Materials Sci. Forum*, **43** (1989), 207.

3. K. Shimizu, R. C. Furneaux, G. E. Thompson, G. C. Wood, A. Gotoh and K. Kobayashi, *Oxid. Met.*, **35** (1991), 427.
4. C. Wagner, *J. Electrochem. Soc.*, **103** (1956), 571.
5. D. P. Whittle, D. J. Young and W. W. Smeltzer, *J. Electrochem. Soc.*, **123** (1976), 1073.
6. R. Prescott and M. J. Graham, *Oxid. Met.*, **3** (1992), 233.
7. M. Bobeth, W. Pompe, M. Rühle and E. Schumann, *Acta Metall. Mater* , **40** (1992), 2669.
8. P. G. Shewmon, *Trans. Met. Soc.* AIME **233** (1965), 736.
9. W. W. Mullins and R. F. Sekerka, *J. Appl. Phys.*, **35** (1964), 444.
10. S. Shankar and L. L. Seigle, *Met. Trans.*, **A9** (1978), 1467.

AEM Study of the Duplex Scale Formed on an ODS Ni₃Al Alloy

B. A. PINT, A. J. GARRATT-REED and L. W. HOBBS

H. H. Uhlig Corrosion Laboratory, Massachusetts Institute of Technology,
Cambridge, MA 02139, USA

ABSTRACT

The oxidation behaviour of ODS Ni₃Al with a Y_2O_3-Al_2O_3 dispersion was investigated at 1000°C and 1200°C. The addition of Y to this mechanically alloyed material reduced the steady-state oxidation rate and produced a duplex scale with an adherent inner α-Al_2O_3 layer and an outer non-adherent Ni-rich oxide layer. Using FEG-STEM/XEDS, the segregation of Y and tramp Zr was detected on α-Al_2O_3 grain boundaries. This reactive element segregation explains the reduction in the oxidation rate at 1000°C and 1200°C. A transverse section of the Al_2O_3/Ni-rich oxide interface revealed large voids which explain the poor adhesion of the outer layer. During cyclic testing at 1200°C, the α-Al_2O_3 layer remained fully adherent; however, the outer Ni-rich layer was observed to spall and re-grow. A model has been developed to explain this behaviour whereby cracks in the inner Al_2O_3 layer occur during cooldown and during the following cycle allow a pathway for the outward transport of Ni. AEM analysis of the scale after cyclic oxidation at 1200°C appears to confirm this model. Ni-rich oxides were found near the metal interface in some areas along with numerous voids. In other areas, the scale was observed to be nearly pure Al_2O_3 and was almost fully dense.

1. Introduction

Ni₃Al is a potentially versatile engineering material because of its combination of mechanical and oxidation-resistant properties. Previous studies[1-4] have examined the oxidation behaviour of Ni₃Al, focusing on oxidation kinetics, reaction products and the effect of various alloy additions (Y, Zr, Ti, Hf and Cr) on the oxidation behaviour. It has been clearly established that some type of oxygen-active or 'reactive' element (RE) such as Y, Zr or Hf is required to produce an adherent alumina scale. More recent studies[5,6] have begun to use higher-resolution techniques to examine the oxide microstructure and the effect of the duplex scale on the oxidation behaviour. Using cross-sectional and parallel TEM specimens, analytical electron microscopy analysis identified the presence of Zr-rich oxides in the outer Ni-rich scale on Zr-doped Ni_3Al[5] and Y segregation to α-Al_2O_3 grain boundaries on Y-doped Ni_3Al.[6]

The current study focused on the oxide scale formed on yttrium oxide dispersion strengthened (ODS) Ni_3Al in order to avoid previously observed problems with the uneven distribution of alloy additions of Y[1] and the limited effect of ion-implanted Y.[7,8] The study of the oxidation kinetics of ODS Ni_3Al at 1000° and 1200°C has shown only beneficial effects of the Y_2O_3 dispersion.[6] Measured parabolic rate constants were similar or smaller than those measured for other RE-doped alumina formers[6,7] and 4–5 times smaller than those measured for undoped Ni_3Al.[1]

Based on this initial characterisation of the oxidation kinetics and cyclic oxidation behaviour of ODS Ni_3Al, further characterisation was undertaken. The most unusual observation was that during cyclic testing at 1200°C (2 h cycles) the outer Ni-rich scale (primarily $NiAl_2O_4$) spalled almost completely after the 14th cycle. Bare metal was not exposed indicating that the inner α-Al_2O_3 layer was largely adherent. However, during the next cycle, the outer Ni-rich scale was observed to re-form, indicating that Ni was able to penetrate through the inner layer. A mechanism has been proposed to explain the breakdown of the scale on ODS Ni_3Al, whereby cracks in the inner α-Al_2O_3 layer allow for the rapid outward transport of Ni, allowing an outer Ni-rich layer to re-grow.[6] Various characterisation techniques have been employed to characterise the growth and breakdown of the duplex scale formed on ODS Ni_3Al.

2. Experimental Procedure

Y_2O_3-dispersed Ni_3Al was prepared using a powder metallurgical process. A prealloyed ingot of Ni-12.1 wt%Al-0.02%B was ultrasonically gas atomised to form a powder which was then wet-attrited. Powders between 38 and 180 μm were next ball milled with 2 vol% Y_2O_3 to form composite powders. These powders were then canned, hot degassed, and extruded at 1175°C to form a 12.7 mm diameter bar.[9,10] A final chemical analysis was performed using inductively coupled plasma analysis (Table 1). The Zr is apparently present as a result of using a zirconia crucible for atomisation.[11] The oxidation behaviour of ODS Ni_3Al was compared with that of doped β-NiAl (0.23wt%Zr), undoped Fe-20 wt%Cr-10%Al, and two commercial ODS FeCrAl alloys: Inco alloy MA956 (Fe-20 wt%Cr-4.5%Al-0.36%Ti + 3 vol%Y_2O_3-Al_2O_3) and Kanthal alloy APM (Fe-20%Cr-5.5%Al-0.23%Si-0.03%Ti + $\approx$2 vol%ZrO_2-Al_2O_3).

Oxidation coupons ($\approx$1 cm diameter and 0.1 cm thick) of the various alloys were polished through 0.3 μm alumina and ultrasonically cleaned in acetone and alcohol prior to oxidation. Isothermal weight gain was measured with a Cahn Instruments model 1000 microbalance. Cyclic weight changes were measured after cooling to room temperature with a Mettler model AE163 balance. Isothermal experiments were performed at 1200°C in a dry, flowing atmosphere of pure O_2. Cyclic experiments were performed in the same

Element	Weight (%)	Element	Weight (%)
Ni	86.00	Fe	0.40
Al	11.67	Mn	0.18
B	0.018	Cr	0.15
O	0.2460	Co	0.10
Y	0.98	Si	0.05
Zr	0.10	Cu	0.04
Hf	0.01	Ti	0.02
C	0.20	Mo	0.01
S	0.0085	–	–

Table 1 Inductively coupled plasma analysis of the chemical composition of ODS Ni₃Al.

apparatus with a cycle time of 2 h. Cold samples were inserted into the hot furnace and rapidly removed (2 s) for both isothermal and cyclic experiments.

After oxidation, specimens were characterised using glancing angle X-ray diffraction (GAXRD), scanning electron microscopy with X-ray energy dispersive spectroscopy (SEM/XEDS), transmission electron microscopy (TEM) and field-emission gun, scanning-transmission electron microscopy (FEG-STEM) equiped with a windowless XEDS detector. The STEM work was done with a VG Microscopes Ltd model HB5 instrument with a <2 nm electron probe. Specimens were sectioned parallel to the alloy-oxide interface and thinned by a combination of jet electropolishing and ion-milling.[12] Transverse sections were prepared using a standard technique.[7,13]

3. Results

Based on the initial study of the oxidation behaviour of ODS Ni₃Al,[6] further characterisation focused on the oxidation behaviour at 1200°C. The cyclic oxidation test (2 h cycles) was repeated to confirm the massive outer scale breakdown, and essentially the same behaviour was observed (Fig. 1). One interesting aspect of the spallation was that it took place gradually over 5–10 min after the sample was removed from the furnace. The spalled pieces were very small, <100 μm in diameter, and again no spalling to bare metal could be observed.

After 20 2 h cycles, the scale surface was examined using SEM/XEDS; the cracked and spalled outer layer, however, prevented examination of the inner layer. Fracture and polished cross-sections of the scale were also not particularly revealing because of questions involving sample preparation. For example, the crack observed in the fracture cross-section (Fig. 2) could have occurred during oxidation or during the cooldown period, or it could have formed during sample preparation.

The scale surface of a specimen isothermally oxidised at 1200°C for 50 h was also examined for comparison. In this case, almost the entire outer scale has

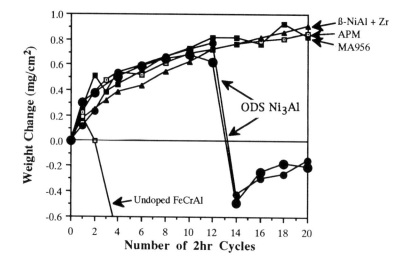

Fig. 1 Weight change during 2 h cyclic tests at 1200°C. ODS Ni₃Al performed similarly to several other doped alumina-formers until massive spallation of the outer scale after the 14th cycle.

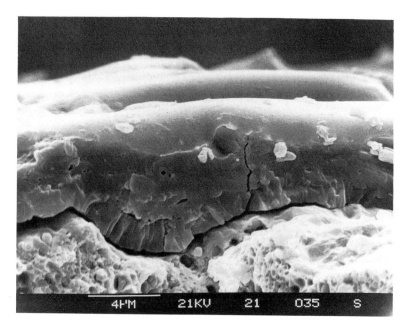

Fig. 2 SEM secondary electron image of a fracture cross-section of the scale formed on ODS Ni₃Al after 20 2 h cycles at 1200°C.

Fig. 3 SEM secondary electron image of the scale on the ODS Ni₃Al after oxidation at 1200°C for 50 h. Near the edge of the sample, a piece of the outer Ni-rich oxide layer is loosely attached. No cracking is observed in the inner α-Al$_2$O$_3$ layer.

spalled away. Figure 3 shows the scale near the sample edge, where a piece of the outer scale is partially attached to the side of the specimen. The surface of the inner scale has areas which appear smooth, with clearly defined grains, and others which appear rough. The latter areas presumably were still attached to the outer scale at temperature, while the smooth areas lost contact. Limited contact between the inner α-Al$_2$O$_3$ scale and the outer Ni-rich layer would explain the poor adhesion of the outer layer. Very few cracks were observed in the alumina scale.

Because of the limited information available from SEM/XEDS analysis of the cycled specimens, TEM specimens were made both parallel and perpendicular to the metal–oxide interface after oxidation for 20 2 h cycles. In the Ni-rich outer scale, voids were present and variations were detected in the Ni-to-Al ratio among grains, but otherwise there were few clues about the nature of the scale breakdown. Likewise, at the interface between the outer and inner (α-Al$_2$O$_3$) layer, the structure was too complicated to reach any definitive conclusions. For example, Ni-rich grains observed among Al$_2$O$_3$ grains could indicate the growth of Ni-rich oxide in cracks or could reflect a rough interface between the two layers.

The scale near the metal interface, while also microstructurally complicated, produced more concrete evidence of the scale breakdown mechanism. In this sample, some areas were found to be nominally pure Al$_2$O$_3$ with no signs of

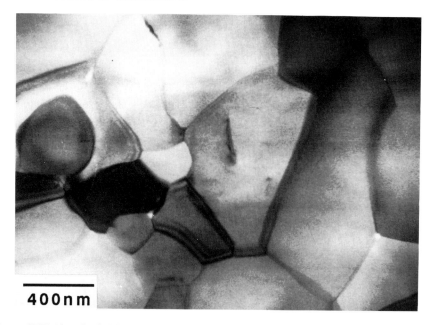

Fig. 4 STEM bright field image of the Al$_2$O$_3$ scale on ODS Ni$_3$Al after oxidation at
1200°C for 20 2 h cycles. Parallel section near the metal interface.

cracking or Ni-rich oxide formation (Fig. 4). However, other regions were a
jumble of Al$_2$O$_3$, mixed (Ni,Al)O$_x$ and voids. An X-ray map of one such region
is shown in Fig. 5. The formation of Ni-rich oxide near the metal–oxide
interface is an indication that some type of through-scale cracking may be
occurring. The observation of thin metal regions confirmed that the parallel
section was close to the metal–oxide interface.

Cross-sections were also made of the scale after oxidation for 20 2 h cycles at
1200°C. The best thin regions were found near the interface between the inner
and outer layers. Similar to the findings of Doychak and Ruhle,[5] the inner
α-Al$_2$O$_3$ layer appears to have a columnar-type structure, and voids are
observed at the interface between the inner and outer layers (Fig. 6). Since this
specimen was cyclically oxidised, the Ni-rich outer scale observed is presumably the re-grown outer scale formed after the 14th cycle. However, it could be
a small area that did not spall during the entire cyclic exposure. Numerous
Y-rich oxides were observed in various locations in the outer scale (Fig. 7),
similar to the Zr-rich particles observed in the outer scale grown on Zr-doped
Ni$_3$Al.[5] The Y-rich particles appear bright in the annular dark-field STEM
images because of atomic number contrast.

As was observed at 1000°C,[5] Y segregation was observed throughout the
scale, even in the Ni-rich scale. The apparent Y/Al weight percentage ratio
determined by XEDS varied considerably throughout the inner scale, from a
high of 0.159 near the Ni-rich oxide interface to a low of 0.015 near the metal
interface. Tramp Zr was also found to be segregated to some but not all
boundaries.

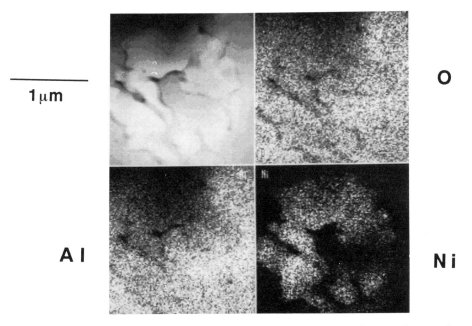

Fig. 5 High resolution X-ray map of the scale on ODS Ni₃Al near the metal interface after oxidation at 1200°C for 20 2 h cycles. (1) STEM annular dark field image; (2) O X-ray map; (3) Al X-ray map; (4) Ni X-ray map.

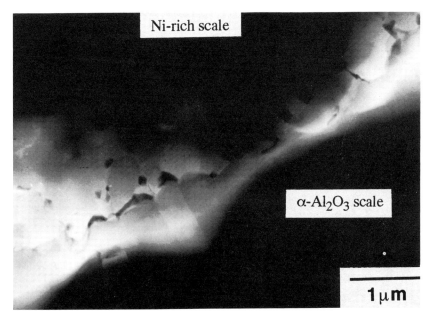

Fig. 6 STEM annular dark field image of the scale formed on ODS Ni₃Al after oxidation for 20 2 h cycles at 1200°C. Transverse section showing voids at the interface between the outer Ni-rich layer (top) and the inner α-Al₂O₃ layer.

Ni-rich scale

400nm

Fig. 7 STEM annular dark field image of the scale formed on ODS Ni₃Al after oxidation for 20 2 h cycles at 1200°C. Transverse section showing Y-rich oxide particles in the outer Ni-rich scale. The Y-rich oxides appear bright in the image because of atomic number contrast.

It has been proposed that RE elements are detected on the scale grain boundaries because they are diffusing outward in the oxygen potential gradient.[14]. The eventual buildup of RE at the gas interface leads to the formation and growth of Re-rich particles. In order to test this hypothesis, the grain boundary segregation was measured in the inner scale as a function of the distance from the outer scale (Fig. 8). These data are a compilation of several diffeerent Al_2O_3 boundaries (similar to those shown in Fig. 6) because no boundary remained properly aligned (parallel to the beam) over its entire length. The general trend is an increasing amount of both Y and Zr segregant on the boundaries closer to the Ni-rich oxide interface, including segregation to the inner-outer scale interface. The total inner-scale thickness is over $2\,\mu m$, so the plot in Fig. 8 is not complete. Closer to the metal interface, the scale had either milled away or was too thick for analysis. However, it was noted in the thicker regions that no voids could be observed at the metal–oxide interface.

Parallel and transverse TEM specimens of isothermally oxidised ODS Ni₃Al were also attempted but with very little success to date. The difficulty experienced with sample preparation compared to the cyclically oxidised specimens is unusual and may be an indication that the α-Al_2O_3 scales grown on ODS Ni₃Al are not as adherent as scales found on other doped alumina-formers such as MA956, APM and Zr-doped β-NiAl.[7]

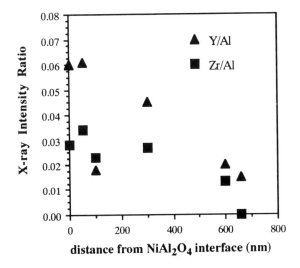

Fig. 8 Plot of the apparent XEDS weight percentage ratios for Zr and Y in the Al$_2$O$_3$ scale as a function of distance from the outer Ni-rich interface after oxidation for 20 2 h cycles at 1200°C.

4. Discussion

The observation of Ni-rich oxides and voids in the α-Al$_2$O$_3$ scale near the metal interface is not fully conclusive but is a strong indication that some type of cracking and re-growth is occurring in the scale. Similar AEM observations of the scale near the metal interface for MA956 after the same cyclic test (20 2 h cycles at 1200°C) did not locate any transient Fe- or Cr-rich oxides.[15] The lack of Cr-rich oxides in the scale on MA956 indicates that the role of Cr in gettering in the FeCrAl alloy (compared to Ni$_3$Al) is probably not significant for the cyclic oxidation resistance. Also, no large voids or evidence of scale cracking were observed near the metal interface. However, cracks were observed at the gas-scale interface using SEM. These cracks apparently did not have a large effect on the cyclic behaviour, because MA956 performed much better in cyclic testing than ODS Ni$_3$Al (Fig. 1).

One significant difference between the scale microstructures on the two materials is that on ODS Ni$_3$Al the α-Al$_2$O$_3$ scale appears to be almost fully columnar beneath the outer Ni-rich scale. The scale on single-layer RE-doped alumina formers is typically finer-grained near the gas interface and columnar near the metal interface.[7] Cracking in the alumina scale on MA956 may be more easily arrested in the finer-grained outer scale and not extend through the entire scale. A crack in the columnar alumina scale on ODS Ni$_3$Al may more easily propagate through the entire scale, leading to the formation of Ni-rich oxide near the gas interface.

Ni-rich oxide has previously been observed to grow in scale cracks on Ni$_3$Al + 2 wt%Cr.[4] This behaviour was observed isothermally when the scale 'blistered' at temperature, an event which does not usually occur on RE-doped alumina-formers. The isothermal behaviour of ODS Ni$_3$Al was parabolic with a measured rate constant very close to that of other doped alumina-formers.[6,7] Spalling of the outer Ni-rich scale occurred during cooldown but not at temperature, and no inner-scale cracking or spalling was observed using SEM. The breakdown during cyclic testing is thus related to stresses during cooldown. The repeatable massive spallation after the 14th cycle (28 h) indicates that this is a critical point. Based on the isothermal kinetic data, the outer scale is not increasing in thickness (which would create a greater stress during cooling).[6] The most likely explanation for the spalling is the presence of voids at the interface between the inner and outer layers (Fig. 6). If these voids increase in size with time, thus limiting contact between the layers, this could eventually lead to spalling in the presence of cooldown stresses. The source of the voids is uncertain. They could be a result of the formation of NiAl$_2$O$_4$ from NiO and Al$_2$O$_3$ or a result of the diffusion of Al from the inner layer to the outer layer.

The voids in the scale near the metal interface also may be a result of a similar reaction of first-formed NiO becoming NiAl$_2$O$_4$. One of the complications with the observations on ODS Ni$_3$Al is that during 20 cycles repeated cracking and re-growth may occur. This is bound to produce a very complex microstructure and also makes it difficult to determine the origins of the Ni-rich oxide formation. Future work will focus on earlier stages of the cyclic failure.

One of the significant observations of this AEM study is the detection of tramp Zr grain boundary segregation in the scale. The segregation of REs to the scale grain boundaries has been attributed to a change in the growth mechanism of α-Al$_2$O$_3$.[16] Tramp Zr has been shown to improve the oxidation adhesion on NiCrAl alloys.[11,17] The small amount of tramp Zr in the alloy may be sufficient to improve the oxidation behaviour without the yttria dispersion. For example, an Al$_2$O$_3$ dispersion could be used to improve the mechanical properties, but improvements in the oxidation behaviour would be due to the tramp Zr, not the Al$_2$O$_3$ dispersion. The detected quantity of Zr segregation was generally less than the Y segregation, a circumstance most likely due to the lower amount of Zr in the alloy (0.10 wt%) compared with that of Y (0.98 wt%). Also, the observation that Zr is not found on every boundary analysed (as is Y) may simply reflect that the tramp Zr is not distributed evenly in the alloy and therefore is not found uniformly in the scale.

Perhaps the most unusual observation was the presence of Y-rich particles in the outer scale after 20 2 h cycles. RE-rich particles in the scale are not an unusual observation,[5,7,18,19] but in this case the initial outer Ni-rich scale was thought to have spalled and re-formed. If the Y-rich particles are also in the outer scale, it would indicate that Y is also rapidly diffusing outward. However, it is also possible that the particular area shown in Fig. 7 is an atypical area where the initial outer scale has been retained. Thus, the Y-rich oxides could be remnants of the initial transient period and not indicative of the rapid transport of Y along with Ni.

In summary, the use of AEM has been very useful in chracterising the growth and breakdown of the duplex scale on ODS Ni$_3$Al. The observation of the spalling and re-growth of the outer Ni-rich scale likely indicates problems with the long-term oxidation resistance of ODS Ni$_3$Al. Further study of the early stages of the scale breakdown is warranted in order to confirm the breakdown mechanism. Understanding the failure may aid in the development of more oxidation-resistant ODS Ni$_3$Al alloys.

Conclusions

1. Voids and Ni-rich oxides were observed in the α-Al$_2$O$_3$ scale near the metal-oxide interface after oxidation of ODS Ni$_3$Al at 1200°C for 20 2 h cycles. These may provide pathways for Ni to penetrate the scale.
2. Y-rich oxides were observed in the outer Ni-rich scale and in the inner α-Al$_2$O$_3$ scale after oxidation at 1200°C. Grain boundary segregation of Y was detected in the inner and outer layers.
3. Tramp Zr segregation was observed on the grain boundaries in the inner and outer scales after oxidation at 1200°C.

Acknowledgments

The authors are very grateful to R. P. Mason at MIT for the preparation of the ODS Ni$_3$Al and to Oak Ridge National Laboratory for supplying the raw ingot. Among our colleagues at MIT, thanks goes to Mr J. Adario for the GAXRD measurements, and to Mr G. Arndt for assistance with the oxidation equipment. The Electric Power Research Institute (contract no. RP242644) provided financial support.

References

1. J. D. Kuenzly and D. L. Douglass, *Oxid. Met.*, **8** (1974), 139–178.
2. S. Taniguchi and T. Shibata, *Oxid. Met.*, **25** (1986) 201–16.
3. S. Taniguchi, T. Shibata and H. Tsuruoka, *Oxid Met.*, **26** (1986), 1–17.
4. S. Taniguchi and T. Shibata, *Oxid. Met.*, **28** (1987), 155–63.
5. J. Doychak and M. Ruhle, *Oxid. Met.*, **31** (1989), 431–52.
6. B. A. Pint and L. W. Hobbs, in *Oxide Films on Metals and Alloys*, B. R. MacDougall, *et al.* (eds), Electrochemical Society, Pennington, NJ, 92-22 (1992), 92–100.
7. B. A. Pint, Ph.D. Thesis, Massachusetts Institute of Technology, Cambridge, MA, 1992.
8. B. A. Pint and L. W. Hobbs, Limitations on the use of ion implantation for the study of the reactive element effect, to be presented at the Spring meeting of the Electrochemical Society, Honolulu, Hawaii, May, 1993, *Electrochem. Soc. Extended Abstracts*, 93-1 (1993), in press.
9. R. P. Mason and N. J. Grant, to be submitted to *Met. Trans. A* (1993).
10. R. P. Mason (1992), Private communication.

11. C. A. Barrett and C. E. Lowell, *Oxid. Met.*, **11** (1977), 199–223.
12. L. W. Hobbs and T. E. Mitchell, in *High Temperature Corrosion*, NACE-6, R. A. Rapp (ed.), NACE, Houston, TX (1983), 76–83.
13. W. E. King, N. L. Peterson and K. Reddy, *Proc. 9th Int. Cong. Met. Corr.*, **4** (1984), 28–35.
14. B. A. Pint and L. W. Hobbs, The dynamic role of interfacial segregants in high temperature oxidation, to be presented at the Spring meeting of the Electrochemical Society, Honolulu, Hawaii, May, 1993, *Electrochem. Soc. Extended Abstracts*, 93-1 (1993), in press.
15. B. A. Pint, A. J. Garratt-Reed and L. W. Hobbs, unpublished research, 1993.
16. B. A. Pint, J. R. Martin and L. W. Hobbs, *Oxid. Met.*, **39** (1993), 167–95.
17. A. S. Khan, C. E. Lowell and C. Barrett, *J. Electrochem. Soc.*, **127** (1980), 670–9.
18. T. A. Ramanarayanan, M. Raghavan and R. Petkovic-Luton, *J. Electrochem. Soc.*, **131** (1984), 923–31.
19. B. A. Pint, A. J. Garratt-Reed and L. W. Hobbs (these Proceedings).

SEM- and TEM-Observations on the Development of the Oxide Scale on Y Implanted Single Crystalline β-NiAl under Low Oxygen Partial Pressure

K. PRÜSSNER, J. BRULEY, U. SALZBERGER, H. ZWEYGART, E. SCHUMANN and M. RÜHLE

Max-Planck-Institut für Metallforschung, Institut für Werkstoffwissenschaft, Seestrasse 92, D-7000 Stuttgart 1, Germany

ABSTRACT

Single crystals of β-NiAl and Y implanted β-NiAl (dose: 5×10^{16} ions cm^{-2}, energy: 70 keV) with surfaces oriented parallel to (100) planes were oxidised under an oxygen partial pressure of 4×10^{-14} Pa in an H_2/H_2O atmosphere at 1223 K for times ranging between 0.1 and 50 h. The oxide scales were examined in plan view and cross-section by SEM. Al concentration profiles were measured as a function of the depth by electron microprobe. Cross-sectional TEM specimen preparation involved the development of a modified technique. First results on TEM and STEM examinations are reported for Y doped specimens. Ion implantation of Y leads to the formation of a very fine grained layer of alumina-containing the Y after oxidation for 0.1 h. Platelets of transition aluminas (most likely θ-Al_2O_3) develop on the fine grained strip after 1 h. After oxidation for 50 h the fine-grained layer is transformed to coarse-grained α-Al_2O_3. Y was detected only (i) at grain boundaries of α-Al_2O_3 in the recrystallised strip, (ii) in an Al, Y and O-containing layer between the recrystallized strip and the platelets, and (iii) at the scale/metal interface. No Y could be identified in the platelets or in the bulk metal.

1. Introduction

The intermetallic compound β-NiAl is a promising material for high temperature applications owing to its high melting point, low specific weight and good oxidation behaviour. Y, or other 'reactive elements' like Ce, Hf, Ta, Zr, added either as an alloying element or by ion implantation or the surface application of oxide dispersions are known to improve oxidation resistance.[1] Different theories have been proposed,[2-6] but the mechanism is not yet fully understood. Therefore, studies of the oxidation of Y implanted β-NiAl and comparisons to the undoped material have been made. TEM studies of thinned cross-sections allow the characterisation of the composition and microstructure of the oxide scales as a function of the distance to the scale/metal interface.

2. Experimental Procedures

Oriented single crystals of β-NiAl (composition: 50.4% Al, 49.6% Ni, <10 ppm S) were cut into slices of $1.5 \times 5 \times 0.8 \, mm^3$ by spark erosion. Surfaces parallel to (100) planes were ground with SiC paper and polished to a final roughness of $0.25 \, \mu m$ with diamond paste. Several crystals were implanted with Y ions (dose: 5×10^{16} ions/cm^2, energy: 70 keV). Y was thus introduced in the form of a gaussian shaped layer with its maximum concentration at a depth of 21.5 nm and a straggling of 2×8.7 nm (FWHM). The oxidation was performed at 1223 K in an H_2/H_2O atmosphere corresponding to an oxygen partial pressure of 4×10^{-14} Pa for time of between 0.1 and 50 h.

The surface morphology of the resulting scale was characterised by scanning electron microscopy (SEM) in plan view. Polished cross-sections of the samples were also examined in the SEM. The concentration profile of Al from the interface into the bulk metal was measured using an electron microprobe operated at 15 kV (JEOL JSM 6400). The distance between two measurements was $1 \, \mu m$. For calibration, β-NiAl of defined composition (50.2% Al, 49.8% Ni) was used. A correction for Z-number, absorption and fluorescence (ZAF) was applied to the raw data.

A modified technique, based on the method suggested by Newcomb et al.,[7] was developed to allow the reproducible preparation of thin cross-sections of β-NiAl and the oxide scale, suitable for TEM. Two specimens of β-NiAl oxidised under the same conditions were glued together with epoxy resin M-Bond 610 such that the oxide scales faced each other. The sandwich was embedded in a brass tube of 3 mm diameter with epoxy resin M-Bond AE 15 enriched with carbon for improved thermal conductivity. The remaining space in the tube was filled with brass. Disks of approximately 0.5 mm thickness were cut from the tube with a wire saw. These disks were ground to ~100 μm thickness on SiC paper. Dimples of ~40 μm depth were mechanically ground into both sides of the disks. The remaining material in the centre of the disks, i.e. near the scale/metal interface, was thus ~20 μm thick. Finally, the specimens were ion-milled to perforation in the thinnest region. The ion-milling was performed in a Bal-Tec RES 010, which allows the very low thinning angles (4–6°) necessary to produce electron transparent foils of β-NiAl and its oxide scale. In conventional ion-mills, only angles down to 8° can be used. For lower angles, the Ar beam cannot be focused properly on the specimen, thereby prolonging preparation times. During ion-milling, care was taken that the specimen was thinned only from the metal side and not from the side of the oxide, because the metal is more resistant to the ion beam.

Examination of the microstructure, chemical composition and crystallography of the oxide scale was carried out on thinned cross-sections by conventional transmission electron microscopy (CTEM), energy-dispersive X-ray spectros-copy (EDS) and selected area diffraction (SAD) using a JEOL JEM 2000 FX operating at 200 kV. The localisation of Y within the scale after 50 h oxidation was determined using the EDS detector on a dedicated scanning transmission electron microscope (VG-STEM HB 501). The effective spatial resolution was limited to 2 nm due to beam broadening. The minimum detection limit for Y in Al_2O_3 was determined to be 0.2 at.%. At grain boundaries of α-Al_2O_3 this corresponds to about ¹⁄₁₀ of a monolayer.

3. Results and Discussion

All the specimens (with and without Y) show continuous oxide scales of variable thickness. The scale/metal interface is rough. The highest number of oxide protrusions into the metal occurs on the Y implanted specimens after oxidation for 50 h. The scales show no spalling or buckling and no pores are observed at the scale/metal interface formed under the chosen experimental conditions. For a general description of the morphologies of similar oxide scales see also, e.g., Rybicki and Smialek[8] or Doychak et al.[9]

After oxidation of pure β-NiAl for 0.1 h, there has been an epitactic growth of platelets consisting of transition aluminas (Fig. 1a). These platelets are aligned along the <110> directions of the metal, as reported by Doychak et al.[9] After oxidation for 50 h, the platelets have grown sufficiently to cover the surface pores present at the beginning (Fig. 1c).

Implantation of Y alters the oxide scale growth. After oxidation for 0.1 h, the platelets of transition aluminas have not yet started to grow. A nearly featureless surface can be seen (Fig. 1b). The nucleation of the platelets occurs only after ~1 h. Initially, needles grow instead of platelets (Fig. 1d), whereas after 50 h the surface is covered by platelets similar to the undoped specimens. No orientation relationship between the platelets and the metal was observed.

SEM examination of polished cross-sections of specimens oxidised for 50 h (Fig. 2a,b) show that the scale/metal interface is rough with oxide protrusions growing inwards. There are more finer protrusions for the Y implanted material and scales are thicker than for the pure metal. Near to the interface of the Y implanted specimens a layer is visible which has a lighter contrast than the nearby oxide (Fig. 2a). This contrast is caused by high concentrations of Y.

A depletion of Al was measured to extend for ~20 μm into the metal before a constant bulk value is reached. At the metal/oxide interface, the concentration of Al is only ~29.5 wt.% compared to ~32.8 wt.% in the bulk (Fig. 2c). Calculation of composition profiles using the composition-dependent diffusion coefficients of Ni and Al have been reported by Brumm.[10] It would be of interest to compare these results to computed profiles in a similar fashion.

TEM examination shows that the oxide scale formed on Y doped β-NiAl after oxidation for 0.1 h consists only of a very fine grained strip of alumina containing Y (Fig. 3a). So far, the structure of the alumina has not been determined and the distribution of Y within the scale is unknown. The thickness of the strip is ~100 nm. The upper half of the strip contains small

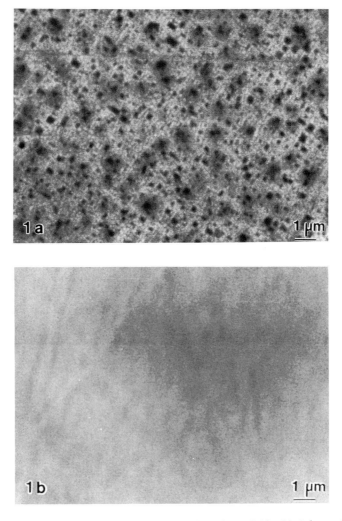

Fig. 1 SEM micrographs of oxidisd β-NiAl and β-NiAl + Y (plan view). (a) −Y/0.1 h, (b) +Y/0.1 h, (c) −Y/10 h, (d) +Y/10 h.

precipitates. On top of the strip, platelets of transition aluminas have not yet nucleated and this is in agreement with the SEM observations. After 10 h oxidation, the width of the Y-containing fine grained strip has broadened to 150 nm (Fig. 3b). On top of the strip platelets of transition aluminas (most likely θ-Al$_2$O$_3$) have grown. These aluminas are heavily twinned. Between the Y containing strip and the metal there is another narrow layer consisting of transition aluminas. After 50 h oxidation the fine grained strip has vanished, apparently recrystallising to form α-Al$_2$O$_3$ (Fig. 3c). The protrusions seen in the SEM were identified to be α-Al$_2$O$_3$ growing into the metal. The upper part of the scale still consists of platelets of transition aluminas. The transition aluminas and the recrystallised strip are separated by a row of particles

Fig. 1—contd.

(30–50 nm in size) which contain Al, Y and O and are probably Y-Al-garnet (YAG).

   In the recrystallised strip (Fig. 4a,b,c) five oxide grain boundaries and oxide grains besides the grain boundaries (Fig. 4d) were analysed. All grain boundaries show an enrichment of Y. The oxide grains are free of Y. The amount of Y segregated to the grain boundaries is not uniform and probably depends on the orientation of the grains, i.e. on the open space provided. Particles with dark contrast found in the strip were found to be rich in Ni and Al and also contain Y. In areas where there are no pegs of α-Al$_2$O$_3$ growing into the metal, i.e. when the former fine crystalline strip is in direct contact with the metal, six spectra were recorded along the scale/metal interface. It was

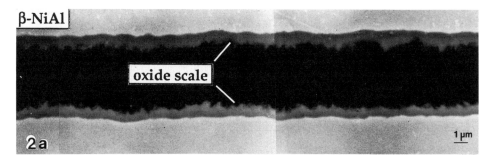

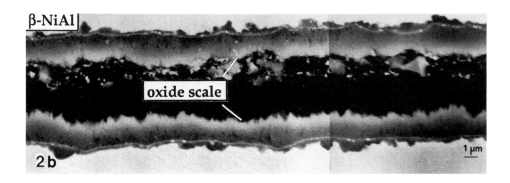

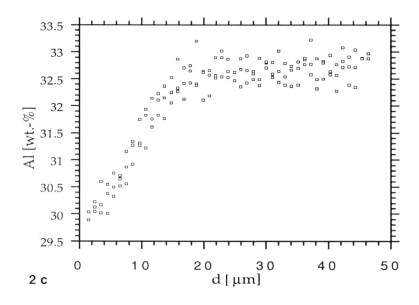

Fig. 2 (a) SEM micrographs of polished cross-sections of β-NiAl (50 h oxidised); (b) SEM micrographs of polished cross-sections of β-NiAl + Y (50 h oxidised); (c) Al concentration profile of β-NiAl (50 h oxidised).

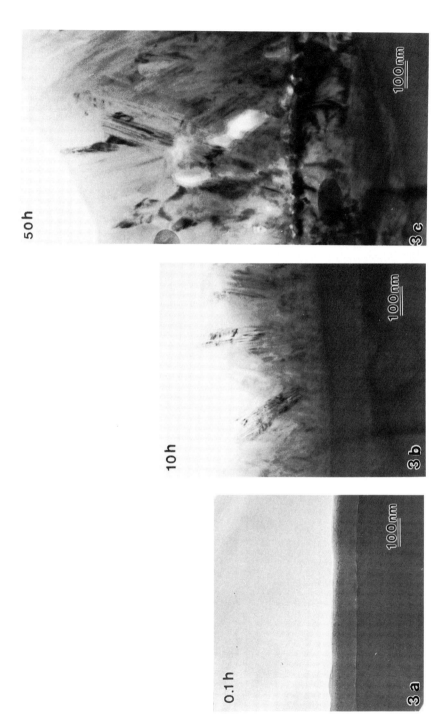

Fig. 3 TEM micrographs of cross sections of β-NiAl + Y. (a) 0.1 h oxidised; (b) 10 h oxidised; (c) 50 h oxidised.

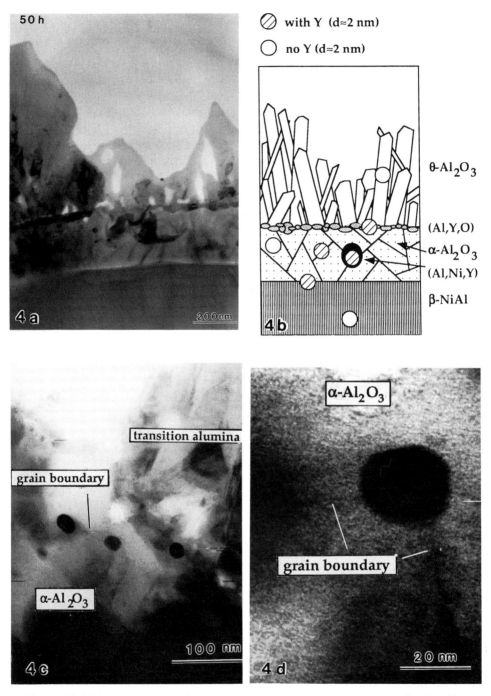

Fig. 4 (a) TEM micrograph of analysed area on β-NiAl + Y (50 h oxidised); (b) schematic representation of oxide scale on β-NiAl + Y (50 h oxidised); (c) STEM micrograph of analysed area; (d) STEM micrograph of analysed grain boundary in oxide scale.

found that Y segregation occurs although its concentration is lower than at grain boundaries. The Y concentration at the interface between α-Al$_2$O$_3$ protrusions and the β-NiAl has not been analysed. The bulk metal and the platelets of transition aluminas formed on top were free of Y. Particles rich in Ni or Fe were found in the transition aluminas. S was not found to be segregated at the scale/metal interface or at grain boundaries and no sulphide particles could be detected.

The observations reported so far are not yet sufficient to suggest a solution to the problem of the Y effect on β-NiAl. Further investigations have to be carried out. So far, there are indications that no single mechanism proposed in the literature[2] occurs, but several of them are operating.

For short oxidation times, there is an additional layer formed in the oxide scale on specimens that had been Y implanted. The extremely fine grain size in this case is probably caused by radiation damage or by the high local Y content. To examine this further, TEM examinations on Y implanted but unoxidised specimens should be carried out, along with oxidaton experiments on β-NiAl alloyed with Y. Because of recrystallisation of the layer after 50 h, this layer may not be of significance for long-term oxidation. After extended times α-Al$_2$O$_3$ is the major phase in the scales on pure and implanted materials.

After recrystallisation of the fine grained strip, Y segregates to grain boundaries of α-Al$_2$O$_3$. Certainly the transport properties of the scale are affected as has been established recently.[11,12] The enhanced formation of inward growing α-Al$_2$O$_3$ pegs might suggest the promotion of oxygen transport. Prevention of pore formation may be another consequence of modified transport properties. Experiments for longer oxidation times may address this issue.

There is also segregation of Y at the scale/metal interface of Y doped specimens after recrystallisation of the strip, which might promote the bonding at the interface. However, the concentrations of Y are quite low at the interface. To maintain a layer of Y during further inward growth of oxide a constant supply for Y from the alloy would be necessary. In ion implanted specimens this cannot occur. β-NiAl alloyed with Y has to be oxidised for a comparison.

No sulphide particles and no S segregation was detected at the interface on Y doped β-NiAl oxidised for 50 h. It is unknown if any S is present at the interface of undoped specimens. Also it is unknown if S or sulphide would be detected in specimens of higher S content. For comparison, the Y free specimens have to be analysed in the STEM. β-NiAl with high S content has to be oxidised under the same conditions.

4. Conclusions

The first results reported here on the oxidation of β-NiAl ion implanted with Y show that a better understanding of the factors causing the so-called 'Y effect' can be obtained by examining thinned cross-sections of oxidised specimens in the TEM and STEM. In this way, the distribution of Y can be localised within the scale. In contrast to the frequently used SIMS technique, the exact location of the probe is known. Further investigations into the oxidation behaviour of β-NiAl alloyed with Y, and β-NiAl with high S content have to be done to collect more data before a mechanism can be suggested.

Acknowledgement

We would like to thank Professor Dr H. J. Grabke (MPI für Eisenforschung, Düsseldorf) for providing us a single crystal of β-NiAl and Dr W. Jäger and Mr Gebauer (Forschungszentrum Jülich) for the Y ion implantation. One author (K. P.) is very much indebted to Mr P. Kopold for teaching us the use of the JEOL JEM 2000 FX. The present work is being funded by the Deutsche Forschungsgemeinschaft.

References

1. L. B. Pfeil, Patent No. 459848 (1937).
2. J. Jedlinski, *Sol. St. Phen.* **21+22** (1992), 335–390.
3. P. A. van Manen, E. W. A. Young, D. Schalkoord, C. J. van der Wekken and J. H. W. deWit, *Surf. Interface Anal.*, **12** (1988), 391–396.
4. H. J. Grabke, D. Wiemer and H. Viefhaus, *Appl. Surf. Sci.*, **47** (1991), 243–250.
5. H. J. Grabke, M. Steinhorst, M. Brumm and D. Wiemer, *Ox. Met.*, **35**(3/4) (1991), 199–222.
6. P. Fox, D. G. Lees and G. W. Lorimer, *Ox. Met.* **35**(5/6) (1991), 491–503.
7. S. B. Newcomb, C. S. Baxter and E. G. Bithell, *Inst. Phys. Conf. Ser.*, **93**(1), 43–48.
8. G. C. Rybicki and J. L. Smialek, *Ox. Met.*, **31**(3/4) (1989), 275–304.
9. J. Doychak, J. L. Smialek and T. E. Mitchell, *Met. Trans.*, **20A**(3) (1989), 499–518.
10. M. W. Brumm, Oxidationsverhalten von β-NiAl und von NiAl-Cr-Legierungen, Diss., MPI f. Eisenforschung, Düsseldorf (1991).
11. M. Harmer, private communication.
12. J. Philibert, private communication.

Redistribution of Major and Minor Alloy Components in Scales Formed During Early Stages of Oxidation on FeCrAl Alloys Studied by Means of SIMS and SNMS

J. JEDLIŃSKI,[1]* G. BORCHARDT,[1] A. BERNASIK,[2]* S. SCHERRER,[2]
R. AMBOS,[3] and B. RAJCHEL[4]

[1]*FB Metallurgie und Werkstoffwissenschaften und SFB 180, Technische Universität
Clausthal, Robert-Koch-Str. 42, D-3392 Clausthal-Zellerfeld, Germany*
[2]*Laboratoire de Physique du Solide (U.A. 155), Ecole des Mines de Nancy, Parc de
Saurupt, F-54042 Nancy Cedex, France*
[3]*Institut für Nichtmetallische Werkstoffe, TU Clausthal, Postfach 1253, D-3392
Clausthal-Zellerfeld, Germany*
[4]*Institute of Nuclear Physics, ul. Radzikowskiego 152, PL-31342 Kraków, Poland*
**Now: Joint University Centre for Chemical Analysis and Microstructural Research,
Surface Spectroscopy Laboratory, ul. Reymonta 23, PL-30-060 Kraków, Poland*

ABSTRACT

The distribution of elements across the scales formed on Fe-23Cr-5Al in oxygen and air during early oxidation stages alloys at temperatures ranging from 1273 to 1473 K was determined using SIMS and SNMS. The effect of reactive element additions either alloyed (yttrium and hafnium) or implanted (yttrium) was studied as well.

Both the major and the minor alloy components were incorporated into the oxide scales from the very beginning of oxidation. The iron and chromium surface maxima were initially observed on the FeCrAl alloys according to the thermodynamic stability of their oxides. Despite their low concentrations in these alloys hafnium and manganese were significantly incorporated into the scales. The distribution of yttrium depended on its form in the substrate and on the reaction conditions. Attempts were made to apply SNMS to determine the sulphur distribution. The redistribution of elements is discussed in terms of the phase changes in the scale.

1. Introduction

Performance of materials for high-temperature applications strongly depends on the protective properties of the oxide scales growing during reaction and on their adherence to the substrate. Both an improvement and deterioration of the oxidation resistance were frequently related to compositional changes occurring in the scales, substrate or at the scale/substrate interface. The most spectacular examples are the so-called reactive-element effect and the sulphur-effect observed for chromia and alumina formers, as reviewed elsewhere.[1-4] Modern surface analytical tools such as secondary ion mass spectrometry (SIMS), sputter neutral mass spectrometry (SNMS) or scanning auger microscopy (SAM) enable investigation of the elemental distributions with good detection limits and in-depth resolutions.

This paper briefly reports the compositional changes occurring across oxide scales growing on FeCrAl alloys and the effect of reactive elements on these changes.

2. Materials and Experimental

Two Fe-23Cr-5Al alloys, unmodified and containing additions of hafnium and yttrium, were obtained by melting the respective metals under vacuum. The ingots were annealed at 1373 K in vacuum for 24 h. The unmodified FeCrAl alloy contained the following elements (in wt. %): Cr-22.7, Al-4.80, Mn-0.68, Si-1.20, Ti-0.025, Ni-0.19, Cu-0.07, C-0.06, S-0.011, P-0.017. The reactive-element-bearing alloy contained: Cr-23.18, Al-4.47, Mn-0.60, Si-0.81, Ti-0.009, Ni-0.11, Cu-0.02, C-0.05, S-0.007, P-0.008, as well as Hf-0.1, Y-0.04, Zr-0.02, and 'Mischmetal'-0.05.

The test samples were prepared using a standard procedure and finished by polishing with 1 μm diamond paste and careful cleaning. Some samples of both alloys were implanted with yttrium ions (Y^+) with an ion beam energy of 70 keV and a dose of 2×10^{16} ion/cm^2.

The oxidation experiments were carried out in the temperature range of 1273 k–1473 K mostly in oxygen and, sometimes, in air. The oxidised samples were analysed by means of SIMS and/or SNMS. A Cameca SMI-300 spectrometer was used for SIMS analysis. In order to reduce the charging effects resulting from the interaction between the bombarding beam and the insulating alumina layer a neutral primary beam of argon atoms with an energy of 10 keV and angle of incidence of 45°, was used. Principles of this approach are described elsewhere.[5]

A Leybold AG INA 3 spectrometer was employed for SNMS analysis in which the ionisation of neutral particles emitted after bombardment is achieved in low-pressure and high-frequency plasma. The following measurement parameters were used: direct bombardment mode, Ar^+ primary beam, accelerating potential 375 V, analysed area diameter – 5 mm, plasma-sample distance 4 mm. A 100 mesh tungsten grid was placed on the sample surface in order to reduce charging effects.

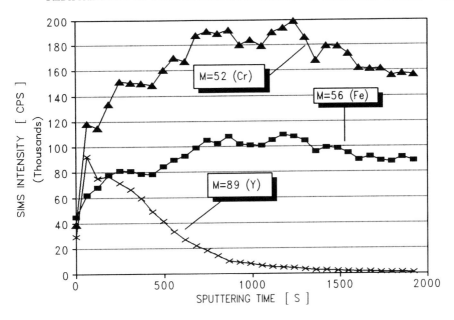

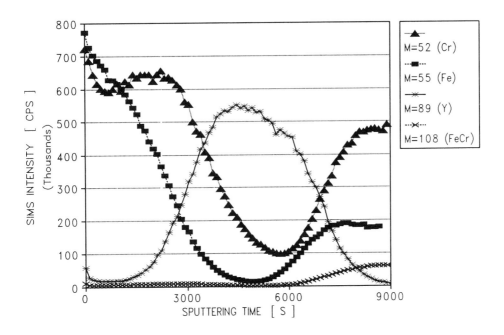

Fig. 1 SIMS in-depth profiles of elemental distributions in yttrium-implanted $(2 \times 10^{16} \, Y^+/cm^2)$ Fe-23Cr-5Al alloy. A, starting material (before oxidation); B, oxidised material (early stages: 1273 K, 2.75 min, oxygen).

3. Results

The distribution of the implanted yttrium was similar in both alloys and is shown in the FeCrAl alloy together with those of Fe and Cr in Fig. 1(A). It should be noted that because of the high SIMS intensity and negligible change observed across the scale, the aluminium distributon was not measured. A somewhat different situation occurred during SNMS analysis. In this case the Al signal was always monitored. Figures 1(B), 2 and 3 present the representative distributions of the elements at various stages of oxidation of the unimplanted and the yttrium-implanted FeCrAl alloy. Similar results for the hafnium and yttrium-containing alloy, unimplanted and yttrium-implanted, are shown in Figs 4 and 5.

In all cases the alloy components were incorporated into the scales from the beginning of the reaction, the amount and sequence being related to their concentration in the alloy and to the thermodynamic stability of their oxides respectively. As the oxidation proceeded the iron and chromium maxima disappeared, indicating that the oxide scale essentially consisted of alumina. Implanted yttrium retarded the disappearance of the Fe- and Cr-maxima (Figs 2B and 3A).

The manganese distribution was difficult to determine since it overlapped with that of iron. However, Figs 2(B) and 3(B) clearly demonstrate that at least during the later stages of oxidation this element tended to enrich in the outer parts of the scales on unimplanted as well as on the yttrium-implanted Fe-23Cr-5Al alloys. In contrast, apparently no irregular shapes were found in silicon distributions. It exhibited a narrow maximum at the outer surface of the scale and was rather uniformly distributed across the oxide layer.

Alloyed yttrium and hafnium were present beneath the outermost Fe-rich part of the scale (Fig. 4). Only slight enrichment of yttrium was detected in the scale with respect to its amount in the alloy. However, the hafnium signal in the scale was higher than that in the alloy.

Yttrium implanted to both alloys initially exhibited only small and narrow maxima at the outer surface of the scale and essentially remained within the scale (Figs 1B, 4B). In the later oxidation stages it tended to migrate towards the outermost part of the scale (Figs 3B, 5). However, this effect was observed only after prolonged oxidation at lower temperatures. Twelve hours at 1273 K were not sufficient for this purpose, while at 1473 K the change of the yttrium distribution was found after 26 min on the unmodified alloy (Fig. 3B) but a considerable amount of yttrium remained within the scale on the yttrium- and hafnium-containing alloy even after 1 h oxidation at this temperature (Fig. 5).

An attempt was made using SNMS results to study the corresponding distribution of sulphur. In order to avoid misleading conclusions the signals ascribed to mass numbers 16 (O) and 32 (S and O_2) were simultaneously monitored. No significant effect of implanted yttrium on the shapes of profiles was observed in either alloy. The results are shown in Fig. 6 for the reactive element-free and the yttrium- and hafnium-bearing alloys.

Remarkable enrichments of sulphur but not of oxygen were observed at the external surfaces of the scales (small and sharp peaks) and near the scale–

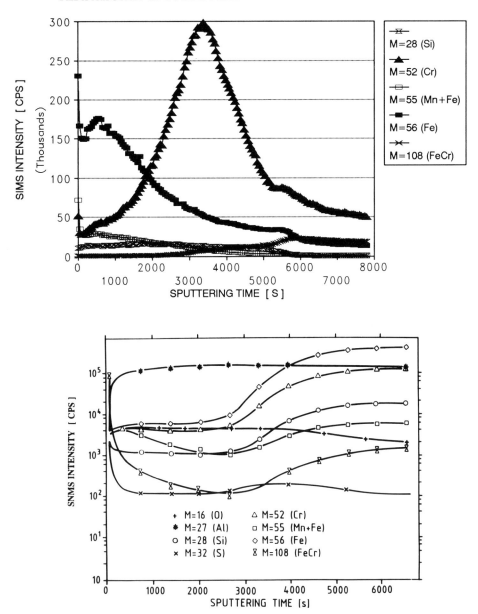

Fig. 2 Distribution of elements across the scales formed on unmodified Fe-23Cr-5Al alloy at 1273 K in oxygen. A, 2.75 min (SIMS profiles); B, 326 min (SNMS profiles).

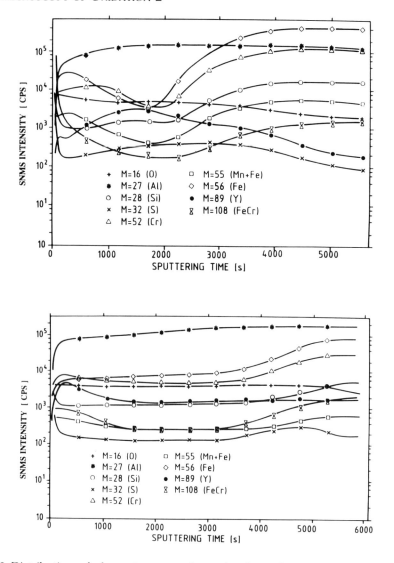

Fig. 3 Distribution of elements across the scales formed on yttrium-implanted (2×10^{16} Y$^+$/cm^2) Fe-23Cr-5Al alloy in oxygen. A, 1273 K, 326 min (SNMS profiles); B, 1473 K, 26 min (SNMS profiles).

substrate interface (broad and rather flat maxima) after relatively prolonged oxidation of the Fe-23Cr-5Al alloy, no matter whether implanted or not (Fig. 6A). In contrast, only rather shallow sulphur-rich regions at the outer surface of the scale were found on the yttrium- and hafnium-containing alloy (Fig. 6B). Virtually no sulphur enrichment was observed at the scale/substrate interface.

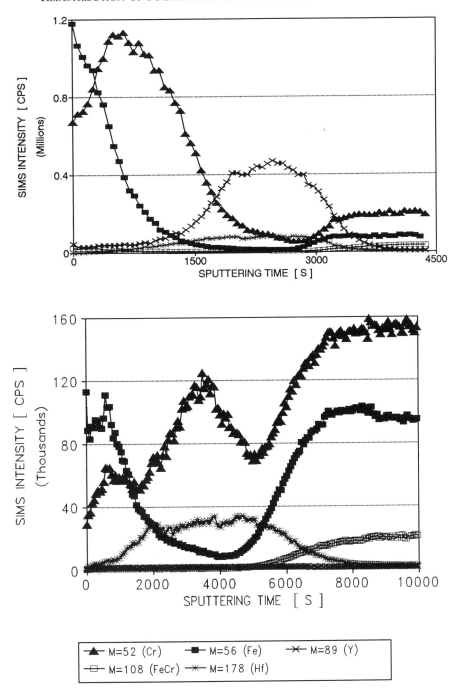

Fig. 4 Distribution of elements across the scales (SIMS profiles) formed on (A) unimplanted and (B) yttrium-implanted (2×10^{16} Y^+/cm^2) Fe-23Cr-5Al + (Hf + Y) alloy (1373 K, 2.75 min, oxygen).

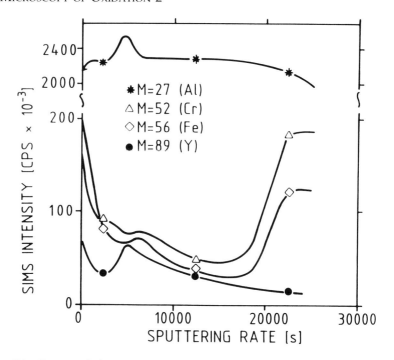

Fig. 5 Distribution of elements across the scale (SIMS profiles) formed on yttrium implanted (2×10^{16} Y$^+$/cm^2) Fe-23Cr-5Al + (Hf + Y) alloy (1473 K, 1 h, oxygen).

4. Discussion

The distributions of the major alloy components are in agreement with those reported in the literature.[6–8] It should be noted however that the corresponding distribution of reactive elements differed considerably from that reported for scales growing on an yttria-containing FeCrAl-type MA956 ODS alloy.[8] The tendency that implanted or alloyed reactive elements incorporated into the oxide layer at the inception of oxidation remain within the scale for extended oxidation periods and migrate towards its outer surface at the later oxidation stages is consistent with the change of the predominant diffusion species in the scales from outward metal transport to inward oxygen transport. The latter change was proposed to be closely related to the phase transformation of the unstable transition-type aluminas into α-Al$_2$O$_3$, as discussed in more detail elsewhere.[9] The unstable aluminas have a defective cation sublattice and are believed to admit much higher dopant concentrations in the cation sublattice than anion-deficient α-Al$_2$O$_3$ with its extremely low point defect concentrations. Due to the lattice rearrangement necessary for the phase transformation to occur the excess of yttrium should be released and either forms precipitates or migrates towards the outermost part of the scale. In the case of the ODS alloys the change of the predominant transport mode occurs very quickly and was observed after between 2 min and 7 min oxidation at 1173 K.[10] High

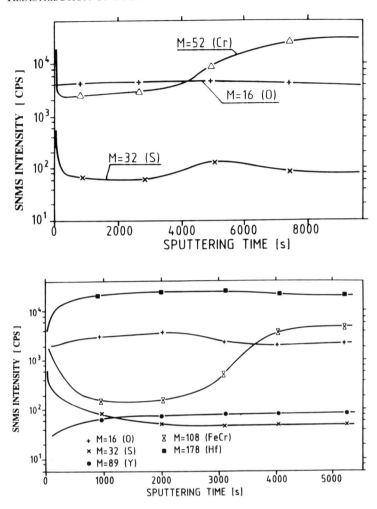

Fig. 6 Shapes of sulphur- and oxygen-related SNMS signals across the scales formed in oxygen on: A, unmodified Fe-23Cr-5Al alloy (1473 K, 9.5 min); B, hafnium- and yttrium-containing Fe-23Cr-5Al alloy (1273 K, 1 h). (Other spectra are shown for the sake of comparison and to indicate the scale–metal interface.)

amounts of yttrium were proposed to retard the phase transformation,[9] which is consistent with the observed element distribution differences between the unimplanted and implanted alloys.

The sulphur enrichment at the scale–substrate interface observed on the Fe-23Cr-5Al alloy suggests that the SNMS method can be used to obtain a qualitative information on the distribution of trace elements across the oxidised samples. The disadvantage of this method is its relatively low spatial resolution, which makes it impossible to study any local compositional inhomogeneities. Moreover, the interaction between the primary ion beam and the target prevents observation of the expected sharp (ca. 10 Å) sulphur concentration maxima at the scale–substrate interface.[3,4] Further experiments should verify the capability of SNMS of detecting such enrichments.

Finally, it should be noted that both alloys exhibit good oxidation resistance under the conditions studied. However, they do degrade under more prolonged oxidation and/or higher reaction temperatures than studied here.[10]

5. Conclusions

1. Both the major and the minor alloy components were incorporated into the oxide scales from the very beginning of oxidation.
2. The iron and chromium surface maxima were initially observed on the FeCrAl alloys according to the thermodynamic stability of their oxides and disappeared after prolonged oxidation.
3. Despite their low concentrations in the alloys hafnium and manganese were significantly incorporated into the scales. The latter tended to enrich in the outermost part of the scale.
4. The distribution of yttrium depended on its form in the substrate and on the reaction conditions.
5. The SNMS technique was helpful in the rough determination of sulphur distribution.
6. The redistribution of elements during the early stages of reaction can be related to the phase composition of the scale.

Acknowledgements

The assistance of E. Ebeling and J. Slowik (TU Clausthal) in preparation of this paper as well as the financial support of the Deutsche Forschungsgemeinschaft and the Polish National Science Foundation 'Komitet Badań Naukowych' (contract no. 7-7326-9203) are gratefully acknowledged.

References

1. D. P. Moon, *Mater. Sci. Techn.*, **5** (1988), 754.
2. J. Jedliński, *Solid State Phenomena*, **21&22** (1992), 335.
3. J. G. Smeggil, N. S. Bornstein and M. A. de Crescente, *Oxid. Met.*, **30** (1988), 259.
4. P. Y. Hou and J. Stringer, *Oxid. Met.*, **38** (1992), 323.
5. G. Borchardt, S. Scherrer and S. Weber, *Microchim. Acta*, **II** (1981), 421.
6. G. Ben Abderrazik, G. Moulin, A. M. Huntz, E. W. A. Young and J. H. W. de Wit, *Solid State Ionics*, **22** (1987), 285.
7. W. J. Quadakkers, J. Jedliński, K. Schmidt, M. Krasovec, G. Borchardt and H. Nickel, *Applied Surface Science*, **47** (1991), 261.
8. W. J. Quadakkers, A. Elschner, W. Speier and H. Nickel, *Applied Surface Science*, **52** (1991), 271.
9. J. Jedliński, *Oxid. Met.*, **39** (1993), 55.
10. J. Jedliński, G. Borchardt, to be published.

A SIMS Study of the Effect of Y and Zr on the Growth of Oxide on β-NiAl

R. PRESCOTT, D. F. MITCHELL and M. J. GRAHAM

Institute for Microstructural Sciences, National Research Council, Ottawa, Canada
K1A 0R6

ABSTRACT

This paper reports on a study of the growth of α-Al_2O_3 on β-NiAl at 1200°C. The influence of alloyed yttrium and zirconium on transport through the oxide has been assessed. Scales have been formed in two-stage experiments using $^{16}O_2$ and $^{18}O_2$ gases, and the various isotopic species have been located by using high-resolution imaging secondary ion mass spectrometry (SIMS). Supplementary information on oxide morphologies has been obtained by SEM. SIMS data show that α-Al_2O_3 scales on β-NiAl develop by both outward transport of aluminium and inward transport of oxygen. Yttrium and zirconium limit the outward transport of aluminium.

1. Introduction

Nickel aluminide, β-NiAl, is one of the many high-temperature materials being considered for use as structural materials in advanced turbine engines. Its oxidation behaviour has been widely characterised, and has been determined to be controlled by growth of a protective Al_2O_3 scale. However, the mechanism of oxide growth is still not completely understood. Oxide growth on NiAl seems to be a special case in which the early stages of growth are characterised by the development of ridges of oxide described by Doychak *et al.*[1] as a lace network which eventually thickens to develop the same type of grain structure as other alumina formers, e.g. NiCrAl or FeCrAl alloys.

Increased understanding of these complex degradation processes may be brought about by advances in analytical instrumentation and techniques. The combination of secondary ion mass spectrometry (SIMS) and ^{18}O tracer experiments has proved to be a valuable tool in the study of oxide growth processes at high temperatures. This paper illustrates the use of high-resolution imaging SIMS to study oxide growth on β-NiAl, and also shows how the growth process is modified by the presence of reactive elements, such as yttrium and zirconium.

2. Experimental

The materials used in this study were β-NiAl, β-NiAl sputter-coated with 4 nm of Y_2O_3, NiAl-0.1%Y and NiAl-0.2%Zr. All concentrations are in weight percent. Two-stage oxidation experiments were conducted in a constant volume apparatus at 1200°C. $^{16}O_2$ was used in the first stage, then the system was briefly evacuated and $^{18}O_2$ was introduced for the second stage. The oxygen pressure was 1.33 kPa.

SIMS analysis was performed in a Perkin-Elmer, Physical Electronics Industries (PHI) model 5500 system using a 25 keV, 400 pA liquid gallium ion source, with a beam diameter ~120 nm. High-resolution SIMS images were acquired in a 256 × 256 pixel format with dwell times of 1 or 2 ms per pixel. Signals for ^{16}O and ^{18}O were collected as the singly charged negative species, and for yttrium and zirconium, the singly charged positive species were detected. Imaging SIMS data have been complemented by sputter depth profiles and morphological examination in a JEOL 840 scanning electron microscope (SEM).

3. Results and Discussion

3.1 Oxidation of β-NiAl

The oxidation of β-NiAl, particularly the transient stages, has been the subject of several thorough studies.[1–3] A characteristic feature of this system is the growth of oxide ridges in the α-Al_2O_3 scale. The SEM micrograph (Fig. 1a) illustrates this ridged morphology. Oxide between the ridges is found to be much thinner than in the ridges. Figures 1(b) and (c) are planar SIMS images of the ridged oxide. Figure 1(b) is an image in which the $^{18}O^-$ species has been detected in the surface of the oxide. The white lines indicate regions of high ^{18}O content and they clearly correspond to the oxide ridges. The dark areas, between the white lines, indicate regions of scale with low ^{18}O content. Figure 1(c) is the complementary $^{16}O^-$ image for the same area of oxide. In this image, the ridges appear dark, which shows that they contain little ^{16}O. The areas between the ridges are brighter, which is indicative of higher ^{16}O levels. From these planar images of the outer surface of the oxide it is clear that after the formation of an α-Al_2O_3 layer, the bulk of the new surface oxide growth is confined to the ridges. This was also found to be the case at 1100°C.[4] From a series of planar images obtained from sputtering the oxide from the outer surface down to the substrate it is possible to observe how the ^{18}O distribution changes with depth.[5] The highest ^{18}O levels are found to be associated with the outer region of the oxide ridges, which shows that the ridges continue to develop throughout the period of oxidation in $^{18}O_2$. Since ^{18}O is detected on the outside of the α-Al_2O_3 ridges, it can be concluded that growth is occurring by outward diffusion of aluminium.

As the time of oxidation at 1200°C increases, the ridged structure coarsens. The ridges become taller and wider and the flat areas between the ridges

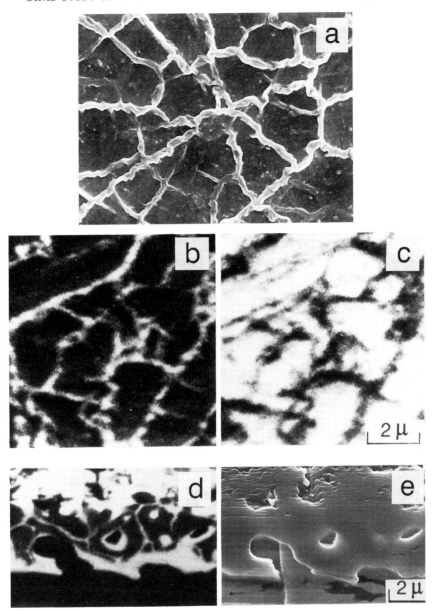

Fig. 1 Oxide formed on β-NiAl at 1200°C. (i) After 4 h; 2 h in $^{16}O_2$ plus 2 h in $^{18}O_2$. (a) SEM micrograph, (b) planar $^{18}O^-$ SIMS image, and (c) complementary $^{16}O^-$ image. (ii) After 12 h; 6 h in $^{16}O_2$ plus 6 h in $^{18}O_2$. (d) $^8O^-$ image, and (e) complementary SEM image of taper section (4×). In SIMS images, regions rich in the particular isotope appear white.

gradually disappear. After 12 h the scale is sufficiently thick to be analysed as a taper section, as shown in Fig. 1(d) and (e). Figure 1(d) shows the ^{18}O distribution in the scale after sequential oxidation in $^{16}O_2$ followed by $^{18}O_2$. (Fig. 1e is a SEM micrograph of the same area for comparative purposes.) A high concentration of ^{18}O is observed in the outer portion of the scale and along the scale-substrate interface as well as in channels or boundaries in the inner part of the oxide. (It should be noted, however, that in this taper section and for the others in Figs 3 and 4 the unpolished top surface of the oxide appears in the image.) The image in Fig. 1(d) clearly shows that inward diffusion of oxygen along short-circuit paths is now beginning to play a significant role in the growth process. This change in growth mechanism appears to coincide with the coalescence or merging of the oxide ridges. Scales formed after 100 h oxidation[6] confirm that the growth mechanism of thicker α-Al_2O_3 scale involves both inward transport of oxygen and outward transport of aluminium.

3.2 Effect of Yttrium

The effects of yttrium on the oxidation behaviour of NiAl have been studied,[7–10] although, according to the literature, their effects are not widely agreed upon. The scale formed on NiAl at 1200°C, after coating with a 4 nm-thick layer of Y_2O_3, is found to consist of two layers, each of which can be analysed by SIMS depth profiling. Figure 2 shows two craters in the two-layered scale from which SIMS analyses were obtained. In each case the area analysed was approximately $5\,\mu m \times 5\,\mu m$. The depth profiles show that the inner layer of oxide (top profile), where the outer layer containing yttrium has spalled off early in the oxidation, has a high ^{18}O content. ^{18}O images of this inner layer[11] are similar to those observed for the uncoated alloy (Fig. 1d) in the substrate interface region. As seen from the depth profile of the complete oxide (bottom profile), yttrium is detected in the outer portion of the outer layer. While the Y_2O_3 coating has not been able to produce the typical effects of reactive elements, the example illustrates the technique of high-resolution SIMS depth profiling and its correlation with microstructural features.

The effects of alloyed yttrium have been studied using an alloy containing 0.1 %Y. Yttrium is found to have a significant effect in slowing the develop-ment of the ridged morphology. SIMS images of a taper section are presented in Fig. 3. Here the bulk of the $^{18}O^-$ signal (Fig. 3b) arises from the vicinity of the scale–substrate interface. The presence of yttrium appears to minimise growth of new oxide in the outer part of the scale. The Y^+ image (Fig. 3c) indicates the distribution of yttrium in the scale. Yttrium is in oxidised form and is observed to be combined with ^{18}O rather than ^{16}O.

3.3 Effect of Zirconium

Additions of zirconium are known to improve the cyclic oxidation resistance of NiAl.[12,13] Development of the ridged morphology is slowed down by the presence of 0.2% Zr. From depth profiles of thin scales[5] formed sequentially in

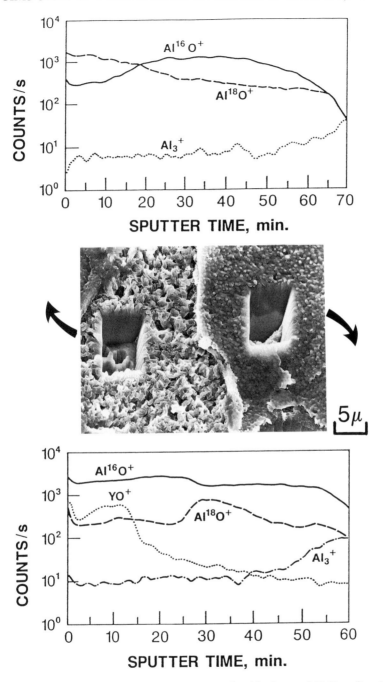

Fig. 2 Oxide formed on β-NiAl, sputter-coated with 4 nm of Y_2O_3, after 24 h at 1200°C; 12 h in $^{16}O_2$ plus 12 h in $^{18}O_2$. The SEM micrograph shows SIMS sputter craters in the two-layered scale and the corresponding SIMS depth profiles (top) from the inner oxide layer (left crater) and (bottom) from the outer oxide layer (right crater).

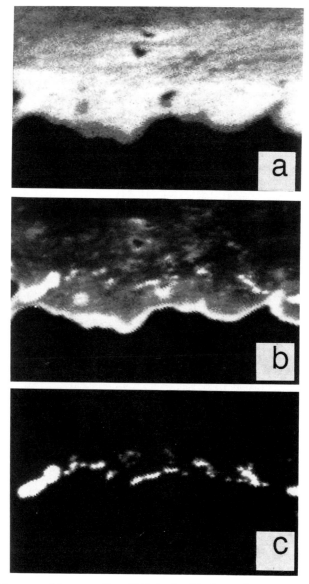

Fig. 3 SIMS images of taper section (2.5×) of oxide formed on NiAl + 0.1%Y after 24 h oxidation at 1200°C; 12 h in $^{16}O_2$ plus 12 h in $^{18}O_2$. (a) $^{16}O^-$ image, (b) $^{18}O^-$ image, and (c) Y^+ image. Regions rich in the particular element or isotope appear white.

$^{16}O_2$ and then in $^{18}O_2$ it is found that much of the ^{18}O is below the ^{16}O although some enrichment of ^{18}O at the surface is interpreted as being due to cation transport along the ridges. The zirconium concentrates and becomes oxidised at the scale–substrate interface. Figure 4 shows a taper section of a mature scale formed after 200 h on NiAl + 0.2% Zr. In this case it can be seen that most of the ^{18}O is close to the scale-substrate interface or within channels

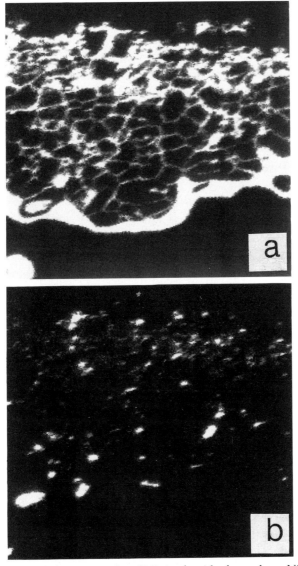

Fig. 4 SIMS images of taper section (2.5×) of oxide formed on NiAl + 0.2% Zr after 200 h oxidation at 1200°C; 100 h in $^{16}O_2$ plus 100 h in $^{18}O_2$. (a) $^{18}O^-$ image, and (b) Zr^+ image. Regions rich in the particular isotope or element appear white.

or boundaries (Fig. 4a), indicating inward transport of oxygen and growth within the scale. There is a lesser amount of ^{18}O on the outside of the scale which shows that the presence of zirconium has not completely stopped the outward diffusion of aluminium. The SIMS image in Fig. 4(b) indicates the distribution of the zirconium. It is now dispersed through the scale as ZrO_2 particles. The particles are finer in the outer portion of the scale and are coarser and more widely spaced close to the scale–substrate interface, probably reflecting the reduced rate of growth of the oxide as the film thickens.

4. Summary

This paper has demonstrated the use of secondary ion mass spectrometry (SIMS) in high-temperature oxidation studies. The techniques illustrated include high-resolution depth profiling, planar imaging and imaging of taper sections. The SIMS results indicate that the early stages of growth of α-Al$_2$O$_3$ on β-NiAl occur by outward transport of aluminium through the ridges of oxide. The later stages of growth occur by a combination of outward aluminium transport and inward oxygen transport along short-circuit paths. Reactive element additions of yttrium and zirconium produce dispersions of oxidised reactive element particles in the α-Al$_2$O$_3$ layer. The major effect of the reactive elements is to limit the outward transport of aluminium.

Acknowledgment

The authors thank J. Doychak of the NASA Lewis Research Center for helpful discussions and for kindly providing the β-NiAl, NiAl + 0.1%Y and NiAl + 0.2% Zr alloys, J. W. Fraser for his assistance with the SEM examination and R. J. Hussey for preparing the taper sections.

References

1. J. Doychak, T. E. Mitchell and J. L. Smialek, *Mat. Res. Soc. Symp. Proc.*, **39** (1985), 475.
2. J. Doychak, J. L. Smialek and T. E. Mitchell, *Met. Trans.*, **20A** (1989), 499.
3. J. Doychak and M. Rühle, *Oxid. Met.*, **31** (1989), 431.
4. R. Prescott, D. F. Mitchell, G. I. Sproule and M. J. Graham, *Solid State Ionics*, **53** (1992), 229.
5. D. F. Mitchell, R. Prescott, M. J. Graham and J. Doychak, *Proc. 3rd Int. SAMPE Metals Conference* (October 1992), Toronto, Canada, M 78.
6. R. Prescott, D. F. Mitchell and M. J. Graham, Paper 133, *Corrosion 92*, Nashville, USA. Published by the National Association of Corrosion Engineers.
7. E. W. A. Young and J. H. W. de Wit, *Oxid. Met.*, **26** (1986), 351.
8. P. A. van Manen, E. W. A. Young, D. Schalkoard, C. J. Van Der Wekken and J. H. W. de Wit, *Surf. Int. Anal.*, **12** (1988), 391.
9. J. Jedlinski and S. Mrowec, *Mater. Sci. Eng.*, **87** (1987), 281.
10. S. Mrowec and J. Jedlinski, *Trans. Metall. Soc. AIME*, **247** (1988), 1099.
11. R. Prescott, D. F. Mitchell, J. W. Fraser and M. J. Graham, *Proc. 3rd Int. Symp. on High Temperature Corrosion* (May 1992), Les Embiez, France (in press).
12. J. Doychak, J. L. Smialek and C. A. Barrett, NASA Tech. Memo. 101455 (1988).
13. C. A. Barrett, *Oxid. Met.*, **30** (1988), 361.

The Effect of a Zr Alloy Addition on the Oxidation Behaviour of β-NiAl: The Transition from Benefit to Breakdown

B. A. PINT, A. J. GARRATT-REED and L. W. HOBBS

H. H. Uhlig Corrosion Laboratory, Massachusetts Institute of Technology, Cambridge, MA 02139, USA

ABSTRACT

The oxidation behaviour of β-NiAl + 0.23wt%Zr was investigated from 1000°–1400°C in 1atm O_2 using a wide range of characterisation techniques. At 1200°C, the addition of Zr appears to significantly improve the oxidation behaviour, reducing the isothermal scale growth rate and improving scale adhesion. However, at 1400°C the addition of Zr causes isothermal breakdown and spalling during cyclic testing. Using FEG-STEM/XEDS, Zr has been found to segregate to oxide grain boundaries over the entire temperature range. This segregation is believed to cause a change in the oxidation mechanism and hence a reduction in the oxidation rate. Zr has also been found in the oxide scale near the gas interface as ZrO_2 particles. Both the amount of Zr segregation and the volume fraction of ZrO_2 increase with temperature. It is proposed that with a fixed Zr alloy content, by increasing the oxidation temperature, Zr is able to diffuse faster within the alloy and to diffuse faster along the grain boundaries in the scale, leading to a buildup of Zr-rich oxides at the scale–gas interface. At 1400°C, this enrichment leads to an eventual breakdown in both the isothermal and cyclic protectiveness of the $α$-Al_2O_3 scale. It is proposed that the optimum reactive element doping level is a function of the oxidation temperature.

1. Introduction

The serendipitous discovery that zirconia crucibles and moulds added enough Zr to Ni-based alloys to improve their oxidation behaviour[1] has led to its common use as a reactive element (RE) addition.[2–10] In β-NiAl it is preferable to other RE additions such as Y because of its higher solubility. Studies on the effect of Zr on the oxidation behaviour of chromia- and alumina-formers have not always demonstrated a benefit.[11–13] However, in direct comparisons of alumina-formers (β-NiAl[14] and FeCrAl alloys,[15,16] Zr appears to provide the same beneficial effects as Y in terms of improving the scale adhesion and slightly reducing the oxidation rate. [18]O tracers have shown that the effect of Zr is analogous to that of Y in changing the growth mechanism of α-Al$_2$O$_3$ by reducing the outward transport of Al.[17,18]

Numerous hypotheses have been put forth to explain the RE effect, but few have been supported by characterisations of the scale microstructure. High-resolution microstructural and microchemical characterisation of RE-doped scales has consistently pointed to the grain boundary segregation of REs, such as Y and Zr, as a necessary condition for a suppression of cation transport in both α-Al$_2$O$_3$[18] and α-Cr$_2$O$_3$[19] scales and that adherent doped scales have an RE segregant at the metal oxide interface.[4-6] For alumina-formers, early characterisation studies concentrated on the segregation of Y in scales grown on Y-doped allows,[20-23] but it has also been observed that Zr segregates to grain boundaries in sintered bulk alumina[23] and in alumina scales grown on Zr-doped FeCrAl alloys,[14-16] β-NiAl[4,8] and Ni$_3$Al.[25]

Based on the initial segregation theory developed for chromia-formers,[26] an amended segregation theory has been proposed which recognises that the RE on the scale grain boundaries is not stagnant but at temperature is diffusing outward from the alloy, to the metal–oxide interface, to the oxide grain boundaries and finally concentrates at the scale–gas interface where RE-rich oxides are formed.[27] This study concentrates on the characterisation of the grain boundary segregation of Zr and the Zr-rich particles formed at the gas interface of α-Al$_2$O$_3$ scales grown at 1000°, 1200° and 1400°C. The goal is to improve the understanding of the RE effect and the role of grain boundary segregation but also to explore the possibility that the formation of a large volume fraction of RE-rich particles is detrimental to the oxidation behaviour.

2. Experimental Procedure

β-NiAl ingots were obtained from the NASA Lewis Research Center. Two compositions were cast in a water-chilled copper mould, undoped and Zr-doped (0.23wt%) NiAl (50at%Al $\approx$ 31.5wt%Al). Alloy coupons ($\approx$15 $\times$ 15 $\times$ 0.8 mm) were polished through 0.3 μm alumina and ultrasonically cleaned in acetone and alcohol prior to oxidation. Isothermal weight gain was measured with a Cahn model 1000 microbalance. Cyclic weight changes were measured after cooling to room temperature with a Mettler model AE163 balance. Isothermal experiments were performed at 1000°, 1200° and 1400°C in a dry, flowing atmosphere of pure O$_2$. Cyclic experiments were performed in the same apparatus. Samples were quickly (2 s) removed from the furnace for cooling to room temperature.

After oxidation, specimens were characterised using glancing angle X-ray diffraction (GAXRD), scanning electron microscopy with X-ray energy-dispersive spectroscopy (SEM/XEDS), transmission electron microscopy (TEM) and field-emission gun scanning-transmission electron microscopy (FEG-STEM) equipped with XEDS. The STEM work was done with a VG Microscopes Ltd

HB5 instrument at 100 kV and a VG HB603 at 200 kV. Both instruments have a ≈2 nm electron probe and a windowless XEDS detector. Specimens were sectioned parallel to the metal–oxide interface, ion-milled so as to examine the oxide either near the gas interface or near the metal interface.[28] In order to examine the metal–oxide interface, samples were also made perpendicular to the metal–oxide interface.[14,29]

To simplify the issue of grain boundary segregant quantification, a basic method was chosen. The probe size and the grain boundary width were assumed to be constant and the beam spreading ignored.[30] Thus, the ratio of grain boundary volume to grain volume is constant, independent of sample thickness. Weight percentages were calculated from the XEDS spectra using the Cliff and Lorimer[31] method of thin film analysis. Weight percentages were ignored when the net count was less than twice the square root of the gross count. The ratio of the cation weight percentages was compared on and away from the boundary, e.g. Zr wt%/Al wt%.

In order to quantify the volume fraction of ZrO_2 particles in the scale, X-ray maps of Zr and Al were used. The area of particles on the Zr X-ray map was divided by the area detected in the Al map. The Al map was used so as not to count the areas due to voids and/or milling.

3. Results

The oxidation behaviour of Zr-doped NiAl at 1000°C is complicated by the transient formation of θ-Al_2O_3.[32] After 1 h at 1000°C, the scale formed on Zr-doped NiAl analysed using GAXRD is approximately 90% θ and 10% α, while that on undoped NiAl averages 25% θ.[33] [The percentage was obtained by a simple primary peak height comparison of θ-Al_2O_3 ($2\theta = 32.8°$ with Cu Kα) and α-Al_2O_3 (57.5°).] Thus in the early stages of oxidation, a Zr addition appears to stabilise the first-forming θ phase. The early formation of the fast-growing θ phase is reflected in the kinetic data (Fig. 1). However, the steady-state growth appears to be controlled by the formation of α-Al_2O_3. Using GAXRD, the scale after 50 h at 1000°C is fully α-Al_2O_3. The RE effect is not clearly observed at 1000°C because the Zr addition has little effect on the oxidation rate. Also, limited cracking and spallation were observed on both doped and undoped samples. Thus there is no clear improvement in oxide adhesion because presumably some of this cracking is a result of the phase transformation.[8]

In order to avoid the complications of the θ–α phase transformation, TEM parallel sections were made from samples oxidised for 50 h at 1000°C where the scale is fully α-Al_2O_3. Zr-rich particles were not uniformly distributed in the scale, large areas contained virtually no particles while others contained clusters of particles (Fig. 2). Voids are present throughout the sample, most likely a result of the θ–α phase transformation which involves a volume reduction of 13%.[8] Some small Ni-rich oxide particles were also observed, presumably formed during the transient stage. The amount of grain boundary segregation and size and fraction of the Zr-rich particles measured after 50 h at

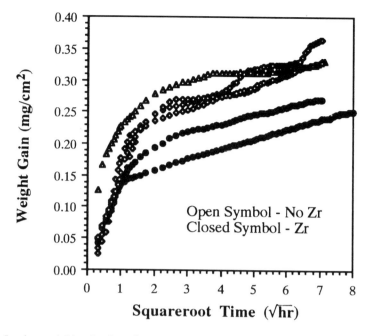

Fig. 1 Isothermal kinetic data from Zr-doped and undoped β-NiAl oxidised at 1000°C in 1atm O_2 showing two regimes, the first for the growth of θ-Al_2O_3 and the second for α-Al_2O_3.

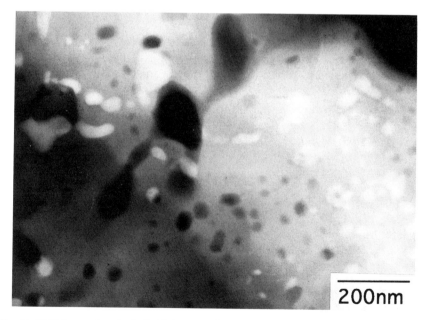

Fig. 2 STEM annular dark field image of the scale formed on β-NiAl + Zr after oxidation for 50 h at 1000°C. Parallel section of the scale near the gas interface.

Location/conditions	Gas interface	Metal interface
1000°C, 50 h	0.030	n/a
1200°C, 50 h	0.048	n/a
1400°C, 1 h	0.110	0.155

Table 1 The average apparent XEDS weight percentage ratio (Zr/Al) on the α-Al$_2$O$_3$ grain boundaries in scales grown on Zr-doped NiAl at several different temperatures. Measurements were taken on parallel sections near either the gas or metal interface.

Temperature	Average particle area (nm^2)	Average particle fraction (%)
1000°C	165	2.9
1200°C	459	4.4
1400°C	1089	14.7

Table 2 The average ZrO$_2$ particle area and particle fraction measured in the scale grown on Zr-doped NiAl at several different temperatures. Measurements were taken on parallel sections near the gas interface.

1000°C are given in Tables 1 and 2. The average particle fraction calculated at 1000°C is the average of several areas ranging from 0% to 7%.

The RE effect was more easily observable at 1200°C. In isothermal tests, a clear reduction in the oxidation rate was evident when Zr was added. The reduction in the parabolic rate constant is by a factor of ≈3, comparable to the reduction observed in Zr-doped FeCrAl alloys, and for Y-doped alumina-formers.[14,15,18] The isothermal weight gains are plotted in Fig. 3 with the weight change over 20 2-h cycles. The Zr addition appears to significantly improve the adherence. More extensive cyclic testing has shown that this benefit lasts for thousands of hours.[7]

Parallel sections were made of the scale after oxidation for 50 h at 1200°C. The scale near the gas interface contained a fairly uniform dispersion of Zr-rich particles in a dense α-Al$_2$O$_3$ matrix (Fig. 4). The volume fraction of Zr-rich particles was large enough to be detected using GAXRD; however, the peaks were too small to identify a phase of ZrO$_2$ conclusively. (The peaks most closely matched the tetragonal phase, JCPDS card 17,923.) Compared to 1000°C, the amount of Zr in the scale has increased with more Zr segregated to the grain boundaries and a larger fraction of Zr particles in the scale (Tables 1 and 2).

The oxidation behaviour was significantly different at 1400°C. Undoped β-NiAl exhibited parabolic behaviour throughout a 50 h isothermal exposure. However, the scale on Zr-doped β-NiAl appeared to break down after ≈20 h (Fig. 5a). This breakdown was also observed for Y-implanted β-NiAl samples,

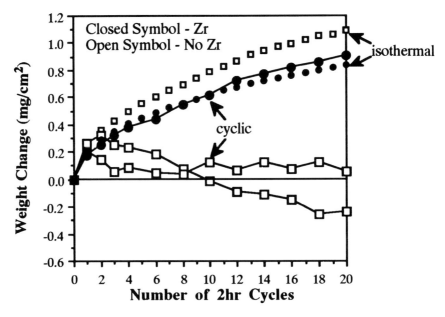

Fig. 3 Isothermal weight gains plotted with the weight change over 20 2-h cycles at 1200°C in 1atm O_2. The Zr addition appears to significantly improve the oxide adherence.

Fig. 4 STEM bright field image of the Al_2O_3 scale on NiAl + Zr after oxidation at 1200°C for 50 h. Parallel section near the gas interface. The dark particles are rich in Zr by XEDS.

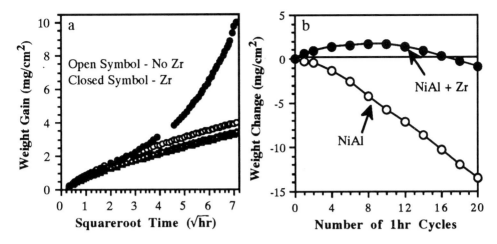

Fig. 5 Oxidation behaviour of Zr-doped and undoped NiAl at 1400°C in 1 atm of O$_2$. (a) Isothermal kinetic data; (b) Cyclic weight changes during 20 1-h cycles. While the Zr alloy addition improves the oxide adherence for the first 10 cycles, thereafter it begins to break down.

and involved the formation of Ni-rich oxides.[14] In cyclic testing of 20 1-h cycles, the undoped β-NiAl sample lost weight after every cycle (Fig. 5b). In the case of NiAl + Zr, the oxide appeared to be somewhat protective for the first ≈10 cycles. However, during the next 10 cycles, a significant breakdown occurred, similar to the breakdown observed in the isothermal exposure. This type of early breakdown had not been observed during similar cyclic tests of β-NiAl + 0.04at% Zr.[9]

At 1400°C, the scale was characterised after a 1 h oxidation in order to avoid the complication of Ni-rich oxides formed at longer times and to look for a source of the eventual breakdown. In this case, the Zr particles were large enough to be easily identified using SEM/XEDS (Fig. 6). As at 1200°C, GAXRD peaks most closely matched the tetragonal ZrO$_2$ phase. The amount of Zr segregant and the particle fraction in the scale again increased with temperature (Tables 1 and 2). The scale contained almost 15% ZrO$_2$ as well as numerous large voids, most likely a result of the rough gas interface. A typical X-ray map of the scale formed at 1400°C is shown in Fig. 7. Scale grain boundaries near the metal interface contained an even higher amount of segregant, with a Zr/Al ratio of 0.155. With this high level of segregant, the grain boundaries appear bright in the annular dark-field image due to atomic number contrast. However, no Zr-rich particles were observed near the metal interface.

The brittle β-NiAl substrate makes transverse section preparation very difficult. To date, the best success has been obtained with a sample oxidised for 50 h at 1000°C. As was detected for a ZrO$_2$-dispersed FeCrAl alloy (Kanthal alloy APM),[14–16] Zr was found to segregate to the metal–oxide interface. In this case, a high-resolution X-ray map was able to detect the Zr along the interface

Fig. 6 SEM secondary electron image of the scale on NiAl + Zr after oxidation at 1400°C for 1 h. The sub-micron particles on the surface are rich in Zr (by XEDS).

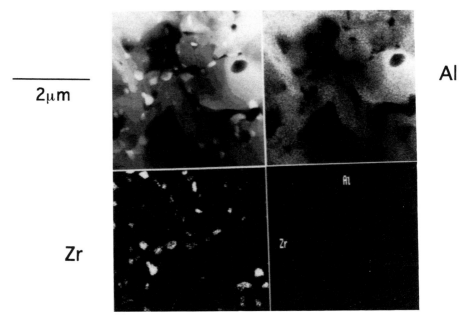

Fig. 7 High-resolution X-ray map of the scale on β-NiAl + Zr near the gas interface after oxidation for 1 h at 1400°C. (1) STEM annular dark field image; (2) Al X-ray map and (3) Zr X-ray map.

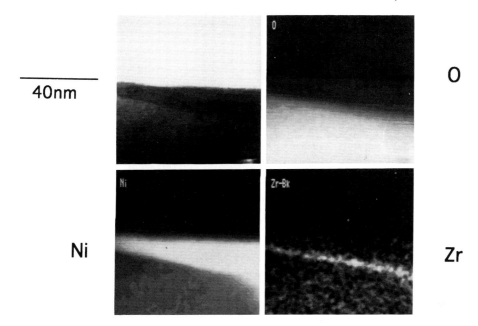

40nm

O

Ni

Zr

Fig. 8 High-resolution X-ray map of the metal–oxide interface after oxidation of NiAl + Zr for 50 h at 1000°C. (1) STEM bright field image, (2) O X-ray map, (3) Ni X-ray map and (4) Zr X-ray map. The background X-ray signal was subtracted from the Zr map.

(Fig. 8). An extensive search was made in the transverse section for S on the metal–oxide interface and for S in the metal but none was detected.

4. Discussion

It appears that while Zr-doping has beneficial effects on the oxidation behaviour of β-NiAl, there are also certain limitations on these improvements. At 1000°C, the Zr addition appeaars to be of some benefit but also detrimental in the stabilisation of the fast-growing θ phase. Scale cracking appears to be exacerbated by the retention of the θ phase and subsequent θ–α transformation. RE additions have been shown to slow the θ–α transformation in both bulk alumina powders (for catalysts)[34,35] and in oxide scales.[33,36]

At 1400°C, the isothermal breakdown only observed in the Zr-doped β-NiAl points to some detrimental role of the RE addition, and not something intrinsic to the oxidation of β-NiAl. There are a number of possible explanations for this breakdown. For example, the RE effect could have some upper temperature bound, above which an RE addition is ineffective. However, similar testing of a Y_2O_3-dispersed FeCrAl alloy (Inco MA956)[14,37] and a ZrO_2-dispersed FeCrAl alloy (Kanthal APM)[14,38] at 1400°C showed a consistent improvement in oxide adhesion in cyclic testing compared to an undoped FeCrAl alloy. Therefore, it is expected that the Zr addition should improve the oxidation behaviour at 1400°C. Another possibility is that a phase transformation in the ZrO_2 particles may cause the detrimental behaviour. However, a phase transformation is not likely to occur during an isothermal exposure, making this an improbable explanation for the observed isothermal breakdown at 1400°C.

Thus, the focus has centred on the amount of Zr in the scale, both in Zr-rich particles and segregated to the oxide grain boundaries. The quantity of Zr in the scale after oxidation at 1400°C is significantly higher, compared to the more protective scale formed at 1200°C. The reason for the increased Zr content in the scale is attributed to the outward diffusion of Zr in the metal and in the scale (along oxide grain boundaries). The driving force for the outward diffusion is the oxygen potential gradient in the scale. Thus, as the temperature increases, more Zr is found in the outer scale, possibly due to a higher activation energy for the diffusion of Zr in the scale compared to the rate of scale growth. The scale thickness after 50 h at 1200°C and 1 h at 1400°C is almost equivalent, slightly more than 5 μm. Thus, for a similar amount of scale growth, the ZrO_2 fraction in the outer scale has increased by a factor of more than 3 (Table 2).

RE-rich particles in alumina scales have been observed by other studies with Y on MA956[20] and Zr on Ni_3Al.[39] The formation of the Zr-rich particles is believed to occur due to saturation of the grain boundaries with Zr.[27] Internal Zr oxidation and subsequent incorporation of the oxides by the inward-growing scale is a possible explanation; however, this would result in particles throughout the scale and no particles were observed in a parallel section near the oxide-metal interface. Also, the higher Zr grain boundary segregation near the metal interface compared to that measured near the gas interface implies that Zr is being incorporated into particles rather than staying on the boundaries near the gas interface (Table 1).

The breakdown of the protective α-Al_2O_3 scale may occur due to some critical ZrO_2 fraction in the scale or a sufficient quantity of Zr on the grain boundaries. Wang and Kroger[40] proposed that second-phase particles can speed the transport of oxygen in α-Al_2O_3. Faster O boundary transport could lead to Al depletion in the metal and subsequent fast oxidation of Ni. It is also possible that the ZrO_2 particles ripen with time causing cracking in the scale as they grow.

It has been proposed that a minimum amount of RE segregation is required to affect the oxidation behaviour.[19] The opposite case is that too much segregant may be detrimental to the boundary transport, perhaps enhancing the transport of O by increasing the effective boundary width.[40] The grain boundary segregation detected at 1400°C, averaging 0.155 (Zr/Al) near the metal–oxide interface, could substantially alter the structure of the grain boundary. Depending on the amount of beam broadening, this could represent segregation from $\approx$30% to 100% of the cation sites on the boundary. Based on the observations from 1000°–1400°C, it appears that a Zr content which is effective at lower temperatures may be detrimental at higher temperatures. Thus, choosing an optimum amount of Zr may be a function of the application temperature.

Location/conditions	Gas interface	Metal interface
1000°C, 100 h	0.018	0.006
1200°C, 100 h	0.130	n/a
1400°C, 25 h	0.038	0.018

Table 3 The average apparent XEDS weight percentage ratio (Zr/Al) on the α-Al$_2$O$_3$ grain boundaries in scales grown on Kanthal APM (FeCrAl + ZrO$_2$) at several different temperatures. Measurements were taken on parallel sections near either the gas or metal interface.[38]

A comparison can be made with Kanthal alloy APM, a Fe-2wt%-4.5%Al alloy with a ZrO$_2$ oxide dispersion of approximately 0.2vol% (0.1wt%Zr). In the case of APM, the Zr addition appears to be beneficial throughout the temperature range, with no breakdown or loss of adhesion at 1400°C.[14,38] In this case, Zr is also observed to segregate to the scale grain boundaries over the temperature range of 1000°–1400°C (Table 3)[38] but no quantification has been made of the particle fractions in the scale. While the amount of segregation varies widely (possibly differing due to different exposure times,[27] the most striking difference is the lower amount of Zr grain boundary segregation at 1400°C, perhaps explaining why no detrimental behaviour is observed with APM.

There are two possible explanations for the lower segregation in the scale on APM. First, the Zr outward diffusion may be restricted in APM. The Zr in APM is not 'free' in the alloy like Zr in NiAl. Instead, it is bound in very stable oxides, possibly limiting the diffusion from the alloy into the oxide. Second, there is significantly less Zr in APM (0.1wt%) than in the β-NiAl (0.23wt%), which could also reduce the amount of Zr diffusing to the gas interface.

Further observations are required to fully establish a relationship among Zr content, temperature and oxidation behaviour. However, the current observations do illustrate the capabilities of high-resolution analytical electron microscopy and the need for further work in this area. Currently, the same characterisation techniques are being applied to studying the effect of time and Zr alloy content on the grain boundary segregation and Zr-rich particle formation. Also this effect will be further studied at 1500°C. The microstructural relationship between Zr and S[5] has not been established but these techniques appear to be the most useful in determining the role of RE additions on scale adhesion.

5. Conclusions

1. For the oxidation of Zr-doped β-NiAl, Zr is found in the scale as Zr-rich oxides and as ions segregated to the α-Al$_2$O$_3$ grain boundaries and to the metal–oxide interface.
2. The quantity of Zr in the α-Al$_2$O$_3$ scale increases with temperature, especially in the volume of Zr-rich oxide particles. The apparent XEDS

weight percentage ratio Zr/Al on the oxide grain boundaries also increases with temperature.

Acknowledgments

The authors are very grateful to Dr J. Doychak at the NASA Lewis Research Center for providing the undoped NiAl and Zr-doped NiAl. Among our colleagues at MIT, thanks goes to Mr J. Adario for the GAXRD measurements, and to Mr G. Arndt for assistance with the oxidation equipment. The Electric Power Research Institute (contract No. RP242644) provided financial support.

References

1. C. A. Barrett and C. E. Lowell, *Oxid. Met.*, **11** (1977), 199–223.
2. A. S. Kahn, C. E. Lowell and C. Barrett, *J. Electrochem. Soc.*, **127** (1980), 670–679.
3. J. Doychak, T. Mitchell and J. L. Smialek, *MRS Symp. Proc.*, **39** (1985), 475–484.
4. S. Taniguchi and T. Shibata, *Oxid. Met.*, **25** (1986), 201–216.
5. J. L. Smialek, in *High Temp. Mat. Chem. IV* (Munir *et al.*, eds), Electrochemical Society, Pennington, NJ (1988), pp. 241–249.
6. J. L. Smialek, J. Doychak and D. J Gaydosh, in *Oxidation of High-Temperature Intermetallics* (T. Grobstein and J. Doychak, eds), TMS, Warrendale, PA, 1989, pp. 83–95.
7. C. S. Barrett, *Oxid. Met.*, **30** (1988), 361–390.
8. J. Doychak, J. L. Smialek and T. E. Mitchell, *Met. Trans.*, **20A** (1989), 499–518.
9. J. Doychak, C. A. Barrett and J. L. Smialek, in *Corrosion & Particle Erosion at High Temperatures* (V. Srinivasan and K. Vedula, eds), TMS-AIME, Metals Park, OH, 1989, pp. 487–514.
10. D. R. Sigler, *Oxid. Met.*, **32** (1989), 337–355.
11. M. H. Lagrange, A. M. Huntz and J. H. Davidson, *Corr. Sci.*, **24** (1984), 613–627.
12. M. Landkof, A. V. Levy, D. H. Boone, R. Gray and E. Yaniv, *Corr. Sci.*, **41** (1985), 344–362.
13. P. Y. Hou and J. Stringer, *Oxid. Met.*, **29** (1988), 45–73.
14. B. A. Pint, Ph.D. Thesis, Massachusetts Institute of Technology, Cambridge, MA (1992).
15. B. A. Pint, K. C. Wills, A. J. Garratt-Reed and L. W. Hobbs, *Electrochem. Soc. Extended Abstracts*, **91-2** (1991), 913–914.
16. B. A. Pint and L. W. Hobbs, submitted to *J. Electrochem. Soc.* (1993).
17. K. P. R. Reddy, J. L. Smialek and A. R. Cooper, *Oxid. Met.*, **17** (1982), 429–449.
18. B. A. Pint, J. R. Martin and L. W. Hobbs, *Oxid. Met.*, **39** (1993), 167–195.

19. C. M. Cotell, G. J. Yurek, R. J. Hussey, D. F. Mitchell and M. J. Graham, *Oxid. Met.*, **34** (1990), 173–200 and 201–216.
20. T. A. Ramanarayanan, M. Raghavan and R. Petkovic-Luton, *J. Electrochem. Soc.*, **13** (1984), 923.
21. K. Przybylski, A. J. Garratt-Reed, B. A. Pint, E. P. Katz and G. J. Yurek, *J. Electrochem. Soc.*, **134** (1987), 3207–3208.
22. B. A. Pint, A. J. Garratt-Reed and G. J. Yurek, unpublished research (1988).
23. B. A. Pint, A. J. Garratt-Reed and L. W. Hobbs, in 'Proceedings of the 3rd International Symposium on the High Temperature Corrosion and Protection of Materials', *R.J. Physique* (1992), in press.
24. C.-W. Li and W. D. Kingery, in 'Structure and properties of MgO and Al_2O_3 ceramics', *Advances in Ceramics*, **10**, Amer. Cer. Soc., Columbus, OH (1984), 368–378.
25. B. A. Pint, A. J. Garratt-Reed and L. W. Hobbs (these Proceedings).
26. K. Przybylski and G. J. Yurek, *Materials Science Forum* **43** (1989), 1–74.
27. B. A. Pint and L. W. Hobbs, The dynamic role of interfacial segregants in high temperature oxidation, to be presented at the Spring meeting of the Electrochemical Society, Honolulu, Hawaii (May 1993), *Electrochem. Soc. Extended Abstracts*, **93-1** (1993), in press.
28. L. W. Hobbs and T. E. Mitchell, in *High Temperature Corrosion*, NACE-6 (R. A. Rapp, ed.), NACE, Houston, TX (1983), 76–83.
29. W. E. King, N. L. Peterson and K. Reddy, *Proc. 9th Int. Cong. Met. Corr.*, **4** (1984), 28–35.
30. W. Furdanowicz, A. J. Garratt-Reed and J. B. Vander Sande, *Inst. Phys. Conf.*, **119** (10), (1991), 437–441.
31. G. Cliff and G. W. Lorimer, *J. Microscopy*, **103** (1975), 203–207.
32. G. C. Rybicki and J. L. Smialek, *Oxid. Met.*, **31** (1989), 276–304.
33. B. A. Pint, A. Jain and L. W. Hobbs, *MRS Symp. Proc.* (1993), in press.
34. H. Schaper, E. B. M. Doesburg and L. L. Van Reijen, *App. Catalysis*, **7** (1983), 211–220.
35. P. Burtin, J. P. Brunelle, M. Pijolat and M. Soustelle, *App. Catalysis*, **34** (1987), 225–38.
36. B. A. Pint, A. Jain and L. W. Hobbs, *MRS Symp. Proc.*, **213** (1991), 981–986.
37. B. A. Pint, A. J. Garratt-Reed and L. W. Hobbs, in 'Proc. 3rd International Symp. on High Temp. Corrosion and Protection of Materials', *R. J. Physique* (1992), in press.
38. B. A. Pint, A. J. Garratt-Reed and L. W. Hobbs, unpublished research (1993).
39. J. Doychak and M. Ruhle, *Oxid. Met.*, **31** (1989), 431–452.
40. H. A. Wang and F. A. Kroger, *J. Amer. Cer. Soc.*, **63** (1980), 613–619.

The Oxidation and Embrittlement of α_2 (Ti$_3$Al) Titanium Aluminides

J. RAKOWSKI, D. MONCEAU, F. S. PETTIT and G. H. MEIER

University of Pittsburgh, Pittsburgh, PA, USA

R. A. PERKINS

Lockheed Palo Alto Research Laboratory, Palo Alto, CA, USA

ABSTRACT

The oxidation behaviour of binary and Nb-modified α_2 alloys has been studied as a function of temperature (700–900°C) and atmosphere. The oxidation kinetics have been characterised by thermogravimetry and the oxidation morphologies have been characterised by a wide variety of metallographic techniques. Additionally, the penetration of interstitials into the alloys has been followed using hardness measurements and the effect of interstitial penetration on mechanical properties has been evaluated. The addition of Nb slows the rate of oxidation as the result of doping of the TiO$_2$ in mixed titania–alumina scales which form on the alloys. The oxidation rates are generally lower in air than oxygen because of nitrides formed at the oxide/alloy interface during oxidation in air. The formation of nitride layers, which is more profuse for alloys containing Nb, also limits interstitial hardening of the underlying alloy. Interstitial penetration severely embrittles Nb-modified α_2, particularly in atmospheres containing water vapour. The mechanisms of oxide formation and interstitial embrittlement are discussed.

1. Introduction

Alloys in the Ti–Al system are of interest for high-temperature systems such as aircraft engines because they have low density and substantial high-temperature strengths. However, their resistance to oxidation and interstitial embrittlement is a concern. The oxidation behaviour of α_2-Ti$_3$Al is the subject of this paper.

1.2 Thermodynamics

An important aspect of the oxidation of Ti-aluminides, compared to the aluminides of Ni and Fe, is the small difference in standard free energy of

476

formation between alumina and the oxides of titanium, which is accentuated by a negative deviation from ideal solution behaviour in the Ti-Al system. The aluminum activity is much smaller than unity in Ti$_3$Al and TiAl. In fact, combining the activities with standard free energy data for the oxides indicates that TiO is more stable in contact with the alloy than is Al$_2$O$_3$ for atom fractions of Al much less than about 0.5.[1,2] Thus, Al$_2$O$_3$ is unstable in contact with binary α_2 and is only marginally more stable than TiO in contact with γ-TiAl.

1.3 Oxidation of Ti$_3$Al (α_2)

The oxidation of Ti$_3$Al alloys would not be expected, in light of the above thermodynamic considerations, to form continuous alumina scales. Instead they form mixed rutile-alumina scales.[3] The oxidation kinetics of Ti$_3$Al between 600 and 960°C are reported to be those expected for rutile growth.[4,5]

Alloying of Ti$_3$Al with β-stabilising elements, particularly Nb, reduces the oxidation rate.[6] The scales developed on α_2 alloys containing multiple additions of β-stabilisers have been described by Wallace *et al.*[7] and Schaeffer.[8]

An additional aspect of the oxidation of Ti$_3$Al alloys is dissolution of oxygen into the alloy at the scale/alloy interface. The embrittlement associated with this phenomenon can be more damaging to the mechanical properties than the surface recession caused by scale formation in the temperature range where Ti$_3$Al will likely be used (<700°C).[9]

2. Experimental

Alloys with compositions Ti-21Al and Ti-21Al-11Nb(at%) were supplied by Alta Group. Large ingots were vacuum-arc melted in water-cooled copper moulds and were forged and rolled to 12.7 mm (Ti$_3$Al) and 6.35 mm (Ti,Nb)$_3$Al plates. Samples for oxidation experiments were cut from the as-rolled plates. A portion of the (Ti,Nb)$_3$Al plate was re-rolled at 800°C to a 2.0 mm thick sheet followed by surface conditioning on both sides to a thickness of 1.5 mm to provide mechanical property specimens. All oxidation specimens were polished through 600 grit SiC. Oxidation experiments were carried out in pure O$_2$ and air at temperatures from 700 to 900°C. The oxidation kinetics were studied using a Cahn Model 2000 microbalance, the oxide morphologies were studied using scanning electron microscopy with energy- and wavelength-dispersive X-ray analysis (EDS and WDS) and Auger electron spectroscopy (AES), and phases were identified by X-ray diffraction (XRD).

The effect of environmental exposure on the mechanical properties of (Ti,Nb)$_3$Al was evaluated in 3-point bend tests at room temperature. Specimens 7.6 mm wide × 38.1 mm long × 1.5 mm thick were annealed in vacuum (10^{-5} torr) for 2 h at 900°C. Surface contamination from the vacuum anneal was removed by acid pickling 0.127 mm from each side. The surfaces were then polished through 600 grit and specimens were exposed to a variety of atmospheres at temperatures between 600 and 900°C. The environmental exposures were conducted in two different test rigs. Most of the exposures were conducted in a closed-end vertical tube furnace using a 2-h flow of the test gas to displace air (purge). The specimens were lowered into the heated test zone, held for the desired time, and withdrawn to cool in the test atmosphere. The test gas was flowed continuously at a rate of 1 litre/min. The test gases were commercial purity bottled gases. A second series of exposures was conducted with ultra-high-purity bottled gases containing <3 or <1 ppm of moisture. The exposures were conducted in a vertical quartz tube furnace heated by radiant lamps. The furnace was evacuated to 10^{-5} torr and heated to 900°C to outgas all surfaces. The atmosphere was then introduced to 1 atm pressure and the specimen was lowered into the hot zone. Gas was flowed through the hot zone continuously at a rate of 0.1 litre/min. After exposure the specimens were tested for strength and ductility by a 3-point bend test using a 2t radius punch and a 25.4 mm wide span. All mechanical tests were conducted at room temperature with a ram speed of 25.4 mm/min. The tests were stopped at 90° residual bend angle if fracture had not occurred. The strength and elongation were calculated from the load and deflection at the time of failure or the maximum bend angle without failure.

3. Results and Discussion

3.1 Oxidation Rates and Morphologies

Figure 1 presents the results of oxidation experiments for both alloys in air and 1 atm. oxygen at 700, 800, and 900°C. The results may be summarised as follows:

(a) At 700°C the rates for both alloys are substantially lower in air than oxygen.
(b) At 800 and 900°C the rates for the binary alloy become comparable in air and oxygen while those for the ternary alloy remain lower in air than oxygen.
(c) For any temperature and atmosphere the oxidation rate is considerably lower for the Nb-containing alloy than for the binary.

The oxidation morphologies of the exposed alloys were carefully evaluated in an attempt to understand the above phenomena. Figure 2(a, b) presents a cross-section of the binary alloy after 165 h oxidation in oxygen at 900°C. The scale consists of an outer layer of single-phase TiO_2 (rutile), which has grown outward, and an inner layer containing rutile and discontinuous alumina particles, which has grown inward. These two layers are separated by a layer of large alumina crystals. The growth of the scale in both directions is consistent with the reported presence of both Ti-interstitials and O-vacancies as significant defects in rutile.[10] The presence of the large alumina crystals indicates that Al is rapidly transported through the inner rutile layer since their size precludes transport through alumina, i.e. continuous alumina layers for these times and temperatures are typically in the 0.1 μm thickness range. Diffusion measurements using a colour-boundary migration technique in Al-doped rutile have been interpreted in terms of a small fraction of the Al being dissolved as interstitials with a very high diffusivity along channels in

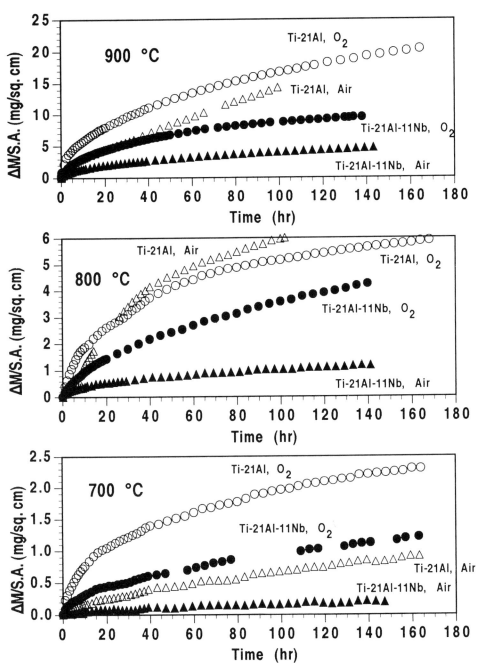

Fig. 1 Mass change versus time data for the oxidation of Ti-21Al and Ti-21Al-11Nb in air and oxygen at 700, 800, and 900°C.

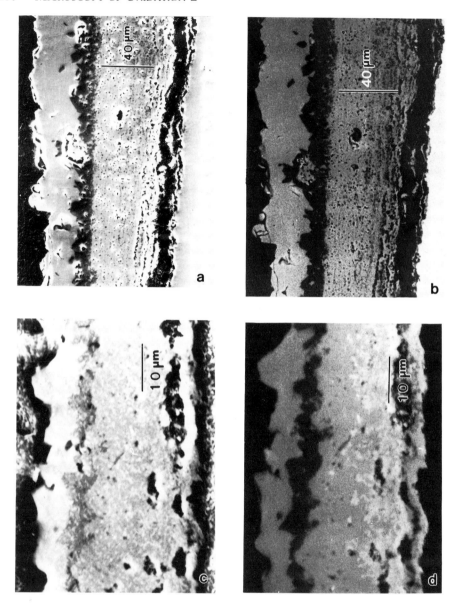

Fig. 2 Cross-sections of specimens oxidised in 1 atm. oxygen at 900°C. (a) is a secondary electron image and (b) a backscattered electron image for Ti-21Al oxidised for 165 hs. (c) is a secondary electron image and (d) is a backscattered electron image for Ti-21Al-11Nb oxidised for 140 h.

the [001] direction in the rutile lattice.[11] It is suggested that a similar process is operative in the growth of mixed scales on α_2.

Figure 2 (c, d) presents a cross-section of the Ti-Al-Nb alloy after 140 h oxidation in oxygen at 900°C. The oxide morphology is qualitatively the same as that for the binary but the scale is substantially thinner, which is consistent with the slower oxidation kinetics for the ternary. In addition, a substantial amount of Nb is found in the inward-growing portion of the scale. These data suggest that the effect of the Nb in slowing the kinetics in oxygen results from a doping effect. The dissolution of Nb^{+5} into TiO$_2$ would be expected to reduce the concentration of both Ti-interstitials and O-vacancies.

Figure 3 (a, b) presents a cross-section of the binary alloy after 100 h oxidation in air at 900°C. The scale morphology is essentially the same as that observed in oxygen except that there is some indication of the presence of nitrogen at the scale/alloy interface. However, the nitrogen-containing phase could not be identified.

The scale on the ternary alloy oxidised in air (Fig. 3c) was rather different from the scales described above. This scale consisted of only one layer which was mainly TiO$_2$ containing a small amount of Al throughout and Nb near the alloy/oxide interface. Discrete particles of alumina were not resolved. In addition a continuous layer of TiN was found at the alloy/oxide interface. This phase was identified by XRD and by a distinct gold colour under the optical microscope. A high concentration of nitrogen was confirmed in this layer by WDS profiles (Fig. 3d) and by scanning Auger line profiles. A thin layer denuded of the β-phase (white phase in Fig. 3c) formed below the nitride layer. The origin of this layer is not clear since no depletion of Nb was observed but may be the result of dissolved oxygen. The significant point, however, is that the presence of Nb has resulted in the formation of a continuous titanium nitride layer which has apparently blocked Al transport and, coupled with the doping effect of Nb in the TiO$_2$, caused the overall rate of oxidation to be greatly reduced. The mechanism whereby Nb affects the nitride formation is still under study.

Figure 4 shows the results of microhardness trace taken on the cross-sections of both alloys after oxidation in air and oxygen for 1 week. The area under the curve is much smaller for both alloys oxidized in air rather than oxygen. This suggests that oxygen has a much higher permeability in α than does nitrogen and that the presence of nitrogen at the metal/oxide interface reduces the dissolution of oxygen.

The major features in the growth of mixed oxides on Ti$_3$Al and (Ti,Nb)$_3$Al are summarised schematically in Fig. 5. Becker et al.[12] have identified similar processes for the growth of mixed oxides on γ-TiAl.

3.2 Mechanical Properties

There is an increasing amount of data which indicates that many intermetallic compounds are susceptible to embrittlement in oxidising atmospheres. The ductility of FeAl, Fe$_3$Al, and Co$_3$Ti have been found to be much lower when tested in air at room temperature as compared with testing in vacuum.[13] This

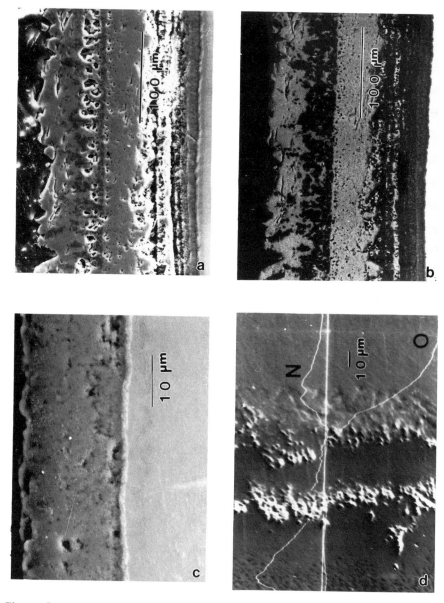

Fig. 3 Cross-sections of specimens oxidised in 1 atm. air at 900°C. (a) is a secondary electron image and (b) a backscattered electron image for Ti-21Al oxidised for 100 h. (c) is a secondary electron image for Ti-21Al-11Nb oxidised for 145 h. (d) is a taper section of Ti-21Al-Nb oxidised for 145 h with WDS traces for oxygen and nitrogen superposed.

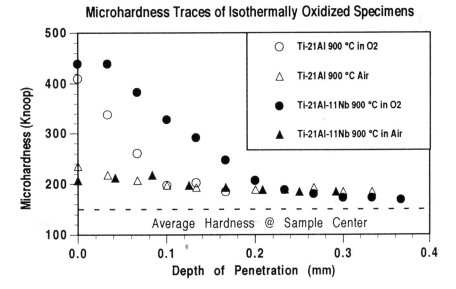

Microhardness Traces of Isothermally Oxidized Specimens

Legend:
- ○ Ti-21Al 900 °C in O2
- △ Ti-21Al 900 °C Air
- ● Ti-21Al-11Nb 900 °C in O2
- ▲ Ti-21Al-11Nb 900 °C in Air

Y-axis: Microhardness (Knoop), 100 to 500

Average Hardness @ Sample Center

X-axis: Depth of Penetration (mm), 0.0 to 0.4

Fig. 4 Microhardness traces across the zone just below the oxide for the specimens presented in Figs. 2 and 3.

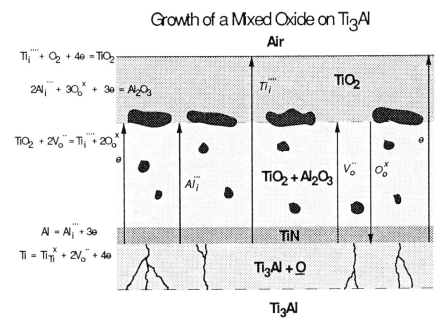

Growth of a Mixed Oxide on Ti$_3$Al

Air

$Ti_i^{\cdots} + O_2 + 4\theta = TiO_2$

$2Al_i^{\cdots} + 3O_o^x + 3\theta = Al_2O_3$

$TiO_2 + 2V_o^{\cdots} = Ti_i^{\cdots} + 2O_o^x$

$Al = Al_i^{\cdots} + 3\theta$

$Ti = Ti_{Ti}^x + 2V_o^{\cdots} + 4\theta$

TiO_2 $Ti_i^{\cdots}$

$TiO_2 + Al_2O_3$ $V_o^{\cdots}$ O_o^x θ

$Al_i^{\cdots}$

TiN

$Ti_3Al + \underline{O}$

Ti$_3$Al

Fig. 5 Schematic diagram of the transport processes and point defect reactions proposed to occur in the growth of mixed titania–alumina scales on α_2-Ti$_3$Al alloys.

Exposure conditions	Elongation (%)
Vacuum, 2 h at 900°C	
As-annealed	1.9
0.127 mm removed	>18.0, >26.5, >22.2
0.254 mm removed	>20.6, >27.9, >24.3
Vacuum, 24 h at 900°C	
As-annealed	3.4
0.076 mm removed	12.2
0.127 mm removed	12.3

Table 1 Contamination embrittlement in vacuum.

effect has been ascribed to the introduction of atomic hydrogen into the compound from water vapour in the test atmosphere. Similarly, Ni_3Al[13,14] and Ti_3Al[9] have been shown to be embrittled by oxygen at elevated temperatures.

The $(Ti,Nb)_3Al$ plate was re-rolled at 800°C to produce a fine-grain, two-phase ($\alpha_2 + \beta$) microstructure which is known to have high ductility at room temperature. The as-rolled and surface-conditioned plate was ductile in bending. A 2t bend with a 90° residual angle (>20% elongation) was attained readily without fracture. However, Table 1 shows the sheet was brittle after a 2 h anneal in vacuum. It fractured at a 1.9% elongation and greatly reduced strength level. When 0.127 mm was removed from each side by acid pickling the annealed sheet was ductile in bending and did not fail on bending to a 135° angle (>26.5% elongation). This experiment shows that superficial contamination of the surface of <0.127 mm occurred during the 2 h anneal in vacuum. The basic alloy is ductile in the annealed condition but was embrittled by a thin layer of contamination, most likely from residual air in the vacuum system. When the vacuum exposure was increased to 24 h at 900°C the depth of contamination increased such that ductility was only partially restored when 0.127 was removed from each side by acid pickling (Table 1).

A series of exposures was then conducted in air, O_2, N_2, CO_2, CO-18%CO_2, and argon for 24 h at 900°C in an attempt to determine which species were responsible for the surface contamination and embrittlement. Exposures were also conducted in air, O_2, and N_2 mixed with 31–46% H_2O as steam to determine the effect of moisture on embrittlement. The results are summarised in Table 2. The elongation was decreased to 6.87%, 3.06%, and 4.37% in exposures to air, O_2, and CO–CO_2, respectively. This indicates that oxygen diffusion into the alloy results in severe embrittlement. Furthermore, those atmospheres which contained water vapour (in the range of 30 to 40 vol%) showed the lowest ductilities. The specimens failed at the proportional limit without yielding. Metallographic studies showed the first crack to form on bending propagated through the specimen. This further suggests that the release of hydrogen when water vapour reacts with the alloy at elevated temperatures may result in embrittlement similar to that proposed for other compounds at room temperature.[13]

Exposure conditions	Elongation (%)	
	24 h-900°C Purged system	20 h-900°C Evacuate and backfill
Argon <1 ppm H$_2$O <10 ppm H$_2$O	 3.60	 >20.3
Nitrogen <1 ppm H$_2$O <32 ppm H$_2$O	 1.55	 17.0
Oxygen <3 ppm <50 ppm	 3.06	 3.60
Air <3 ppm <50 ppm	 6.87	 2.70
CO$_2$ Liquified 82% CO – 18% CO$_2$	 1.42 4.37	 – –
Air – 46% H$_2$O Oxygen – 37% H$_2$O (82% CO – 18% CO$_2$) – 31% H$_2$O H$_2$ – 39% H$_2$O	1.30 1.91 1.35 1.11	– – – –

Table 2 Contamination embrittlement in various gases.

Similar contamination embrittlement was observed at lower temperatures (600, 700, and 800°C) if the alloy was exposed for a sufficiently long time.

The foregoing results were obtained by the use of a flowing gas to displace air from the furnace before exposure (purged). Analysis of the data indicates that either not all of the air was removed by the purge or that small amounts of moisture in the atmosphere was a source of contamination. Samples exposed to flowing N$_2$ with 32 ppm H$_2$O formed a rutile scale over a thin TiN interface layer and were embrittled. It could not be determined which of the atmosphere's components was the major source of contamination embrittlement. Therefore, a second series of exposures was conducted in ultra-pure gases (<1–3 ppm H$_2$O) in a system that could be evacuated and outgassed to remove residual air and moisture. Results are summarised in Table 2. The alloy was not embrittled by exposure to argon or nitrogen with <1–3 ppm H$_2$O for 24 h at 900°C. A thin TiN scale without surface oxides was formed in the ultra-pure nitrogen. The samples were embrittled on exposure to CO$_2$ and air with 1–3 ppm H$_2$O. This indicates strongly that oxygen contamination is the primary source of embrittlement. The addition of water in large amounts appears to be harmful as a result of H$_2$ embrittlement effects. Moisture also may be harmful at lower concentrations (10–50 ppm) but results are inconclusive. Further work on the effect of moisture is needed.

4. Summary

Mixed titania–alumina scales were formed on Ti_3Al and $(Ti,Nb)_3Al$ under all conditions studied. The addition of Nb slows the rate of oxidation as the result of doping of the TiO_2 in the scales. The oxidation rates are generally lower in air than oxygen because of nitrides formed at the oxide/alloy interface during oxidation in air. The formation of nitride layers, which is more profuse for alloys containing Nb, also limits interstitial hardening of the underlying alloy. Interstitial penetration severely embrittles Nb-modified α_2. Oxygen clearly embrittles the alloy and hydrogen (from H_2O)) may also play a role. Nitrogen was found not to cause embrittlement.

References

1. A. Rahmel and P. J. Spencer, *Oxid. Met.*, **35** (1991), 53–68.
2. K. L. Luthra, *Oxid. Met.*, **36** (1991), 475–490.
3. J. M. Rakowski, Senior Thesis, University of Pittsburgh (1992).
4. N. S. Choudhury, H. C. Graham, and J. W. Hinze, in *Properties of High Temperature Alloys* (Z. A. Foroulis and F. S. Pettit, eds.), pp. 668–680 (1976), The Electrochemical Society.
5. M. G. Mendiratta and N. S. Choudhury, 'Properties and Microstructure of High-Temperature Materials', AFML-TR-78-112, Contract No. F33615-75-C-1005 (Systems Research Laboratories, Inc., Ohio, August 1978).
6. M. Khobaib and F. W. Vahldiek, 'High temperature oxidation behavior of Ti_3Al alloys', *Second International SAMPE Metals Conference* (F. H. Froes and R. A. Cull, eds.), Covina, CA (1988), pp. 262–270.
7. T. A. Wallace, R. K. Clark, K. E. Wiedemann, and S. K. Sankaran, *Oxid. Met.*, **37** (1992), 111–124.
8. J. C. Schaeffer, *Scripta Met.*, **28** (1993), 791–796.
9. S. J. Balsone, 'The effect of elevated temperature exposure on the tensile and creep properties of Ti-24Al-11Nb', in *Oxidation of High Temperature Intermetallics* (T. Grobstein and J. Doychak, j. eds.), pp. 219–234 (1989), TMS.
10. P. Kofstad, *High Temperature Corrosion* (1988), p. 295, Elsevier, London.
11. J. A. S. Ikeda, Y. Chiang, and B. D. Fabes, *J. Amer. Ceram. Soc.*, **73** (1990), 1633–1640.
12. S. Becker, A. Rahmel, M. Schorr, and M. Schutze, *Oxid. Met.*, **38** (1992), 425–464.
13. C. T. Liu and C. G. McKamey, 'Environmental embrittlement – a major cause for low ductility of ordered intermetallics', in *High Temperature Aluminides and Intermetallics* (S. H. Whang, C. T. Liu, D. P. Pope, and J. O. Stiegler, eds.), p. 133 (1989), TMS.

14. J. H. DeVan and C. A. Hippsley, 'Oxidation of Ni$_3$Al below 850°C and its effect on fracture behavior', in *Oxidation of High Temperature Intermetallics* (T. Grobstein and J. Doychak, j. eds.), pp. 31–40 (1989), TMS.

SNMS Investigations Concerning the Effect of Niobium Additions on the Oxidation Behaviour of Titanium Aluminides

W. J. QUADAKKERS, A. ELSCHNER,[1] N. ZHENG, H. SCHUSTER, H. NICKEL

Research Centre Jülich, Institute for Reactor Materials, PO Box 1913, 5170 Jülich, Germany

[1] *Bayer AG, 4150 Krefeld, Germany*

ABSTRACT

The effect of niobium additions on the oxidation behaviour of TiAl-base intermetallics at 800°C in air was investigated. For this purpose titanium-aluminium alloys with and without niobium addition were studied during cyclic oxidation up to exposure times of 3000 h. The scale growth mechanisms during the early stages of oxidation were studied by a two-stage oxidation technique using ^{18}O and ^{15}N tracer. The scales formed during this oxidation process were analysed by SNMS. Nitrogen was found to form titanium and titanium–aluminium nitrides at the alloy/oxide interface. Niobium was mainly incorporated in the titanium dioxide and in the nitride-containing layer at the oxide/metal interface. The positive effect of niobium on the oxidation resistance is believed to be caused by a decreased oxygen vacancy concentration in the titanium dioxide due to niobium doping of the rutile lattice. The ^{15}N-tracer showed that the nitrogen is transported through the oxide layer as well as the nitride-containing sub-scale. It could not be established whether the transport occurs in ionic or molecular form.

1. Introduction

The potential applications of intermetallic phases on the basis of γ-TiAl as construction materials are significantly limited by their poor high-temperature oxidation resistance.[1,2,3] A special property of TiAl-based intermetallics is that

488

the oxidation rates in air considerably differ from those in oxygen.[4] It is generally accepted that additions of the element niobium significantly increase the oxidation resistance of γ-TiAl.[4,5,6,7] Several possible explanations have been given for this effect; however, the exact mechanisms are not yet clarified.

In the present study, the formation and growth of oxide scales on γ-TiAl are being studied at 800°C. Emphasis was placed on the effect of niobium additions on the early stages of oxidation. For this purpose the oxide scales were analysed by secondary neutrals mass spectrometry (SNMS) with the aim to determine the niobium distribution in the various oxide phases and to analyse the incorporation of nitrogen in the corrosion scales.

2. Experimental

The alloys used in the present investigation were Ti50Al, Ti48Al2Nb, Ti47.5Al5Nb and Ti45Al10Nb (additions given in at. %). The alloys were produced by induction melting in an argon atmosphere. Specimens of 15 mm diameter and 2 mm thickness were prepared from the cast ingots and then ground and polished up to a 1 μm diamond finish. The oxidation and spalling kinetics were investigated in still air at 800°C up to exposure times of 3000 h. Specimens were examined before and after oxidation using optical metallography, scanning electron microscopy (SEM) with energy-dispersive X-ray analysis (EDA), electron probe microanalysis (EPMA) and X-ray diffraction (XRD).

The early stages of oxidation of alloy Ti47.5Al5Nb were studied in a two-stage oxidation method at 800°C. In this technique[8] the materials were first oxidised in air for 10 min, 0.5 h, 2 h, 5 h or 24 h, and subsequently in a gas consisting of 16% ^{16}O, 4% ^{18}O, 64% ^{14}N and 16% ^{15}N for 20 min, 1 h, 4 h, 10 h or 48 h. The scales formed during this oxidation process were then analysed by secondary neutrals mass spectrometry (SNMS) (Leybold, INA3). The primary energy of the Ar-plasma was 800 V.[9] The quantified oxygen and nitrogen tracer profiles were recalculated to data which would have been obtained if the gas in the second oxidation stage had contained 20% of ^{18}O and 80% ^{15}N instead of an enriched gas.[9]

3. Results

3.1 Effect of Niobium on Cyclic Oxidation Behaviour at 800°C

Figure 1 shows the oxidation behaviour of Ti50Al and the three niobium-containing alloys during oxidation at 800°C in air. The scale on the binary alloy already showed spallation at the first interruption of the experiment for weight measurement, i.e. the scale growth which actually occurred in the period up to the first interruption for weight measurement was significantly larger than indicated by the first data point in Fig. 1. All niobium-containing alloys showed smaller growth rates and a significantly better scale adherence. The scale growth rates decreased with increasing alloy niobium content.

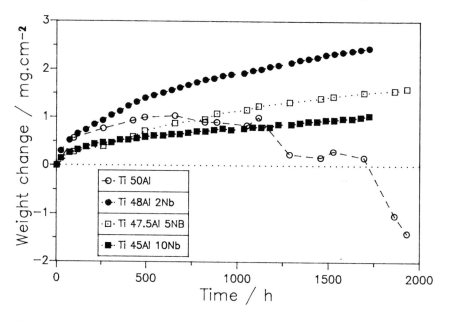

Fig. 1 Weight change as a function of time for Ti50Al and three niobium-containing titanium aluminides during oxidation at 800°C in air. Weight changes were measured after cooling to room temperature.

3.2 Mechanisms of Scale Formation

Figure 2 shows an SNMS depth profile of the oxide layer on the alloy with 5% Nb after two-stage oxidation for 10 min and 20 min at 800°C. The results show the formation of an alumina-based scale at the surface and at the interface with the metal. Between these a titanium-rich scale is formed. At the scale/metal interface a titanium-rich layer has been formed. Niobium is present only in the inner part of the scale. The corresponding oxygen tracer profiles (Fig. 3) illustrate that during the second stage of oxidation the scale has grown by outward cation diffusion and short circuit oxygen diffusion.[8]

Figure 4 shows the depth profiles after 0.5 h and 1 h two-stage oxidation. The aluminium oxide is still present at the scale surface whereas the aluminium peak in the inner part of the scale disappeared. An interesting difference with the results in Fig. 2 is that the nitrogen enrichment and the dip in the aluminium profile coincide after 30 min exposure whereas after 1.5 h they are separated.

After a total of 72 h oxidation the formation of a titanium oxide becomes visible on top of the initially formed outer alumina layer (Fig. 5). The middle part of the oxide scale contains smaller amounts of oxygen than the outer and

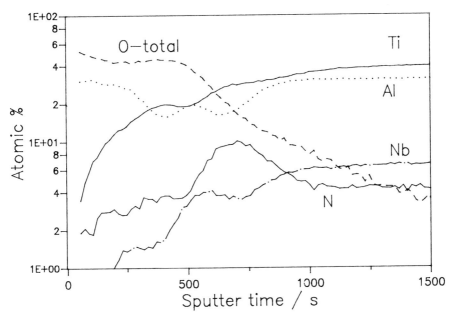

Fig. 2 SNMS analysis of Ti47.5Al5Nb after two-stage oxidation (10 min and 20 min) in air and air + $^{18}O/^{15}N$ at 800°C.

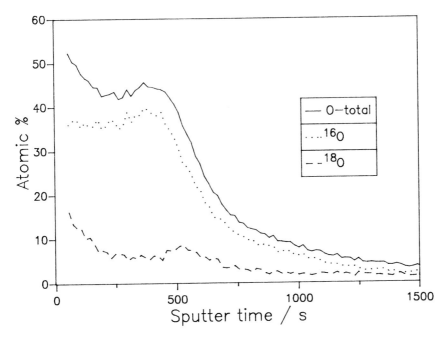

Fig. 3 As Fig. 2, showing oxygen isotope distributions.

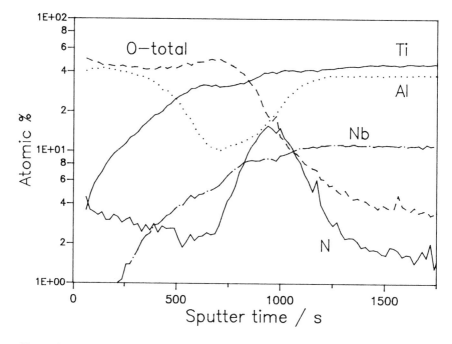

Fig. 4 SNMS analysis of Ti47.5Al5Nb after two-stage oxidation (0.5 h and 1 h) in air and air + ^{18}O/^{15}N at 800°C.

inner part (Fig. 6). In the latter areas of the scale the oxide is mainly titania which possesses a larger oxygen/metal ratio than alumina. The ^{18}O profile (Fig. 8) shows a more pronounced peak at the outer scale than after the shorter oxidation times (Fig. 3), apparently due to outward diffusion of titanium. An interesting observation is that the enrichment of ^{18}O in the inner part of the scale does not occur at the scale/metal interface but near the inner boundary between the titania and the alumina-rich part of the scale (Fig. 6): the relative minimum in the ^{16}O profile at a sputter time of around 1500 s is not visible in the ^{18}O profile. Analysis of the ^{15}N and ^{14}N profiles reveals that the two isotopes are not clearly separated although some indication was present that the ^{15}N is slightly more shifted to the inner part of the nitride-containing sub-scale (Fig. 8). The niobium is present only in the inner titanium-rich oxide and in the nitride-containing layer, whereas in the outer alumina and titania no niobium is detected. An interesting feature is that the niobium becomes enriched in the alloy immediately beneath the nitrogen-rich layer (Fig. 5) in agreement with recently published data after 15 h oxidation.[10]

4. Discussion and Conclusions

The long-term experiments confirm that the niobium-containing titanium aluminides possess considerably better oxidation resistance at 800°C than

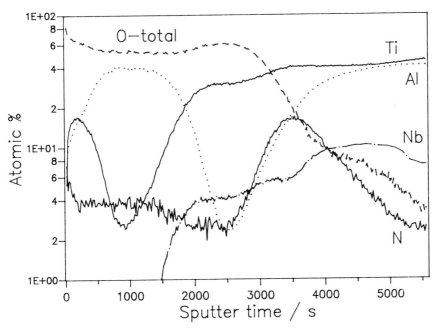

Fig. 5 SNMS analysis of Ti47.5Al5Nb after two-stage oxidation (24 h and 48 h) in air and air + $^{18}O/^{15}N$ at 800°C.

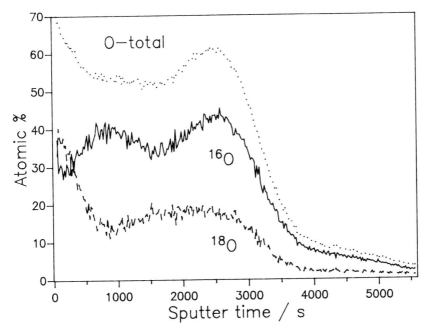

Fig. 6 As Fig. 5, showing total oxygen and oxygen isotope distributions.

Ti50Al (Fig. 1). As in the case of binary titanium aluminides[2,7] the surface scales are multiphase titania/alumina-based mixtures, with, however, a significantly lower growth rate. In the early stages of oxidation at 800°C, the alloy containing 5 at.% niobium forms an alumina-based surface scale (Fig. 2). At this stage of oxidation the scale formation is very similar to that of Ti50Al.[10] Beneath the outer alumina scale, a layer rich in titanium nitride is formed. Alumina formation leads to aluminium depletion and consequently to a decrease in aluminium activity in the sub-surface zone. The resulting increase in oxygen activity causes the alumina to become unstable at the oxide/alloy interface relative to titania[2] and therefore this latter oxide is being formed (Figs. 2 and 4). Initially this formation of titania beneath the alumina in turn leads to a relative increase in aluminium activity and subsequently to formation of aluminium-rich oxide between the titania and the nitrogen-containing layer (Fig. 2). On further exposure the alumina and titania at the inner scale side become intermixed (Fig. 4).

According to Becker *et al.*[11] titania has a significant solubility for alumina which increases with decreasing oxygen partial pressure, i.e. under the prevailing conditions the solubility would be larger in the inner scale side than near the scale/gas interface. This means that the initially formed alumina would be 'attacked' at the interface between the two oxides after longer exposure times. By this effect, the barrier function of the alumina is deteriorated and titanium cations can diffuse outward (Fig. 5). In this context it is interesting to note that the highest ^{18}O enrichment in the inner part of the scale is not present at the scale/metal interface but near the boundary between the alumina-rich scale and the inner titania-rich layer (Fig. 6). This might be related to the dissolution of alumina in titania; however, this could not yet be derived with certainty from the present results.

Niobium is present in the titania in the inner part of the scale and in the titanium nitride. A possible explanation for the slower growth rates of the oxide scales on the niobium-containing alloys compared to that on Ti50Al is that niobium decreases the oxygen vacancy concentration in the rutile lattice due to its higher valency compared to that of titanium. Consequently the oxygen inward diffusion decreases and so the scale growth is slower than in the case of Ti50Al.

For a correct interpretation of the presented depth profiles it is important to note that in SNMS studies nitrogen is an element which is difficult to quantify.[12] Therefore it is not absolutely sure that the maximum concentration of the nitrogen peak at the scale/metal interface is really around 10–20 at.% as measured in the presented SNMS profiles (Figs. 2, 4, 5, 7). Recent results of detailed microstructural studies using SEM and TEM[13] indicate that the nitrogen-containing layer consists of an alumina/nitride mixture. XRD studies showed that both TiN and $TiAl_2N$ were present in agreement with recent results from Becker *et al.*[11] The occurrence of a mixed oxide/nitride layer would be in agreement with the present data in which the nitrogen content was always around only 20 at.%. This result also explains why the ^{15}N and ^{14}N are nearly completely intermixed in the nitride-rich layer at the oxide/metal interface (Fig. 7). Whether the nitrogen is transported through the oxide scale

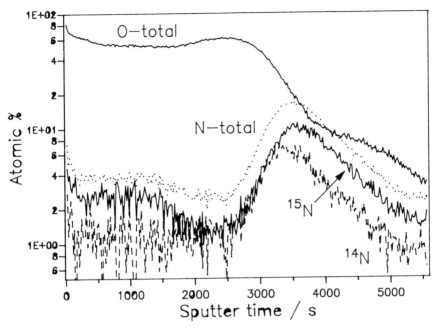

Fig. 7 As Fig. 5, showing total nitrogen content and nitrogen isotope distributions compared with total oxygen concentration.

in ionic or molecular form could not be derived with certainty. Previous SNMS studies of surface scales on $Ti50Al^{10}$ indicated some solubility for nitrogen in TiO_2; however, due to the difficulties in quantification of nitrogen in SNMS analyses this could not be concluded with certainty.

Acknowledgements

The authors are grateful to the 'Deutsche Forschungsgemeinschaft which has partly funded the present investigations. Thanks are due to Mr Beyess, KFA/IFF, and Dr Beavan, GKSS for supplying the alloys and Mr Baumanns, KFA/IRW, for carrying out the oxidation experiments.

References

1. M. Yamaguchi, High temperature Intermetallics, *Proc. Roy. Soc. Lond.* (1991), 15.
2. G. H. Meier, D. Apollonia, R. A. Perkins, and K. T. Chiang, in *Oxidation of High Temperature Intermetallics*, (T. Grobstein and J. Doychak, eds.), The Minerals Metals & Materials Society (1989), p. 185.

3. A. K. Misra, *Metall. Trans.*, **22A** (1991), 715.
4. M. S. Choudhury, H. C. Graham, and J. W. Hinze, in *Properties of High Temperature Alloys* (Z. A. Foroulis, F. S. Pettit, eds.), The Electrochemical Society (1976), p. 668.
5. J. Subrahmanyam, *J. Mat. Sci.*, **23** (1988), 1906.
6. G. Welsch and A. I. Kahveci, *Oxidation of High Temperature Intermetallics* (T. Grobstein and J. Doychak, eds.), The Minerals Metals & Materials Society (1989), p. 207.
7. U. Figge, W. J. Quadakkers, H. Schuster, and F. Schubert, EUROCORR 92 (Espoo, Finland, 31 May–4 June, 1992), Proceedings, Vol. 1, p. 591.
8. W. J. Quadakkers, H. Holzbrecher, K. G. Briefs, and H. Beske, *Oxid. Met.*, **32** (1989), p. 67.
9. W. J. Quadakkers, A. Elschner, W. Speier, and H. Nickel, *Applied Surface Science*, **52** (1991), 271.
10. U. Figge, A. Elschner, N. Zheng, H. Schuster, and W. J. Quadakkers, Arbeitstagung Angewandte Oberflächenanalytik, 22–25 June, 1992, Jülich FRG, Proceedings in Fresenius, *J. Anal. Chem.*, in print.
11. S. Becker, A. Rahmel, M. Schorr, and M. Schütze, *Oxid. Met.*, **38** (1992), 425.
12. K. H. Müller, K. Seifert, and M. Wilmer, *J. Vac. Sci. Technol.*, **A3** (1985), 1367.
13. W. J. Quadakkers and N. Zheng, Research Centre Jülich, Germany, to be published.

In situ Observation of Oxidation in Oxygen and Other Gases of Nb₃Al-based Alloys by Hot-stage Light Microscopy

ISAO TOMIZUKA, MIEKO OKAMOTO and AKIMITSU MIYAZAKI

National Research Institute for Metals, 2-3-12 Nakameguro, Meguro, Tokyo, Japan

ABSTRACT

Reaction of as-melted Nb-19Al alloy, a less brittle material among Nb₃Al-based alloys, was observed by heating it at a constant rate of 2 K/min in oxygen, nitrogen, Ar-20%O₂ and N₂-20%O₂ gases in a small furnace under a light microscope. The observed morphological changes were discussed with reference to information from SEM, EPMA and X-ray diffractometry.

In oxygen, apparent variation of the surface started at about 630 K, while weight increase was detected only at a temperature as high as about 1070 K. At the stage of the oxidation which was accompanied by the weight increase scale formation was pronounced at some places, while it was limited at other places. Subsequent SEM observation revealed a unique scale structure in the regions with a thick scale. A model is suggested to explain the formation of this peculiar structure, based on its cell structure of A15-phase enmeshed in a network of A2-phase.

In other gases, the process was qualitatively similar, but the reaction was slow. A slight effect of nitrogen gas was observed, especially on general appearance.

1. Introduction

The Japanese government is promoting a research project to develop a lightweight metallic material to be applied at 2073 K. The participants of the project have selected Nb₃Al-based intermetallics as suitable materials. A series of experimental alloys has been prepared, and these are being characterised from various points of view. Of these alloys, Nb-19at%Al was relatively easy to handle, because it is least brittle. The current presentation is concerned with its reactions in atmospheres containing oxygen and/or nitrogen in the course of heating up to 1273°K at a rate of 2 K/min.

2. Experimental

The original alloy was prepared as an ingot of about 10 kg by an induction skull-melting process. Experiments were carried out on it without further heat treatment. The ingot was roughly cut to a square rod of about 13 mm by an electro-spark machine, followed by slicing to sizes appropriate to thermomicroscopy or thermogravimetry by a wheel of silicon carbide in a wet condition. The specimen was a square tablet of 5 mm × 5 mm × 2 mm for thermomicroscopy and of 10 mm × 10 mm × 2 mm for thermogravimetry. These were polished by Buehler's 0000 polishing paper in a dry condition and rinsed with acetone immediately before the experiments. The microstructure of the alloy was not thoroughly homogeneous. It contained voids and other types of flaws. Some of the flaws might have been caused by the polishing, since niobium aluminide of a higher aluminium content is fairly brittle. The size of the crystal grains was approximately 0.8 mm. The alloy was mono-phase, as far as it was observed by SEM. Whether it is in equilibrium or in supersaturation is, however, not clear, since the phase diagram for the Nb–Al system is still a matter of controversy. The alloy should be in supersaturation according to the phase diagram suggested within our project,[1] while it should be in equilibrium according to that reported in reference 2. In any case, however, Nb-19Al alloy is located in a position not far from the Nb-side border of A-15 phase.

Reaction atmospheres were oxygen, nitrogen, Ar-20%O_2 or N_2-20%O_2. The oxygen and nitrogen were of ordinary Japanese technical grade. According to the specification from the supplier, oxygen content and the dew point of the nitrogen gas were 30 ppm and 213 K, respectively. The other gases were purchased from an outside supplier. They were compressed in steel bottles and reported to be purposely prepared from high-purity gases.

Thermomicroscopy was performed in a Leitz Microscope Heating Stage 1350. The image was observed through a TV camera and recorded on videotape for 2 s every 30 s. The heating was performed at a rate of 2 K/min except below 573 K, to which temperature the specimen was heated from room temperature in 10 min.

Thermogravimetry was performed under the same conditions as the microscopy, except for the initial heating to 573 K, which took 20 min. SEM observations and X-ray diffractometry were performed on the specimens which were prepared in the furnace employed for the thermomicroscopy.

3. Results and Discussion

3.1 Reaction in Oxygen

In case of oxygen atmosphere, weight changes in a thermobalance were detected only above 1070 K. The change between 1070 and 1273 K implied at least two cycles of formation-and-break-down of protective layers.

The colour or image of the specimen as observed through the TV system varied in the following 6 stages before temperature reached 1273 K (see Fig. 1).

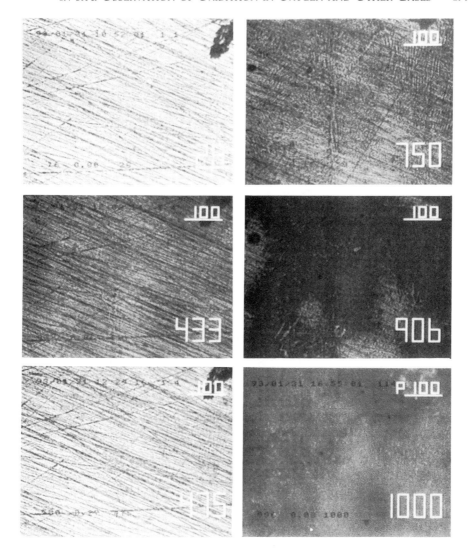

Fig. 1 Variation of surface observed by thermomicroscopy in oxygen (bars stand for 100 micrometer).

1. RT–590 K: As at room temperature. Flat image of magenta colour. A faint image of a lamellar structure was discerned on the screen. Since this type of structure was rare when the specimen was mirror-polished in a wet condition, it is believed to have originated (or at least intensified) by dry polishing with a polishing paper.
2. 590–720 K: TV screen was dark. The surface appeared to be covered with a very thin layer of fine downy material, which is likely to be an open mat of crystals of oxide(s).

3. 720–770 K: TV screen returned to the appearance of the first stage. This is likely because the downy crystals transformed to a very thin transparent film.
4. 770–950 K: Whole TV screen darkened gradually and a lamellar image was gradually obvious. A new scale appeared to be forming.
5. 950–1223 K: Whole TV screen darkened further. The lamellar region was so dark that almost nothing was visible. The dark region expanded rapidly.
6. 1223–1273 K: Whole TV screen was completely dark under normal illumination. Under polarised illumination, the former lamellar region appeared as if it were covered by a fog. This was due to protrusion of parts of the surface.

The surface of the specimen once heated to 1273 K appeared to be whitish-black to the naked eye at room temperature.

Peaks of theta-alumina and several niobium oxides were identified from X-ray diffractograms. A similar diffractogram was obtained from a specimen heated up to 1148 K.

SEM observation of the surface of the specimen previously heated to 1273 K revealed that the surface was composed of two types of structure (see Fig. 2). One (marked R) is rough, containing crevices running almost parallel with each other. Based on analogy of the configurations, they are supposed to be connected with the lamellar structure. The crevices appeared to be created by eruption from inside rather than erosion or corrosion from outside. The other (marked F) appeared flat at low magnification. Observation at a higher magnification revealed, however, a number of small eruptions being about to take place especially near the edges of regions with the rough structure. This type of structure suggests a process in which an initially formed scale was ruptured from within. An observation of a specimen heated to 1148 K showed a similar image.

SEM observation across a plane normal to the specimen surface revealed that the thickness of the scale was not even. On some parts of the surface, scale was as thick as 100 micrometer (see Fig. 3). Most of the thick scale had a unique structure which resembled woven textile fabric. The 'warps' were straight from the scale surface inwards and continued as bright bands into the base alloy. They disappeared gradually deep in the alloy.

Figure 4 illustrates maps of elements of a scale structure similar to that in Fig. 3. The map for aluminium reveals that the 'warps' are relatively low in aluminium content. This means that it is A2-phase. Since this type of bright strip was extremely rare in the alloy in the as-supplied condition, precipitation of A2-phase might have been facilitated by dissolution of oxygen in the A15-phase, accompanied by internal oxidation of aluminium. This view is supported by a considerable amount of oxygen in the sub-scale metal layer shown in the map of oxygen in Fig. 4. In Fig. 3, the oxidation front advanced a little deeper into the alloy in the 'warp' area than in the other areas. This agrees well with the supposition that the 'warp' is A2-phase, because it should be less protective due to its lesser aluminium content. The 'woofs' are, on the other hand, less straight and less continuous. The distance between the 'woofs' increased towards the outside. This is considered to be due to expansion of the

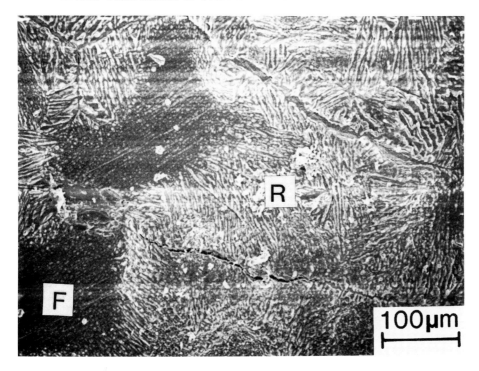

Fig. 2 An example of the surface after heating in oxygen to 1273 K.

'warp' in the process of further oxidation of suboxides of niobium to higher ones. This consideration agrees with the fact that niobium oxides of various degree of oxidation have been confirmed in X-ray diffractograms. The textile-fabric structure is maintained in the alloy side of the scale/alloy border. This suggests that the composition of the alloy between the 'warps' is not homogeneous and that there are channels connecting the 'warps', along which preferential oxidation proceeds by the oxygen transported via the 'warps'.

Based on the observation made so far, we can construct a model of the oxidation process proceeding in the thick-scale region. The model is schematically illustrated in Fig. 5. In this figure, left- and right-hand columns illustrate variations respectively of aluminium activity and of the phase structure in the scale/A15-phase cross-section.

Stage (a) stands for A15-phase enmeshed in a network of A2-phase. The network is formed likely because A2-phase has a high oxygen solubility,[3] which enables internal oxidation, followed by growth of A2-phase, deep in A15-phase. Stage (b) stands for further decrease in activity of metallic Al in the A15-phase adjacent to the scale due to Al-internal oxidation (see map of O in Fig. 4). As a result, the A15-phase in the sub-scale region is again supersaturated with Nb. Stage (c) stands for splitting of the A15-phase highly supersaturated with Nb into an A2-phase and a new A15-phase which is higher in activity of metallic Al than before. (A likely reason why the new A2-phase grows in parallel to the alloy surface is presumed to lie in the difference in the

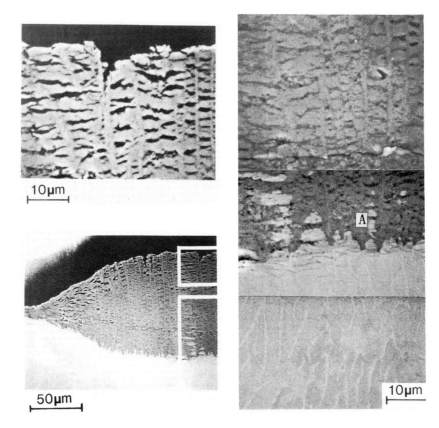

Fig. 3 Cross-section of a thick portion of the scale formed on a specimen heated in oxygen to 1273 K.

rate of splitting. Stage (a) has started at a much lower temperature.) This results in a trough of Al-activity between the newly formed and the original A15-phase regions, as is illustrated in Fig. 5(c).

Stage (d) stands for oxidation after the split. The A2-phase immediately below the top scale is oxidised rapidly because of its lower Al-activity. The A15-phase in the trough region is oxidised more rapidly than the A15-phase immediately above it for the same reason. This results in an enclosed metallic region of A15-phase. Their examples are specified by A in Fig. 3. Stage (e) stands for the enclosed A15-phase being finally oxidised. The situation near the scale/metal interface returns to Stage (a), leaving an advanced oxidation front.

The scale was thin in other parts of the specimen surface. In these parts, no fabric-weave scale was formed. Although the bright network of strips of A2-phase might be observed in the base alloy under the thin scale as well, their density was low. Casting flaws were seen more frequently in the alloy where the scale was thin. This evidence gives a clue to explain why the scale was thin there. As the casting flaws were formed in the region where consolidation proceeded in the last stage of cooling, this region has the lowest liquidus

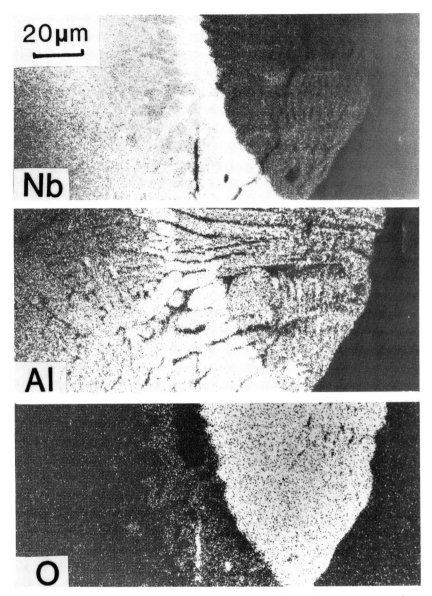

Fig. 4 Element maps of a thick portion of the scale formed on a specimen heated in oxygen to 1273 K.

temperature. Because the liquidus temperature decreases with increase in aluminium content in the region near Nb-19Al, this means that the aluminium content is relatively high in the region where casting flaws are abundant. This higher aluminium content caused a more protective scale, which retarded oxidation in later stages.

In addition to this straightforward effect of the higher aluminium content forming a more protective alumina layer, a locally higher aluminium content

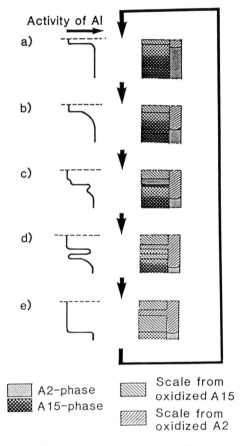

Activity of Al

a)

b)

c)

d)

e)

A2–phase

A15–phase

Scale from
oxidized A15

Scale from
oxidized A2

Fig. 5 Schematic model of a mechanism accounting for the structure observed in
Figs. 3 and 4.

leads for another reason to a thinner scale: the composition is located farther
from the A15-phase border (unless the alloy is in supersaturation). This results
in being less liable to precipitate the A2-phase which offers a preferred path for
oxygen transport.

3.2 Reactions in Other Gases

Reactions in other gases observed by thermomicroscopy were qualitatively
almost identical with that in oxygen. Every step from the first to the fifth was
observed during the heating, while the sixth step was missing in these cases.
Weight gain started at 1073 K as in oxygen. The increase was, however, smaller
as compared to that in oxygen. Relative weight gains were in the order of
nitrogen $<<$ N_2-20%O_2 $\leq$ Ar-20%O_2 $<<$ oxygen. Formation and break-down of
a protective layer was not obvious in the case of nitrogen.

The surfaces as observed by naked eye appeared slightly different, depending on whether the gas contained nitrogen or not. When gas did not contain

nitrogen, the specimen was whitish or exhibited a white powder over the surface. With nitrogen present, the surface was almost completely black, while something white deposited over the side face of the specimen (where convection was relatively common).

X-ray diffractogram from the specimen heated in gases containing nitrogen suggested the presence of some nitrides on the specimen surface, together with oxides. More detailed characterisation will be required, however.

SEM images of the specimen surfaces after heating up to 1273 K were little different from those for oxygen gas exposure.

The scale thicknesses were not uniform in these other gases. Thicker portions were, however, far less thick than for the case of oxygen. The 'textile-fabric weave' scale and narrow bright strips of A2-phase were seen in every gas, most likely because every gas contained oxygen, although the content was very small in the case of nitrogen. The scale thickness was almost negligible in some parts of the specimens. In these parts, the specimen has cracks or voids, implying these parts being more brittle than the other parts.

4. Conclusion

When Nb-19Al alloy, prepared by an induction skull melting, is subjected to a high-temperature atmosphere containing oxygen (even to a commercial nitrogen containing a slight amount of oxygen), a scale of unique structure is formed on part of its surface, presumably because the alloy is easily transformed to a cell structure of an A15-phase enmeshed by a network of A2-phase. The A2-phase suffers preferential oxidation. Through channels of the oxide of the A2-phase, oxygen is transported into the alloy. This causes oxidation of the A15-phase from within. Due to this cell structure, the alloy may be more vulnerable to oxidation than is expected from its aluminium content alone.

Acknowledgment

The authors acknowledge Dr G. Borchardt of Institute fuer Allgemeine Metallurgie, Technische Universitaet Clausthal, for his supply of useful information.

References

1. R. Suyama *et al.*, 'Microstructure and high temperature properties of Nb-Al intermetallic compounds', *Proc. of the 3rd Symposium on High-performance Material for Severe Environments*, pp. 141–149 (1992), Japan Ind. Tech. Assoc.
2. J. L. Jorda, R. Fluekiger and J. Muller, 'A new metallurgical investigation of the niobium aluminium system', *J. Less Comm. Met.*, **75** (1980), 227–239.
3. *Binary Alloy Phase Diagrams*, Vol. 2, p. 1684, American Society for Metals.

7

OXIDATION OF SILICON AND TITANIUM BASED CERAMICS

Oxidation of Powdered and Sintered Titanium Nitride

YU. G. GOGOTSI, G. DRANSFIELD* and F. PORZ[†]

Tokyo Institute of Technology, Research Laboratory of Engineering Materials, 4259 Nagatsuta, Midori-ku, Yokohama 227, Japan
*Tioxide Group Ltd, Billingham, Cleveland TS23 1PS, UK
[†]Institut für Keramik im Maschinenbau, Universität Karlsruhe, D-76128 Karlsruhe 1, Germany

ABSTRACT

Oxidation of three TiN powders prepared by plasma vapour phase synthesis and having the specific surface area of 33, 46 and 75 $m^2 g^{-1}$, as well as oxidation of pressureless-sintered bodies produced from these powders, was studied. The experiments were performed in flowing air under programmed heating up to 1500°C and under isothermal conditions. Oxidation of ultrafine powders starts above ~250°C and results in the formation of TiO_2 as anatase and brookite at lower temperatures, and as rutile at higher temperatures. The grain size can considerably affect the oxidation rate of TiN powders. The surface of the oxidised sintered TiN samples in most cases was covered by a layer of highly textured rutile crystals. The formation of multi-layered oxide films was observed above 1000°C. The growth of such rutile scales may be based on the increased grain boundary or volume diffusion of titanium ions, recrystallisation and sintering processes in the outer scale at high temperatures, as well as on the gas (N_2, NO_x) evolution at the oxide/substrate interface or exfoliation of the oxide layers due to internal stresses.

1. Introduction

TiN is used as a metallurgical coating, a hard refractory material and as a component of ceramic matrix composites. In the last cases, fine particles are needed to improve the sinterability, mechanical and electrophysical properties. A number of studies of the processing, sinterability and properties of ultrafine TiN powders have been conducted.[1,2] There exists also information on the oxidation of powdered[3–6] and dense[7–11] titanium nitride. Studies of a TiN powder[6] show that its oxidation starts at about 500°C. The oxidation of an ultrafine TiN can occur even at room temperature[3] and is accelerated above 250°C.[4] Pronounced oxidation of dense TiN starts at ~700°C, but an attack at grain boundaries has already started at 400°C.[11]

509

However, the information on the oxidation of ultrafine powders is extremely limited. The influence of the particles size on the oxidation resistance of the powders and dense TiN material has not been reported. There are wide variations in the observed morphology of oxidation products and explanations of the oxidation mechanisms.

The purpose of this study was to investigate the oxidation behaviour of ultrafine TiN powders and sintered materials, produced from the same powders. This is accomplished by discussing the possible oxidation mechanisms and their influences on the structure of the oxide scale.

2. Materials and Experimental

In the present work the oxidation behaviour of three TiN powders (Table 1) and sintered specimens produced from these powders (Table 2) was studied. The ultrafine TiN powders were produced by DC-plasma synthesis based on the vapour phase reaction of the titanium chloride with ammonia:

$$TiCl_4 + 2NH_3 = TiN + 4HCl + 1/2N_+ H_2 \quad (1)$$

They were single phase with an NaCl-like structure. The lattice constants (Table 1) were close to that of stoichiometric TiN (0.4240 nm) in spite of a relatively high oxygen content (~6%). Pressureless-sintered TiN samples were prepared from ultra-fine plasma synthesised powders with various surface

Powder	Lot	Specific surface (m^2/g)	Lattice constant (Å)	Nitrogen content (%)
TTN 25	NP 92/274	33	4.236 ± 0.002	19.26
TTN 50	NP 92/275	46	4.237 ± 0.002	N/A[b]
TTn 70	NP 92/276	75	4.237 ± 0.001	N/A[b]

[a]Tioxide Group PLC.
[b]N/A, not analysed.

Table 1 Characterisation of powders[a]

Material	Initial powder	Density (g/cm^3)	Oxygen content (%)	Nitrogen content (%)
TAJ/1	TTN 25	5.15	6.3	19.17
TAJ/5	TTN 50	5.03	7.4	N/A[a]
TAJ/7	TTN 70	5.00	N/A	N/A

[a]N/A, not analysed.

Table 2 Characterisation of sintered materials

areas by sintering without additives under flowing nitrogen at 1600°C, with a soak time of 1 h.[1,2] Ramp rate up and down was 10°C/min from/to 1000°C. The properties of sintered materials are described in Table 2. All samples possessed residual porosity after sintering, the pores mostly had a regular shape (Fig. 1a), corresponding to that of the TiN crystals. The highest density and optimum properties (strength ~500 MPa, hardness ~15 GPa and fracture toughness ~4 MPa) were reached for TAJ/1 at the smallest specific surface of the initial powder. The grain size of this material was less than 5 μm.

Oxidation was studied in flowing air under programmed heating up to 1500°C and under isothermal conditions (the heating rate up to oxidation temperature was 2–20°C/min, cooling rate was ~10°C/min) with 1.55 mm * ~4 mm * ~18 mm specimens using a Netzsch thermobalance capable of $2*10^{-5}$ g resolution. All samples were ultrasonically cleaned in acetone and weighed with a microbalance before and after oxidation. For the investigation of the diffusion processes during TiN oxidation, Pt-marker was put onto the surface of TAJ/1 samples (Fig. 1b) by magnetron sputtering.

Samples before and after tests were examined by X-ray diffraction (XRD) using Cu K_α-radiation (Siemens D500) and scanning electron microscopy (SEM) (Cambridge Stereoscan S4-10).

3. Results and Discussion

3.1 Oxidation of Ultrafine Powders

As can be seen in Fig. 2, the mass gain of TiN powders starts at 240–260°C. At lower temperatures a mass loss due to the desorption of water vapours and adsorbed gases from the surface of the powders occurred. At the heating rate of 2°C/min the oxidation was completed at 520–580°C. The oxidation rates for powders increased with increasing specific surface area. The full transformation to TiO_2 occurred at lower temperatures for TiN with a higher specific surface.

XRD investigations of the oxidised powders show the formation of TiO_2, thus, confirming the reaction

$$2 \text{ TiN} + 2 \text{ O}_2 = 2 \text{ TiO}_2 + \text{N}_2 \tag{2}$$

over the whole temperature range studied. Other titanium oxides were not found. However, different modifications of TiO_2 were formed depending on the oxidation temperature. Only brookite and anatase were found at 300°C. Rutile appears at 350°C and its content grows with increasing temperature. The content of brookite and anatase in the oxidation products decreases with increasing temperature. The diffraction lines of brookite disappear from the diffraction patterns at 400°C, and the lines of anatase at 1000°C. The increase of the size of TiO_2 particles with increasing temperature results in a sharpening of the diffraction lines. After heating to 1300°C with a rate of 2°C/min the size of rutile particles reached 1 μm and sintering of the powder began (Fig. 3a). After heating to 1500°C the powder was sintered into a very hard agglomerate consisting of dense granules ~100 μm in diameter (Fig. 3b).

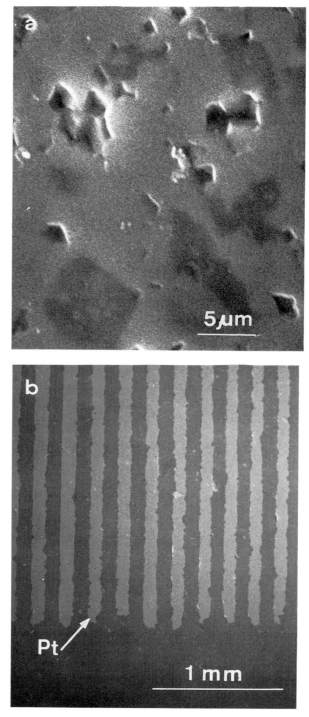

Fig. 1 Electron micrographs of the microstructure (a) and Pt marker (b) on the surface of a TAJ/1 specimen.

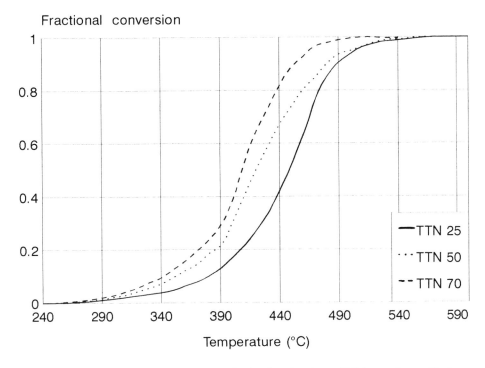

Fig. 2 The thermogravimetric curves obtained at heating of TiN powders with the rate of 2°C/min.

3.2 Oxidation of Sintered TiN

As can be seen in Fig. 4, a noticeable mass gain of sintered TiN under non-isothermal conditions starts at ~700°C and up to 1000°C its rate is very slow. Above 1000°C the mass gain increases and at 1150–1500°C shoulders on the TG-curves and some exothermal peaks on DTA curves in conjunction with the peaks on DTG curves were observed (Fig. 4). Thus, an acceleration of the oxidation occurred repeatedly. These changes in the oxidation behaviour can be explained by changes of oxidation mechanisms and/or by cracking of the oxide layer. The formation of TiO_2 as rutile confirms reaction (2). The differences in structure, porosity and composition of the samples prepared from different powders resulted in different oxidation behaviour (Fig. 5). The TAJ/7 sample heated with the rate of 2°C/min was disrupted during heating (sharp increase of the mass gain above 1400°C in Fig. 4) and completely oxidised (Fig. 6). Perhaps inner oxidation of this specimen, having a higher porosity than other materials and strong internal stresses arising on oxidation, caused the rupture of the sample. However, the differences in the shape of the TG-curves cannot be simply explained on the basis of the available information on the specimens' structure and properties. Further work is needed to resolve the effect of impurities, oxygen content and grain size on TiN oxidation.

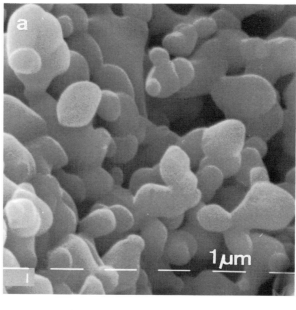

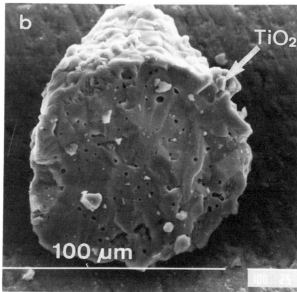

Fig. 3 Electron micrographs of the TTN 25 powder after heating up to 1300°C (a) and 1500°C (b).

Analysis of the oxidation kinetics was not the subject of this work. However, it is necessary to note that the kinetic curves obtained at 1200°C followed a parabolic law. Thus, we can assume that the reaction rate is limited by diffusion processes. On the other hand, gradual changes of oxidation kinetics depending on the temperature and time were found[9,11] and the kinetic curves

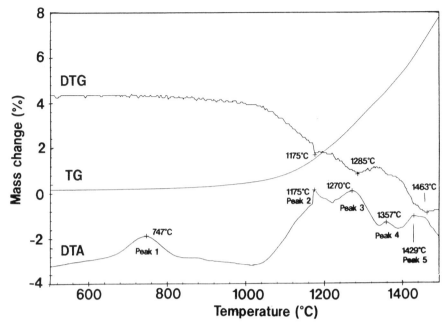

Fig. 4 Thermogram obtained at heating of a TAJ/1 sample in air with the rate of 20°C/min.

followed a logarithmic or power law. This suggests changes in the oxidation mechanisms.

After oxidation of a marked specimen during 15 h at 800°C the Pt-marker was found on the outer surface of the oxide layer (Fig. 7). This position indicates that the diffusion of oxygen through the oxide layer to the rutile/TiN interface caused the growth of the rutile scale at relatively low temperatures. The apparent activation energy E was calculated using the Ozawa equation

$$E = 2.19R^*d(\log \beta)/d(1/T_p) \tag{3}$$

where R is the universal gas constant, β is the heating rate and T_p is the temperature of the peak on the DTA curve. A series of four sets of program rate and peak temperature data was generated and plotted as seen in Fig. 8. The E value (Fig. 8) for the first stage of oxidation (corresponding to the peak 1 in Fig. 4) is in agreement with the value of 254 kJ/mol reported for the oxidation of TiN at 700–800°C.[11] This value is comparable with the activation energy of the oxygen diffusion into TiO_2 (189 kJ/mol).[10] This fact suggests that the oxygen diffusion through the TiO_2 film is a limiting factor at the first stage of oxidation (below 1000°C).

It is more complicated to interpret the oxidation behaviour of TiN above 1000°C. Four maxima are present on the DTA curve in the high-temperature range (Fig. 4). The dependence of the 2nd and 3rd exothermal effects on the heating rate (Fig. 8) suggests that they correspond to thermally activated processes. The calculated E values are relatively high and may correspond, for

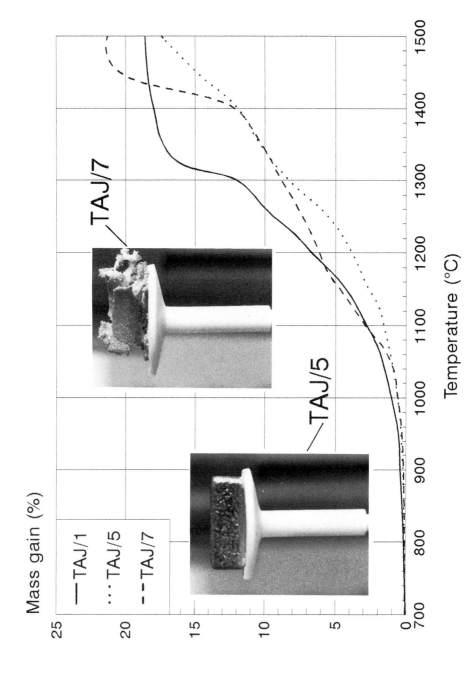

Fig. 5. Thermogravimetric curves obtained at heating of different TiN samples with the rate of 2°C/min and appearance of the samples after heating and cooling to room temperature.

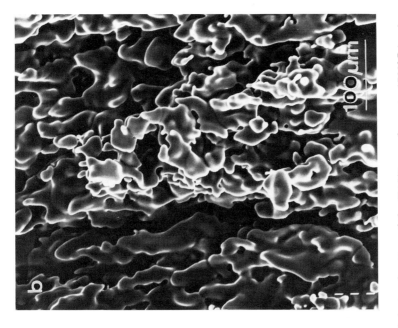

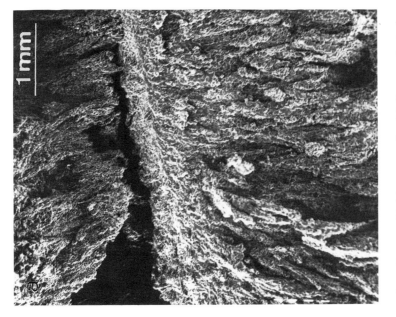

Fig. 6 Porous, highly textured oxidation products formed during heating of the TAJ/7 sample up to 1500°C in air.

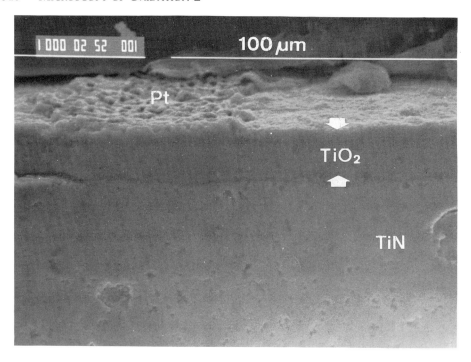

Fig. 7 Electron micrograph of the Pt marker on the surface of the TAJ/1 specimen after oxidation during 15 h at 800°C.

example, to the diffusion of titanium ions through TiO_2. After oxidation at 1100°C and 1200°C the marker was found within the oxide scale. This position indicates that the diffusion of titanium in rutile, as well as diffusion of oxygen through the oxide layer are important at the high-temperature oxidation of TiN. As is known,[10] the diffusion rate of oxygen ions in TiO_2 is much higher than that of Ti ions. But the diffusion mobility of the interstitial titanium cations increases above 950°C[10] and begins to play a more important role than oxygen diffusion at the formation of a rutile scale during oxidation of pure titanium. Dissolution of nitrogen in the rutile lattice in the case of TiN oxidation should increase the concentration and diffusion mobility of the interstitial titanium cations (in comparison with the case of Ti oxidation), thus increasing the role of the titanium diffusion and decreasing the role of oxygen diffusion.

Microstructural investigations of the oxidised samples showed that the surface of the oxidised samples was completely covered by a scale of highly textured TiO_2 crystals (Fig. 9). The growth of such a scale should be related to reaction (2) and based on the diffusion of titanium and/or oxygen to the reaction interface. After heating up to 1500°C a relatively smooth, sintered oxide layer with rutile grain size of −100 μm was formed (Fig. 9a). Cracking and spallation of the oxide scales were also observed. After oxidation above 1000°C multi-layered oxide scales were found. Beneath the dense, surface rutile layer

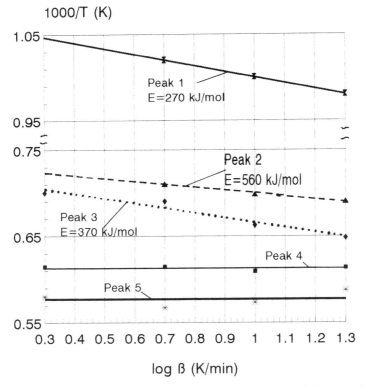

Fig. 8 Inverse temperature of the DTA peak as a function of the logarithm of heating rate for TAJ/1 TiN.

one or several fine-grained porous rutile layers were observed (Fig. 9b). Such a structure of the oxide scale may be based on the recrystallisation and sintering processes in the outer scale at high temperatures, as well as on the gas (N_2, NO_x) evolution at the oxide/substrate interface or exfoliation of the oxide layers due to internal stresses. On the other hand, different crystal structure of outer and inner oxide layers can be a result of different growth directions of these layers. Perhaps the outer layer grows outside due to the diffusion of Ti ions and the inner layer grows inside the material due to oxygen ions diffusion through the oxygen vacancies in the TiO_2 layer. At lower temperatures (e.g. 800°C), when the diffusion mobility of Ti ions is negligible, the marker was found on the surface of the oxide scale and only a single-layered rutile scale covered the surface of the specimen (Fig. 7). At higher temperatures (>1000°C), when the marker was found within the oxide scale, multi-layered scales were formed (Figs 6, 9).

A rounded, barrel-like form of the specimens after oxidation above 1000°C, as well as the sintering of the outer oxide film and cavities formation between the inner oxide layers and along the rutile/TiN interface confirm the increasing role of Ti diffusion at high temperatures. Diffusion of titanium to the surface leading to the formation of porosity in the sub-surface layer was found also at oxidation of Si_3N_4-TiN composites.[12]

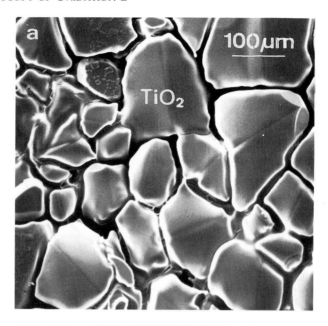

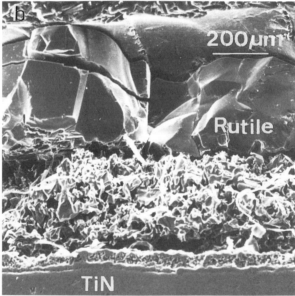

Fig. 9 Surface (a) and cross-section (b) of the TAJ/1 sample after heating up to
1500°C with the heating rate of 2°C/min.

High stresses and strains resulting from the TiN-TiO$_2$ transformation can
lead to cracking and exfoliation of the outer oxide layer. In the case of the TAJ/7
specimen, produced from the finest TTN 70 powder, these stresses caused its
failure (Fig. 5) and promoted its complete oxidation under heating to 1500°C
(Fig. 6). Repeated separation of a thin rutile layer and its subsequent fracture

can also be a reason for the formation of the layered structures. The mechanism of the process can be described as follows. The movement of atoms from the bulk material into the film as ions leaves vacancies at the interface and these may coalesce to form cavities so that the film becomes locally detached from the ceramics (Fig. 9b). The coalescence of vacancies can occur in the presence of concentration gradients of cation and anion vacancies, which is possible due to the diffusion of titanium and oxygen ions in different directions. However, the pore formation can also occur as a result of N_2 or NO evolution. Furthermore, the removal of Ti from the sub-surface layer into the surface oxide film may upset the equilibrium between compressional and tensional internal stresses, so that a resultant stress arises which may rupture the film at places where it is unsupported. The 4th and 5th peaks on the DTA curves (Fig. 4), which do not show a pronounced temperature dependence (Fig. 8), can correspond to cracking of the oxide layer promoting the oxidation.

4. Conclusions

1. Oxidation behaviour of three ultrafine TiN powders and sintered materials produced from these powders was studied.
2. Mass gain of the powders at heating in air starts at 240–260°C. The oxidation rate increases with increasing the specific surface area of the powder. The oxidation products were identified as brookite and anatase at lower temperatures and as rutile at high temperatures.
3. Oxidation of sintered TiN results in different behaviour depending on the temperature and initial powder. The oxidation of TiN is not strictly oxygen diffusion controlled in the whole temperature range.
4. Single-layered rutile scales covered the surface of the specimens below 1000°C. At higher temperatures multi-layered scales were formed.
5. Spallation and cracking of the oxide layers on TiN specimens occurred. The formation of non-protective lamellar oxide films leads to a low oxidation resistance of sintered TiN above 1000°C.
6. Cracking of the specimen produced from the finest powder promoted its complete oxidation under heating to 1500°C.
7. The microstructural investigations of the oxide layer suggest that both the diffusion of titanium in rutile and diffusion of oxygen through the oxide layer may be important at the oxidation of TiN.

Acknowledgements

Thanks are due to Professor M. Yoshimura (Tokyo Institute of Technology, Japan) for a useful discussion, to Brian Bennett for estimation of the lattice constants of TiN, Tony Jones, Tim Jannett (all with Tioxide Group PLC, UK) and Mrs Christine Taut (IKTS, Dresden, Germany) for the experimental help. Dr Yury Gogotsi gratefully acknowledges the receipt of a Research Fellowship from the Japan Society for the Promotion of Science.

References

1. G. Dransfield, in *Ceramic Technology International 1992* (ed. I. Birkby), pp. 71–74 (1991), Sterling, London.
2. G. P. Dransfield and A. G. Jones, in *Proc. 2nd European Ceramic Society Conference*, Augsburg (Sept. 1991) (to be published).
3. Y. Sakka, S. Ohno and M. Uda, *J. Am. Ceram. Soc.*, **75** (1) (1992), 244–248.
4. A. A. Kochergina and D. V. Fedoseev, *Inorganic Materials*, **3** (1990), 549–550.
5. V. A. Lavrenko, A. F. Alexeev, V. S. Neshpor *et al.*, *Sverkhtverdye Materialy* (4) (1988), 40–44 (in Russian).
6. A. Bellosi, A. Tampieri and Yu-Zh. Liu, *Mater. Sci. Eng.*, **A127** (1990), 115–122.
7. J. Desmaison, P. Lefort and M. Billy, *Oxid. Metals*, **13** (6) (1979), 505–517.
8. J. Desmaison, P. Lefort and M. Billy, *Oxid. Metals*, **13** (3) (1979), 203–222.
9. Yu. G. Gogotsi, F. Porz and V. P. Yaroshenko, *J. Am. Ceram. Soc.*, **75** (8) (1992), 2251–2259.
10. R. F. Voitovich, 'Oxidation of carbides and nitrides' (1981), *Naukova Dumka*, Kiev (in Russian).
11. A. Tampieri, E. Landi and A. Bellosi, *Brit. Ceram. Trans. J.*, **90** (6) (1991), 194–196.
12. Yu. G. Gogotsi, V. A. Lavrenko, *Corrosion of High-Performance Ceramics* (1992), Springer, Berlin–Heidelberg, p. 56.

Oxidation Kinetics of a Silicon Carbide Fibre-Reinforced Ceramic Matrix Composite

V. GUÉNON, F. LEGENDRE and G. GAUTHIER

SNECMA, Materials and Processes Department B.P.81, 91003 Evry Cedex, France

ABSTRACT

The oxidation kinetics of a silicon carbide fibre-reinforced silicon carbide matrix (SiC/SiC) composite were studied. Weight variation versus time curves for the Nicalon SiC fibres and the SiC/SiC composite were experimentally determined using thermogravimetric analysis (TGA) for temperatures ranging from 600°C to 1200°C. Two competing oxidation reactions occur at the same time: (1) carbon interphase oxidation leading to a weight loss and (2) fibre and – to a lesser extent – matrix oxidation by formation of a silica layer leading to a weight increase. Experimental work was completed with TEM observation and EELS analysis of the fibre/matrix interphase. Based on experimental and microscopic observations, a simple incremental model is proposed to describe the basic oxidation mechanisms and their relative contributions to the overall kinetics as a function of temperature. Weight variation was calculated using this model. Good agreement between experimental results and model prediction was obtained for temperatures under 1000°C.

1. Introduction

Ceramic matrix composites (CMCs) have been of growing importance in turbine engine technology development. Improvement in turbine engine performance occurs by decreasing weight and allowing higher use temperatures which decreases the need for energy-consuming cooling air. Because CMCs have low density and high temperature resistance, they are expected to contribute in great part to the increase of turbine engine performance. For several years, SNECMA have been conducting CMC material and engine parts development studies in collaboration with SEP.[1] Using SEP's carbon/SiC and SiC/SiC materials, SNECMA have been developing hot parts, such as afterburner elements for military applications. Such parts are subjected to use temperatures up to 1000°C, and engine tests have shown that they have to withstand severe thermomechanical loads and long-term ageing in an oxidising environment. Coupling between oxidation and mechanical loading was found to be extremely damaging to the SiC/SiC material.[2] It is therefore of great importance to understand thoroughly the oxidation mechanisms of these materials, in order to:

(a) be able to modify the materials and improve their oxidation resistance,
(b) understand the coupling between oxidation and mechanical loading,
(c) provide information to set up the bases of design and analysis of CMC parts and
(d) extend the use of these materials to more structural parts.

The behaviour of ceramic matrix composites is highly dependent on the fibre/matrix interphase. This interphase is the key factor in consuming crack energy, but at the same time it is very sensitive to oxidation. Therefore, primary attention must be paid to the interphase when studying oxidation of CMCs. High-temperature oxidation behaviour of SEP's SiC/SiC material has been studied experimentally through TGA analysis. Electron microscopy has allowed the observation of the interphase and provided useful information for understanding the oxidation kinetics of this material. A simple model which takes into account the oxidation mechanisms of the carbon interphase, of the SiC matrix and SiC fibre along the interphase has been set up. Oxidation kinetics at temperatures ranging from 600°C to 1200°C have been calculated in terms of weight change with this model and compared to experimental observations. Very good agreement has been obtained for medium temperatures (600–850°C). For higher temperatures (1000–1200°C), experimental and model-based weight change curves do not match, but they show the same general features.

2. Material

The material under study is a CMC manufactured by SEP (France). It is composed of 18 layers of silicon carbide Nicalon fibres woven in a plain-weave fabric, embedded in a silicon carbide matrix deposited through a chemical vapour infiltration (CVI) technique. The SiC fibres are covered with a turbostratic CVD carbon layer which creates a weak interface between the fibres and the matrix. Such a weak interface is needed in CMCs to achieve good mechanical properties through crack deviation. The fibre volume fraction is 42%. Optical microscopic analysis shows large pores around the fibre bundles, as well as smaller pores inside the bundles, and size and shape differences between individual fibres (see Fig. 1).

3. Thermogravimetric Analysis

3.1 Experimental Procedure

In order to obtain experimental data on the behaviour of the SiC/C/SiC material under oxidising environment, thermogravimetric analysis was performed. The

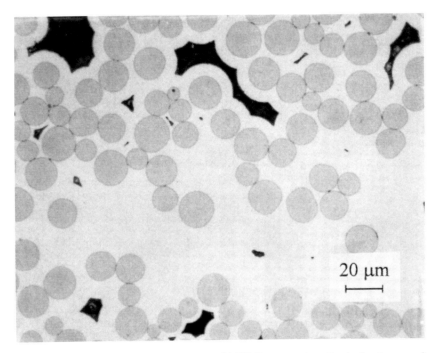

Fig. 1 Transverse cross-section of the SiC/C/SiC composite. Optical micrograph.

tests were performed on the Nicalon fibres, then on the composite itself. The samples were put in the furnace after the desired test temperature was reached. Tests were run at the following tmperatures: 600°C, 750°C, 850°C, 1000°C and 1200°C. Test duration was 100 h, and the atmosphere was 20% O_2, 80% N_2 gas flowing in the furnace at $0.31\,min^{-1}$. A thermal balance was used for continuous recording of the weight variations undergone by the material during thermal exposure. Nicalon fibres were desized before TGA. Composite samples were 18 mm wide and 7 mm thick squares. The sides were machined, thus allowing preferential oxidation through the side faces.

3.2 Results and Analysis

The test results were recorded as relative weight change versus exposure time curves. The relative weight change is defined as:

$$\delta m/m_0 = (m - m_0)/m_0$$

m_0 being the initial sample weight and m the current sample weight.

Experimental TGA for desized Nicalon fibres resulted in a weight gain, which stems from the formation of an oxide layer, SiO_2, from the SiC of the fibre, according to:

$$SiC + 3/2\ O_2 => SiO_2 + CO \tag{1}$$
or
$$SiC + 2O_2 => SiO_2 + CO_2 \tag{1'}$$

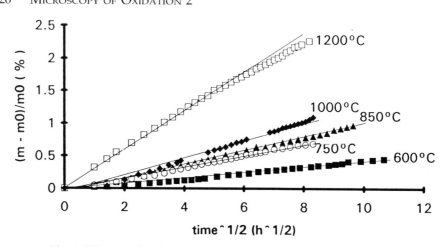

Fig. 2 TGA weight change curves for the desized Nicalon fibres.

Whether the oxidation product is CO or CO_2, the weight variation due to replacement of SiC by SiO_2 is the same. When plotted against square root of time (Fig. 2), the weight change curve is linear. Therefore, the oxidation kinetics of Nicalon fibres can be described by the equation:

$$\delta m/m_0 = kt^{1/2} \qquad (2)$$

where t is the thermal exposure time and k is a constant at a given temperature. For each temperature, k was determined from Fig. 2. Equation (2) defines a parabolic weight change curve, which is representative of a diffusion-limited process. Here, diffusion of the oxygen takes place through the silica layer in order to reach the SiC fibre. The relation between the silica layer thickness, e_f and the weight gain was calculated using fibre geometry, and SiC and SiO_2's respective densities ($\rho_{SiC} = 3.15$ and $\rho_{SiO2} = 2.32$) and molecular weights. The relation is:

$$e_f = 12462 \; \delta m/m_0 \quad \text{(nm)} \qquad (3)$$

This relation leads to the kinetics of silica layer formation, which can be described by an equation similar to equation (2):

$$e_f = k_f t^{1/2} \qquad (4)$$

where k_f can be determined in the same way as k.

Experimental TGA curves for the SiC/C/SiC composite are shown in Fig. 7. All curves, except for the 600°C curve, show a weight loss, followed by a weight gain. The 600°C curve exhibits a continuous weight loss. As the temperature rises, the weight loss stage shortens, and the weight gain starts more quickly. At $T = 750$°C, the maximum weight loss is 0.35% after 40 h, while at $T = 1200$°C, the sample loses at most 0.03% of its weight after 5 h. However, for the higher temperatures (1000 and 1200°C), the weight gain seems to reach a maximum and remains constant after only 15 h for the 1200°C curve.

These curves show that the composite oxidation is a combination of at least two competing processes. Weight loss is the result of the carbon interphase burn-out by the oxygen. Weight gain is the result of equation (1), i.e. formation of a silica layer after the loss of carbon has allowed the oxygen to reach the SiC. Reaction (1) was shown to occur with the fibres, but it can also occur with the SiC matrix. The two competing mechanisms are supposed to occur almost simultaneously, as soon as enough carbon has been burnt out to allow reaction (1) to take place. In the descending part of the curve, carbon oxidation prevails over reaction (1), and the curve reaches a minimum when this tendency starts to reverse. As temperature increases, the rate of the oxidation reactions increases such that these mechanisms become faster.

4. Microscopy

The aim of microscopic analysis was to provide information on the fibre/matrix interphase oxidation mechanisms. The carbon interphase thickness before oxidation is about 100 nm. The oxidation product, SiO_2, contains a light element, O, which is difficult to detect with a classical energy selection system. Electron energy loss spectrometry (EELS) was used in order to obtain an image of the silica formed at the interphase. This method is able to identify light elements (with an atomic number smaller than 11) such as oxygen, carbon or nitrogen, and also the form under which this element is combined (SiO, SiO_2, SiC, SiN . . .). The following equipment was used: transmission microscope,

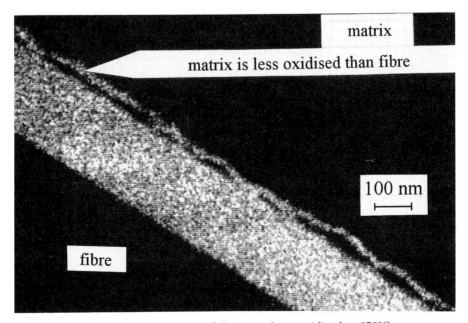

Fig. 3 TEM micrograph of the interphase oxidised at 850°C.

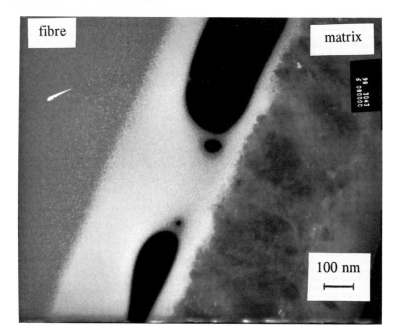

Fig. 4 TEM micrograph of the interphase oxidised at 1000°C.

EELS, STEM unit, EDAX 9900. In order to obtain an image of the silica layers, an image subtraction method was used. This method consists of setting two windows on the EELS spectrum: one on the oxygen peak, and another one right before the peak. In this way, two images are obtained: one image of the background, and one image of the background plus oxygen information. The difference between the two digitalised images is obtained with the EDAX 9900 computer. The result is a third, digitalised image, showing the oxygen. Figure 3 shows the image of the silica formed at the interphase after thermal treatment at 850°C. The picture clearly shows the layers formed on the Nicalon fibre on one side, and on the SiC matrix on the other side. The total silica thickness on the fibre side is about 200 nm. The silica layer formed on the matrix is about ten times thinner.

Figure 4 shows a TEM micrograph of the interphase oxidised at 1000°C. As in the previous discussion, the white layers are silica layers. The silica layer formed on the matrix is about three times thinner than that formed on the fibre.

By showing the silica layers thicknesses, micrographs gave important information for oxidation modelling.

5. Weight Change Model

As was shown by the experimental results, several different mechanisms contribute to the SiC/C/SiC composite oxidation. These mechanisms were also identified by Filipuzzi[3] as:

(a) carbon interphase oxidation,
(b) SiC fibres oxidation and
(c) CVD SiC matrix oxidation.

These three mechanisms were taken into account in order to model the oxidation kinetics of the composite.

5.1 Interphase Oxidation Kinetics

According to Bernstein and Koger,[4] the oxidation kinetics of a carbon layer between two plates are parabolic, which is characteristic of a diffusion-limited process. It was assumed that in the 600–1200°C temperature range, the length of burnt-out carbon between the fibre and matrix, L_c, follows the same law as equation (2):

$$L_c(t) = k_c t^{1/2} \tag{5}$$

where k_c is a constant at a given temperature.

Burning out of the carbon interphase leaves a cylinder-shaped pore around the fibre. As explained earlier, SiC oxidation creates a silica layer on the fibre and matrix, i.e. on the walls of the cylinder-shaped pore. This layer can grow until the pore closes. This mechanism is illustrated in Fig. 5. In order to take into account the effect of pore entrance closure, a correction factor, $F(t)$, was determined:

$$F(t) = e_e(t)/e_0 \tag{6}$$

where e_0 is the initial pore thickness, i.e., the carbon interphase thickness, and $e_e(t)$ is the pore entrance thickness, as shown in Fig. 5. Their calculation is explained in the next paragraph. At $t = 0$, $F(t) = 1$, and when the pore is totally closed by the silica, $F(t) = 0$. The burnt-out carbon length as a function of time becomes:

$$L_c(t) = F(t) k_c t^{1/2} \tag{7}$$

In the numerical model, the length of burnt-out carbon within a small time interval, dL_c, was calculated using equation (7).

5.2 SiC Oxidation Kinetics

The oxidation kinetics of the SiC fibres were described by equations (1) and (4), and by Fig. 2. It was assumed that the SiC matrix follows the same type of law as the SiC fibre. Therefore, a silica layer opposite to the fibre grows on the SiC matrix. Its thickness, e_m, is given by:

$$e_m = k_m t^{1/2} \tag{8}$$

where k_m is a constant at a given temperature. The value of k_m was determined on the basis of the microscopic observations of section 4. The silica layers

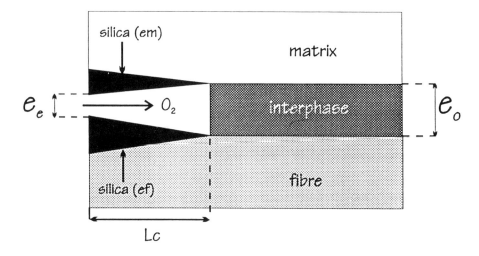

Fig. 5 Representation of SiC/C/SiC oxidation mechanisms.

thicknesses e_m and e_f were measured directly from the micrographs for 850°C and 1000°C (see Figs 3 and 4). The value of k_m for each of these temperatures was calculated as:

$$k_m = (e_m/e_f)k_f \qquad (9)$$

The matrix was found to be less subject to oxidation than the fibres: at 1000°C, the silica layer formed on the matrix is three times thinner than the silica layer formed on the fibre (see Fig. 4); at 850°C, the ratio becomes 10 (see Fig. 3). Therefore, matrix oxidation was considered to be negligible at 600 and 750°C, and the value of k_m at 1200°C was extrapolated. These observations are in agreement with Filipuzzi[5] who found that the parabolic constant of Nicalon fibres is higher than that of pure SiC, especially at low temperature.

In order to simplify the model, the total silica layer thickness is considered as $e = e_f + e_m$, such as:

$$e(t) = (k_f + k_m) \, t^{1/2} \qquad (10)$$

As can be seen from Figs 5 and 6, the pore entrance thickness, $e_e(t)$, used in equation (6), is given by:

$$e_e(t) = e_0 - e(t)/2 \qquad (11)$$

The oxidation kinetics were modelled using small time intervals, as illustrated in Fig. 6. At time t_n, the formed silica longitudinal surface section is given by:

$$S(t_n) = \sum_{i=0}^{n-1} [L_c(t_{i+1}) - L_c(t_i)] e(t_n - t_i) \qquad (12)$$

where $L_c(t_i)$ is given by equation (7).

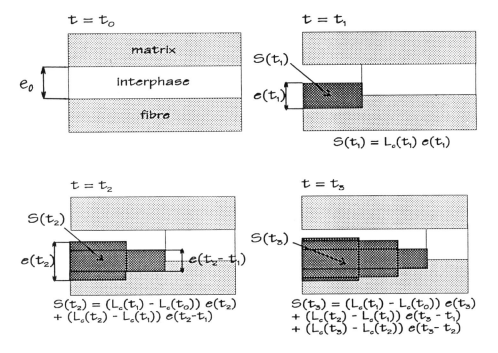

Fig. 6 Incremental modelling of the SiC/C/SiC oxidation mechanisms.

5.3 Weight Change Calculation

At a given time t_n, the thickness of formed silica at the pore entrance is known by equation (4). This allows the calculation of the closure factor, $F(t_n)$, through equation (6). The length of burnt-out carbon can then be calculated through (7), which gives way to the weight loss due to carbon oxidation.

Knowing that the volume occupied by the formed silica is about twice that of the SiC used in the reaction, the weight gained by SiC oxidation can be derived using the value of $S(t_n)$ from equation (12), as:

$$Pm_{SiO_2}(t_n) = \left(\rho_{SiO_2} - \frac{\rho_{SiC}}{2} \right) S(t_n) 2\pi r_f \tag{13}$$

where r_f is the average fibre radius. The program stops if:

- the cylinder-shaped pore closes,
- the carbon is totally burnt out,
- the testing time length is over.

As the value of k_c was not experimentally determined, the calculations were performed for various values of k_c, until a good agreement between the model and the experimental results were obtained.

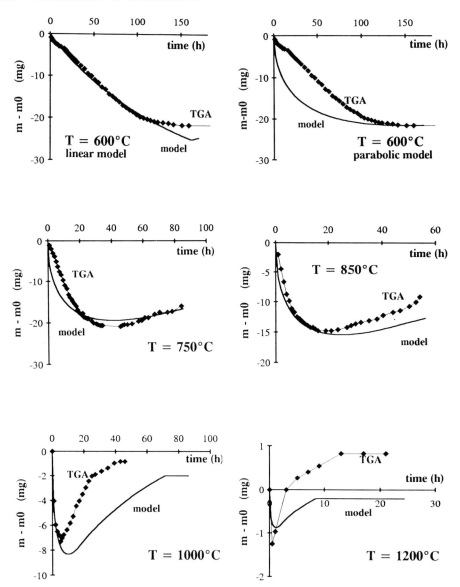

Fig. 7 Comparison of TGA and calculated weight change curves.

6. Results and Discussion

The described model was applied to calculate weight change at several temperatures. The theoretical curves obtained were compared to the TGA results. Figure 7 shows the theoretical curves and the experimental data. At 600°C, the model does not fit the experimental data. Carbon oxidation is faster on the TGA curve. Overall oxidation kinetics are controlled by the slowest step of the mechanism. According to Medford,[6] at low temperature, oxidation is controlled by chemical reaction, while at high temperature, diffusion is the rate-limiting factor. Therefore, a linear law, such as

$$L_c(t) = F(t)k'_c t \tag{14}$$

was used in the model for 600°C. It gives a satisfactory agreement with the experimental data, as shown in Fig. 7. At this temperature, the rate-limiting step is the reaction of carbon with the oxygen. The parabolic law of equation (7) gave a very good agreement with experimental data for 750°C and 850°C. At higher temperatures, theoretical curves do not fit experimental data after a certain time, although they exhibit the same overall shape: a quick weight loss followed by weight gain, and then a stabilisation. The weight gain is faster in the experimental data than calculated. Another oxidation mechanism has not been taken into account in the model, because it was supposed to be negligible: it is the oxidation of the SiC matrix through the surface and internal porosity. It would be useful to carry out TGA on the CVI SiC alone in order to assess its influence on the overall mechanism. Moreover, such measurements would provide experimental assessment of the k_m constants at various temperatures. The curves in Fig. 7 show that weight change stops quickly at elevated temperatures. As temperature increases, SiO_2 formation prevails over carbon burn-out. Silica quickly closes the pores created around the fibres by carbon loss and stops the oxidation mechanism after a short time (15 h at 1200°C). Therefore, in the long term, the oxidation kinetics of SiC/C/SiC are such that temperatures around 600°C lead to deeper progression of carbon oxidation than higher temperatures.

7. Conclusions

The oxidation mechanisms of a SiC/C/SiC ceramic composite material were observed through thermogravimetric analysis and electron microscopy. A simple numerical model was set up and applied for comparison with experimental data. Good agreement was obtained at temperatures up to 850°C and, at 1000°C and 1200°C, the experimental and theoretical curves exhibit a similar overall shape. Oxidation is the result of at least two competing phenomena: interphase carbon burn-out, which creates a cylinder-shaped pore, and silica formation on the fibre and the matrix, which tends to close the pore and stops the mechanism. At low temperature (600°C), carbon oxidation is the major mechanism, while at high temperature (1200°C), silica formation prevails. More experimental evaluations are needed to improve the model.

8. Acknowledgements

The authors wish to thank Mr Gérard Simon for conducting the microscopic analyses. This work was realised for the *Groupement Scientifique GS 4C*, financially supported by CNES, CNRS, DRET, MRE, Aérospatiale, SEP and SNECMA.

9. References

1. G. Gauthier, G. Bessenay and Y. Honnorat, 'CMC evaluation for use in military aircraft engines', *Proceedings of 4th International Symposium on Ceramic Materials and Components for Engines*, Goteborg, Sweden (10–12 June 1991).
2. B. Dambrine and G. Gauthier, 'Fatigue behaviour of a SiC/SiC composite: influence of oxidation', *Revue des Composites et Matériaux Avancés*, **3**, (out of series), J. L. Chermant and G. Fantozzi eds, Hermès, France (1993).
3. L. Filipuzzi, R. Naslain and J. Thebault, 'Etude de l'oxydation de composites SiC (Nicalon)/SiC (CVI) à interphase de pyrocarbone', *Matériaux composites pour applications à hautes températures*, proceedings of conference organised by AMAC/CODEMAC Bordeaux, France (March 1990), pp. 289–301.
4. J. Bernstein and T. B. Koger, 'Carbon film oxidation-undercut kinetics', *J. Electrochem. Soc.: Solid State Science and Technology* **135**(8) (1988), 2086–2090.
5. L. Filipuzzi, 'Oxydation des Composites SiC/SiC et de leurs constituants: approche expérimentale, modélisation et influence sur le comportement mécanique, doctoral thesis, Université de Bordeaux I, France (1991).
6. J. E. Medford, 'Prediction of in-depth oxidation distribution', *AIAA* (1977), 77–783.

Microscopic Studies of Oxidised Si₃N₄-based Ceramics

YU. G. GOGOTSI,[1] G. GRATHWOHL,[2] F. PORZ,[2]
V. V. KOVYLYAEV,[3] A. D. VASIL'EV[3]

[1]Tokyo Institute of Technology, Research Laboratory of Engineering Materials, 4259
Nagatsuta, Midori-ku, Yokohama 227, Japan
[2]Institut für Keramik im Maschinenbau, Universität Karlsruhe, D-76128 Karlsruhe,
Germany
[3]Institute for Problems of Materials Science, 252680 Kiev, Ukraine

ABSTRACT

Scanning electron microscopy (SEM), energy- and wavelength-dispersive micro-analysis (EDAX and WDS), light microscopy and X-ray diffraction (XRD) techniques were used to analyse the oxide scales formed on silicon nitride-based ceramics during oxidation in air at temperatures up to 1500°C. A group of hot-pressed Si₃N₄-based ceramics with 1–3 mass % Al₂O₃, 5–9 mass % Y₂O₃ and 0–50 mass % TiN was used for oxidation experiments. The effects of the content of sintering aids (Al₂O₃ and Y₂O₃) and the content and grain size of the particulate addition (TiN) on the composition, structure and protective properties of oxide layers were investigated.

1. Introduction

Today, high-strength ceramic materials are widely used for engine components, turbochargers, cutting tools, etc. It is well known that silicon nitride-based ceramics are attractive candidates for use at high temperatures in oxidising environments. In the case of high-temperature engineering applications of Si₃N₄, one of the most important factors affecting the performance is its oxidation behaviour.

Thermochemical analysis provided by Luthra[1] indicated that the reaction rates of high-purity Si₃N₄ should be influenced both by diffusion of oxygen and nitrogen through SiO₂ and an interface reaction. The formation of a very thin Si₂N₂O layer at the Si₃N₄/SiO₂ boundary may be responsible for the superior oxidation resistance of pure silicon nitride.[2] The process of oxidation of pure Si₃N₄ is schematically shown in Fig. 1.

Silicon nitride ceramics for engineering applications usually contain sintering aids and other additives which can change the structure and composition of protective oxide layers and affect the oxidation resistance of the materials.[3] The most favoured interpretation of oxidation phenomena in polyphase Si₃N₄

535

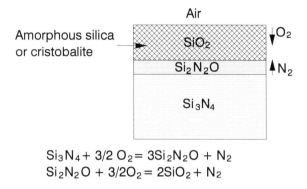

$$Si_3N_4 + 3/2\,O_2 = 3Si_2N_2O + N_2$$
$$Si_2N_2O + 3/2O_2 = 2SiO_2 + N_2$$

Fig. 1 Schematic presentation of the oxidation of pure Si_3N_4.

materials considers that the oxidation rate-controlling process is the sequential inward diffusion of oxygen (to generate silicon oxynitride and silica), and the outward diffusion of intergranular cations as a result of ionic chemical potential gradients.[4] Diffusion processes are indicated as the rate-controlling step for the parabolic oxidation behaviour observed in a range of testing conditions.

Recently increased attention has been devoted to particulate additions for the production of structural Si_3N_4-based ceramics. TiN is most frequently used for particulate reinforcement of the silicon nitride matrix.[5] The alteration of the Si_3N_4 microstructure by the dispersed additions may then offer a potential for improvement of the composite's strength at room temperature, but the effect at elevated temperature may be adverse. The addition of compounds that are easily oxidised in air may result in lower oxidation resistance of ceramic composites and can considerably influence both the oxidation mechanisms of the material, and the composition and protective properties of the oxide layer.[3]

The purpose of this study is to compare and discuss the structure, composition and protective properties of oxide films formed after oxidation of various Si_3N_4-based materials.

2. Materials and Experimental

Five model compositions containing various amounts of yttria and alumina (Fig. 2) were used for the experiments. The samples were hot pressed at 1840°C (0.5 h) in graphite dies at a pressure of 30 MPa (processed at the Institute for Problems of Materials Science, Kiev, Ukraine). Up to 50 mass% TiN with different grain sizes were added to the ceramic with 5 mass% Y_2O_3 and 2 mass% Al_2O_3. From the billets obtained, the test samples ($2 \times 5 \times 15$ mm) were cut and ground with a 100 μm diamond wheel. The preparation and properties of the materials used are described in more detail elsewhere.[5,6]

Oxidation products were examined by X-ray diffractometry (XRD) (Siemens D 500), scanning electron microscopy (Cambridge Stereoscan S4-10 and Superprobe 733) and metallography (Zeiss, Axiomat). Energy-dispersive X-ray

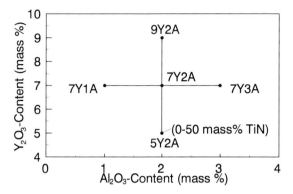

Fig. 2 Compositions under study.

microanalysis (EDX) and wavelength-dispersive microanalysis (WDS) were used to identify the composition of the materials and element distribution in the oxide layers. XRD peak intensity measurements of the phases were taken for the strongest diffraction signals: (101) for α-cristobalite (SiO$_2$), (200) for β-Si$_3$N$_4$ and (021) for β-Y$_2$Si$_2$O$_7$. Relationships between the intensities of α-cristobalite or yttrium silicate peaks and the intensity of the strongest β-Si$_3$N$_4$ peak were used to normalize the XRD data. Assuming a uniform distribution of oxide phases in the surface layer and the formation of thin ($<$100 μm) oxide films, no other special corrections to the measured XRD peak heights were made.

3. Results and Discussion

3.1 Oxide-containing Ceramics

The oxidation kinetics of silicon nitride-based ceramics are determined by O$_2$ diffusivity in the heterogeneous oxide scale, the composition of which is in turn controlled by the diffusion of additive and impurity cations from the bulk ceramics to the reaction interface. The oxide scales usually have a non-uniform complex structure and composition, depending on the oxidation temperature. XRD analysis of the surfaces of samples showed the presence of α-cristobalite and yttrium disilicate β-Y$_2$Si$_2$O$_7$, which is in accordance with the published data.[3,4,7,8] A glassy silicate phase in the Si-Al-Y-O system was always present in the oxide film. During oxidation, the SiO$_2$ produced as the oxidation product is brought to chemical equilibrium with those intergranular phases not already existing in equilibrium with SiO$_2$ by the outward diffusion of Y and Al cations. The equilibrium condition changes as oxidation proceeds. Precipitation of crystalline Y$_2$Si$_2$O$_7$ and SiO$_2$ from the silicate melt maintains the equilibrium with the changing composition of the grain-boundary phase.

The dependence of the oxide layer phase composition on the additives content is strongly pronounced (Fig. 3). The higher amount of additives contributed to a lower cristobalite content in the oxide scale above 1350°C (Fig.

3) due to diffusion of Y and Al cations to the surface and formation of glassy and crystalline silicate phases. The higher content of yttria in the ceramics increases the quantity of β-$Y_2S_2O_7$ in the oxide layer (Fig. 3). A considerable amount of the Al_2O_3 addition in the material is incorporated during sintering into the Si_3N_4 according to the equation:

$$z/2(Al_2O_3) + (6 - z/4)Si_3N_4 = Si_{6-z}Al_zN_{8-z}O_z + 3/4_z(SiO_2)$$

According to this equation a higher alumina content in the initial composition leads to a higher SiO_2 content in the grain boundary phase. This leads to decreasing oxidation below 1200°C with increasing Al content. This change in the SiO_2 content can also explain changes in $Y_2Si_2O_7/SiO_2$ ratio in the surface layer of the materials oxidised at 1200°C (Fig. 3). Al_2O_3 dissolved in the silicate phase hinders crystallisation, thereby promoting glass formation. At relatively low temperatures it can improve the oxidation resistance due to the protection of the unstable oxynitride grain-boundary phases from catastrophic oxidation. At higher temperatures the effect can be reversed. It is known that the diffusion rate of oxygen in silicates is much higher than in pure SiO_2.[9] An increase of the oxygen transport rate will increase the overall oxidation rate. Thus the more strongly fluxed the surface silicate layer by dissolved constituents from the substrate (Y,Al), the lower the viscosity of the glass and the material's resistance to oxidation above 1300°C. Hot-pressed ceramics with a lower content of additives were more stable under oxidation above 1350–1400°C than additive-rich ceramics, probably because of the protective effect of the cristobalite layer formed. The measurement of the mass gain[6] offers clear evidence that an increase in the total additives content increases oxidation rates at high temperatures.

The oblique, polished surface of a sample 7Y1A oxidised at 1500°C (Fig. 4) shows an example of the oxide scale, containing preferentially oriented, elongated and platelike crystals of β-$Y_2Si_2O_7$ (white in Fig. 4a,b) with different morphologies in a glassy phase (dark) and bubbles. A distribution of elements in the oxide layer shows that the silicate phase was enriched with Al.

Oxide scales, in general, were formed at >1200°C with important microstructural variations between their surface and the reaction interface (Fig. 5) irrespective of composition and nature of the material. The β-$Y_2Si_2O_7$ crystals are present on the oxide surface and at the oxide-substrate interface (white in Fig. 5), but usually do not appear in the mid-zone of the oxide film, which is only α-cristobalite and a glassy phase. A pure cristobalite layer at the boundary between the silicate oxide film and Si_3N_4 substrate was found only at 1200–1300°C. At higher temperatures dissolution of pure SiO_2 in a silicate phase probably occurred, and a glassy phase with relatively uniformly distributed aluminium (Fig. 4) was found. The formation of a porous layer between Y-rich layers and/or on the oxide/bulk material interface was observed after the oxidation above 1300°C (Fig. 5). This leads to a partial spallation of the scale.

Two different oxidation regimes, based on the features of oxidation at different temperatures, can be identified: I, low ($T<1300$°C), and II, high ($T>1350$°C) temperature. Oxidation below 1050°C was not found for the

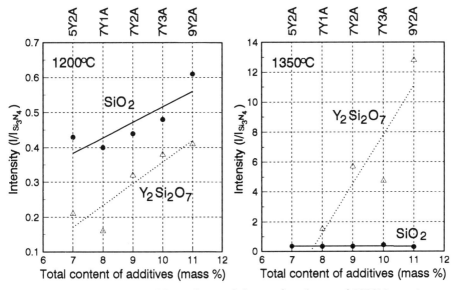

Fig. 3 Semiquantitative XRD analysis of the surface layer of HPSN specimens oxidised 100 h at 1200 and 1350°C as function of the total additives content.

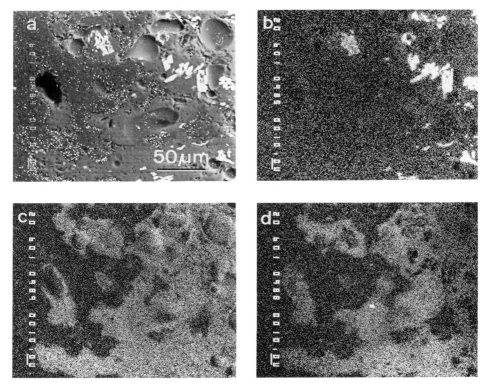

Fig. 4 The oblique polished surface of the 7Y1A sample after heating to 1500°C. Back-scattered electron image (a) and X-ray maps of Y (b), O (c) and Al (d).

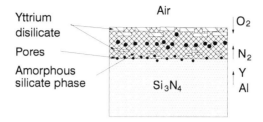

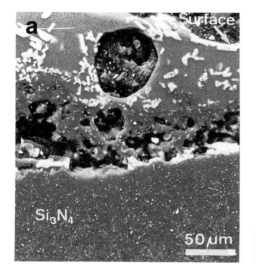 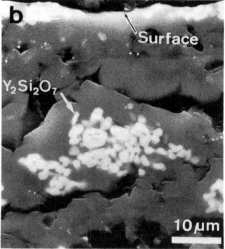

Fig. 5 Schematic diagram of the oxide film after oxidation above 1400°C and cross-sections of the oxide films on the 7Y1A (a) and 7Y3A (b) samples oxidised during 100 h (a) and 25 h (b) at 1500°C.

hot-pressed ceramics under study. However, sintered materials of the same composition[6] demonstrated instability at lower temperatures, because of a lower oxygen content in the intergranular phase and its oxidation.

The formation of a protective SiO_2 scale on the surface of the samples after a few minutes of oxidation above 1100°C leads to a high resistance of silicon nitride ceramics, for which inward diffusion of oxygen is rate limiting (Fig. 6). Below 1350°C the formation of a uniformly thick 'fish scale' type of cristobalite film on HPSN samples was observed (Fig. 6).

Above 1400°C (this temperature is probably near to the eutectic temperature for the silicate layer formed) the viscosity of the oxide scale decreases strongly, large pores and bubbles form (Figs 4 and 5). After a short time the concentration of impurities in the oxide scale reached a sufficient level to dissolve the protective cristobalite layer, formed at lower temperatures (Fig. 6).

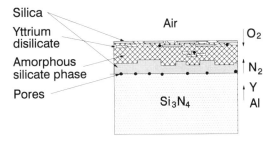

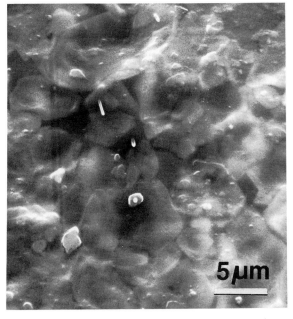

Fig. 6 Schematic diagram of the oxide film after oxidation below 1350°C and surface morphology of the oxide film on the 7Y2A HPSN oxidised 75 h at 1300°C.

The low viscosity silicate film then became evidently less protective and the materials oxidised almost linearly.

Thus, the content of additives in silicon nitride ceramics (and consequently in the oxide layer) determines the oxidation behaviour at high temperatures.

3.2 TiN-containing Ceramics

The detectable mass gain of TiN-containing ceramics starts at $>600°C$ but up to 1100°C oxidation is not very pronounced.[10] The mass gain of the composites in this temperature range is connected with the oxidation of TiN. In the whole temperature range studied the mass gain related to the amount of TiN in the composite (Fig. 7). Furthermore, there is a marked decrease of oxidative attack with decreasing TiN grain size.

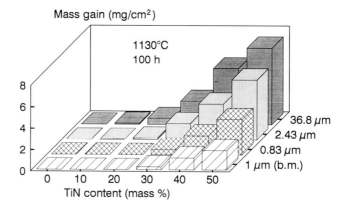

Fig. 7 Mass gain of the TiN-containing HPSN samples oxidised 100 h at 1130°C vs the content and grain size of TiN.

XRD analysis of the surfaces of samples oxidised at >1000°C showed the presence of rutile (TiO_2), α-cristobalite (SiO_2), yttrium disilicate (β-$Y_2Si_2O_7$), mullite, yttrialite (y-phase, $R_xY_{2-x}Si_2O_7$, R-Th,Zr,Ti, etc.), yttrium dititanate ($Y_2Ti_2O_7$) and Si_2N_2O. At high TiN contents in ceramics these phases are present in the sub-surface layer beneath the rutile scale. Microstructural investigations showed also numerous inclusions of other phases in rutile. On the surface of samples with 40 or 50% TiN the formation of a continuous rutile layer was found (Fig. 8a). The growth of such a scale should be related to the diffusion of titanium to the surface of the composite. A correlation between the size of TiN particles in the ceramics and TiO_2 crystals in the oxide layer was observed. With a decreasing amount of TiN only those groups of grains that exist near the surface can be oxidised since the TiN grains are present as isolated inclusions in a Si_3N_4 matrix. Rutile islands protruding from the oxide layer are formed in these areas only. At high temperatures, the formation of a viscous silicate film on the surface of the composites can protect them from further oxidation even at relatively high TiN contents (Fig. 8b). In this case the presence of small TiN particles is also desirable, because coarse TiN grains prevent the formation of a continuous silicate layer on the surface and decrease its protective properties.

4. Conclusions

1. The oxidation resistance of silicon nitride ceramics is strongly dependent on the content of additives, resulting in different phase composition, structure and protective properties of the oxide layers.
2. Samples with the higher amount of yttria and alumina were more stable under oxidation below 1350°C, the reverse was the case at higher temperatures.

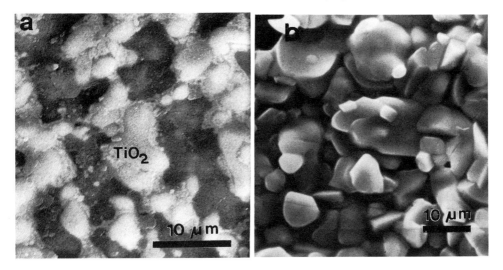

Fig. 8 Oxide layers on the surface of samples with 30 mass% TiN (a) and 50 mass% TiN (b) after heating to 1500°C (a) and 1300°C (b).

3. In the range 1100–1350°C oxygen diffusion through a protective SiO$_2$ or silicate layer should be the rate-limiting step for parabolic oxidation of Y$_2$O$_3$+Al$_2$O$_3$-doped ceramics; above ~1350°C the formation of a liquid silicate phase and dissolution of the protective SiO$_2$ film in this liquid are responsible for the markedly faster oxidation rate.

4. The content of α-cristobalite and β-Y$_2$Si$_2$O$_7$ in the oxide scale is dependent on ceramic composition.

5. The mass gain due to oxidation increases with the increase of TiN grain size and content.

6. The presence of fine isolated inclusions of TiN in the silicon nitride matrix does not decrease the oxidation resistance of the composite. For materials containing a continuous skeleton of TiN, a noticeable increase of the oxidation rate was observed.

7. Formation of a viscous silicate film on the surface of the composites can protect them from further oxidation, even at high TiN contents.

8. Oxide scales on both TiN-containing and TiN-free ceramics exhibit a layered structure with microstructural variations from their surface to the reaction interface.

Acknowledgements

Thanks are due to Dr. V. P. Yaroshenko (Institute for Problems of Materials Science, Kiev, Ukraine) for hot pressing of the specimens.

References

1. K. L. Luthra, *J. Am. Ceram. Soc.*, **74** (1991), 1095–103.
2. L. U. Ogbuji, *J. Am. Ceram. Soc.*, **75** (1992), 2995–3000.
3. Yu. G. Gogotsi and V. A. Lavrenko, *Corrosion of High-Performance Ceramics*, Springer, Berlin-Heidelberg, 1992.
4. P. Andrews and F. L. Riley, *J. Europ. Ceram. Soc.*, **7** (1991), 125–32.
5. Yu. G. Gogotsi and F. Porz, *Corrosion* Sci., **33** (1992), 627–40.
6. Yu. G. Gogotsi, G. Grathwohl, F. Thümmler, V. P. Yaroshenko, M. Herrmann and Ch. Taut, *J. Europ. Ceram. Soc.*, **11** (1993) 375–386.
7. A. Bellosi, G. N. Babini, H. Li-Ping and F. Xi-Ren, *Mat. Chem. Phys.*, **26** (1990), 21–33.
8. P. Andrews and F. L. Riley, *J. Europ. Ceram. Soc.*, **5** (1989), 245–56.
9. W. D. Kingery, H. K. Bowen and D. R. Uhlmann, *Introduction to Ceramics*, Wiley, New York (1976).
10. Yu. G. Gogotsi, F. Porz and G. Dransfield, *Oxidation of Metals*, **39** (1993), 69–91.

In Situ Studies of the Surface Changes Occurring During High-Temperature Oxidation of Silicon Nitride

R. J. FORDHAM, J. F. NORTON, S. CANETOLI and J. F. COSTE*

Commission of the European Communities, Institute for Advanced Materials,
JRC-Petten Establishment, Postbus 2, 1755 ZG Petten, The Netherlands
*Ecole Nationale Supérieure des Mines de Saint-Etienne, Département Sciences des
Matériaux, 158 Cours Fauriel, 42023 Saint-Etienne 2, France

ABSTRACT

The air oxidation of a hot-pressed silicon nitride ceramic containing yttrium has been studied by hot-stage microscopy at temperatures up to 1160°C. Previous studies have shown that silicon nitrides exhibit reduced oxidation resistance at relatively low temperatures if densification has been achieved using yttrium oxide additions compared with other metallic oxides. Evidence supports the hypothesis that this accelerated oxidation is associated with low-temperature oxidation in the grain boundaries of yttrium silicates containing nitrogen. The hot-stage micros-copy was complemented by post-exposure examinations using SEM-EDX, EPMA and XRD.

Between 750°C and 920°C oxidation of the grain boundary phases [mainly Y-N-apatite] progressed by an oxygen diffusion-controlled mechanism to form Y-O-apatite and a silicate glass. Between 920°C and 950°C the glass became liquid and a vigorous evolution of nitrogen bubbles was observed. Prolonged exposure at 950°C and above leads ultimately to the formation and crystallisation of Y_2SiO_5 as well as SiO_2.

This paper concentrates on the visual observations made and confirms the previously established oxidation model.

1. Introduction

Silicon nitride is one of the more important 'new' or advanced ceramic materials which show great promise for high temperature and strength applications at the limit of, or beyond, the capabilities of 'conventional' high-performance alloys. Good oxidation or corrosion resistance is vitally important for the extended lifetimes demanded. As with metallic alloys and other non-oxide ceramics, this long-term resistance is reliant upon the formation of a

545

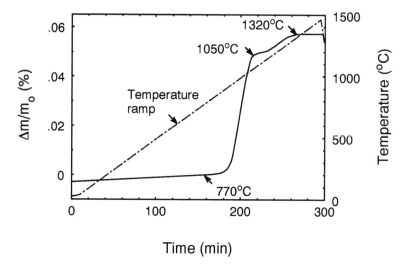

Fig. 1 Thermogravimetric oxidation of HPSN[Y] in air; RT → 1450°C at 5°C/min ramp (after Veyret[1]).

stable oxide layer. In the case of silicon nitride, the oxide is silica which is an efficient barrier to oxygen diffusion and further oxidation. However, silicon nitrides must be fabricated with additions of metallic oxides (up to 15wt%) and are not monophase materials; the oxide additives lead to the formation of secondary phases which are to be found in grain boundaries or inclusions. The high-temperature performance of these materials is often critically limited by the properties of the secondary phases. Ideally, they are refractory crystalline silicates so that the high-temperature properties of the silicon nitride itself may be exploited.

A typical example is given by silicon nitrides fabricated using additions of yttria. The pseudo-ternary phases $Y_{10}(SiO_4)_6N_2$ (known as H-phase or Y-N-apatite) and $YSiO_2N$ (known as K-phase or Y-N-α-wollastonite) occur frequently in materials densified with Y_2O_3. These phases have been known for some time to oxidise rapidly at temperatures below 1000°C with an accompanying volume change which is nearly always detrimental to the material. At higher temperatures the formation of silica, from the oxidation of silicon nitride itself, is believed to be beneficial and to prevent or reduce the otherwise adverse effects of the oxidation of the secondary phase(s). Figure 1[1] illustrates the air oxidation on a thermobalance of a commercial grade of hot-pressed silicon nitride containing both H- (mainly) and K- (minor) phases as well as $FeSi_2$ in inclusions. The first sign of oxidation is at 770°C. Si_3N_4 itself only begins to oxidise significantly above 1050°C, as shown by the change in slope of the thermogram. The oxidative weight gain in the range 770–1050°C represents primarily the reaction of the H-phase with oxygen and a model has been developed to explain the behaviour of the material via the formation first of a Y-O-apatite analogue of H-phase followed by further oxidation to $Y_2Si_2O_7$ accompanied by the release of molecular nitrogen.[2]

The present work illustrates the use of hot-stage microscopy to observe the oxidation in real time. It will be seen that the model, with minor variations, is fully supported.

2. Experimental

The material chosen was Ceranox NH209 (*ex*. Feldmühle Ag, BRD), a dense (3.32 Mg m^{-3}) hot-pressed (40 MPa at 1750°C) silicon nitride containing the equivalent of ~9.3wt% Y_2O_3 and ~1.7wt% Fe_2O_3 as well as traces of MgO (0.01wt%), CaO (0.05wt%) and Al_2O_3 (0.12wt%). X-ray diffraction analysis shows the presence of β-Si_3N_4, H-phase, $FeSi_2$ and, possibly, a small amount of K-phase. Transmission electron microscopy identified only the H-phase and $FeSi_2$ as secondary phases and found no evidence of a vitreous phase. The microstructure of the material is illustrated in Fig. 2. From the optical micrograph four phases are identifiable, in order of importance:

1. *Mid-grey*, which fills the majority of the surface and is therefore the β-Si_3N_4; this appears black in the SEM micrograph.
2. *Dark grey*, the second most abundant phase which is identified by EPMA as H-phase and is bright in the SEM.
3. *Brilliant white*, which represents a phase present in isolated spots to about 2% of the surface area and which are sometimes found locally in great concentration. From their metallic appearance and high refractive index, they are believed to be $FeSi_2$, and

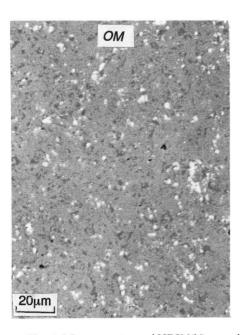

OM

20μm

SEM

10μm

Fig. 2 Microstructure of HPSN(Y) revealed by (a) optical microscopy and (b) SEM.

4. *Intermediate grey*, only visible in isolated places and representing only a very small proportion of the surface and believed to be K-phase.

The oxidation behaviour of the material was studied using a home-built hot-stage microscope with a video recording attachment to furnish a permanent record for detailed image analysis. Individual frames were grabbed and printed using a VIDAS 21 Image Processing End Analysis System [Zeiss Kontron Elektronik GmBH]. The hot-stage apparatus is described in detail elsewhere in these proceedings.[3] The specimens were prepared by cutting thin blanks [4 mm ϕ] which were then lapped and polished on the upper face to a finish of 1 μm using an automatic machine to ensure reproducibility (Struers Abramin). The final thickness of the specimens varied between 200 and 400 μm. The environment used for the hot-stage microscopy was static laboratory air. At first, a ramp from RT to 1160°C at 50°C. min^{-1} was followed to record the temperature of the onset of visible changes and to establish the overall behaviour of the material under these conditions. Thereafter a series of isothermal exposures at various temperatures for up to 3 h was carried out after rapid heating [80°C.min^{-1}] to the dwell temperature. Optical magnifications of $\times 320$ were used which after processing through the video were enhanced to apparent magnifications of $\times 1000$.

After cooling to RT the surfaces of the specimens were examined by XRD, SEM-EDX and EPMA to identify as far as possible the products of the oxidation and to relate them to the features visible during the HT microscopy.

3. Results and Discussion

3.1 Ramp from RT → 1160°C at 50°C min^{-1}

A series of video images at a range of temperatures from this experiment is shown in Fig. 3. Until 750°C there were no visible changes to the material. At about 750°C very dark spots appeared on the surface which grew and spread until nearly the whole surface was covered at about 900°C. At 920°C there was the first appearance of liquid in very isolated spots. This was evident due to the bursting of a bubble, presumably containing nitrogen. As the temperature continued to rise the liquid phase accompanied by the bursting of bubbles became more and more evident until by 1000°C the whole surface appeared to 'boil'. Further heating to 1060°C showed the first appearance of solid, presumably crystalline material in the 'boiling' liquid and the viscosity of the liquid was seen to increase as the 'boiling' slowly subsided by 1160°C, the end of the experiment.

3.2 Isothermal test at 950°C

This experiment is illustrated by a series of video frames over a period of 180 min at temperature [Fig. 4]. The first frame when the temperature had just reached 950°C shows still a preponderance of the very dark phase which was spreading rapidly over the surface. After 1 min liquid was seen to be spreading

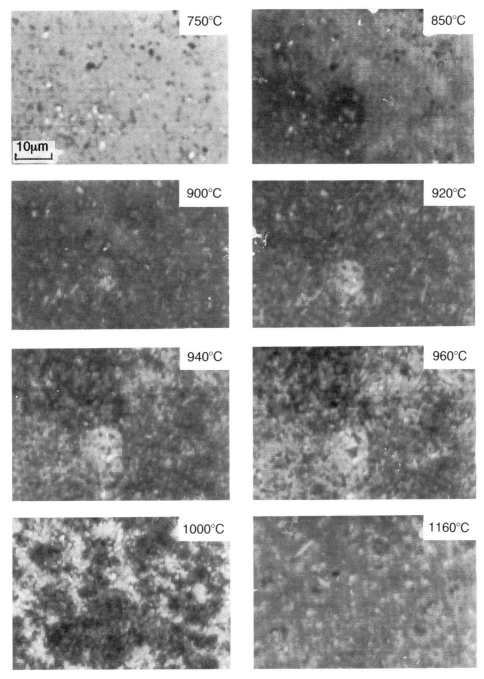

Fig. 3 Hot-stage microscopy video frames from RT → 1160°C ramp at 50°C/min in air.

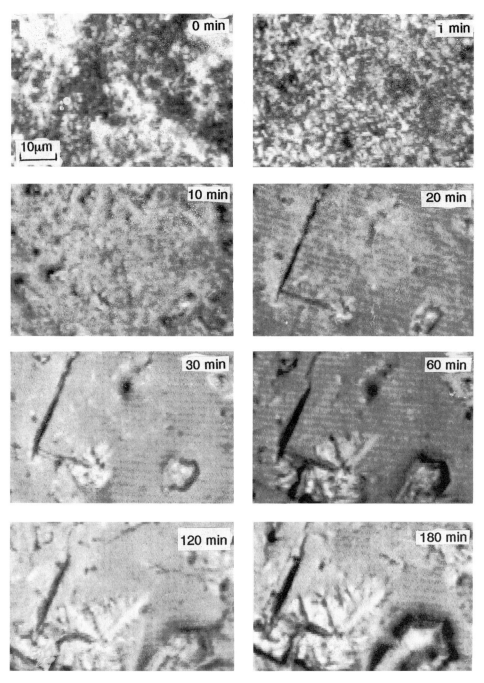

Fig. 4 Hot-stage microscopy video frames from isothermal exposure at 950°C in air.

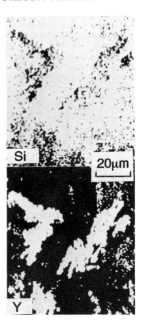

Fig. 5 Surface morphology after 180 min at 950°C revealed by (a) SEM and (b) elemental maps for Si and Y from EPMA analysis.

over the surface which was then bubbling vigorously after 10 mins. The first crystals were seen to form after 20 mins at temperature and the remaining frames show the growth of the crystalline product in the viscous liquid. The surface continued to bubble, although less vigorously, even after 180 min at temperature. After cooling to RT, Y_2SiO_5 was identified by XRD as one oxidation product in addition to SiO_2. K-phase was also evident by XRD. Figure 5 shows both an SEM micrograph and two element maps (for Si and Y) from the EPMA. The concentration of the Y-rich crystalline material is clearly seen.

3.3 Isothermal test at 980°C

Figure 6 illustrates three video frames after 0, 20 and 120 min exposure. The contrast between these frames is very strong. After 20 min exposure the surface is boiling vigorously but after 120 min, although nitrogen gas is still being evolved, crystallites are clearly evident and the viscosity of the liquid has significantly increased.

In the context of the model developed by Veyret and Billy,[2] these observations can be explained by the oxidation first of the intergranular Y-N-apatite to Y-O-apatite with a small volume expansion and extrusion and spreading of the O-apatite over the surface. Nitrogen is largely retained within this phase until the formation of the liquid when it is released violently. Experiments with pure Y-N-apatite do not show the formation of a liquid

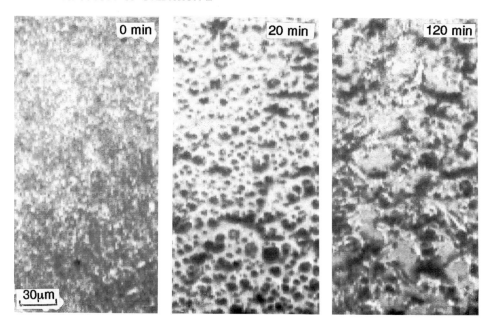

Fig. 6 Hot-stage microscopy video frames from isothermal exposure at 980°C in air.

during oxidation to Y-O-apatite and then yttrium silicates. It was observed that the liquid and N_2 bubbling appeared first at regions which had originally shown the presence of the white phase $FeSi_2$. It is proposed that the liquid results from the formation of a low melting eutectic glass containing Fe and other impurity elements such as Al. After cooling, some Fe was still found to be present as $FeSi_2$ in isolated pockets although at a lower surface concentration than before exposure.

Two modifications of the earlier model seem therefore to be appropriate, as illustrated in Fig. 7: the formation of a low-viscosity molten glass covering the whole surface at a relatively low temperature and the violent release of nitrogen gas are remarkable and were unexpected phenomena, as was the formation after prolonged exposure of Y_2SiO_5 rather than $Y_2Si_2O_7$, which is the silicate fully compatible with SiO_2. Longer exposure times may be expected to lead to oxidation to the higher silicate.

The modified model shown in Fig. 7 represents a summary of the observations from these hot-stage microscopy experiments in the context of the earlier model.

4. Conclusions

1. Surface phenomena on a silicon nitride ceramic exposed to air between 750 and 1160°C have been successfully observed in real time.

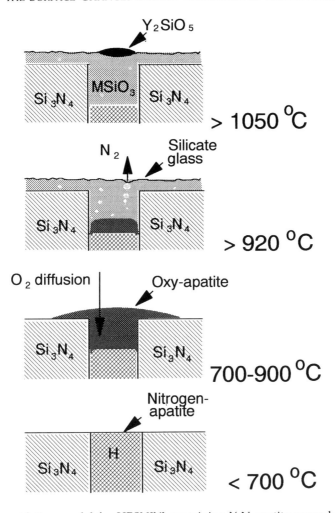

Fig. 7 Air oxidation model for HPSN[Y] containing Y-N-apatite secondary phase (after Veyret and Billy[2]) revised by present work.

2. The observations have led us to propose minor modifications to an existing oxidation model. These are that:
3. The oxide film is liquid at very low temperatures and outgassing of N_2 occurs violently.
4. Crystalline products (initially Y_2SiO_5) separate from the liquid as temperature or time is increased.

Acknowledgement

The contribution of Dr J. B. Veyret to the thermogravimetry of this material and the development of the oxidation model is gratefully acknowledged.

Thanks are due also to M. Spreij for some of the hot-stage experiments and to K. Schuster and M. Moulaert for the SEM and microprobe examinations. Finally, one of the authors (J. F. Coste) wishes to thank the Commission of the European Communities for the award of a grant to execute this study.

References

1. J. B. Veyret, Comportement à l'oxydation de céramiques à base de nitrure de silicium en atmosphère complexe air/SO$_2$, doctoral thesis, Université de Limoges, France (1989).
2. J. B. Veyret and M. Billy, Oxidation of hot-pressed silicon nitride: modelling, in *Euro-Ceramics*, Vol. 3; *Engineering Ceramics*, Proceedings of 1st European Ceramic Society Conference, Maastricht (Netherlands) (June 1989), (G. de With, R. A. Terpstra and R. Metselaar, eds), Elsevier Applied Science (1989), pp. 512–6.
3. J. F. Norton, S. Canetoli and P. Pex, The use of a hot stage microscope in high-temperature corrosion studies (these Proceedings).

The Hot Salt Corrosion of Silicon Nitride in a Burner Rig

D. J. BAXTER, P. MORETTO, A. BURKE and J. F. NORTON

Commission of the European Communities Institute for Advanced Materials Joint Research Centre Petten Establishment 1755 ZG Petten, The Netherlands

ABSTRACT

A commercially produced monolithic silicon nitride was corrosion tested in a low-velocity burner rig in the temperature range 800°C to 1100°C at ambient pressure. A marine-type gas turbine environment was created by burning a high-sulphur fuel in air contaminated with sea salt. At temperatures less than 1000°C deposition of sodium sulphate on the surfaces of the ceramic occurred following reaction of sodium chloride with SO_2/SO_3. Above the dew point of Na_2SO_4 (approximately 950°C) the main sodium-containing species was NaOH vapour. In the 50-h tests, a predominantly SiO_2 surface layer formed which provided protection to the underlying ceramic. Both Na from the salt contaminant and Y, present as a sintering agent in the Si_3N_4, contributed to the corrosion process. The mechanism of corrosion is discussed and a preliminary assessment of the effect of corrosion on the flexural strength of the material is presented.

1. Introduction

Increasing energy efficiency and reducing environmental emissions are major considerations in the design of future power plants. Greater efficiency can often be obtained by increasing the process operating temperature which has the consequence of placing more stringent demands on the materials of construction. Superalloys have for a number of years satisfied the lifetime requirements of critical components in various types of plant operating at elevated temperatures. These materials have, however, virtually reached their limit of exploitation in terms of both metallurgical development and engineering design. Maintenance of mechanical properties to very high temperatures, coupled with low density, makes ceramic materials attractive alternatives to the currently used metal counterparts. Hence, ceramic-base structural materials are being developed in an attempt to provide the possibility of achieving the efficiency and environmental objectives.

Considerable work has been done over the last two decades to improve mechanical behaviour and identify failure mechanisms of engineering ceramics. Exposure at elevated temperatures clearly leads to the possibility of environmentally induced degradation, with corrosion becoming a potentially serious concern in some systems.[1,2] Non-oxide ceramics such as silicon carbide (SiC) and silicon nitride (Si_3N_4) form a surface oxide layer composed of silica (SiO_2) upon exposure to an oxidising environment.[3,4,5] In principle, the slow-growing oxide is capable of providing a very high degree of protection to the underlying ceramic up to temperatures of around 1500°C.[6] Contaminants, such as the salts NaCl and Na_2SO_4, typically present in gas turbine atmospheres, can cause accelerated rates of corrosion of SiC and Si_3N_4.[7,8,9] In addition, corrosion not only affects component surfaces, causing a loss of load-bearing section, but can induce significant changes in the strength of the bulk material.[10,11] Experimental methods for studying salt-induced corrosion range from simple laboratory crucible tests to actual engine tests. High cost precludes the wide use of engine tests to study corrosion, while the static nature of most laboratory methods renders the results obtained difficult to relate to real operating conditions. An attractive alternative is the burner rig, which provides a dynamic corrosive environment in which condensed phase contaminants can be continuously added in a controlled manner.

Using a low-velocity burner rig, a commercially produced hot-pressed Si_3N_4 was corrosion tested in the temperature range 800°C to 1100°C in the environment produced by combusting a marine-grade high-sulphur fuel contaminated with artificial ocean water. The temperature range used in the investigation covered regimes in which solid Na_2SO_4 would be expected to deposit on exposed surfaces of the ceramic up to the melting point (884°C), with liquid deposition up to the dew point (approximately 950°C). The higher temperatures of test are above the dew point of Na_2SO_4.

2. Materials and Test Conditions

2.1 Test Material

The Si_3N_4 used in the study was a commercially available grade (Feldmuhle AG, Germany), produced by hot pressing at 1750°C under a pressure of 40 MPa using Y_2O_3 as the sintering additive. The resulting ceramic had a density of 3.33 g cm^{-3} with 2% open porosity. The microstructure contained nitride grains up to 20 μm long and an intergranular phase with an apatite structure, identified as $Y_5(SiO_4)_3N$ by X-ray diffraction. Precipitates of $FeSi_2$ were also present. The full chemical composition of the ceramic was 1.7% Fe_2O_3, 0.12% Al_2O_3, 0.05% CaO, 0.01% MgO and 0.01 ZrO_2, with 9.0% Y_2O_3 (in weight %). Test specimens were diamond cut to 3.5 × 4.5 × 55 mm in size, the edges chamfered and faces polished longitudinally to minimise transverse defects that might adversely affect four-point bend test strength determination.

Specimens were cleaned, measured and weighed prior to exposure in the burner rig.

2.2 Test Conditions

The specimens were placed vertically in a holder, allowing 45 mm of their length to be exposed to the gas stream. (The four-point bend test required a uniform length of 40 mm.) The burner rig was preheated to the required test temperature before the specimens were inserted. After a period of 15 min for the specimens to reach the test temperature, the fuel and salt supplies were turned on.

A 1% (by weight) sulphur marine-grade diesel fuel was added to the furnace in the form of an atomised spray at a rate of 72 ml h^{-1}. Artificial ocean water conforming to the standard, ASTM D1141-86 was added in a similar manner at a rate of 12.5 ml h^{-1}. The ratio of atomising air to fuel was 28:1 by weight, providing excess oxygen to the combustion process. The partial pressure of SO_2 was calculated, assuming thermodynamic equilibrium, to be almost constant at 2×10^{-4} bar, while the partial pressure of SO_3 decreased from 10^{-4} to 10^{-5} bar over the test temperature range. However, owing to the relative slowness that an SO_2/SO_3 equilibrium is reached, it is possible that the above levels of SO_3 are not attained under the dynamic conditions prevailing in the burner rig. The residence time of gases in the rig is approximately 3 s. Nevertheless, reaction of NaCl in the ocean water with SO_2 and/or SO_3 produces Na_2SO_4. Deposition of Na_2SO_4 on exposed surfaces should be solid below the melting point (884°C), although the additional presence of NaCl in the deposit can contribute to a eutectic phase with a melting point below 700°C. Evidence of a molten product on the surfaces of specimens exposed at 800°C was indeed found, but post-exposure analysis failed to confirm the presence of NaCl. Liquid Na_2SO_4 was deposited at 900°C. The dew point of Na_2SO_4 is dependent on both system pressure and the concentrations of Na and S in the gas stream. At the ambient operating pressure of the burner rig, the dew point was calculated to be approximately 950°C. Sodium sulphate vapour is formed above the dew point, but most Na is present as NaOH, which increases in concentration with temperature.

Eight cycles of 6.25 h made up a total exposure time of 50 h at each temperature. After each cycle the specimens were removed from the hot furnace and cooled in laboratory air. Before reaching 100°C, the warm specimens were placed in a desiccator to prevent the pick-up of moisture. Included in each test were 'inert' alumina blanks placed alongside the Si_3N_4 specimens. By measuring weight changes, these blanks enabled monitoring of the steadiness of conditions in the burner rig to be undertaken. Flexural strength of the silicon nitride was determined from four-point bend tests. The corroded specimens were analysed using a combination of X-ray diffraction, scanning electron microscopy (SEM), with energy-dispersive analysis, and electron probe microanalysis (EPMA), with wavelength-dispersive analysis methods.

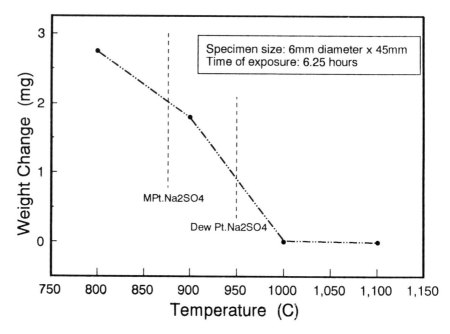

Fig. 1 Weights of salt deposited on 'inert' alumina specimens during 6.25-h exposure cycles.

3. Results and Discussion

3.1 Salt Deposition

The alumina blanks, exposed simultaneously with the Si_3N_4, gained weight due to the deposition of Na_2SO_4 in a consistent manner at temperatures of 800°C and 900°C. The amount of salt deposition at 1000 and 1100°C on the alumina blanks was negligible (Fig. 1). The gross weight changes of the Si_3N_4 after each cycle of exposure are shown in Fig. 2. Analysis of the Si_3N_4 specimen surfaces by X-ray diffraction revealed the major phase present to be Na_2SO_4 after exposure at 800°C and 900°C. Examination of fracture sections in the SEM showed the salt layer to be up to $30\,\mu m$ thick at the centre of specimen faces, thinning to 5-$10\,\mu m$ at the edges of the 800°C-exposed specimens. The shape of the weight change curve (Fig. 2) actually suggests that salt spalling occurred after the second and third cycles at 800°C. At 900°C, the salt layer was discontinuous with a maximum thickness of only $3\,\mu m$. Sodium sulphate was not detected on the specimens exposed at 1000°C and 1100°C.

3.2 Corrosion Behaviour

As described above, the major surface phase on the 800°C specimens was the deposited Na_2SO_4. The actual corrosion product, revealed by washing the specimens in distilled water, was very thin (less than $1\,\mu m$), composed of SiO_2,

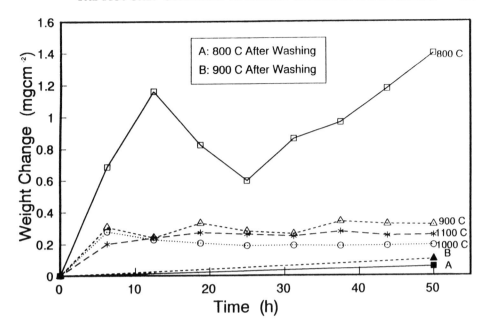

Fig. 2 Weight changes of the Si_3N_4 specimens directly after removal from the burner rig.

and discrete particles of a phase containing Y, Si and O (Fig. 3a). Previously,[12] it was found that short time exposure of the same ceramic to air/1%SO_2 at 800°C results in the formation of SiO_2, an yttrium silicate phase and $Y_2(SO_4)_3$. With the aid of kinetic data, it has been proposed[3] that, for air oxidation, a very thin layer of SiO_2 forms only on Si_3N_4 grains, while $Y_{4.67}(SiO_4)_3O$ is the product of oxidation of the intergranular oxynitride apatite phase. In the present work at 800°C, Y sulphate was not detected despite the availability of sulphur/sulphate in the deposited Na_2SO_4. However, Na_2SO_4 should be relatively stable at this temperature and the partial pressure of SO_2 in the burner rig was low (300 ppm) compared with the 1% level in the previous work.[12]

After exposure at 900°C, coverage of SiO_2 on the original Si_3N_4 grains was more apparent, with voids present at the oxide grain boundaries (indicated by arrows in Fig. 3b). The voids are associated with the production of nitrogen gas during the oxidation of Si_3N_4. Also shown in Fig. 3b are the extended islands of salt (Na_2SO_4), and protruding through the salt layer, larger particles of the oxidised apatite phase. At 800°C and 900°C weight changes due to corrosion were overwhelmed by the effects of salt deposition and therefore different stages in the corrosion process could not be identified. Most of the weight gained could be removed by washing the specimens. From the net weight gains, and assuming parabolic kinetics, apparent reaction rate constants of 3×10^{-8} and 6×10^{-8} mg^2 $cm^4 s^{-1}$ at 800°C and 900°C, respectively, were calculated.

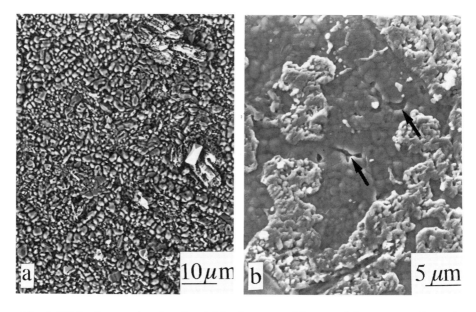

Fig. 3 SEM micrographs showing (a) back-scattered image of the outer surface of the scale formed at 800°C, after washing, and (b) the silica-rich scale partially covered with salt deposit after exposure at 900°C.

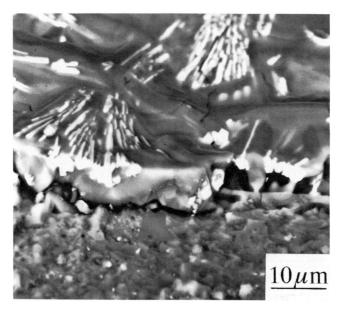

Fig. 4 Back-scattered image of a fracture section through the thin scale formed at 1000°C.

Approximately 50% of specimen surface areas were covered by a scale 4–7 μm thick and composed of a continuous, void-containing layer of SiO_2 with needles of $Y_2Si_2O_7$ embedded in the outer region after exposure at 1000°C (Fig. 4). While this scale morphology would normally be associated with protective corrosion kinetics, a much thicker (15 μm) layer with more complex morphology was formed in patches up to 400 μm in diameter over the remainder of the specimen surfaces. The X-ray maps in Fig. 5 show the distribution of the elements forming the main phases in the thicker scale. These phases were SiO_2 (tridymite) doped with Na, $Y_2Si_2O_7$ as well as Na and Mg (the latter not shown in the figure) doped silicate, possibly miserite, previously reported by Babini *et al.*[13] and a phase virtually free of silicon and containing Y, S and O, deduced to be $Y_2(SO_4)_3$. It was evident that within the 50 h of exposure, formation of a continuous layer of SiO_2, undercutting the silicate and sulphate phases, had not occurred. In a previous study[14] it was observed that at 1000°C sodium silicate is the major phase in the first hour of exposure, but SiO_2 soon develops and effectively seals the silicate off from the base ceramic.

Corrosion at 1100°C was characterised by an initial period of rapid weight gain, followed by very slow scale growth at an approximate parabolic rate of 10^{-7} mg^2 cm^{-4} s^{-1}. The dramatic slowdown in corrosion rate can be attributed to the completion and subsequent thickening of the continuous SiO_2 layer.[3] The micrograph in Fig. 6 shows that the scale is not only thinner than the scale formed at 1000°C, but that the $Y_2Si_2O_7$ (white material) has a different, more compact morphology and a second SiO_2 phase is present in needle form (arrow). The needles were an almost pure form of SiO_2. The remaining SiO_2 contained a significant concentration of sodium. In the temperature range 900°C to 1100°C, tridymite is the main stable form of SiO_2 whose formation is not only favoured by monovalent metal ions,[15] but can accommodate up to 9 mol% Na.[16]

3.3 Mechanical Effects

Corrosion of silicon-base ceramics can either strengthen or weaken the bulk material depending upon their microstructure and the mode of corrosive attack. Since intergranular sintering phases to a large extent determine the mechanical properties of engineering ceramics, the performance of the inter-granular phase under corrosive attack would be expected to be a major factor determining changes in mechanical behaviour. The intergranular phase was fully crystalline in the as-received material and its structure in the bulk of the specimens did not change measurably during corrosion exposure at elevated temperatures. The room temperature flexural strength of the 9% Y_2O_3, Si_3N_4 was found to increase after exposure at 800°C, with negligible change in strength after 900°C and 1000°C, before showing an 18% reduction after exposure at 1100°C (Fig. 7). Surface scales may or may not contribute to the strength of the 'composite' and loss of section due to scale formation can usually only account for a very small decrease in strength. More importantly, the strength of brittle materials such as Si_3N_4 is strongly dependent upon the

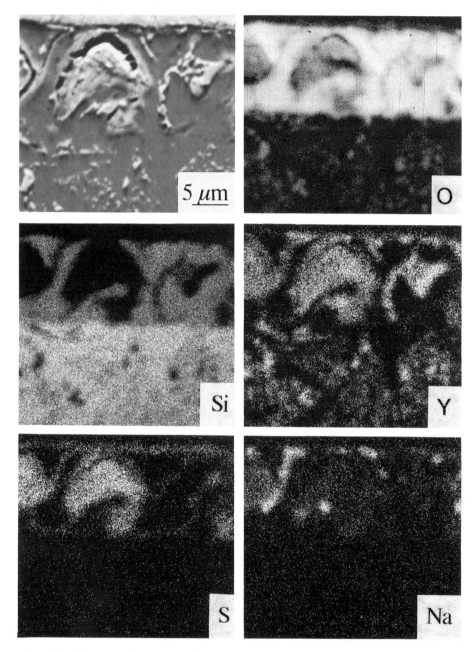

Fig. 5 Back-scattered image with X-ray maps of the thick scale formed at 1000°C.

Fig. 6 Back-scattered image of a fracture section through the scale formed at 1100°C.

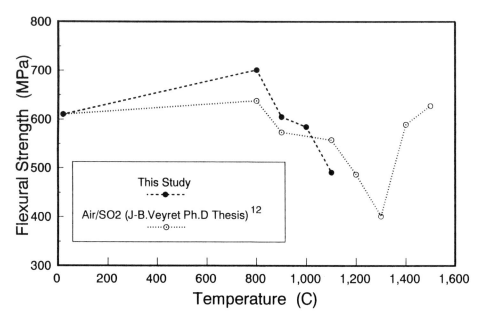

Fig. 7 The effect of corrosion test temperature on flexural strength.

presence of defects. Corrosion can, in particular, affect the size and population of surface defects and therefore influence mechanical behaviour.

After 50 h exposure at 800°C, corrosion had hardly changed the morphology of the machined surfaces of the specimens (Fig. 3a). In this case the strength increase can be attributed to the very thin surface scale smoothing over the pre-existing machining flaws.[11] While the specimen surfaces were not marked-ly roughened through exposure at 1100°C, surface corrosion pits of the order 30–50 μm in size have previously been correlated with loss of strength of Si_3N_4.[8] Further work is required to establish the mechanism affecting the change in strength observed in the material used in this investigation.

4. Conclusions

A low-velocity burner rig permits close control of corrosion experiments in which the test environment contains both gaseous and condensed phases. From a study of the corrosion behaviour of Si_3N_4, hot-pressed with 9% Y_2O_3, in an environment created by burning a high-sulphur marine-grade diesel fuel in air contaminated with artificial ocean water, the following conclusions can be drawn.

1. The Y_2O_3-modified Si_3N_4 forms a thin Si-rich surface scale at temperatures of 800°C, 900°C and 1100°C, while at 1000°C, sulphur plays a significant role in the formation of a mixed, relatively thick scale.
2. Temperature dictates the rate of Na_2SO_4 deposition, from zero above the dew point (approximately 950°C) to a small amount of liquid at 900°C and a 30-μm-thick layer in 50 h at 800°C. The deposit was not a significant contributor to the corrosion process.
3. The participation of Na increases with increasing temperature, but does not lead to breakdown of the silica layer within 50 h at 1100°C.
4. Room temperature flexural strength is increased after exposure at 800°C, unchanged after 900°C and 1000°C and reduced by 18% after exposure at 1100°C.

Acknowledgements

The authors wish to express their thanks for the contributions made by S. Canetoli, P. Frampton, Ph. Glaude, M. Moulaert, K. Schuster and K. Teipel throughout the course of the work.

References

1. F. Costa Oliveira, R. A. H. Edwards and R. J. Fordham, in *High Temperature Corrosion of Technical Ceramics* (R. J. Fordham, ed.), Elsevier Applied Science (1990), pp. 53–67.

2. J. Luyten, P. Lemaitre and A. Stalios, ibid, pp. 161–168.
3. J. B. Veyret and M. Billy, *Euro-Ceramics*, Vol. 3 (G. de With *et al.*, eds.), Elsevier Applied Science (1989), pp. 512–516.
4. M. Billy and J. G. Desmaison, *High Temp. Technol.*, **4** (1986), 131–139.
5. S. C. Singhal, *J. Mater. Sci.*, **11** (1976), 500–509.
6. D. Cubicciotti and K. H. Lau, *J. Am. Ceram. Soc.*, **61** (1978), 512–517.
7. W. C. Bourne and R. E. Tressler, *Ceram. Bull.*, **59** (1980), 443–452.
8. D. S. Fox and J. L. Smialek, *J. Am. Ceram. Soc.*, **73** (1990), 303–311.
9. N. A. Jacobson and J. L. Smialek, *J. Am. Ceram. Soc.*, **68** (1985), 432–439.
10. G. A. Gogotsi, Y. U. Gogotsi, V. P. Zavada and V. V. Traskovsky, *Ceram. Int.*, **15** (1989), 305–310.
11. K. Jakus, J. E. Ritter and W. P. Rogers, *J. Am. Ceram. Soc.* **67** (1984), 471–475.
12. J. B. Veyret, Ph.D. Thesis, JRC Petten (1989).
13. G. N. Babini, A. Bellosi and P. Vincenzini, *J. Mater. Sci.*, **19** (1984), 3487–3497.
14. D. S. Fox and N. S. Jacobson, *J. Am. Ceram. Soc.*, **71**, 1988, 128–138.
15. G. D. Rieck and J. M. Stevels, *J. Soc. Glass Technol.*, **35** (1951), 284–288.
16. S. H. Garofalini and A. D. Miller, *J. Cryst. Growth*, **78** (1986), 85–96.

Understanding the Morphological Development of Oxidation Products Formed on β-sialon Materials with Different z-values

M. J. POMEROY and R. RAMESH

University of Limerick, Limerick, Ireland

ABSTRACT

The oxidation behaviour of β-sialons ($Si_{(6-z)}Al_zO_zN_{(8-z)}$) with z-values of 0.2, 0.5, 1.0, 1.5, 2.0 and 3.0 at 1350°C is reported for times up to 256 h. The sialons can be grouped into two categories with respect to the predominant crystalline phases formed in the surface layers. For the z = 0.2 to 1.0 materials, yttrium disilicate or cristobalite are the major surface phases. For the higher z-value materials, mullite is the predominant phase. The development of surface morphology is explained in terms of Y and Al diffusion effects which stabilise a liquid phase and the manner in which excesses of silicon, aluminium and/or yttrium are precipitated from this liquid.

1. Introduction

The literature[1–5] shows that z-value has a significant effect on the nature of crystalline phases formed in oxide scales developed on β-sialons during oxidation. The references cited have also drawn attention to the dependence of oxidation rates on additive diffusion effects and have reported the formation of liquid phases during oxidation. The work reported here attempts to define more closely the effect of z-value, additive diffusion effects and liquid formation on the phases arising in the surface layers after oxidation for various time periods.

566

2. Experimental procedure

β-sialons corresponding to nominal z-values of 0.2, 0.5, 1.0, 1.5, 2.0 and 3.0 containing, theoretically, 11-v/o Y-Si-Al-O-N glass were prepared by ball milling Si_3N_4, Y_2O_3, Al_2O_3 and AlN powders for 24 h in isopropanol using sialon milling media. The mixtures were then homogenised for 30 min, dried and sieved to <212 µm. They were then cold isostatically pressed into $40 \times 15 \times 15$ mm bars and pressureless sintered in a powder bed (50 w/o boron nitride + 50 w/o silicon nitride) at 1700°C for 2 h in flowing nitrogen. The relevant compositions are given in Table 1. The oxidation experiments were carried out in a tube furnace at a temperature of 1350°C for 1 h, 32 h, 64 h, 128 h and 256 h using specimens ($10 \times 10 \times 2$ mm) of each of the compositions, polished to a 1 µm finish. The phase assemblage of the surface products was determined using X-ray diffraction (XRD) while the morphology of the surfaces was examined using back scattered image (BEI) techniques. Energy-dispersive

	Base material	1 h	32 h	64 h	128 h	256 h
z = 0.2						
Y	2.3	24	20	30	35	30
Si	93.8	70	75	67	61	57
Al	3.9	6	6	3	4	13
z = 0.5						
Y	2.4	9	4	18	19	15
Si	90.0	86	92	77	76	82
Al	7.6	5	4	5	6	3
z = 1.0						
Y	2.4	6	9	10	6	6
Si	83.6	79	65	52	75	73
Al	14.0	15	26	38	19	22
z = 1.5						
Y	2.5	5	11	8	9	8
Si	76.9	69	47	45	50	46
Al	20.6	26	42	46	41	47
z = 2.0						
Y	2.5	8	6	6	10	7
Si	70.1	59	53	52	50	47
Al	27.4	33	41	42	40	46
z = 3.0						
Y	2.6	11	7	7	8	7
Si	55.7	59	46	53	47	45
Al	41.7	30	47	40	45	48

Table 1 Cation compositions of base materials and surface layers (equivalent %)

analyses (EDA) for Si, Y and Al were also conducted over large areas (2.3 × 1.6 mm) of the surface layers in order to establish the cation content in equivalent percentages. The values quoted are the mean of three analyses and the range observed was typically 5% of the mean value.

3. Results

3.1 Phase Assemblage of Surface Products

Table 2 shows the phase assemblages observed in the external layers of the oxide scales formed. It can be seen from this table that, in general, after 64 h oxidation the phase assemblage becomes effectively constant. The materials may be classified in terms of the predominant phase formed, thus the $z = 0.2$ material is a silicate former, the $z = 0.5$ and 1.0 silica formers and the $z = 1.0$, 1.5 and 3.0 mullite formers. Table 2 shows the occurrence of the silicon oxynitride in the external layers formed on the $z = 0.2$, 0.5, 1.0 and 1.5 materials after various times of oxidation. The only other nitrogen-containing phases observed were for the $z = 3.0$ material after 1 h when trace amounts of X-phase (nitrogen mullite) and Y_2SiAlO_5N were detected.

Two further results of note arising from Table 2 relate to the type of yttrium disilicate formed in the surface layers of the $z = 0.2$, 0.5 and 1.0 materials. It is seen that γ-yttrium disilicate only arises after 64 h oxidation. Furthermore, for the $z = 0.2$ and 1.0 materials the predominant silicate formed is γ-yttrium disilicate whilst for the $z = 0.5$ material β-ytrium disilicate is the predominant form.

3.2 Elemental Analysis of Surface Layers

Table 1 compares cation compositions in equivalent percent derived from the large area analyses after the various oxidation times with those of the starting composition based on the weights of powders used to fabricate the materials. For each of the materials it is seen that there is a significant enrichment of yttrium in the surface layers after 1 h. After a period of 64 h yttrium levels become relatively constant and the yttrium levels observed for the $z = 0.2$ and 0.5 materials are significantly greater than those observed for the $z = 1.0$ to 3.0 materials. One anomalous trend is observed for the $z = 0.5$ material where a significant decrease in yttrium content is observed after 32 h; this is followed by an enrichment effect after 64 h. A similar but less significant trend may be observed for the $z = 0.2$ material. In terms of aluminium content, there is an initial enrichment of the surface layers for the $z = 0.2$, 1.0, 1.5 and 2.0 materials after 1 h (Table 1). In contrast the $z = 0.5$ and 3.0 materials show an initial depletion in aluminium. In the case of the mullite forming $z = 1.5$, 2.0 and 3.0 materials, aluminium contents become constant after 32 h oxidation. For the $z = 1.0$ material there is a gradual enrichment in aluminium up to 64 h followed by a significant decrease after 128 h after which the aluminium level is constant. Of particular interest is the low aluminium level observed for the 0.5 material after 256 h oxidation.

	Phase assemblage				
Material	**1 h**	**32 h**	**64 h**	**128 h**	**256 h**
0.2 M	β-Si$_3$N$_4$ silica β-Y$_2$Si$_2$O$_7$ Si$_2$N$_2$O[a]	Silica β-Y$_2$Si$_2$O$_7$ β-Si$_3$N$_4$	Silica β-Y$_2$Si$_2$O$_7$ γ-Y$_2$Si$_2$O$_7$	Silica γ-Y$_2$Si$_2$O$_7$[b] β-Y$_2$Si$_2$O$_7$	γ-Y$_2$Si$_2$O$_7$[b] Silica β-Y$_2$Si$_2$O$_7$ Si$_2$N$_2$O[a]
0.5 M	Silica β-sialon β-Y$_2$Si$_2$O$_7$[a] Mullite[a]	Silica β-Y$_2$Si$_2$O$_7$ β-sialon Mullite γ-Y$_2$Si$_2$O$_7$[a]	Silica β-Y$_2$Si$_2$O$_7$ γ-Y$_2$Si$_2$O$_7$ Mullite	Silica β-Y$_2$Si$_2$O$_7$[c] Mullite γ-Y$_2$Si$_2$O$_7$[a]	Silica β-Y$_2$Si$_2$O$_7$[c] Mullite Si$_2$N$_2$O[a]
1.0 M	Silica β-sialon Mullite	Silica Mullite β-sialon Si$_2$N$_2$O β-Y$_2$Si$_2$O$_7$[a]	Silica Mullite γ-Y$_2$Si$_2$O$_7$[a]	Silica Mullite γ-Y$_2$Si$_2$O$_7$[b]	Silica Mullite Si$_2$N$_2$O[a]
1.5 M	β-sialon Silica Mullite	Mullite Silica β-sialon	Mullite Silica Si$_2$N$_2$O[a]	Mullite Silica Si$_2$N$_2$O[a]	Mullite Silica
2.0 M	Silica Mullite β-sialon	Silica Mullite	Mullite Silica	Mullite Silica	Mullite Silica
3.0 M	β-sialon Silica Mullite Y$_2$SiAlO$_5$N[a] X-phase[a]	Mullite Silica β-sialon	Mullite Silica β-Y$_2$Si$_2$O$_7$	Mullite Silica β-Y$_2$Si$_2$O$_7$	Mullite Silica

[a]trace amount; [b]preferred orientation <010>; [c]preferred orientation <110>.

Table 2 Phase assemblage observed on surface layers of β-sialons after their oxidation at 1350°C for various times

3.3 Surface Texture

Figures 1(a)–1(d) show the surface textures observed for the $z = 0.2$, 1.0, 1.5 and 3.0 materials after 256 h oxidation. The $z = 0.2$ material shows a marked level of crystallinity in terms of bright acicular yttrium disilicate crystals in a darker siliceous background. The $z = 3.0$ material shows a similar texture but with needle-like mullite crystals embedded in a Y-Si-Al containing amorphous phase. The 1.0 and 1.5 materials are typified by the occurrence of needle-like mullite crystals and a grey cristobalite phase located in an amorphous Y-Si-Al-containing phase. It is clear from Figures 1(b) and 1(c) that the cristobalite content for the $z = 1.0$ material is greater than that for the $z = 1.5$ material. This observation is consistent with the XRD data presented in Table 2. The

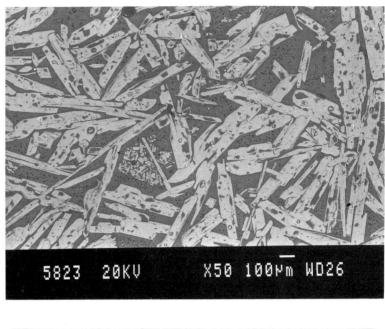

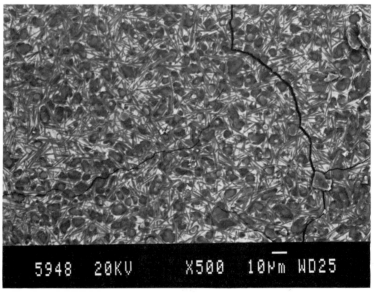

Fig. 1 Surface texture of surface products formed on β-sialon materials after 256 h oxidation at 1350°C. (a) z = 0.2; (b) z = 1.0; (c) z = 1.5; (d) z = 3.0.

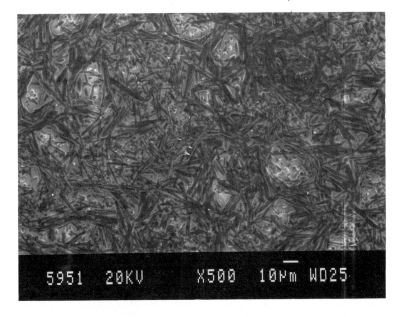

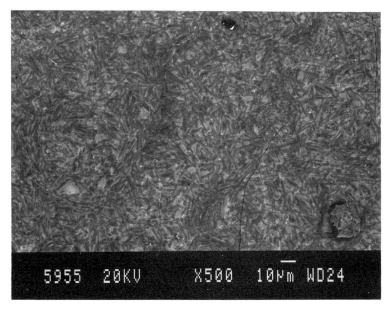

Fig. 1—contd.

surface textures are reasonably typical of those observed after various times, but most importantly they demonstrate the occurrence of an amorphous Y-containing phase.

4. Discussion of Results

With respect to the predominant phase formed in the exterior surface layers, it is seen that there is a critical z-value which must be exceeded before mullite formation occurs. This value is between 1.0 and 1.5. For lower z-values the predominant phase formed is either silica or yttrium disilicate depending on the level of yttrium enrichment in the external layers (see Table 1). Close examination of Table 1 shows that the surface layers of the 0.2 and 1.0 materials are richer in aluminium than those of the 0.5 material and this may explain the predominance of the β-yttrium silicate for the $z = 0.5$ material and the preferred stability of the γ-yttrium disilicate for the $z = 0.2$ and 1.0 materials. The occurrence of nitrogen-containing phases within the exterior layers of the materials confirms that the scales formed during oxidation contain small levels of nitrogen. Furthermore the Y_2SiAlO_5N phase only arises on devitrification of Y-Si-Al-O-N glasses at temperatures less than 1150°C. A further factor of significance is the transformation of β-yttrium disilicate to a textured γ form. Such a reaction is likely to occur via a solution precipitation process which would enable a preferred growth mechanism to operate. The lowest melting-point for an oxynitride liquid in this system would appear to be 1330°C which is close to the lowest oxide eutectic temperature (1340°C) and thus liquid phase formation should certainly occur at the temperature employed in this work. These observations, in conjunction with the surface textures observed for the oxidised materials, confirm previous clear evidence[4,5] of oxynitride liquid phase formation during oxidation.

As stated in the introduction, the exact role played by this liquid during oxidation has not been qualified. Furthermore the reasons for additive diffusion being the rate-controlling step have not been qualified either. Liquid phase cation compositions are in the range 16–24 Y, 56–60 Si and 16–24 Al (equivalent %). Examination of the starting compositions given in Table 1 shows that no compositions approach the liquid compositions. Accordingly if liquid phase formation is to occur then certain surface enrichment effects must also occur. The only manner in which such surface enrichment effects can occur is by the diffusion of aluminium and yttrium through the grain boundary amorphous phase in the ceramic. In this case the rate of liquid formation and the amount of liquid formed must be controlled by additive diffusion rates. A rate dependency on additive diffusion rates would therefore be expected. If the rate and extent of liquid formation are controlled by additive diffusion then the liquid phase must be an integral part of the mechanism of oxidation. Since the β-sialon grains are precipitated from an oxynitride liquid during sintering it would seem logical to assume that they are dissolved by the liquid film during oxidation. Any excess cation concentrations should be precipitated from the

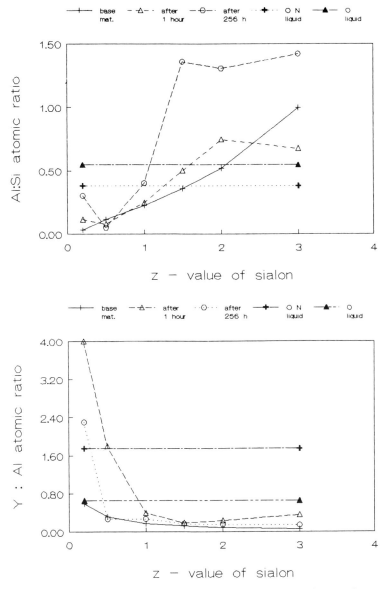

Fig. 2 (a) Variation of atomic ratios of aluminium to silicon observed in surface layers before and after oxidation for 1 and 256 h. (b) Variation of atomic ratios of yttrium to aluminium observed in surface layers before and after oxidation for 1 and 256 h.

liquid as a silica-containing species as the XRD analyses of the surface layers of the oxidised ceramics indicate.

Figure 2(a) shows the variation of aluminium to silicon ratio with z-value and enables comparison with that required for liquid formation. It is seen from Fig. 2(a) that for z-values of 0.2, 0.5 and 1.0 aluminium enrichment effects are required. It would appear that no aluminium enrichment effects are required for the $z = 1.5$, 2.0 and 3.0 materials since the Al:Si ratios for these materials are close to (1.5 and 2.0) or in excess of (3.0) those required for liquid formation. Figure 2(b) shows the variation in Y:Al ratio with z-value and enables comparison with those values required for liquid formation. Figures 2(a) and 2(b) also show the variation of aluminium to silicon and yttrium to aluminium atomic ratio after 1 h oxidation and after 256 h oxidation. If the liquid phase is important in the oxidation process it should be possible to predict crystalline phases formed within the oxide by comparing the Y:Al and Al:Si ratios observed with those required for liquid formation since any element present to excess in the liquid will precipitate as a silica-containing species. In terms of Y:Al and Al:Si ratios it should be possible to define three criteria:

1. If the Y:Al ratio observed is greater than that required for liquid formation and the Al:Si ratio less, yttrium and silicon should be precipitated from the liquid as yttrium disilicate and cristobalite.
2. If the Y:Al ratio observed is less than that required for liquid formation and the Al:Si ratio less, silicon should be precipitated from the liquid as cristobalite.
3. If the Y:Al ratio observed is less than that required for liquid formation and the Al:Si greater, aluminium and silicon should be precipitated from the liquid as mullite and cristobalite.

Comparing the data given in Figs 2(a) and 2(b) with the phase assemblages given in Table 2 shows excellent consistency with the criteria listed above, thus confirming the importance of liquid phase formation. Anomalies are observed in some instances where yttrium disilicate and mullite unexpectedly arise. However, it is to be remembered that the cooling rates employed during the oxidation experiments were relatively slow and the liquid phase is likely to have devitrified as indicated by the presence of Y_2SiAlO_5N.

The stability of any liquid formed will depend solely on cation composition since nitrogen is not essential for liquid formation at the temperature in question. In this instance liquid formation rates and consequently sialon solution rates will be controlled by two factors: (i) rates of diffusion of yttrium and aluminium in the case of the $z = 0.2$, 0.5 and 1.0 materials and yttrium diffusion rates in the case of the $z = 1.5$, 2.0 and 3.0 materials and (ii) rates of precipitation of yttrium and aluminium from the liquid as silicates. The first factor will accelerate the oxidation rate whilst the latter will retard it. Evidence of this latter effect has been reported by Persson et al.[6] who showed that enhanced crystallisation of oxides formed on silicon oxynitrides due to surface contamination by baria significantly reduced oxidation rates. It is also probable that after a certain duration of oxidation the liquid may redissolve crystalline oxide phases in order to maintain its composition and such effects may explain the inconsistent EDA data acquired for the $z = 0.2$ and 0.5 materials oxidised for 32 and 64 h.

From the above arguments it is clear that the morphological development of oxide scales on β-sialon ceramics is controlled by solution-precipitation effects which arise in a liquid phase formed at the ceramic–oxide scale interface.

5. Conclusions

1. The morphological development of oxide scales can be understood in terms of differences between Y : Al and Al : Si atomic ratios in the oxide scale layers and those required for liquid phase formation.
2. The formation of liquid phases at the ceramic–oxide interface and the phase assemblage of oxidation products appear to be determined by yttrium diffusion for materials with z-values of 0.2 to 3.0.
3. For liquid phase formation, aluminium diffusion is required for $z = 0.2$ to 1.0 materials but not for materials with z-values of 1.5 to 3.0. In the latter case mullite is observed as the predominant surface phase.
4. The kinetics and mechanism of oxidation and morphological development of the oxide scale can be more readily rationalised in terms of additive diffusion rates if the formation of a liquid phase is considered to be an integral part of the oxidation process.

Acknowledgements

Thanks are due to Mr P. Byrne, Ms A. Flannery and Mr R. Bartley for specimen fabrication and preparation.

References

1. J. Schlichting, in F. L. Riley (ed.), 'Nitrogen Ceramics', *NATO ASI series E23* (1977), p. 627. Noordhoff, Leyden.
2. J. Persson and M. Nygren, in *Proc. 11th RISO Int. Symp. Structural Ceramics – Processing, Microstructure and Properties*, Roskilde, Denmark (1990), p. 451.
3. M. H. Lewis and P. Bernard, *J. Mat. Sci.*, **15** (1980), 443–448.
4. M. J. Pomeroy and S. Hampshire, *Materials Chemistry & Physics*, **13** (1985), 437–448.
5. L. Themelin, M. Desmaison-Brut, M. Billy and J. Crampon, *Ceramics International*, **18** (1992), 119–130.
6. J. Persson, P.-O. Kall and M. Nygren, in M. Buggy and S. Hampshire (eds), *Materials for Advanced Technology Applications* (1992), pp. 49–60, Trans Tech Publications, Aedermannsdorf, Switzerland.

Oxidation of SiC Whisker-reinforced-Mullite in Dry and Wet Air

F. KARA and J. A. LITTLE

Department of Materials Science and Metallurgy, University of Cambridge, Pembroke Street, Cambridge CB2 3QZ, UK

ABSTRACT

25 vol% SiC whisker-reinforced mullite was oxidised in dry and wet air at temperatures between 1200°C and 1500°C. The oxidation in both cases was parabolic. The presence of water increased the oxidation rate of the composite while the activation energies remained unchanged. The increased rate of oxidation was attributed to the decrease in viscosity of the resulting aluminosilicate glass scale by water.

1. Introduction

Mullite has attracted a great deal of interest as an engineering ceramic due to its unique properties of low thermal expansion coefficient and density, high-temperature strength retention and excellent creep resistance. However, as in other ceramics, its brittleness is a major limiting factor when consideration is given to practical applications.

In order to overcome this problem of brittleness, many engineering ceramics have been fabricated in composite forms and considerable toughness increases have been achieved in alumina and mullite reinforced with SiC whiskers.[1-3]

Although SiC has excellent oxidation resistance due to the formation of a protective silica film, when it is incorporated in alumina or mullite the oxidation rate of the composites is reported to increase by several orders of magnitude due to the reaction of silica with the matrix materials.[4] The oxidation rate of hot-pressed SiC with 4 wt.% Al_2O_3 as a sintering additive in oxygen was reported to increase compared to pure SiC.[5] This was attributed to the formation of an aluminosilicate glass of lower viscosity due to the reaction of alumina with silica. Oxidation of the same material in wet oxygen was observed to be slightly greater than in dry oxygen.[6] However, there have been no studies on the effect of the presence of water vapour on the oxidation behaviour of SiC whisker-reinforced oxide matrix composites.

In this study, the oxidation of hot isostatically pressed 25 vol% SiC whisker-reinforced mullite was studied in dry and wet air at a range of temperatures.

2. Experimental

Mullite containing 25 vol.% SiC whiskers was produced by a colloidal mixing method.[7] The mullite source, boehmite and fumed silica, were dispersed in water at, pH 3.5 and 7, respectively, and mixed for 0.5 h. An SiC whisker dispersion (pH = 2) was added to the mixed sol and stirred over a hot plate until the mixture became viscous. Cold isostatically pressed composite powder compacts were glass-encapsulated at 700°C and hot isostatically pressed (HIPed) at 1600°C and 200 MPa for 5 h. The density of the composite was about 99%.

Samples for oxidation studies were polished to 1 μm and oxidised in a tube furnace at temperatures of 1200°C to 1500°C. At set times, samples were taken out and cooled in air directly to room temperature to retain the high-temperature reaction products. The oxidation atmosphere was respectively dry and wet air. The dry air oxidation runs were carried out after passing the air through silica gel, and the wet air atmosphere was achieved by passing the air through a water bath at a constant temperature of 85°C. The flow rate of air in both runs was 120 ml/min. This resulted in a partial pressure of 8×10^{-4} atm water in the air.

Both before and after oxidation, samples were characterised by X-ray diffraction and microscopical techniques. The scale thicknesses were measured by using light optical and/or scanning electron microscopes on polished cross-sections.

3. Results and Discussion

3.1 Composite Microstructure

Figure 1(a) shows a light optical micrograph of the composite and confirms that a good degree of mixing was achieved between the whiskers and the matrix. TEM investigation of the microstructure showed that there were occasional glassy pockets located especially in the vicinity of SiC whisker clumps (Fig. 1b). The interfaces between the SiC and mullite had a glassy layer of 1–2 nm in thickness.[7]

3.2 Oxidation Kinetics

Figures 2(a) and (b) record the scale thickness as a function of square root of time in dry and wet air oxidation, respectively for the composite at various temperatures. These plots show that the parabolic law is obeyed and indicate that oxidation is controlled by diffusion. The activation energies were 340 and 357 kJ/mol for dry and wet air oxidation, respectively (Fig. 2c). Thus there is

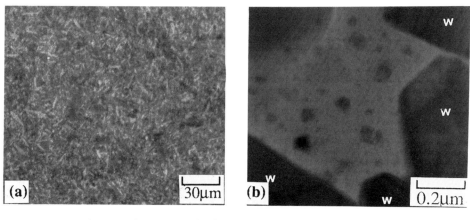

Fig. 1 (a) Light optical micrograph of the composite and (b) TEM dark field image of a glassy region (w=whisker).

almost no difference in the activation energies of dry and wet air oxidation, although the presence of water enhanced the absolute oxidation rate. Singhal reported similar observations in the oxidation of alumina-added SiC in dry and wet oxygen.[6]

The reason for the increased rate of oxidation in the presence of water may be due to the increased rate of transport of oxidants. It has been reported that water is soluble in oxides with the formation of hydroxide groups and, for glassy oxides, the concentration of the hydroxide groups depends linearly on the square root of water vapour partial pressure.[8] The solubility of water in sodium-aluminosilicate glasses at 1 atm partial pressure of water was reported to be ~ 0.39–0.4 mol% between 1250°C and 1500°C.[9] It is also well established that the viscosity of silicate glasses is decreased by an increasing partial pressure of water vapour.[8] This might explain the increased oxidation rate in wet air, since diffusivity would increase with decreasing viscosity. Comparison of scale microstructures after dry and wet air oxidation at 1200°C after 12 h supports this argument (Fig. 3). The viscosity of the glass in wet air oxidation seems to be lower because the glass was spread evenly over the surface (Fig. 3b), while it was not in dry air oxidation due to its apparently higher viscosity (Fig. 3a).

3.3 Characterisation of the Oxide Scale

Between 1325°C and 1500°C, X-ray diffraction showed only mullite as a crystalline phase and an amorphous peak due to the presence of a glass phase both in dry and wet air oxidation. The scale surfaces at 1200°C to 1500°C are shown in Fig. 4 after oxidation in dry air. At 1200°C, the scale surface consisted of glassy humps containing pores. The general appearance of the scale surface was identical at 12 h and after 72 h of oxidation (Fig. 3a and Fig. 4a). As the temperature increases, the scale morphology changes. There were many

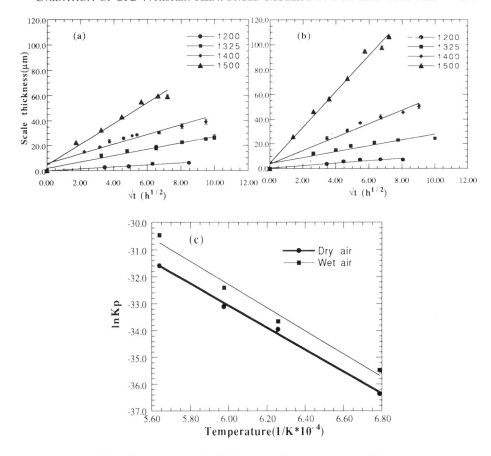

Fig. 2 (a) and (b) Change in scale thickness with square root of time in dry and wet air oxidation, respectively, and (c) activation plot.

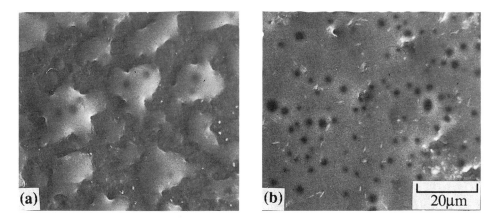

Fig. 3 Scale surface after 12 h of (a) dry air and (b) wet air oxidation.

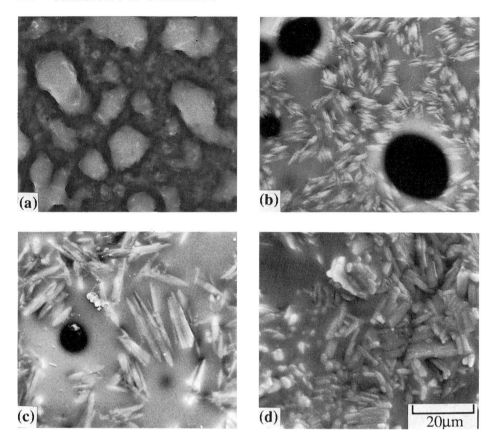

Fig. 4 Scale surfaces after oxidation in dry air: (a) 1200°C, 72 h, (b) 1325°C 90 h, (c) 1400°C, 90 h and (d) 1500°C, 52 h.

elongated mullite precipitates in the scale at the top of the glass at 1325°C, 1400°C and 1500°C which were not observed at 1200°C. At 1500°C, equiaxed mullite grains were also observed together with elongated mullite precipitates.

In wet air oxidation, the scale surfaces at all temperatures looked like that of the 1500°C dry oxidation (Fig. 5). However, initial stage oxidation scales were similar to those observed at 1325 and 1400°C in dry air oxidation, viz. mullite precipitates were present at the top of the glass, together with bubbles and no equiaxed mullite grains (Fig. 6). The size of the reprecipitated mullite crystals formed on the scale surface increases with time and temperature, and their growth was observed to be parabolic (Fig. 7).

The formation of mullite from alumino-silicate mixtures necessitates the dissolution of alumina into silica up to a critical concentration followed by nucleation and growth.[10] Evidence from this work indicates that SiC oxidises to silica which then reacts with mullite to form an alumino-silicate liquid from which mullite crystals precipitate.

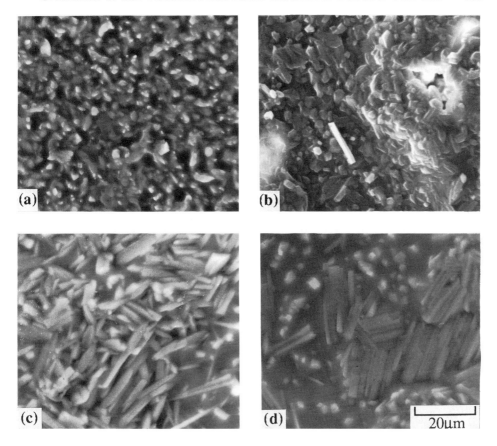

Fig. 5 Scale surfaces after oxidation in wet air: (a) 1200°C, 72 h, (b) 1325°C 90 h, (c) 1400°C, 90 h and (d) 1500°C, 52 h.

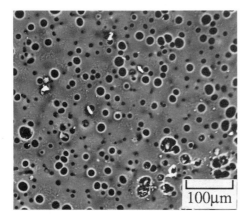

Fig. 6 Scale surface at 1500°C after 2 h of wet air oxidation.

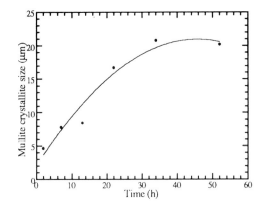

Fig. 7 Change in mullite crystallite size with time (1500°C, wet air).

Cross-section micrographs of wet air oxidation specimens at a particular temperature (1500°C) show the development of the scale microstructure (Fig. 8). Initially, the scale is composed of mullite grains in the glass matrix, above which is a glassy layer with the reprecipitated mullite crystals at the surface. The glassy layer is at the surface, probably due to the ~100% volume expansion which occurs when SiC is oxidised to SiO_2. This volume expansion is accommodated by the viscous glass being pushed from the interface to the top of the scale. The mullite grains in the glassy matrix originate from the mullite matrix itself. As the oxidation proceeds the glassy layer disappears. The time taken for the glassy layer to disappear increases with decreasing temperature. The disappearance of the glassy layer was observed at all temperatures studied in wet air oxidation, while it occured at only 1500°C in dry air oxidation. This indicates that the presence of water and temperature play important roles in this process. We believe that the glassy layer volatilises. The evidence for this is that there is a slight deviation from the parabolic behaviour, viz. the scale was thinner, at the later stages of oxidation, when the glassy layer starts to volatilise (Fig. 2a, 1500°C data and Fig. 2b).

The scale–matrix interface was relatively flat and the scale was relatively porous under all conditions studied. Bubble formation was observed at the scale surface (Fig. 6) as well as at the scale–matrix interface (Fig. 8). Bubble formation is due to the formation of gas products (mainly CO), the pressure of which must exceed at least 1 atm for pore formation.[11] These pores grow due to coalescence as they migrate away from the interface. However, their maximum size is limited by their physical environment, i.e. they do not grow to more than some certain size in the inner part of the scale due to the presence of mullite grains around them. This appears to increase the gas pressure inside these pores to very high levels. The evidence for this is the presence of bulb-shaped pores in the glassy layer, e.g. in Fig. 8 after 7 h. This particular shape indicates that as soon as a pore reaches the glassy layer it can decrease its pressure by increasing its volume due to the lower viscosity of the glass as compared with the layer containing the mullite grains. A similar pore

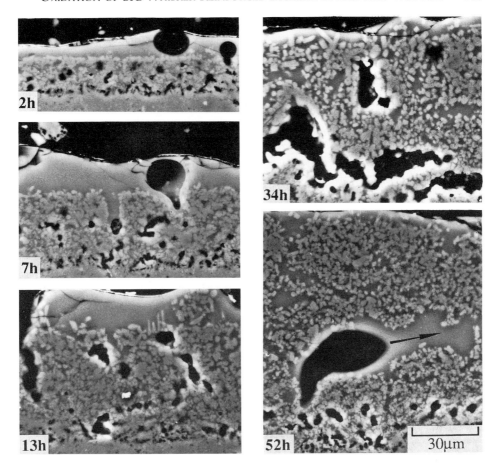

Fig. 8 Cross-section micrographs of the scale development at 1500°C in wet air oxidation.

development is shown in Fig. 8 after 52 h which also reveals that the large pores adopt their shape according to the shape of the glassy region around them. The pore in this micrograph will grow in the direction of the arrow.

Development of large pores lying parallel to the interface was also observed (Fig. 8 after 34 h and Fig. 9). This type of pore development was seen only at 1500°C at the later stages of dry air oxidation, i.e. after volatilisation of the top glassy layer. However, in wet air oxidation they were present at all temperatures except at 1200°C at the later stages of oxidation, i.e. after volatilisation of the glassy layer. These observations suggest that there is an interaction between the development of these pores and the top glassy layer. The development of these pores is probably due to the fact that, as the glassy layer disappears, the reprecipitated mullite crystals impinge on the mullite grains from the mullite matrix (Fig. 10) and form a rather impervious barrier to gas escape. Gases forming near the interface cannot escape easily to the

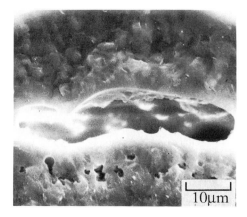

Fig. 9 Development of a large pore parallel to the interface (1325°C, 65 h, wet air).

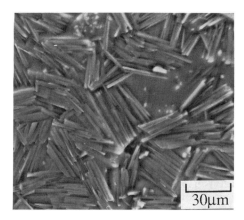

Fig. 10 Precipitated mullite crystals on the scale surface (1500°C, 34 h, wet air).

scale–gas interface and thus develop these large pores by coalescence. However, when the glassy layer is present, gases diffusing out meet the glass rather than the impervious barrier of precipitated mullite crystals and can form large bubbles in the glass, thus relaxing the gas pressure near the interface and thus preventing the development of large pores. The absence of such pore development at 1200°C after wet air oxidation can be attributed to the volume of reprecipitated mullite crystals being not high enough (Fig. 3b) to form such a barrier.

3.4. Composition of the Glass

EDX analysis of the glass after dry and wet air oxidation showed that it had an alumino-silicate composition. The Al_2O_3 concentration in the glass decreased almost parabolically with time. For a given time, the higher the temperature the lower was the Al_2O_3 concentration. Also, the Al_2O_3 concentration in the glass was always lower in wet air oxidation for the same temperature and time. As mentioned earlier, the reprecipitated mullite crystallite size increased parabolically (Fig. 7). This suggests that Al_2O_3 in the glass is consumed by mullite crystals, and their growth is governed by the glass composition. The presence of water was observed to promote the precipitation of mullite crystals in the oxidation of hot pressed SiC with Al_2O_3 additive.[12] A similar observation was also noted in this study (compare Fig. 3a and 3b).

4. Conclusions

The presence of water in air increased the oxidation rates of the SiC whisker-reinforced mullite composite, while the activation energies remained almost unchanged. The oxidation in both cases was parabolic. The reason for increased oxidation in wet air was attributed to the decrease in the viscosity of the glass by the formation of hydroxyl groups and, as a result, increased transport of oxidant.

The top glassy layer in the scale volatilised at higher temperatures (1500°C) in dry air and at all temperatures studied in wet air oxidation. After volatilisation, reprecipitated mullite crystals impinged on the mullite grains from the matrix and formed a layer impervious to escape of the gases and this resulted in development of large pores lying parallel to the scale/matrix interface.

The amount of Al_2O_3 in the glass decreased with time and temperature as it was consumed by reprecipitated mullite crystals. In wet air, the consumption of Al_2O_3 was higher. The presence of water promotes the formation of mullite precipitates.

Acknowledgement

The financial support of The Education Ministry of Turkey is gratefully acknowledged. Thanks are due to the School of Industrial and Manufacturing Science of Cranfield Institute of Technology for allowing us to use the HIPing facilities and John Hedge for conducting the HIPing experiments.

References

1. G. C. Wei and P. F. Becher, *Am. Ceram. Soc. Bull.*, **64** (1985), 298–304.
2. T. N. Tiegs and P. F. Becher, *Am. Ceram. Soc. Bull.*, **66** (1987), 330–342.
3. J. Homeny and W. L. Vaughn, *MRS Bull.*, October 1/November 15 (1985), 66–71.
4. M. P. Borom, M. K. Brun and L. E. Szala, *Ceram. Eng. Sci. Proc.* (1987), 654–670.
5. S. C. Singhal, *J. Mater. Sci.*, **11** (1976), 1246–1253.

6. S. C. Singhal, *J. Am. Ceram. Soc.*, **59** (1976), 81–82.
7. F. Kara and J. A. Little, submitted to HT-CMC 93 Conference of EACM.
8. I. B. Cutler, *J. Am. Ceram. Soc.*, **52** (1969), 11–13.
9. N. P. Bansal and R. H. Doremus, *Handbook of Glass Properties* (1986), Academic Press, Florida.
10. I. A. Aksay and D. M. Dabs and M. Sarikaya, *J. Am. Ceram. Soc.*, **74** (1991), 2343–2358.
11. D. M. Mieskowski, T. E. Mitchell and A. H. Heuer, *Com. Am. Ceram. Soc.*, **57** (1984), C17–C18.
12. M. Maeda, K. Nakamura and T. Ohkubo, *J. Mater. Sci.*, **23** (1988), 3933–3938.

Author Index

Subject Index